Actuators for Intelligent Electric Vehicles

Actuators for Intelligent Electric Vehicles

Editors

Peng Hang
Xin Xia
Xinbo Chen

MDPI • Basel • Beijing • Wuhan • Barcelona • Belgrade • Manchester • Tokyo • Cluj • Tianjin

Editors
Peng Hang
School of Mechanical and
Aerospace Engineering
Nanyang Technological
University
Singapore

Xin Xia
Department of Civil and
Environmental Engineering
University of California
Los Angeles
United States

Xinbo Chen
School of Automotive Studies
Tongji University
Shanghai
China

Editorial Office
MDPI
St. Alban-Anlage 66
4052 Basel, Switzerland

This is a reprint of articles from the Special Issue published online in the open access journal *Actuators* (ISSN 2076-0825) (available at: www.mdpi.com/journal/actuators/special_issues/ Actuators_Intelligent_Electric_Vehicles).

For citation purposes, cite each article independently as indicated on the article page online and as indicated below:

LastName, A.A.; LastName, B.B.; LastName, C.C. Article Title. *Journal Name* **Year**, *Volume Number*, Page Range.

ISBN 978-3-0365-3512-8 (Hbk)
ISBN 978-3-0365-3511-1 (PDF)

Contents

About the Editors

Peng Hang

Peng Hang is a Research Fellow with RRIS, Nanyang Technological University, Singapore. His research interests include decision making, motion planning and control for connected autonomous vehicles. He received the Ph.D. degree with the School of Automotive Studies, Tongji University, in 2019. He has been a Visiting Researcher with the Department of Electrical and Computer Engineering, National University of Singapore, and a Software Engineer in Research and Advanced Technology Dept., SAIC Motor, China. He has contributed more than 50 academic papers in the field of autonomous driving, and applied for more than 20 patents. He serves as an Associate Editor of SAE International Journal of Vehicle Dynamics, Stability, and NVH, a Section Chair of CAA International Conference on Vehicular Control and Intelligent, a Guest Editor of Actuators, IET Intelligent Transport Systems, Autonomous Intelligent Systems, and Journal of Advanced Transportation, and an active reviewer for more than 40 international journals and conferences.

Xin Xia

Xin Xia received the B.E. degree in vehicle engineering from the School of Mechanical and Automotive Studies, South China University of Technology, Guangzhou, China, in 2014, and the Ph.D. degree in vehicle engineering from the School of Automotive Studies, Tongji University, Shanghai, China, in 2019. He was a Postdoctoral Fellow associated with Dr. Amir Khajepour with the Department of Mechanical and Mechatronics Engineering, University of Waterloo, Waterloo, ON, Canada, from January 2020 to March 2021. He is currently a Postdoctoral researcher associated with Dr. Jiaqi Ma with the Department of Civil and Environmental Engineering, University of California, Los Angeles, CA, USA. His research interest includes state estimation, cooperative localization, cooperative perception, and dynamics control of the autonomous vehicle.

Xinbo Chen

Xinbo Chen received the B.S. degree in mechanical engineering from Zhejiang University, Hangzhou, China in 1982. He received the M.S. degree in mechanical engineering from Tongji University, Shanghai, China, in 1985 and the Ph.D. degree in mechanical engineering from Tohoku University, Japan, in 1995. From 1988 to 1995, he was an Assistant Professor with the School of Mechanical Engineering, Tongji University. From 1996 to 2002, he was an Associate Professor with the School of Mechanical Engineering, Tongji University. Since 2002, he has been a Professor with the School of Automotive Studies, Tongji University. He is the author of more than 200 articles and more than 70 patents. His research interests include dynamic control of omni-directional electric vehicle, design and control of active/semi-active suspension system.

Review

Towards Autonomous Driving: Review and Perspectives on Configuration and Control of Four-Wheel Independent Drive/Steering Electric Vehicles

Peng Hang [1,*] and Xinbo Chen [2]

[1] School of Mechanical and Aerospace Engineering, Nanyang Technological University, Singapore 639798, Singapore
[2] School of Automotive Studies, Tongji University, Shanghai 201804, China; chenxinbo@tongji.edu.cn
*   Correspondence: peng.hang@ntu.edu.sg

Abstract: In this paper, the related studies of chassis configurations and control systems for four-wheel independent drive/steering electric vehicles (4WID-4WIS EV) are reviewed and discussed. Firstly, some prototypes and integrated X-by-wire modules of 4WID-4WIS EV are introduced, and the chassis configuration of 4WID-4WIS EV is analyzed. Then, common control models of 4WID-4WIS EV, i.e., the dynamic model, kinematic model, and path tracking model, are summarized. Furthermore, the control frameworks, strategies, and algorithms of 4WID-4WIS EV are introduced and discussed, including the handling of stability control, rollover prevention control, path tracking control and active fault-tolerate control. Finally, with a view towards autonomous driving, some challenges, and perspectives for 4WID-4WIS EV are discussed.

Keywords: autonomous driving; four-wheel independent drive; four-wheel independent steering; path tracking; handling stability; active safety control; electric vehicle

Citation: Hang, P.; Chen, X. Towards Autonomous Driving: Review and Perspectives on Configuration and Control of Four-Wheel Independent Drive/Steering Electric Vehicles. *Actuators* **2021**, *10*, 184. https://doi.org/10.3390/act10080184

Academic Editor: Hai Wang

Received: 5 July 2021
Accepted: 1 August 2021
Published: 5 August 2021

Publisher's Note: MDPI stays neutral with regard to jurisdictional claims in published maps and institutional affiliations.

1. Introduction

Autonomous driving techniques can not only reduce human drivers' driving burden, but also advance driving safety and reduce traffic accidents. In addition to realizing zero emissions targets and reducing air pollution, electric vehicles (EVs) have better control performance than traditional fuel vehicles. Therefore, autonomous vehicles (AVs) and EVs have been a popular issue in vehicle development [1–3].

In recent years, most AVs have been studied and developed based on the traditional fuel vehicle platform, e.g., those used by Baidu, Waymo, Uber, etc. These so-called AVs are designed by applying advanced perception sensors, decision-making and control systems to the existing commercial vehicles [4]. Most autonomous driving companies are not automobile manufacturers and cannot integrate autonomous driving technology into the autonomous driving platform design, which restricts the commercial development of AVs [5]. In fact, traditional fuel vehicles are not the best autonomous driving platform. Their complex drive and transmission systems, i.e., the internal combustion engine, torque converter, etc., have slow response rates and the low control accuracy [6]. In contrast, EVs are preferred by many researchers. Without the complex drive and transmission systems, accurate control is easier to achieve [7]. As a result, the decision-making commands from the autonomous driving system can be better executed [8]. Therefore, towards future autonomous driving, autonomous mobile platforms have been widely studied, including those of Schaeffler, Protean, etc. [9–11]. In the autonomous mobile platforms, the X-by-wire chassis technique is a critical issue for accurate control [12,13].

Traditional vehicles usually adopt the centralized drive system and the front-wheel steering (FWS) system, which is a common chassis configuration. With the development of chassis modularization and electrification, the integrated X-by-wire module has been widely studied, in which the steering system, drive system and braking system are all controlled by wire [14]. They are integrated with the vehicle suspension and make up an integrated chassis module, which is beneficial to the chassis reconstruction for different demands [15]. Due to the X-by-wire module, vehicles can easily realize accurate dynamic control to advance active safety [16]. Four X-by-wire modules make up a four-wheel independent drive/steering electric vehicle (4WID-4WIS EV). Due to the application of X-by-wire modules, the steering angle and drive/braking torque of each wheel can be controlled independently [17]. As a result, 4WID-4WIS EV can easily realize multi-objective optimization control, e.g., handling stability control, rollover prevention control and path tracking control [18]. Therefore, 4WID-4WIS EV is regarded as an ideal EV development platform by many researchers.

4WID-4WIS EV has been widely studied in recent years. Some prototypes have been designed and developed by vehicle companies and universities. Moreover, various control frameworks, algorithms and strategies have been studied as well. However, some critical issues of 4WID-4WIS EV have not been completely resolved, which prevents its commercial application. Towards autonomous driving, this paper aims to review the chassis configuration and control technique of 4WID-4WIS EV. Focusing on certain technical difficulties of 4WID-4WIS EV, some perspectives are given at the end of this paper.

The rest of this paper is organized as follows. In Section 2, the chassis configuration of 4WID-4WIS EV is introduced and analyzed. Section 3 presents the typical control models of 4WID-4WIS EV. In Section 4, control frameworks and control algorithms of 4WID-4WIS EV are reviewed. Section 5 gives the challenges and perspectives of 4WID-4WIS EVs' future development. Finally, Section 6 concludes this paper.

2. Chassis Configuration of 4WID-4WIS EV

This section mainly focuses on the chassis configuration analysis of the 4WID-4WIS EV. Firstly, the typical prototypes of 4WID-4WIS EV are introduced and the configuration analysis is conducted. Then, the key component of 4WID-4WIS EV, i.e., the X-by-wire module, is reviewed, and the comparative study of different modules is carried out. Finally, the steering modes of 4WID-4WIS EV are analyzed and the switching logic between different steering modes is introduced.

2.1. Configuration Analysis of 4WID-4WIS EV

As shown in Figure 1, the chassis of 4WID-4WIS EV is made up of four X-by-wire modules that integrate the steering, drive, braking and suspension systems. Three actuators are included in the X-by-wire module, i.e., the steering-by-wire actuator, drive-by-wire actuator, and braking-by-wire actuator. The steering-by-wire actuator is usually integrated with the steering kingpin, which can be a virtual kingpin or a component of the suspension system. The in-wheel motor is usually taken as a drive-by-wire actuator, which can be integrated with the wheel rim. Compared with the conventional centralized drive system, the drive shaft, the differential mechanism, and reducers are cancelled. An electronic hydraulic braking (EHB) system and an electronic mechanical braking (EMB) system are usually adopted as the braking-by-wire actuator [19–21].

Due to the application of X-by-wire modules, the steering angle and drive/braking torque of each wheel can be controlled independently. As a result, 4WID-4WIS EV has more degrees of freedom (DOF) in terms of control than conventional vehicles, which leads to more steering and motion modes.

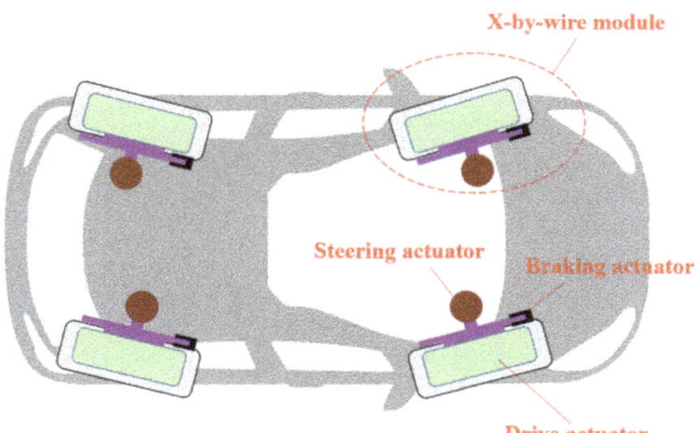

Figure 1. Chassis configuration of 4WID-4WIS EV.

2.2. Prototypes of 4WID-4WIS EV

In recent years, 4WID-4WIS EV has been widely studied by many companies and universities. Some prototypes of 4WID-4WIS EV are shown in Figure 2. As a futuristic looking vehicle, Fine-T is proposed by Toyota, which is equipped with a 4WID-4WIS technique that can realize pivot steering in favor of parking in a tighter area [22]. Additionally, Nissan also designed three generations of 4WID-4WIS concept cars, i.e., PIVO1, PIVO2 and PIVO3 [23]. ROboMObil is an autonomous 4WID-4WIS EV. With the application of the 4WID-4WIS technique, it not only shows strong maneuverability at low-speed conditions, e.g., parking, but also has good handling stability at high-speed conditions [24,25]. DFKI EO Smart 2 is a highly flexible micro-car designed for mega-cities, which is also an autonomous concept car. Besides the 4WID-4WIS technique, it can change the morphology of its height and length to further improve the maneuverability. In addition to single-vehicle autonomous driving, platooning autonomous driving can be realized with EO Smart 2 [26]. With the intelligent corner module, Schaeffler proposed the 4WID-4WIS EV Mover that is the solution to the autonomous and sustainable mobility in urban spaces [27]. With the reconstruction of the chassis configuration, Schaeffler Mover can be applied to different types of vehicles. In addition to the vehicle companies, some universities also developed some 4WID-4WIS EV prototypes, including Jilin University [28,29], The Chinese University of Hong Kong (CUHK) [30–32], Massachusetts Institute of Technology (MIT) [33], Universiti Teknologi Malaysia (UTM) [34], Tongji University [35–37], Pusan National University [38], and Iowa State University [39].

Figure 2. Protypes of 4WID-4WIS EV: (**a**) Toyota Fine-T; (**b**) Nissan PIVO3; (**c**) ROboMObil; (**d**) DFKI EO Smart 2; (**e**) Schaeffler Mover; (**f**) Jilin University; (**g**) CUHK OK-1; (**h**) MIT Hiriko; (**i**) UTM; (**j**) Tongji University.

Table 1 shows the performance analysis of the 4WID-4WIS EV prototypes. Most of them have a 180° steering angle range, which is in favor of high maneuverability. Compared with the prototypes designed by universities, the prototypes developed by vehicle companies have higher speed, which is closer to the performance requirements of passenger cars. Some 4WID-4WIS EV prototypes can realize simple autonomous driving functions, e.g., automatic parking. ROboMObil and DFKI EO Smart 2 can realize high-level autonomous driving.

Table 1. Configuration analysis of 4WID-4WIS EV.

Protype	Steering Angle	Speed	Autonomous Driving	Reference
Toyota Fine-T	$\pm90^\circ$	-	×	[22]
Nissan PIVO3	$\pm90^\circ$	-	√	[23]
ROboMObil	$-25^\circ \sim 95^\circ$	100 km/h	√	[24,25]
DFKI EO Smart 2	$\pm90^\circ$	65 km/h	√	[26]
Schaeffler Mover	$\pm90^\circ$	60 km/h	√	[27]
Jilin University	$\pm90^\circ$	8 km/h	×	[28,29]
CUHK OK-1	$\pm90^\circ$	10 km/h	√	[30–32]
MIT Hiriko	$\pm60^\circ$	50 km/h	×	[33]
UTM	$-60^\circ \sim 30^\circ$	30 km/h	×	[34]
Tongji University	$\pm90^\circ$	10 km/h	√	[35–37]
FABOT	$\pm35^\circ$	3 km/h	×	[38]
AgRover	360°	5 km/h	√	[39]

2.3. Integrated X-by-Wire Module of 4WID-4WIS EV

The key component of the 4WID-4WIS EV is the integrated X-by-wire module that integrates the steering, drive, braking and suspension systems. Four X-by-wire modules make up the chassis of 4WID-4WIS EVs. Figure 3 shows four typical X-by-wire modules, in which the first three are mature product prototypes. The X-by-wire modules (b) and (c) have been applied to the 4WID-4WIS EV Schaeffler Mover and ROboMObil. The last module is designed and developed by the authors.

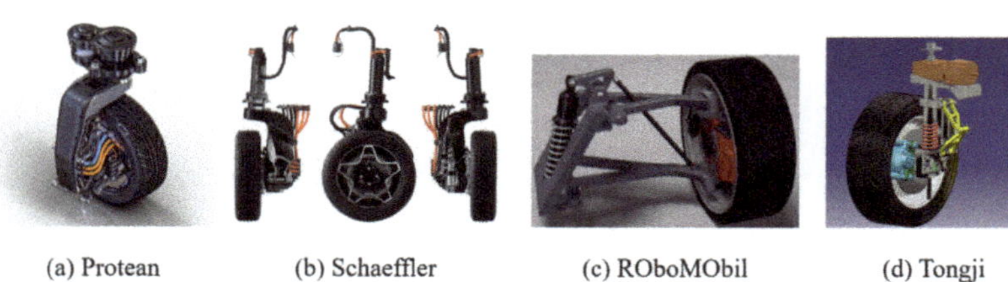

| (a) Protean | (b) Schaeffler | (c) ROboMObil | (d) Tongji |

Figure 3. Integrated X-by-wire module for 4WID-4WIS EV.

Table 2 shows the structure analysis of four integrated X-by-wire modules. The steering actuators of the four X-by-wire modules have a similar structure, i.e., servo motor and reducer. However, the layout positions of the four steering actuators are different, i.e., above the wheel (Protean. Surrey, United Kingdom, and Schaeffler, Herzogenaurach, Germany), inside the wheel (ROboMObil, Wessling, Germany) and beside the wheel (Tongji, Shanghai, China). Due to different layout positions of the steering actuator, it yields various steering ranges and control issues. If the steering actuator is placed above the wheel, it can realize zero steering kingpin offset, which is able to reduce the steering resistance. However, it will increase the vertical size of the X-by-wire module. If the steering actuator is placed beside the wheel, the vertical size of the X-by-wire module can be reduced, but it brings large steering kingpin offset, which brings a challenge to the capability of the steering motor. If the steering actuator is placed inside the wheel, it can reduce both the vertical size of the X-by-wire module and the steering kingpin offset, but it brings challenges to the layout of the in-wheel space.

Table 2. Structure analysis of integrated X-by-wire modules for 4WID-4WIS EV.

Type	Steering	Drive	Braking	Suspension
Protean	$360°$	In-wheel motor (80 kW, 1250 N·m)	HB + Motor	Candle type
Schaeffler	$\pm 90°$	In-wheel motor (13 kW, 250 N·m)	HB + Motor	Trailing arm type
ROboMObil	$-25° \sim 95°$	In-wheel motor (160 N·m)	HB + Motor	Double wishbone type
Tongji	$\pm 90°$	In-wheel motor (180 N·m)	HB + Motor	Candle type

The drive actuators of the four X-by-wire modules all take the in-wheel motor. The Protean X-by-wire module adopts the PD18 in-wheel motor, which has the largest power and torque among the four modules. The braking actuators of the four X-by-wire modules all take the hybrid braking system that integrates hydraulic braking (HB) and motor regenerative braking. The suspension systems of the four X-by-wire modules are different, and can be divided into three types, i.e., the candle type, trailing arm type and the double wishbone type. Compared with the candle suspension and the trailing arm suspension, the double wishbone suspension has better lateral and roll stiffness, which is in favor of safe driving in the condition of the large lateral acceleration. Therefore, it can be found from Table 1 that the design speed of ROboMObil is the largest among all prototypes, i.e., 100 km/h.

2.4. Steering Modes and Switching Logic

As mentioned above, due to the application of X-by-wire modules, the steering angle of each wheel can be controlled independently. As a result, 4WID-4WIS EV has more steering modes than traditional vehicles. The steering modes of 4WID-4WIS EV are illustrated in Figure 4, including FWS, rear-wheel steering (RWS), 4-wheel steering (4WS), oblique moving, crab moving, and pivot steering. With these steering modes, the maneuverability can be advanced remarkably, e.g., crab moving for side parking, and pivot steering for turning around in narrow spaces [40]. In addition to the maneuverability advancement at low-speed conditions, active 4WS can improve vehicles' handling stability at high-speed conditions [41,42].

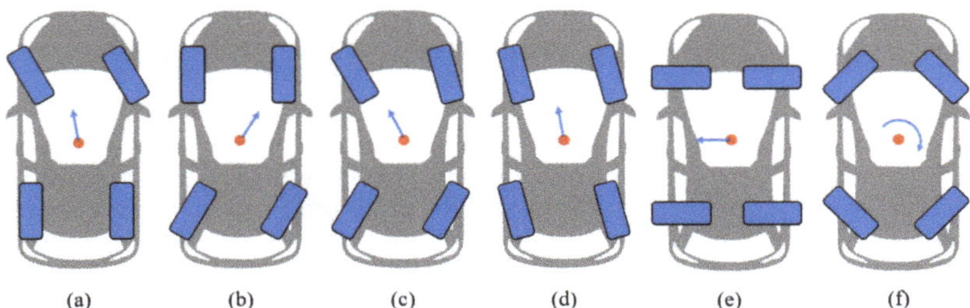

Figure 4. Steering modes of 4WID-4WIS EV: (**a**) FWS; (**b**) RWS; (**c**) 4WS; (**d**) oblique moving; (**e**) crab moving; (**f**) pivot steering.

To deal with different missions, effective switching between steering modes becomes very necessary. Based on the principle that the turning center is continuous, a logic of steering mode switching is proposed, which can realize smooth switching at low-speed conditions without stopping the car [43]. The dynamic switching logic between FWS and RWS, and FWS and 4WS, is studied, which is verified with real vehicle tests [44]. To minimize the sudden change of vehicle dynamic parameters and the energy consumption in the switching process, a B-spline curve is proposed to design the switching trajectory, which is optimized with the multi-objective genetic algorithm [45]. Based on the kinematic model and dynamic model of 4WID-4WIS EV, a steering mode switching strategy is designed and verified [46]. To realize the switching control between FWS and 4WS at high-speed conditions, a robust controller is designed [47], which aims to achieve a smooth transition of sideslip angle and yaw rate.

3. Control Model of 4WID-4WIS EV

This section mainly reviews the common control models of 4WID-4WIS EV, including the vehicle dynamic model, vehicle kinematic model, and path tracking model.

3.1. Vehicle Dynamic Model

Vehicle dynamic model is usually used to describe the dynamics of vehicles, especially at high-speed conditions. It is mainly derived through Newton's Law. According to the number of control DOF, the vehicle dynamic model has various evolutions [48]. A complex vehicle dynamic model can accurately describe the dynamic characteristics of the vehicle. However, it will introduce difficulty to the design of controllers due to the strong nonlinearity and coupling of the complex vehicle dynamic model [49]. Although the simplified vehicle dynamic model is in favor of controller design, some assumptions are made, which are invalid at some conditions. For instance, the assumption of the linear tire model is invalid at extreme conditions [50].

As for the vehicle dynamic control, longitudinal motion, lateral motion, yaw motion and roll motion are commonly considered by researchers. Figure 5 shows the dynamic model of 4WID-4WIS EV. According to Figure 5, the four DOF vehicle dynamic model can be expressed as follows [51,52].

$$
\begin{cases}
m\left(\dot{v}_x - v_x \beta r\right) = \sum F_x - F_w - F_f \\
m v_x \left(\dot{\beta} + r\right) + m_s h_s \ddot{\phi} = \sum F_y \\
I_z \dot{r} - I_{xz} \ddot{\phi} = \sum M_z \\
I_x \ddot{\phi} - I_{xz} \dot{r} = \sum L_x
\end{cases}
\tag{1}
$$

$$
\begin{cases}
\sum F_x = F_{xfl} \cos \delta_{fl} + F_{xfr} \cos \delta_{fr} + F_{xrl} \cos \delta_{rl} + F_{xrr} \cos \delta_{rr} \\
\sum F_y = F_{yfl} \cos \delta_{fl} + F_{yfr} \cos \delta_{fr} + F_{yrl} \cos \delta_{rl} + F_{yrr} \cos \delta_{rr} \\
\sum M_z = \left(F_{yfl} \cos \delta_{fl} + F_{yfr} \cos \delta_{fr}\right) l_f - \left(F_{yrl} \cos \delta_{rl} + F_{yrr} \cos \delta_{rr}\right) l_r + \Delta M_z \\
\sum L_x = m_s g h_s \phi - b_\phi \dot{\phi} - k_\phi \phi
\end{cases}
\tag{2}
$$

where v_x denotes the longitudinal velocity. β and r denote the sideslip angle and yaw rate at the center of gravity (CG), and ϕ is the roll angle. In addition, $\sum F_x$, $\sum F_y$, $\sum M_z$ and $\sum L_x$ denote the total longitudinal tire force, lateral tire force, yaw moment and roll moment acting on the vehicle. F_w and F_f denote the wind resistance and the rolling resistance, respectively. m and m_s denote the vehicle mass and vehicle sprung mass. h_s is the height of sprung mass. I_z, I_{xz} and I_x are the yaw inertia moment, the product of inertia and the roll inertia moment. δ_i ($i = fl, fr, rl, rr$) denotes the steering angle of each wheel (*fl* denotes the front left wheel, *fr* denotes the front right wheel, *rl* denotes the rear left wheel, and *rr* denotes the rear right wheel). F_{xi} and F_{yi} ($i = fl, fr, rl, rr$) denote the longitudinal and lateral forces of each tire. k_ϕ and b_ϕ denote the roll stiffness and damping of the vehicle suspension. ΔM_z is the external yaw moment, which is generated by the torque difference between the left wheel and the right wheel.

$$
\Delta M_z = [-F_{xfl} \cos \delta_{fl} + F_{xfr} \cos \delta_{fr} - F_{xrl} \cos \delta_{rl} + F_{xrr} \cos \delta_{rr}] \frac{B}{2}
\tag{3}
$$

where B is the vehicle track. According to different control objectives, the above 4DOF vehicle model can be simplified as a 3DOF vehicle model or a 2DOF vehicle model.

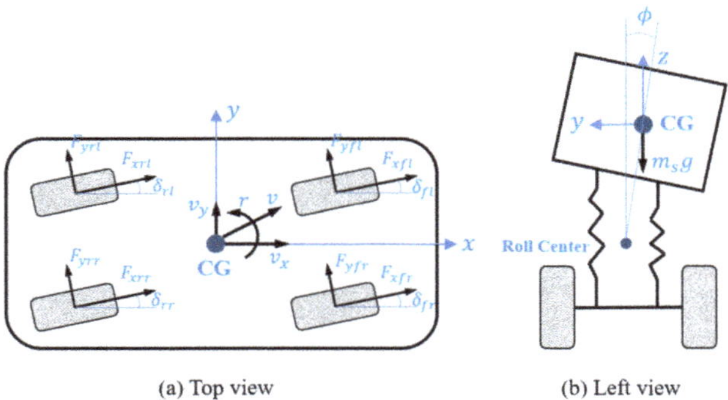

(a) Top view (b) Left view

Figure 5. Dynamic model of 4WID-4WIS EV.

It can be found from Equation (2) that the vehicle dynamic model is mainly determined by the tire force F_{xi} and F_{yi}. The tire is a critical component of the vehicle, and its structural characteristics and mechanical properties (vertical force, longitudinal force, lateral force, and torque of return) have a significant impact on the dynamic performance of the vehicle (ride, handling, stability, and safety) [53]. The mechanical properties of tires are mainly affected by factors such as tire type, cornering angle, slip rate, speed, etc. Tire models describe the relationships between the tire force and these influencing factors [54].

Tire models are mainly divided into three types: theoretical models with analytical formulas obtained by simplifying the mechanics of tires; empirical models obtained by analyzing and fitting tire force characteristic test data; semi-empirical models that combines the theoretical model and the analysis of experimental data [55]. Most of the empirical or semi-empirical models have the advantages of simple representation, easy calculation, and high fitting accuracy for specific tires, e.g., the magic formula [56], Dugoff tire model [57], UniTire model [58], Burckhardt tire model [59], HSRI tire model [60], etc. The theoretical model does not require fitting of experimental parameters and has strong versatility, e.g., the Gim tire model [61], string tire model [62], Fiala tire model [63], etc. The selection of tire models depends on the actual vehicle dynamics problem to be solved, whether it needs a more accurate theoretical model for modeling, or an empirical model towards practical engineering applications.

To reduce the complexity of controller design, the four-wheel vehicle model is usually simplified as a single-track model, as shown in Figure 6. As a result, the four steering control variables are reduced to two. The steering angle transformation relationship between the two models follows the Ackerman steering geometry [64].

$$\tan \delta_{fl} = \frac{\tan \delta_f}{1 - \frac{B}{2l}\left(\tan \delta_f - \tan \delta_r\right)}, \quad \tan \delta_{fr} = \frac{\tan \delta_f}{1 + \frac{B}{2l}\left(\tan \delta_f - \tan \delta_r\right)}$$
$$\tan \delta_{rl} = \frac{\tan \delta_r}{1 - \frac{B}{2l}\left(\tan \delta_f - \tan \delta_r\right)}, \quad \tan \delta_{rr} = \frac{\tan \delta_r}{1 + \frac{B}{2l}\left(\tan \delta_f - \tan \delta_r\right)} \tag{4}$$

where δ_f and δ_r denote the front and rear steering angles. l denotes the wheelbase.

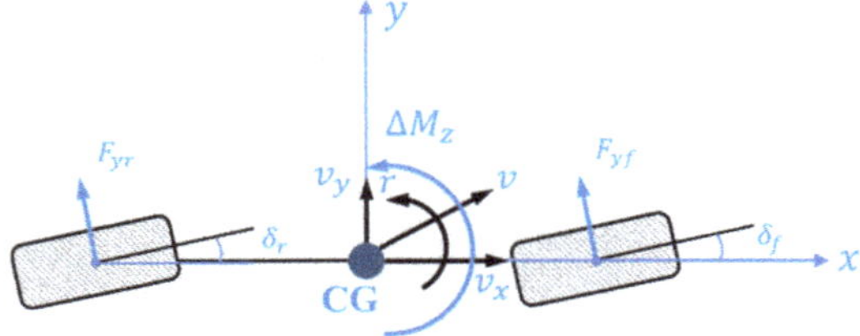

Figure 6. Single-track model for 4WID-4WIS EV.

3.2. Vehicle Kinematic Model

The vehicle kinematic model is usually used to address the motion planning and control of vehicles at low-speed conditions, e.g., automatic parking control [65]. For motion control at high-speed conditions, the vehicle dynamic model is preferred [66].

The single-track kinematic model for 4WID-4WIS EV is derived as follows [67].

$$
\begin{cases}
\dot{v}_x = a_x \\
\dot{\varphi} = v_x\left(\tan\delta_f + \tan\delta_r\right)/l \\
\dot{X} = v_x \cos(\beta + \varphi)/\cos\beta \\
\dot{Y} = v_x \sin(\beta + \varphi)/\cos\beta
\end{cases}
\tag{5}
$$

where a_x denotes the longitudinal acceleration. (X, Y) is the position coordinate of the vehicle.

3.3. Path Tracking Model

According to the information difference of the target path, i.e., the target position coordinate or target path curvature, the path tracking model is divided into two types. The first kind of path tracking model is based on the given information of φ, X and Y, which aims to minimize the following errors [68].

$$
\begin{cases}
\Delta\varphi = \varphi - \varphi_d \\
\Delta X = X - X_d \\
\Delta Y = Y - Y_d
\end{cases}
\tag{6}
$$

$$
\begin{cases}
\dot{\varphi} = r \\
\dot{X} = v_x \cos\varphi - v_y \sin\varphi \\
\dot{Y} = v_x \sin\varphi + v_y \cos\varphi
\end{cases}
\tag{7}
$$

where φ_d, X_d and Y_d denote the desired values for the target path.

The second kind of path tracking model is derived according to the curvature information of the target path, which is illustrated in Figure 7. To make the vehicle track the target path precisely, the path-tracking problem is equivalent to minimizing the yaw angle error $\Delta\varphi$ and the lateral offset Δy, which are derived as follows [69].

$$
\begin{cases}
\Delta\dot{\varphi} = r - \dfrac{v_x}{\rho} \\
\Delta\dot{y} = v_y + v_x\Delta\varphi
\end{cases}
\tag{8}
$$

where ρ denotes the curvature of the target path.

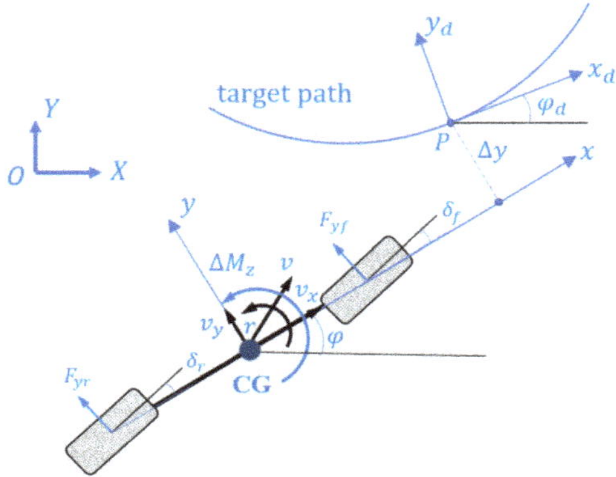

Figure 7. Path-tracking model for 4WID-4WIS EV.

4. Control of 4WID-4WIS EV for Autonomous Driving

In this section, the control framework of 4WID-4WIS EV is introduced. Then, the control algorithms and strategies of handling stability, rollover prevention and path tracking are reviewed and discussed. Finally, active fault-tolerate control algorithms for 4WID-4WIS EVs are introduced.

4.1. Control Framework of 4WID-4WIS EV

The control framework of 4WID-4WIS can be divided into two types, i.e., the coupling control framework [70] and the decoupling control framework [71], which are shown in Figure 8a,b, respectively. In the coupling control framework, the longitudinal motion control is coupled with the lateral motion control, which yields a multi-objective control. It brings a challenge to the control algorithm design. In the decoupling control framework, the longitudinal motion control is decoupled with the lateral motion control, which can reduce the complexity of controller design.

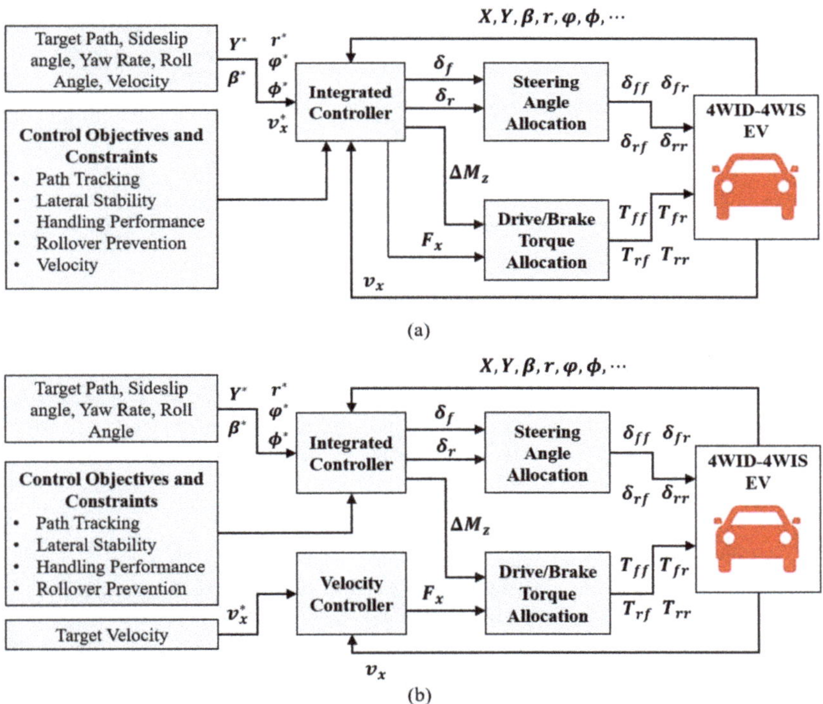

Figure 8. Control framework of 4WID-4WIS EV: (**a**) Coupling control framework; (**b**) decoupling control framework. * denotes the target reference.

From Figure 8, we can find that both the coupling control framework and the decoupling control framework consist of two levels. The high level is the controller design. According to the control objectives of path tracking, lateral stability, handling performance, rollover prevention and velocity tracking, it aims to track various references including the target path, sideslip angle, yaw rate, roll angle and velocity. During the tracking control process, various control constraints must be considered. All the control algorithms are designed with an integrated controller. Then, the integrated controller outputs the control signals to the low-level control system, i.e., the allocation level.

The allocation level includes the steering angle allocation and the torque allocation. The steering angle allocation is based on Equation (4). The torque allocation algorithm is used to adjust the total longitudinal force F_x and the external yaw moment ΔM_z, i.e., direct yaw-moment control (DYC). Various torque allocation algorithms have been studied including the direct allocation approach [72], affine control allocation [73], sequence least squares [74], weighted least squares [75], dynamic allocation [76], model predictive control (MPC) [77], etc. After torque allocation, the target drive/brake torques of four wheels will be worked out.

Finally, the allocation level will output the target steering angles and torques of four wheels to the 4WID-4WIS EV. For the closed-loop control, the vehicle's motion state and position information will be fed back to the integrated controller and velocity controller.

Due to the application of the 4WID-4WIS technique, 4WID-4WIS EV has four kinds of control strategies for dynamic control, which are listed in Table 3, i.e., active front steering (AFS), AFS + DYC, 4WS, and 4WS + DYC. Due to the various control strategies, 4WID-4WIS EVs can achieve superior driving performance compared to conventional vehicles in terms of path tracking, handling stability and rollover prevention.

Table 3. Control Strategies of 4WID-4WIS EV.

Control Strategy	Control Variable
AFS	δ_f
AFS + DYC	$\delta_f, \Delta M_z$
4WS	δ_f, δ_r
4WS + DYC	$\delta_f, \delta_r, \Delta M_z$

4.2. Handling Stability Control

The handling stability control of vehicles is defined to track the desired sideslip angle and yaw rate [78]. For traditional FWS vehicles, only the front-wheel steering angle can be controlled. When conducting the steering maneuver at high-speed conditions, the front tire lateral force may enter the saturation region, which cannot provide enough force to guarantee the lateral stability of vehicles [79]. For 4WID-4WIS EVs, since the braking and drive torque of each wheel can be controlled independently, DYC can be realized easily. As a result, the external yaw moment can make up for the lack of tire lateral force to increase the handling stability. In [80], a BP-PID controller-based multi-model control system is proposed for lateral stability improvement via DYC. In [81], a novel control algorithm of DYC based on the correctional LQR is designed to realize vehicle dynamic stability control. Based on the slide model control (SMC), a DYC-based hierarchical control strategy is proposed to improve lateral stability at driving limits [82]. By calculating the stability boundary with the phase plane method, a new extension coordinated controller is designed to improve the driving stability and handling performance, which can find the best balance between AFS and DYC [83]. To enhance the lateral stability, a robust internal model control method with a modified structure is applied to the integrated controller design of AFS + DYC [84]. The control diagram is illustrated in Figure 9.

Compared with DYC, the 4WS technique makes it easier to realize zero sideslip angle. Meanwhile, it is not necessary to deal with the allocation of the external yaw moment and the total longitudinal force [85]. In [86], the linear-parameter-varying (LPV) model is used to simplify the nonlinear model, and the decoupling control is applied to the velocity tracking control and handling stability control. In [87], considering the velocity-varying motion, a LPV controller is designed for handling stability control of 4WS. In addition, the attenuation of diagonal decoupling (ADD) control is proposed for 4WS vehicles, which shows good robustness to address uncertainties and disturbances [88]. In [89], an internal model control (IMC) strategy is proposed to address the nonlinearity of the stability control system. Additionally, the multi-input-multi-output (MIMO) IMC is adopted for vehicle stability control [90]. In [91], a handling modification method is applied to the handling stability control of 4WS vehicles. Based on SMC, the decentralized control algorithm is robust to arbitrary lateral disturbances and can guarantee that the vehicle converges to reference yaw rate and zero sideslip [92]. Due to the advantage of strong robustness to deal with parametric uncertainties, external disturbances and sensor noise, robust control has been studied by many researchers and applied to the handling stability control in 4WS vehicles, including H2 control, H∞ control, and μ-synthesis control [93–96]. In [97], a H2/H∞ mixed robust controller is designed for stability control. In [98], pre-compensation decoupling control with H∞ performance is applied to the longitudinal motion control and handling stability control. In [99], the handling stability and system robustness are advanced with the μ-synthesis robust controller. In [100], varying parameters are considered in the vehicle model and the μ-synthesis controller is designed for 4WS. Although robust control approaches show strong robustness to deal with parametric perturbations, a large range of perturbation will lead to a high-order controller, which brings large amounts of calculation to the hardware. We need to find a good balance between control performance and calculation efficiency in the controller design.

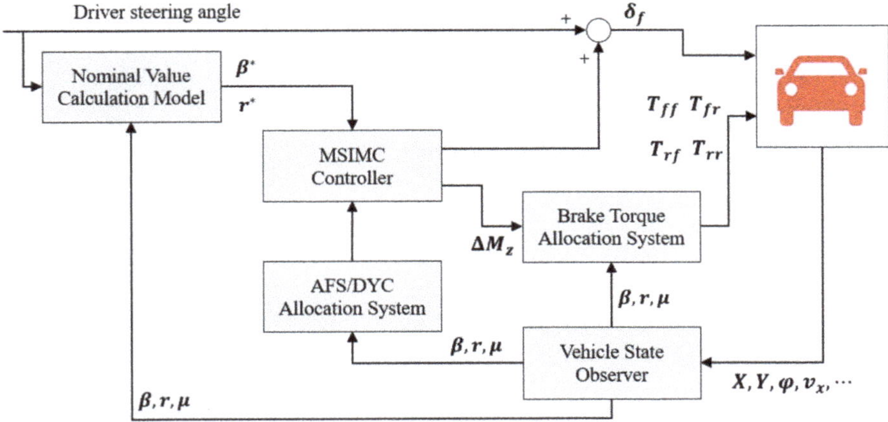

Figure 9. Control diagram of the AFS + DYC control system in [84].

With the advantages of 4WS and DYC, the combination of 4WS and DYC yields the superior handling stability for 4WID-4WIS EVs [101]. In [102,103], two feed-forward and feedback controllers are designed to realize zero sideslip angle and target yaw rate tracking with the integrated control of 4WS and DYC. In [104], a robust H∞ control approach is applied to the coordinated control of 4WS and DYC to improve handling stability in extreme conditions. In [105], fuzzy control theory is used to design the feedback controller of 4WS + DYC to improve lateral stability at high-speed conditions. To obtain a gain-scheduled controller, the LPV system is combined with the H∞ optimal control theory for the handling stability controller design of 4WS and DYC [106]. Besides, taking the tire nonlinearity into consideration, 4WS and DYC control are combined with the active suspension control to advance both the handling stability and the ride comfort [107]. Compared with AFS, the coordinate control of 4WS and DYC can advance the active safety of AVs at extreme conditions.

4.3. Rollover Prevention Control

Although the handling stability control can enhance the lateral driving safety at driving limits, for some vehicles with high size, e.g., trucks and buses, it is necessary to consider the rollover prevention performance [108]. The rollover prevention control is usually considered with the handling stability control [109]. The rollover index (RI) is usually used as the control performance index of rollover prevention. In [110], a RI is proposed to evaluate the rollover effect, a roll state estimator is designed, based on RI and the roll state estimator, and an integrated rollover mitigation controller is designed to reduce the danger of rollover without loss of vehicle lateral stability. Furthermore, a multiple-rollover-index (MRI) minimization approach is proposed to realize active rollover prevention control for heavy articulated vehicles [111].

Different control algorithms have been designed for rollover prevention control. In [112], a linear quadratic static output feedback (LQSOF) approach is applied to the preview controller design for vehicle rollover prevention. In [113], a nonlinear control strategy is designed, which can guarantee the handling stability while preventing rollover. In [114,115], a pulsed steering system and a hydraulic-mechanical pulsed steering system are designed, which integrate the handling stability control and rollover prevention control. In [116], linear-time-varying (LTV) MPC is applied to the integrated controller design, which can advance lateral stability, handling performance and rollover prevention via the 4WS technique. In [117], the fuzzy SMC approach is applied to the vehicle dynamic control of 4WS vehicles, which can enhance the dynamic response and deal with system nonlinearity. As Figure 10 shows, in [118], a new type of hierarchical control is proposed for 4WS vehicles, which uses the fractional SMC to obtain good robustness. Although SMC shows good performance in terms of dealing with system nonlinearity, controller chattering is still a critical issue for application.

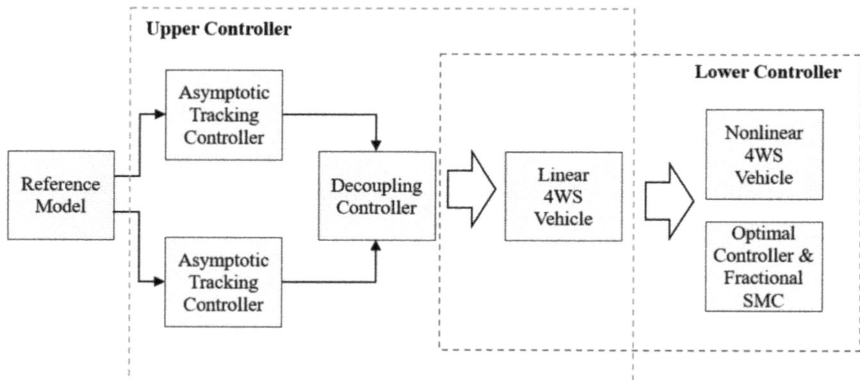

Figure 10. Control diagram of the 4WS control system in [118].

Additionally, 4WS and DYC are usually combined to advance the rollover prevention performance. With 4WS and DYC techniques, an integrated dynamic control with steering (IDCS) system is proposed to improve the handling stability and rollover prevention performance through fuzzy logic [119]. In [120], a switching MPC controller is designed to realize rollover prevention with active steering control and active differential braking control. Based on the SMC approach, a hierarchical coordinated control algorithm for integrating active steering control and driving/braking force distribution is proposed, which can enhance the handling stability and rollover prevention performance [121].

4.4. Path Trakcing Control

Path tracking control is the main control task for AVs [122]. Therefore, it has been widely studied in recent years and various control algorithms have been designed. In [123], DYC is used to advance the path tracking performance, and a robust H∞ control approach is applied to the DYC controller design. In [124], a coupling control framework is proposed based on DYC, and both the velocity tracking control and the path tracking control are considered with LTV MPC. In [125], based on LQR technique, both 4WS and DYC are utilized to improve the path-tracking performance. To improve the robustness of the path-tracking controller, a robust path-tracking controller is designed for the 4WID-4WIS agricultural robotic vehicle with the backstepping SMC theory [126]. To improve the control accuracy of the backstepping SMC, a comprehensive method that combines feedforward and backstepping SMC is applied to the path tracking control of 4WID-4WIS EVs [127]. In [128], a four-wheel SMC steering controller is designed for the path tracking of 4WID-4WIS EVs. Meanwhile, the longitudinal velocity controller is designed with the SMC approach.

For low-speed autonomous driving, it is sufficient to consider the path tracking control. However, with the increase in vehicle speed, the issue of handling stability and rollover prevention becomes more and more prominent. The path tracking issue is needs to be considered together with handling stability at high-speed conditions, especially at extreme conditions [129]. Compared with traditional vehicles, 4WID-4WIS EV has more control DOF; therefore, it is easier to realize the integrated control of path tracking and handling stability. In [130], a LQR feedback controller is applied to the path tracking of 4WS under the condition of high-speed emergency obstacle avoidance. In addition to the path tracking issue, the issue of handling stability control is considered as well. However, LQR approach has poor robustness to deal with the system nonlinearity and uncertainties. A robust LQR controller is designed for path tracking via the integration of AFS and DYC [131]. Based on the SMC theory, an automatic path-tracking controller is designed for 4WS vehicle, which has strong robustness to deal with system uncertainties such as cornering power perturbation, path radius fluctuation, and cross wind disturbance [132]. In [133], Hamilton energy function control theory is applied to the path tracking and lateral stability control of the 4WS + DYC control system. Besides, a robust controller is applied to the integrated 4WS + DYC control system, which can not only improve the path-tracking performance and handling stability but also has good robustness to address parametric perturbation [134]. The control diagram is shown in Figure 11.

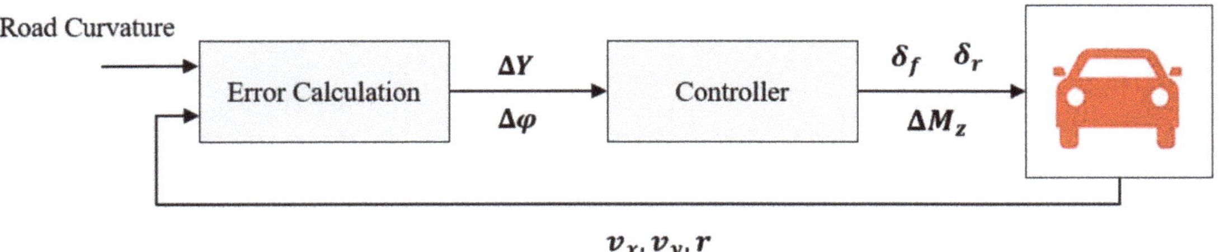

Figure 11. Control diagram of the integrated 4WS + DYC control system in [134].

Moreover, MPC has been widely applied to the path tracking control of AVs [135]. In [136], a coupling control framework is designed based on MPC, which comprehensively considers the velocity tracking control, handling stability control and path tracking control. Besides, the road adhesion coefficient is estimated to improve the control accuracy. Based on the nonlinear 4WS vehicle model, nonlinear model predictive control (NMPC) is used to design an integrated controller that considers handling stability and path tracking [137]. Although MPC has superior control accuracy than other control algorithms, the real-time optimization brings a large amount of calculation to the hardware.

Finally, Table 4 shows the summary of various control instances for 4WID-4WIS EV. It can be found that the 2DOF single track model is the most common control model for 4WID-4WIS EVs. If the longitudinal motion control or rollover prevention control is considered, another control DOF is required, which yields a 3DOF control model. To advance handling stability, rollover prevention performance and path tracking performance, different control strategies, i.e., AFS + DYC, 4WS, and 4WS + DYC, have been widely applied to the dynamic control of 4WID-4WIS EVs. Furthermore, LQR, SMC, robust control and MPC are the common control algorithms for 4WID-4WIS EVs. LQR can only deal with linear systems. SMC and robust control have good robustness to address system uncertainties and disturbances, but their control performances are remarkably affected by the model accuracy. With model prediction and real-time optimization, MPC can realize accurate control, but the real-time optimization also brings a large amount of calculation to the hardware. Simulation, the hardware-in-the-loop (HIL) test and the road test are the three kinds of algorithm verification methods. It can be found that most papers evaluate the control algorithm with simulation. Only few papers conduct the road test. One important reason is that the techniques used for 4WID-4WIS EVs are not vary mature, especially for the X-by-wire technique, and their reliability and safety cannot be guaranteed completely. Road tests involve a degree safety risk.

Table 4. Summary of various control instances.

Reference	Control Objective	Control Model	Control Strategy	Control Algorithm	Test Environment
[80]	HS	2DOF	AFS + DYC	BP PID	Simulation
[81]	HS	2DOF	AFS + DYC	LQR	HIL Test
[82]	HS	2DOF	AFS + DYC	SMC	HIL Test
[83]	HS	2DOF	AFS + DYC	Coordinated control	HIL Test
[84]	HS	2DOF	AFS + DYC	H∞ control	Simulation
[89]	HS	2DOF	4WS	Internal model control	Simulation
[90]	HS	2DOF	4WS	Internal model control	Simulation
[99]	HS	2DOF	4WS	μ-synthesis	Simulation
[100]	HS	2DOF	4WS	μ-synthesis	HIL Test
[92]	HS	2DOF	4WS	Feed-forward control	Simulation
[97]	HS	2DOF	4WS	H2/H∞	Simulation
[95]	HS	2DOF	4WS	LPV H∞	Simulation
[96]	HS	2DOF	4WS	μ-synthesis	Road Test
[87]	HS + VC	3DOF	4WS	LPV	Simulation
[88]	HS + VC	3DOF	4WS	Decoupling control	Simulation
[98]	HS + VC	3DOF	4WS	Decoupling control	Simulation
[102]	HS	2DOF	4WS + DYC	Feed-forward, feedback	Simulation
[104]	HS	2DOF	4WS + DYC	H∞ control	Simulation
[105]	HS	2DOF	4WS + DYC	fuzzy control	Simulation
[111]	HS + RP	3DOF	AFS + DYC	LQR	Simulation
[117]	HS + RP	3DOF	4WS	Fuzzy SMC	Simulation
[118]	HS + RP	3DOF	4WS	Fractional SMC	Simulation
[116]	HS + RP	3DOF	4WS	LTV-MPC	Simulation
[119]	HS + RP + VC	4DOF	4WS	Fuzzy logic	Simulation
[121]	HS + RP	3DOF	4WS + DYC	SMC	Simulation
[123]	PT + HS	2DOF	AFS + DYC	H∞ control	Simulation
[124]	PT + HS + VC	3DOF	AFS + DYC	MPC	Simulation
[129]	PT + HS	2DOF	AFS + DYC	LTV-MPC	Road Test
[127]	PT + HS	2DOF	4WS	SMC	Simulation
[132]	PT + HS	2DOF	4WS	SMC	Simulation
[126]	PT + VC	3DOF	4WS	Backstepping SMC	Road Test
[136]	PT + VC	3DOF	4WS	MPC	Simulation
[137]	PT + VC	3DOF	4WS	MPC	Simulation
[125]	PT	2DOF	4WS + DYC	LQR	Simulation
[133]	PT + HS	2DOF	4WS + DYC	Hamilton	Simulation
[134]	PT + HS	2DOF	4WS + DYC	μ-synthesis	Simulation

Where HS, PT, RP, and VC are the abbreviation of handling stability, path tracking, rollover prevention and velocity control, respectively.

4.5. Active Fault-Tolerant Control

Although X-by-wire modules can bring various control strategies and steering modes to 4WID-4WIS EVs in favor of driving performance advancement, once one X-by-wire module fails, it will increase the risk of vehicle instability [138]. To address this issue, active fault-tolerant control algorithms have been widely studied [139].

In [140], an MPC-based fault tolerant control system is designed, in which one MPC is used for fault tolerant control and another MPC is used as an observer to estimate and compensate for the actuator fault. In [141], a multiple model-based fault-tolerant control system is proposed based on fuzzy logic and MPC. In [142], a dual-loop SMC is used to deal with the fault of in-wheel motor. In [143], an adaptive SMC fault-tolerant controller is designed. Furthermore, a modified SMC is applied to the active fault-tolerant control of 4WID-4WIS EV, in which the steering geometry is re-arranged according to the location of faulty wheels [144]. In [145], a robust adaptive fault-tolerant control scheme is designed with adaptive fast terminal SMC. Moreover, game theory has been applied to the active fault-tolerant control. In [146], a cooperative game-based actuator fault-tolerant control strategy is designed based on a differential game. Additionally, feedback linearization and cooperative game theory are combined to design the fault-tolerant controller [147]. To advance the robustness of the fault-tolerate controller, a model-independent self-tuning fault-tolerant control framework is designed, which can enhance the longitudinal and lateral tracking ability under different failure conditions [148].

To improve the performance of monitor vehicle states, a fault detection and diagnosis algorithm is designed to monitor vehicle states and provide feedback containing fault information to the controller [149]. In [150], an active fault-tolerant control framework is designed, which includes a baseline controller, a set of reconfigurable controllers, a fault detection and diagnosis mechanism, and a decision mechanism.

Furthermore, control allocation methods have been widely used to realize active fault-tolerant control of 4WID-4WIS EVs [151]. In [152], an orientated tire force allocation algorithm is proposed to address the steering system fault in the path tracking process. In [29], based on the pseudo-inverse matrix, a control allocation method is introduced for decoupling of the forces and moment. Based on the LPV framework, reconfiguration control is applied to the torque allocation, which can realize velocity and path tracking even during a fault event of the steering-by-wire system [153]. In [154], based on the fault detection and diagnosis module, a reconfigurable control allocator is designed, which optimally distributes the generalized forces/moments to four wheels.

5. Challenges and Perspectives for 4WID-4WIS EV

Although 4WID-4WIS EV has superior performance than traditional vehicles, some critical technical issues related to machinery and control have not been resolved, which prevents its commercial application.

The first challenge is the high cost of 4WID-4WIS EV. Due to the application of the X-by-wire module, 12 control actuators are included in a 4WID-4WIS EV. Compared with traditional centralized-control vehicles, more actuators lead to higher cost. Therefore, cost reduction is the first consideration. The highly integrated design of the X-by-wire module and the concept of the reconfigurable chassis are good solutions. With the highly integrated X-by-wire module, the reconfigurable chassis can be formed with different numbers of X-by-wire modules according to different demands, and applied to different autonomous mobile platforms, e.g., four X-by-wire modules forming the autonomous passenger car, and eight X-by-wire modules forming the autonomous truck. Once the mission is finished, X-by-wire modules will be separated and ready for reorganization for the next mission.

The second challenge is that the mechanical structure and integration technique of the integrated X-by-wire module are not mature, especially in terms of dealing with extreme conditions. According to the literature review of the integrated X-by-wire module, it can be found that most X-by-wire modules adopt simple suspension structures, which cannot withstand huge lateral force. Therefore, existing 4WID-4WIS EVs can only travel in common conditions; they cannot deal with severe and extreme conditions. Therefore, it is necessary to design an advanced and practical X-by-wire module for the future applications of 4WID-4WIS EVs.

The third challenge is the reliability limitation of the X-by-wire technique. Compared with traditional mechanical systems, the reliability and safety of the X-by-wire technique are worse, and are generally untrustworthy. Since the 4WID-4WIS EV has 12 control actuators involved in steering, drive and braking, the probability of an actuator fault is still a crucial issue. Additionally, considering that the X-by-wire technique, and especially the steering-by-wire technique, is not a mature technique, it is necessary to design an effective active fault-tolerant control system to guarantee the functional safety of the system.

The last challenge is the control technique. For 4WID-4WIS EVs, which have non-linear MIMO control systems, it is not easy to deal with the parametric uncertainties, external disturbances, and sensor noise with simple control algorithms, e.g., PID control. Although some control algorithms can realize accurate dynamic control and have good robustness, e.g., MPC, real-time optimization brings a large amount of calculation, which is a challenge for the hardware platform. Therefore, improving the computing efficiency of control algorithms is an urgent task. Additionally, as for multi-objective control, i.e., handling stability control, rollover prevention control, and path tracking control, there is no good adaptive control strategy to adjust the control priority and weighting to deal with different cases. For instance, at low-speed conditions, path tracking is the main control task. However, handling stability control and rollover prevention control must be given priority in extreme conditions.

6. Conclusions

Focusing on chassis configuration and control techniques, a literature review of—and various perspectives on—4WID-4WIS EVs are presented in this paper. Various prototypes of 4WID-4WIS EVs and integrated X-by-wire modules are introduced. Different chassis configurations and mechanical structures are compared and analyzed. Furthermore, the steering modes and switching logics of 4WID-4WIS EV are discussed. In addition, the common control models of 4WID-4WIS EV are summarized, including the kinematic model, dynamic model, and path tracking model. Based on different control models, different control objectives can be realized, including handling stability control, rollover prevention control, path tracking control, and active fault-tolerant control. For different control objectives, the control algorithms are reviewed and analyzed. Finally, for the development and application of 4WID-4WIS EV, some challenges, and perspectives are discussed, including the cost, mechanical design, control technique, etc.

Author Contributions: Writing—original draft preparation, P.H.; writing—review and editing, X.C. Both authors have read and agreed to the published version of the manuscript.

Funding: This research was funded by the National Key R&D Program of China (Grant No. 2018YFB0104802).

Conflicts of Interest: The authors declare no conflict of interest.

References

1. Chen, G.; Cao, H.; Conradt, J.; Tang, H.; Rohrbein, F.; Knoll, A. Event-based neuromorphic vision for autonomous driving: A paradigm shift for bio-inspired visual sensing and perception. *IEEE Signal Process. Mag.* **2020**, *37*, 34–49. [CrossRef]
2. Liu, Y.; Chen, G.; Knoll, A. Globally optimal vertical direction estimation in Atlanta World. *IEEE Trans. Pattern Anal. Mach. Intell.* **2020**. [CrossRef] [PubMed]
3. Hang, P.; Lv, C.; Huang, C.; Xing, Y.; Hu, Z. Cooperative decision making of connected automated vehicles at multi-lane merging zone: A coalitional game approach. *IEEE Trans. Intell. Transp. Syst.* **2021**. [CrossRef]
4. Xing, Y.; Lv, C.; Cao, D.; Hang, P. Toward human-vehicle collaboration: Review and perspectives on human-centered collaborative automated driving. *Transp. Res. Part C Emerg. Technol.* **2021**, *128*, 103199. [CrossRef]
5. Huang, C.; Huang, H.; Hang, P.; Gao, H.; Wu, J.; Huang, Z.; Lv, C. Personalized Trajectory Planning and Control of Lane-Change Maneuvers for Autonomous Driving. *IEEE Trans. Veh. Technol.* **2021**, *70*, 5511–5523. [CrossRef]
6. Chen, X.; Hang, P.; Wang, W.; Li, Y. Design and analysis of a novel wheel type continuously variable transmission. *Mech. Mach. Theory* **2017**, *107*, 13–26. [CrossRef]
7. Huang, C.; Lv, C.; Hang, P.; Xing, Y. Toward Safe and Personalized Autonomous Driving: Decision-Making and Motion Control with DPF and CDT Techniques. *IEEE/ASME Trans. Mechatron.* **2021**, *26*, 611–620. [CrossRef]
8. Xing, Y.; Lv, C.; Mo, X.; Hu, Z.; Huang, C.; Hang, P. Toward Safe and Smart Mobility: Energy-Aware Deep Learning for Driving Behavior Analysis and Prediction of Connected Vehicles. *IEEE Trans. Intell. Transp. Syst.* **2021**, *22*, 4267–4280. [CrossRef]
9. Van Rensselar, J. New vehicle technology and market innovation. *Tribol. Lubr. Technol.* **2020**, *76*, 26–34.
10. Biček, M.; Pepelnjak, T.; Pušavec, F. Production aspect of direct drive in-wheel motors. *Procedia CIRP* **2019**, *81*, 1278–1283. [CrossRef]
11. Unseld, R. The next generation of vehicles will no longer have mechanical steering. *ATZelectron. Worldw.* **2020**, *15*, 14–17. [CrossRef]
12. Ni, J.; Hu, J.; Xiang, C. Robust control in diagonal move steer mode and experiment on an X-by-wire UGV. *IEEE/ASME Trans. Mechatron.* **2019**, *24*, 572–584. [CrossRef]

13. Ni, J.; Hu, J.; Xiang, C. A review for design and dynamics control of unmanned ground vehicle. *Proc. Inst. Mech. Eng. Part D J. Automob. Eng.* **2021**, *235*, 1084–1100. [CrossRef]
14. Ni, J.; Hu, J.; Xiang, C. An AWID and AWIS X-by-wire UGV: Design and hierarchical chassis dynamics control. *IEEE Trans. Intell. Transp. Syst.* **2018**, *20*, 654–666. [CrossRef]
15. Jiang, D.; Danhua, L.; Wang, S.; Tain, L.; Yang, L. Additional yaw moment control of a 4WIS and 4WID agricultural data acquisition vehicle. *Int. J. Adv. Robot. Syst.* **2015**, *12*, 78. [CrossRef]
16. Demirci, M.; Gokasan, M. Adaptive optimal control allocation using Lagrangian neural networks for stability control of a 4WS–4WD electric vehicle. *Trans. Inst. Meas. Control* **2013**, *35*, 1139–1151. [CrossRef]
17. Fahimi, F. Full drive-by-wire dynamic control for four-wheel-steer all-wheel-drive vehicles. *Veh. Syst. Dyn.* **2013**, *51*, 360–376. [CrossRef]
18. Liang, Y.; Li, Y.; Yu, Y.; Zheng, L. Integrated lateral control for 4WID/4WIS vehicle in high-speed condition considering the magnitude of steering. *Veh. Syst. Dyn.* **2020**, *58*, 1711–1735. [CrossRef]
19. Ko, J.W.; Ko, S.Y.; Kim, I.S.; Hyun, D.Y.; Kim, H.S. Co-operative control for regenerative braking and friction braking to increase energy recovery without wheel lock. *Int. J. Automot. Technol.* **2014**, *15*, 253–262. [CrossRef]
20. Milanés, V.; González, C.; Naranjo, J.E.; Onieva, E.; Pedro, T.D. Electro-hydraulic braking system for autonomous vehicles. *Int. J. Automot. Technol.* **2010**, *11*, 89–95. [CrossRef]
21. Xu, G.; Li, W.; Xu, K.; Song, Z. An intelligent regenerative braking strategy for electric vehicles. *Energies* **2011**, *4*, 1461–1477. [CrossRef]
22. Lam, T.L.; Yan, J.; Qian, H.; Xu, Y. Traction/braking force distribution algorithm for omni-directional all-wheel-independent-drive vehicles. In Proceedings of the 2013 IEEE International Conference on Robotics and Automation, Karlsruhe, Germany, 6–10 May 2013.
23. Lee, M.H.; Li, T.H.S. Kinematics, dynamics and control design of 4WIS4WID mobile robots. *J. Eng.* **2015**, *2015*, 6–16. [CrossRef]
24. Bünte, T.; Ho, L.M.; Satzger, C.; Brembeck, J. Central vehicle dynamics control of the robotic research platform robomobil. *ATZelektron. Worldw.* **2014**, *9*, 58–64. [CrossRef]
25. De Castro, R.; Bünte, T.; Brembeck, J. Design and validation of the second generation of the robomobil's vehicle dynamics controller. In Proceedings of the 24th Symposium of the International Association for Vehicle System Dynamics (IAVSD 2015), Graz, Austria, 17–21 August 2015.
26. Birnschein, T.; Kirchner, F.; Girault, B.; Yuksel, M.; MacHowinski, J. An innovative, comprehensive concept for energy efficient electric mobility-EO smart connecting car. In Proceedings of the 2012 IEEE International Energy Conference and Exhibition (ENERGYCON), Florence, Italy, 9–12 September 2012.
27. Kraus, M.; Harkort, C.; Wuebbolt-Gorbatenko, B.; Laumann, M. Fusion of Drive and Chassis for a People Mover. *ATZ Worldw.* **2018**, *120*, 46–51. [CrossRef]
28. Li, C.; Song, P.; Chen, G.; Zong, C.; Liu, W. *Driving and Steering Coordination Control for 4WID/4WIS Electric Vehicle*; SAE Technical Paper; SAE International: Warrendale, PA, USA, 2015.
29. Liu, Y.; Zong, C.; Zhang, D.; Zheng, H.; Han, X.; Sun, M. Fault-tolerant control approach based on constraint control allocation for 4WIS/4WID vehicles. *Proc. Inst. Mech. Eng. Part D J. Automob. Eng.* **2021**, *235*, 2281–2295. [CrossRef]
30. Lam, T.L.; Qian, H.; Xu, Y. Omnidirectional steering interface and control for a four-wheel independent steering vehicle. *IEEE/ASME Trans. Mechatron.* **2009**, *15*, 329–338.
31. Qian, H.; Lam, T.L.; Li, W.; Xia, C. System and design of an omni-directional vehicle. In Proceedings of the 2008 IEEE International Conference on Robotics and Biomimetics, Bangkok, Thailand, 21–26 February 2009.
32. Lam, T.L.; Qian, H.; Xu, Y.; Xu, G. Omni-directional steer-by-wire interface for four wheel independent steering vehicle. In Proceedings of the 2009 IEEE International Conference on Robotics and Automation, Kobe, Japan, 12–17 May 2009.
33. Ashley, S. Shrink-to-Fit Car for City Parking. In *The New York Times*; The New York Times Company: New York, NY, USA, 2012.
34. Rasul, M.H.; Zamzuri, H.; Mustafa, A.M.A.; Ariff, M.H.M. Development of 4WIS SBW in-wheel drive compact electric vehicle platform. In Proceedings of the 2015 10th Asian Control Conference (ASCC), Sabah, Malaysia, 31 May–3 June 2015.
35. Hang, P.; Chen, X. Path tracking control of 4-wheel-steering autonomous ground vehicles based on linear parameter-varying system with experimental verification. *Proc. Inst. Mech. Eng. Part I J. Syst. Control Eng.* **2021**, *235*, 411–423.
36. Hang, P.; Chen, X.; Luo, F. *Path-Tracking Controller Design for a 4WIS and 4WID Electric Vehicle with Steer-by-Wire System*; SAE Technical Paper; SAE International: Warrendale, PA, USA, 2017.
37. Hang, P.; Han, Y.; Chen, X.; Zhang, B. Design of an active collision avoidance system for a 4WIS-4WID electric vehicle. *IFAC PapersOnLine* **2018**, *51*, 771–777. [CrossRef]
38. Choi, M.W.; Park, J.S.; Lee, B.S.; Lee, M.H. The performance of independent wheels steering vehicle (4WS) applied Ackerman geometry. In Proceedings of the 2008 International Conference on Control, Automation and Systems, Seoul, Korea, 14–17 October 2008.
39. Tu, X. Robust Navigation Control and Headland Turning Optimization of Agricultural Vehicles. Ph.D. Thesis, Iowa State University, Ames, IA, USA, 2013.
40. Jung, S.; Guenther, D.A. *An Examination of the Maneuverability of an All Wheel Steer Vehicle at Low Speed*; SAE Technical Paper; SAE International: Warrendale, PA, USA, 1991.

41. Hang, P.; Chen, X.; Fang, S.; Luo, F. Robust control for four-wheel-independent-steering electric vehicle with steer-by-wire system. *Int. J. Automot. Technol.* **2017**, *18*, 785–797. [CrossRef]
42. Kobayashi, T.; Katsuyama, E.; Sugiura, H.; Ono, E.; Yamamoto, M. Efficient direct yaw moment control: Tyre slip power loss minimisation for four-independent wheel drive vehicle. *Veh. Syst. Dyn.* **2018**, *56*, 719–733. [CrossRef]
43. Chen, X.; Fang, S.; Hang, P.; Luo, F.; Wu, X. The steering modes switching control of four wheels independently steering-by-wire vehicle based on path planning. In Proceedings of the FISITA 2016 World Automotive Congress, Busan, South Korea, 26–30 September 2016.
44. Chen, X.; Luo, F.; Hang, P.; Luo, J. Steering Mode Switch Control of Four-Wheel-Independent-Steering Electric Vehicle. In Proceedings of the 19th Asia Pacific Automotive Engineering Conference & SAE-China Congress 2017, Shanghai, China, 24–26 October 2017; pp. 437–453.
45. Xu, F.; Liu, X.; Chen, W.; Zhou, C. Dynamic switch control of steering modes for four wheel independent steering rescue vehicle. *IEEE Access* **2019**, *7*, 135595–135605. [CrossRef]
46. Lai, X.; Chen, X.B.; Wu, X.J.; Liang, D. A study on control system for four-wheels independent driving and steering electric vehicle. *Appl. Mech. Mater.* **2015**, *701*, 807–811. [CrossRef]
47. Hang, P.; Chen, X.; Fang, S. Controller design of steer-by-wire 4WIS electric vehicle with various steering modes. In Proceedings of the FISITA 2016 World Automotive Congress, Busan, Korea, 26–30 September 2016.
48. Bai, G.; Liu, L.; Meng, Y.; Luo, W.; Gu, Q.; Ma, B. Path tracking of mining vehicles based on nonlinear model predictive control. *Appl. Sci.* **2019**, *9*, 1372. [CrossRef]
49. Chen, X.; Han, Y.; Hang, P. Researches on 4WIS-4WID Stability with LQR Coordinated 4WS and DYC. In *The IAVSD International Symposium on Dynamics of Vehicles on Roads and Tracks*; Springer: Cham, Switzerland, 2019; pp. 1508–1516.
50. Hang, P.; Luo, F.; Fang, S.; Chen, X. Path tracking control of a four-wheel-independent-steering electric vehicle based on model predictive control. In Proceedings of the 2017 36th Chinese Control Conference (CCC), Dalian, China, 26–28 July 2017.
51. Hang, P.; Chen, X. Integrated chassis control algorithm design for path tracking based on four-wheel steering and direct yaw-moment control. *Proc. Inst. Mech. Eng. Part I J. Syst. Control Eng.* **2019**, *233*, 625–641. [CrossRef]
52. Hang, P.; Chen, X.; Zhang, B.; Tang, T. Longitudinal velocity tracking control of a 4WID electric vehicle. *IFAC PapersOnLine* **2018**, *51*, 790–795. [CrossRef]
53. Baffet, G.; Charara, A.; Lechner, D. Estimation of vehicle sideslip, tire force and wheel cornering stiffness. *Control Eng. Pract.* **2009**, *17*, 1255–1264. [CrossRef]
54. Ray, L.R. Nonlinear tire force estimation and road friction identification: Simulation and experiments. *Automatica* **1997**, *33*, 1819–1833. [CrossRef]
55. Wang, R.; Wang, J. Tire–road friction coefficient and tire cornering stiffness estimation based on longitudinal tire force difference generation. *Control Eng. Pract.* **2013**, *21*, 65–75. [CrossRef]
56. Pacejka, H.B.; Bakker, E. The magic formula tyre model. *Veh. Syst. Dyn.* **1992**, *21*, 1–18. [CrossRef]
57. Ding, N.; Taheri, S. A modified Dugoff tire model for combined-slip forces. *Tire Sci. Technol.* **2010**, *38*, 228–244. [CrossRef]
58. Guo, K.; Lu, D.; Chen, S.; Lin, W.C.; Lu, X. The UniTire model: A nonlinear and non-steady-state tyre model for vehicle dynamics simulation. *Veh. Syst. Dyn.* **2005**, *43*, 341–358. [CrossRef]
59. Leng, B.; Jin, D.; Xiong, L.; Yang, X.; Yu, Z. Estimation of tire-road peak adhesion coefficient for intelligent electric vehicles based on camera and tire dynamics information fusion. *Mech. Syst. Signal Process.* **2021**, *150*, 107275. [CrossRef]
60. Feng, Y.; Chen, H.; Zhao, H.; Zhou, H. Road tire friction coefficient estimation for four wheel drive electric vehicle based on moving optimal estimation strategy. *Mech. Syst. Signal Process.* **2020**, *139*, 106416. [CrossRef]
61. Kim, S.; Nikravesh, P.E.; Gim, G. A two-dimensional tire model on uneven roads for vehicle dynamic simulation. *Veh. Syst. Dyn.* **2008**, *46*, 913–930. [CrossRef]
62. Pacejka, H.B. Analysis of the dynamic response of a rolling string-type tire model to lateral wheel-plane vibrations. *Veh. Syst. Dyn.* **1972**, *1*, 37–66. [CrossRef]
63. Hsu, Y.H.J.; Laws, S.M.; Gerdes, J.C. Estimation of tire slip angle and friction limits using steering torque. *IEEE Trans. Control Syst. Technol.* **2009**, *18*, 896–907. [CrossRef]
64. Hang, P.; Chen, X.; Luo, F.; Fang, S. Robust control of a four-wheel-independent-steering electric vehicle for path tracking. *SAE Int. J. Veh. Dyn. Stab. NVH* **2017**, *1*, 307–316. [CrossRef]
65. Hang, P.; Huang, S.; Chen, X.; Tan, K.K. Path planning of collision avoidance for unmanned ground vehicles: A nonlinear model predictive control approach. *Proc. Inst. Mech. Eng. Part I J. Syst. Control Eng.* **2021**, *235*, 222–236.
66. Hang, P.; Lv, C.; Huang, C.; Cai, J.; Hu, Z.; Xing, Y. An integrated framework of decision making and motion planning for autonomous vehicles considering social behaviors. *IEEE Trans. Veh. Technol.* **2020**, *69*, 14458–14469. [CrossRef]
67. Hang, P.; Huang, C.; Hu, Z.; Xing, Y.; Lv, C. Decision Making of Connected Automated Vehicles at An Unsignalized Roundabout Considering Personalized Driving Behaviours. *IEEE Trans. Veh. Technol.* **2021**, *70*, 4051–4064. [CrossRef]
68. Bai, G.; Meng, Y.; Liu, L.; Luo, W.; Gu, Q. Review and comparison of path tracking based on model predictive control. *Electronics* **2019**, *8*, 1077. [CrossRef]
69. Hang, P.; Chen, X.; Luo, F. LPV/H∞ controller design for path tracking of autonomous ground vehicles through four-wheel steering and direct yaw-moment control. *Int. J. Automot. Technol.* **2019**, *20*, 679–691. [CrossRef]

70. Bai, G.; Liu, L.; Meng, Y.; Luo, W.; Gu, Q.; Wang, J. Path tracking of wheeled mobile robots based on dynamic prediction model. *IEEE Access* **2019**, *7*, 39690–39701. [CrossRef]
71. Han, Q.; Dai, L. A non-linear dynamic approach to the motion of four-wheel-steering vehicles under various operation conditions. *Proc. Inst. Mech. Eng. Part D J. Automob. Eng.* **2008**, *222*, 535–549. [CrossRef]
72. Durham, W.C. Constrained control allocation. *J. Guid. Control Dyn.* **1993**, *16*, 717–725. [CrossRef]
73. Oppenheimer, M.W.; Doman, D.B.; Bolender, M.A. Control allocation for over-actuated systems. In Proceedings of the 2006 14th Mediterranean Conference on Control and Automation, Ancona, Italy, 28–30 June 2006.
74. Harkegard, O. Efficient active set algorithms for solving constrained least squares problems in aircraft control allocation. In Proceedings of the 41st IEEE Conference on Decision and Control, Las Vegas, NV, USA, 10–13 December 2002.
75. Xiong, L.; Yu, Z.; Jiang, W.; Jiang, Z. Research on vehicle stability control of 4WD electric vehicle based on longitudinal force control allocation. *J. Tongji Univ. Nat. Sci.* **2010**, *38*, 417–421.
76. Zaccarian, L. Dynamic allocation for input redundant control systems. *Automatica* **2009**, *45*, 1431–1438. [CrossRef]
77. Skjong, S.; Pedersen, E. Nonangular MPC-based thrust allocation algorithm for marine vessels—a study of optimal thruster commands. *IEEE Trans. Transp. Electrif.* **2017**, *3*, 792–807. [CrossRef]
78. Hu, J.; Hu, Z.; Fu, C.; Nan, F. Integrated control of AFS and DYC for in-wheel-motor electric vehicles based on operation region division. *Int. J. Veh. Des.* **2019**, *79*, 221–247. [CrossRef]
79. Feng, T.; Wang, Y.; Li, Q. Coordinated control of active front steering and active disturbance rejection sliding mode-based DYC for 4WID-EV. *Meas. Control* **2020**, *53*, 1870–1882. [CrossRef]
80. Huang, G.; Yuan, X.; Shi, K.; Wu, X. A BP-PID controller-based multi-model control system for lateral stability of distributed drive electric vehicle. *J. Frankl. Inst.* **2019**, *356*, 7290–7311. [CrossRef]
81. Li, L.; Jia, G.; Chen, J.; Zhu, H.; Cao, D.; Song, J. A novel vehicle dynamics stability control algorithm based on the hierarchical strategy with constrain of nonlinear tyre forces. *Veh. Syst. Dyn.* **2015**, *53*, 1093–1116. [CrossRef]
82. Zhao, Y.; Zhang, C. Electronic stability control for improving stability for an eight in-wheel motor-independent drive electric vehicle. *Shock Vib.* **2019**, *2019*, 1–21. [CrossRef]
83. Chen, W.; Liang, X.; Wang, Q.; Zhao, L.; Wang, X. Extension coordinated control of four wheel independent drive electric vehicles by AFS and DYC. *Control Eng. Pract.* **2020**, *101*, 104504. [CrossRef]
84. Wu, J.; Zhao, Y.Q.; Ji, X.W.; Liu, Y.H.; Yin, C.Q. A modified structure internal model robust control method for the integration of active front steering and direct yaw moment control. *Sci. China Technol. Sci.* **2015**, *58*, 75–85. [CrossRef]
85. Canale, M.; Fagiano, L. Stability control of 4WS vehicles using robust IMC techniques. *Veh. Syst. Dyn.* **2008**, *46*, 991–1011. [CrossRef]
86. Li, M.; Jia, Y. Decoupling control in velocity-varying four-wheel steering vehicles with H∞ performance by longitudinal velocity and yaw rate feedback. *Veh. Syst. Dyn.* **2014**, *52*, 1563–1583. [CrossRef]
87. Li, M.; Jia, Y.; Du, J. LPV control with decoupling performance of 4WS vehicles under velocity-varying motion. *IEEE Trans. Control Syst. Technol.* **2014**, *22*, 1708–1724.
88. Li, M.; Jia, Y.; Matsuno, F. Attenuating diagonal decoupling with robustness for velocity-varying 4WS vehicles. *Control Eng. Pract.* **2016**, *56*, 49–59. [CrossRef]
89. Men, J.; Wu, B.; Chen, J.; Zhang, Z. Comparisons of vehicle stability controls based on 4WS, Brake, Brake-FAS and IMC techniques. *Veh. Syst. Dyn.* **2012**, *50*, 1053–1084.
90. Men, J.; Wu, B.; Chen, J. Comparisons of 4WS and Brake-FAS based on IMC for vehicle stability control. *J. Mech. Sci. Technol.* **2011**, *25*, 1265–1277.
91. Russell, H.E.B.; Gerdes, J.C. Design of variable vehicle handling characteristics using four-wheel steer-by-wire. *IEEE Trans. Control Syst. Technol.* **2015**, *24*, 1529–1540. [CrossRef]
92. Wu, Y.; Li, B.; Zhang, N.; Du, H.; Zhang, B. Rear-steering based decentralized control of four-wheel steering vehicle. *IEEE Trans. Veh. Technol.* **2020**, *69*, 10899–10913. [CrossRef]
93. Yaniv, O. Robustness to speed of 4WS vehicles for yaw and lateral dynamics. *Veh. Syst. Dyn.* **1997**, *27*, 221–234. [CrossRef]
94. Canale, M.; Fagiano, L. Comparing rear wheel steering and rear active differential approaches to vehicle yaw control. *Veh. Syst. Dyn.* **2010**, *48*, 529–546. [CrossRef]
95. Yu, H.; Gao, L. Two-degree-of-freedom vehicle steering controllers design based on four-wheel-steering-by-wire. *Int. J. Veh. Auton. Syst.* **2007**, *5*, 47–78. [CrossRef]
96. Zhao, W.; Qin, X.; Wang, C. Yaw and lateral stability control for four-wheel steer-by-wire system. *IEEE/ASME Trans. Mechatron.* **2018**, *23*, 2628–2637. [CrossRef]
97. Zhao, W.Z.; Qin, X.X. Study on mixed H2/H∞ robust control strategy of four wheel steering system. *Sci. China Technol. Sci.* **2017**, *60*, 1831–1840. [CrossRef]
98. Li, M.; Jia, Y. Precompensation decoupling control with H∞ performance for 4WS velocity-varying vehicles. *Int. J. Syst. Sci.* **2016**, *47*, 3864–3875. [CrossRef]
99. Yin, G.; Chen, N.; Li, P. Improving Handling Stability Performance of Four-Wheel Steering Vehicle via μ-Synthesis Robust Control. *IEEE Trans. Veh. Technol.* **2007**, *56*, 2432–2439. [CrossRef]
100. Yin, G.D.; Chen, N.; Wang, J.X.; Wu, L.Y. A study on μ-synthesis control for four-wheel steering system to enhance vehicle lateral stability. *J. Dyn. Syst. Meas. Control* **2011**, *133*, 011002. [CrossRef]

101. Abe, M.; Ohkubo, N.; Kano, Y. Comparison of 4WS and Direct Yaw Moment Control (DYC) for Improvement of Vehicle Handling Performance. *JSAE Rev.* **1995**, *2*, 214.
102. Abe, M.; Ohkubo, N.; Kano, Y. A direct yaw moment control for improving limit performance of vehicle handling-comparison and cooperation with 4WS. *Veh. Syst. Dyn.* **1996**, *25*, 3–23. [CrossRef]
103. Tianjun, Z.; Changfu, Z. Research on control algorithm for DYC and integrated control with 4WS. In Proceedings of the 2009 International Conference on Computational Intelligence and Natural Computing, Wuhan, China, 6–7 June 2009.
104. Nagai, M.; Hirano, Y.; Yamanaka, S. Integrated control of active rear wheel steering and direct yaw moment control. *Veh. Syst. Dyn.* **1997**, *27*, 357–370. [CrossRef]
105. Zhou, L.; Ou, L.; Wang, C. A simulation of the four-wheel steering vehicle stability based on DYC control. In Proceedings of the 2009 International Conference on Measuring Technology and Mechatronics Automation, Zhangjiajie, China, 11–12 April 2009.
106. Hang, P.; Xia, X.; Chen, X. Handling Stability Advancement With 4WS and DYC Coordinated Control: A Gain-Scheduled Robust Control Approach. *IEEE Trans. Veh. Technol.* **2021**, *70*, 3164–3174. [CrossRef]
107. Furukawa, Y.; Abe, M. Advanced chassis control systems for vehicle handling and active safety. *Veh. Syst. Dyn.* **1997**, *28*, 59–86. [CrossRef]
108. Chen, B.C.; Peng, H. Differential-braking-based rollover prevention for sport utility vehicles with human-in-the-loop evaluations. *Veh. Syst. Dyn.* **2001**, *36*, 359–389. [CrossRef]
109. Rajamani, R.; Piyabongkarn, D.N. New paradigms for the integration of yaw stability and rollover prevention functions in vehicle stability control. *IEEE Trans. Intell. Transp. Syst.* **2012**, *14*, 249–261. [CrossRef]
110. Yoon, J.; Cho, W.; Koo, B.; Yi, K. Unified chassis control for rollover prevention and lateral stability. *IEEE Trans. Veh. Technol.* **2008**, *58*, 596–609. [CrossRef]
111. Huang, H.H.; Yedavalli, R.K.; Guenther, D.A. Active roll control for rollover prevention of heavy articulated vehicles with multiple-rollover-index minimisation. *Veh. Syst. Dyn.* **2012**, *50*, 471–493. [CrossRef]
112. Yim, S. Design of a preview controller for vehicle rollover prevention. *IEEE Trans. Veh. Technol.* **2011**, *60*, 4217–4226. [CrossRef]
113. Schofield, B.; Hagglund, T.; Rantzer, A. Vehicle dynamics control and controller allocation for rollover prevention. In Proceedings of the 2006 IEEE Conference on Computer Aided Control System Design, 2006 IEEE International Conference on Control Applications, 2006 IEEE International Symposium on Intelligent Control, Munich, Germany, 4–6 October 2006.
114. Zhang, B.; Khajepour, A.; Goodarzi, A. Vehicle yaw stability control using active rear steering: Development and experimental validation. *Proc. Inst. Mech. Eng. Part K J. Multi-Body Dyn.* **2017**, *231*, 333–345. [CrossRef]
115. Zhang, Y.; Khajepour, A.; Xie, X. Rollover prevention for sport utility vehicles using a pulsed active rear-steering strategy. *Proc. Inst. Mech. Eng. Part D J. Automob. Eng.* **2016**, *230*, 1239–1253. [CrossRef]
116. Hang, P.; Chen, X.; Wang, W. Cooperative Control Framework for Human Driver and Active Rear Steering System to Advance Active Safety. *IEEE Trans. Intell. Veh.* **2020**. [CrossRef]
117. Alfi, A.; Farrokhi, M. Hybrid state-feedback sliding-mode controller using fuzzy logic for four-wheel-steering vehicles. *Veh. Syst. Dyn.* **2009**, *47*, 265–284. [CrossRef]
118. Tian, J.; Ding, J.; Tai, Y.; Chen, N. Hierarchical control of nonlinear active four-wheel-steering vehicles. *Energies* **2018**, *11*, 2930. [CrossRef]
119. Song, J. Integrated control of brake pressure and rear-wheel steering to improve lateral stability with fuzzy logic. *Int. J. Automot. Technol.* **2012**, *13*, 563–570. [CrossRef]
120. Lee, S.; Yakub, F.; Kasahara, M.; Mori, Y. Rollover prevention with predictive control of differential braking and rear wheel steering. In Proceedings of the 2013 6th IEEE Conference on Robotics, Automation and Mechatronics (RAM), Manila, Philippines, 12–15 November 2013.
121. Zhou, Z.; Miaohua, H.; Yachao, Z.; Chen, F. Vehicle Stability Control through Optimized Coordination of Active Rear Steering and Differential Driving/Braking. *SAE Int. J. Passeng. Cars Mech. Syst.* **2018**, *11*, 239–248. [CrossRef]
122. Wu, J.; Cheng, S.; Liu, B.; Liu, C. A human-machine-cooperative-driving controller based on AFS and DYC for vehicle dynamic stability. *Energies* **2017**, *10*, 1737. [CrossRef]
123. He, X.; Yang, K.; Liu, Y.; Ji, X. *A Novel Direct Yaw Moment Control System for Autonomous Vehicle*; SAE Technical Paper; SAE International: Warrendale, PA, USA, 2018.
124. Lin, F.; Zhang, Y.; Zhao, Y.; Yin, G.; Zhang, H.; Wang, K. Trajectory tracking of autonomous vehicle with the fusion of DYC and longitudinal–lateral control. *Chin. J. Mech. Eng.* **2019**, *32*, 1–16. [CrossRef]
125. Mashadi, B.; Ahmadizadeh, P.; Majidi, M. *Integrated Controller Design for Path Following in Autonomous Vehicles*; SAE Technical Paper; SAE International: Warrendale, PA, USA, 2011.
126. Tu, X.; Gai, J.; Tang, L. Robust navigation control of a 4WD/4WS agricultural robotic vehicle. *Comput. Electron. Agric.* **2019**, *164*, 104892. [CrossRef]
127. Lei, Y.; Wen, G.; Fu, Y.; Li, X.; Hou, B.; Geng, X. Trajectory-following of a 4WID-4WIS vehicle via feedforward–backstepping sliding-mode control. *Proc. Inst. Mech. Eng. Part D J. Automob. Eng.* **2021**. [CrossRef]
128. Li, B.; Du, H.; Li, W. Trajectory control for autonomous electric vehicles with in-wheel motors based on a dynamics model approach. *IET Intell. Transp. Syst.* **2016**, *10*, 318–330. [CrossRef]
129. Guo, J.; Luo, Y.; Li, K.; Dai, Y. Coordinated path-following and direct yaw-moment control of autonomous electric vehicles with sideslip angle estimation. *Mech. Syst. Signal Process.* **2018**, *105*, 183–199. [CrossRef]

130. Liu, R.; Wei, M.; Zhao, W. Trajectory tracking control of four wheel steering under high speed emergency obstacle avoidance. *Int. J. Veh. Des.* **2018**, *77*, 1–21. [CrossRef]
131. Hu, C.; Wang, R.; Yan, F.; Chadli, M.; Chen, N. Output constraint control on path following of four-wheel independently actuated autonomous vehicles. In Proceedings of the 2015 American Control Conference (ACC), Chicago, IL, USA, 1–3 July 2015.
132. Hiraoka, T.; Nishihara, O.; Kumamoto, H. Automatic path-tracking controller of a four-wheel steering vehicle. *Veh. Syst. Dyn.* **2009**, *47*, 1205–1227. [CrossRef]
133. Chen, T.; Chen, L.; Xu, X.; Cai, Y.; Sun, X. Simultaneous path following and lateral stability control of 4WD-4WS autonomous electric vehicles with actuator saturation. *Adv. Eng. Softw.* **2019**, *128*, 46–54. [CrossRef]
134. Mashadi, B.; Ahmadizadeh, P.; Majidi, M.; Mahmoodi-Kaleybar, M. Integrated robust controller for vehicle path following. *Multibody Syst. Dyn.* **2015**, *33*, 207–228. [CrossRef]
135. Zhang, H.; Heng, B.; Zhao, W. Path tracking control for active rear steering vehicles considering driver steering characteristics. *IEEE Access* **2020**, *8*, 98009–98017. [CrossRef]
136. Liu, R.; Wei, M.; Sang, N.; Wei, J. Research on curved path tracking control for four-wheel steering vehicle considering road adhesion coefficient. *Math. Probl. Eng.* **2020**, *2020*, 1–18. [CrossRef]
137. Yu, C.; Zheng, Y.; Shyrokau, B.; Ivanov, V. MPC-based path following design for automated vehicles with rear wheel steering. In Proceedings of the 2021 IEEE International Conference on Mechatronics (ICM), Kashiwa, Japan, 7–9 March 2021.
138. Wang, Y.; Zong, C.; Li, K.; Chen, H. Fault-tolerant control for in-wheel-motor-driven electric ground vehicles in discrete time. *Mech. Syst. Signal Process.* **2019**, *121*, 441–454. [CrossRef]
139. Li, C.; Chen, G.; Zong, C. *Fault-Tolerant Control for 4WID/4WIS Electric Vehicles*; SAE Technical Paper; SAE International: Warrendale, PA, USA, 2014. [CrossRef]
140. Shi, K.; Yuan, X. MPC-based fault tolerant control system for yaw stability of distributed drive electric vehicle. In Proceedings of the 2019 3rd Conference on Vehicle Control and Intelligence (CVCI), Hefei, China, 21–22 September 2019.
141. Liu, L.; Shi, K.; Yuan, X.; Li, Q. Multiple model-based fault-tolerant control system for distributed drive electric vehicle. *J. Braz. Soc. Mech. Sci. Eng.* **2019**, *41*, 1–15. [CrossRef]
142. Zhang, H.; Zhao, W.; Wang, J. Fault-tolerant control for electric vehicles with independently driven in-wheel motors considering individual driver steering characteristics. *IEEE Trans. Veh. Technol.* **2019**, *68*, 4527–4536. [CrossRef]
143. Zhang, D.; Liu, G.; Zhou, H.; Zhao, W. Adaptive sliding mode fault-tolerant coordination control for four-wheel independently driven electric vehicles. *IEEE Trans. Ind. Electron.* **2018**, *65*, 9090–9100. [CrossRef]
144. Li, B.; Du, H.; Li, W. Fault-tolerant control of electric vehicles with in-wheel motors using actuator-grouping sliding mode controllers. *Mech. Syst. Signal Process.* **2016**, *72*, 462–485. [CrossRef]
145. Guo, B.; Chen, Y. Robust adaptive fault-tolerant control of four-wheel independently actuated electric vehicles. *IEEE Trans. Ind. Inform.* **2019**, *16*, 2882–2894. [CrossRef]
146. Zhang, B.; Lu, S.; Zhao, L.; Xiao, K. Fault-tolerant control based on 2D game for independent driving electric vehicle suffering actuator failures. *Proc. Inst. Mech. Eng. Part D J. Automob. Eng.* **2020**, *234*, 3011–3025. [CrossRef]
147. Zhang, B.; Lu, S. Fault-tolerant control for four-wheel independent actuated electric vehicle using feedback linearization and cooperative game theory. *Control Eng. Pract.* **2020**, *101*, 104510. [CrossRef]
148. Luo, Y.; Luo, J.; Qin, Z. Model-independent self-tuning fault-tolerant control method for 4WID EV. *Int. J. Automot. Technol.* **2016**, *17*, 1091–1100. [CrossRef]
149. Liu, C.; Zong, C.; He, L.; Li, C.; Liu, M. Actuator Fault Detection and Diagnosis of 4WID/4WIS Electric Vehicles. *SAE Int. J. Passeng. Cars Electron. Electr. Syst.* **2013**, *7*. [CrossRef]
150. Zhang, G.; Zhang, H.; Huang, X.; Wang, J.; Yu, H.; Graaf, R. Active fault-tolerant control for electric vehicles with independently driven rear in-wheel motors against certain actuator faults. *IEEE Trans. Control Syst. Technol.* **2015**, *24*, 1557–1572. [CrossRef]
151. Liu, C.; Zong, C.; He, L.; Liu, J.; Li, C. *Passive Fault-Tolerant Performance of 4WID/4WIS Electric Vehicles Based on MPC and Control Allocation*; SAE Technical Paper; SAE International: Warrendale, PA, USA, 2013.
152. Chen, T.; Chenc, L.; Xu, X.; Cai, Y.; Jiang, H.; Sun, X. Passive fault-tolerant path following control of autonomous distributed drive electric vehicle considering steering system fault. *Mech. Syst. Signal Process.* **2019**, *123*, 298–315. [CrossRef]
153. Mihály, A.; Gáspár, P. Reconfgurable fault-tolerant control of in-wheel electric vehicles with steering system failure. *IFAC PapersOnLine* **2015**, *48*, 49–54. [CrossRef]
154. Zhang, Y.; Zheng, H.; Zhang, J.; Cheng, C. *A Fault-Tolerant Control Method for 4WIS/4WID Electric Vehicles Based on Reconfigurable Control Allocation*; SAE Technical Paper; SAE International: Warrendale, PA, USA, 2018. [CrossRef]

Article

In-Wheel Two-Speed AMT with Selectable One-Way Clutch for Electric Vehicles

Dele Meng [1,2], Fei Wang [1], Yuhai Wang [1,*] and Bingzhao Gao [3]

[1] State Key Laboratory of Automotive Simulation and Control, Jilin University, Changchun 130012, China; mengdl19@mails.jlu.edu.cn (D.M.); wangfei_jlu@jlu.edu.cn (F.W.)
[2] Qingdao Automotive Research Institute, Jilin University, Qingdao 266108, China
[3] Clean Energy Automotive Engineering Center, Tongji University, Shanghai 201804, China; gaobz@tongji.edu.cn
[*] Correspondence: wangyuhai@jlu.edu.cn

Abstract: To improve the efficiency of the electric vehicle (EV) drive systems and EV performance, the use of multi-speed transmissions and distributed drives has been studied extensively. In addition, to develop efficient and compact drive systems, new clutch solutions are needed. In this paper, we propose an in-wheel two-speed automatic mechanical transmission (IW-AMT) with a selectable one-way clutch (SOWC). The IW-AMT consists of a high-speed motor and a mechanical shift actuator, and it can realize shifting without power interruption, thus effectively reducing the unsprung mass and the technical specifications of the motor. We established a virtual prototype model of the IW-AMT to show the shifting process and evaluate the quality of shifting. The simulation results of the upshifting process indicated that the vehicle torque and velocity changed smoothly, and the maximum jerk is less than 10 m/s^3. Furthermore, to improve the jerk induced by the downshifting process, we analyzed the momentary state of the SOWC struts that are dropped and attempted to improve the jerk from two aspects: improving the wet multi-plate clutch (WMPC) combination curve and improving the SOWC structure. The results indicated that the downshift-induced jerk can be reduced to 13 m/s^3.

Keywords: electric vehicles; two-speed AMT; in-wheel-drive; shifting process; selectable one-way clutch

Citation: Meng, D.; Wang, F.; Wang, Y.; Gao, B. In-Wheel Two-Speed AMT with Selectable One-Way Clutch for Electric Vehicles. *Actuators* **2021**, *10*, 220. https://doi.org/10.3390/act10090220

Academic Editor: Hai Wang

Received: 28 July 2021
Accepted: 31 August 2021
Published: 2 September 2021

Publisher's Note: MDPI stays neutral with regard to jurisdictional claims in published maps and institutional affiliations.

1. Introduction

Electric vehicles (EVs), as a promising way to reduce the greenhouse effect and alleviate the problem of climate change, have been extensively studied [1]. As a key component of EVs, the drive system has a direct impact on the energy efficiency of EVs. Therefore, it is extremely important to improve the efficiency of the drive system while satisfying vehicle performance standards [2].

Studies have demonstrated that integrating a multi-speed gearbox into the drive system can effectively help to utilize the high-efficiency range, improve motor efficiency, and reduce energy consumption [3,4]. Continuously variable transmission (CVT), and dual-clutch transmission (DCT) have multiple gears, but they have complex structures and high manufacturing costs [5,6]. By contrast, automatic mechanical transmission (AMT) has a simple structure, low manufacturing cost, and high transmission efficiency [7,8]. Owing to the working characteristics of the motor, EVs do not need many gears. Therefore, a two-speed AMT can serve as an economical and effective solution to improve the efficiency of the EV drive system. In [9], a novel two-speed planetary AMT (PAMT) was proposed, which used a synchronizer and a brake band to achieve two-gear switching. In [10], a two-speed uninterrupted mechanical transmission (UMT) composed of a planetary gear set, brake belt, and centrifugal clutch that realized seamless switching between two gears was proposed. In [11], a fork-less two-speed AMT (I-AMT) with a dry clutch was proposed, without torque interruption during gear shifting.

In addition, to shorten the distance of the transmission chain, reduce the energy loss of the transmission joint, and further improve the efficiency of the drive system, various scholars and manufacturers have conducted extensive research on distributed in-wheel drive systems [12,13]. Protean proposed the concept of integrating the motor and brake and connecting the motor directly to drive the wheels [14]. Schaeffler developed the E-Wheel Drive system by further integrating the drive motor, electrical equipment, and braking and cooling systems [15]. However, the use of in-wheel motors significantly increases the unsprung mass of a vehicle, which degrades vehicle handling and ride quality. Moreover, an in-wheel motor is less efficient when the vehicle is operated under diverse conditions. By installing a miniaturized high-speed motor in the wheel to match a two-speed transmission drive scheme, the work efficiency of the motor can increase, and EV performance can be improved. In [16], a wheel-mounted two-speed configuration composed of two independent high-speed motors and two planetary gear sets was introduced. It was possible to smoothly switch between the two gears. NSK Ltd. optimized the above structure and successfully manufactured a test vehicle [17]. However, the problem of additional unsprung mass due to the dual-motor design was not effectively solved. In [18], a new configuration composed of one motor and an electromechanical shift actuator to achieve two gears in the wheel was proposed. This configuration effectively reduced the unsprung mass and improved system reliability.

Furthermore, to develop an efficient and compact drive system, a new clutch solution is needed [19]. With the advancement of mechatronics technology, a variety of new actuators are being used in drive systems to improve transmission efficiency [20]. General Motors has replaced the original one-way clutch (OWC) and low/reverse friction clutch with a selectable one-way clutch (SOWC) in its GF9 gearbox product series to avoid the use of friction clutches, reduce weight and cost, and improve transmission efficiency [21]. The SOWC has higher efficiency and torque density compared to the OWC, and it requires less packaging space [22]. Moreover, Ford and Honda have used the SOWC in their AT applications owing to its compact dimensions and high-efficiency [23,24]. In addition, Means Industry has introduced a new type of static SOWC for use in hybrid power systems to improve transmission efficiency [25]. It allows for the controllable locking of two independent components and performs motor actions to achieve precise synchronization.

In this paper, we describe an in-wheel two-speed AMT (IW-AMT) that uses the SOWC. Moreover, the proposed IW-AMT unit also consists of two planetary gear sets and a wet multi-plate clutch (WMPC) unit. The IW-AMT uses a single motor that cooperates with a mechanical shift actuator to realize the change of the two gears without power interruption, which effectively reduces the unsprung mass and motor technical specifications, in addition to improving vehicle performance. To demonstrate the proposed IW-AMT with the SOWC shifting process and evaluate the quality of shifting, we perform a simulation by using a virtual prototype simulation model of IW-AMT and improve the WMPC combination curve and the SOWC structure based on the simulation results to achieve superior shift quality.

In Section 2, we describe the structure and characteristics of IW-AMT with the SOWC. In Section 3, the ideal shifting process of IW-AMT is analyzed, which includes the upshift/downshift process. In Section 4, a virtual prototype simulation model of IW-AMT with the SOWC is established to demonstrate the changes in key parameters during gear shifting. In Section 5, the simulation results are discussed. Finally, our conclusions and future work are provided in Section 6.

2. Structure and Characteristics

2.1. Planetary Gear Set

To minimize the unsprung mass and improve the reliability of the drive system, we adopt a miniaturized high-speed drive motor. In addition, given the limited design space in the wheel and the high likelihood of interference between the transmission system and the steering and braking systems, the gear transmission unit must have strict design requirements. Considering the above problems, we use two-stage planetary gear trains as the transmission unit. A block diagram of IW-AMT and the planetary gear train are shown in Figures 1 and 2, respectively.

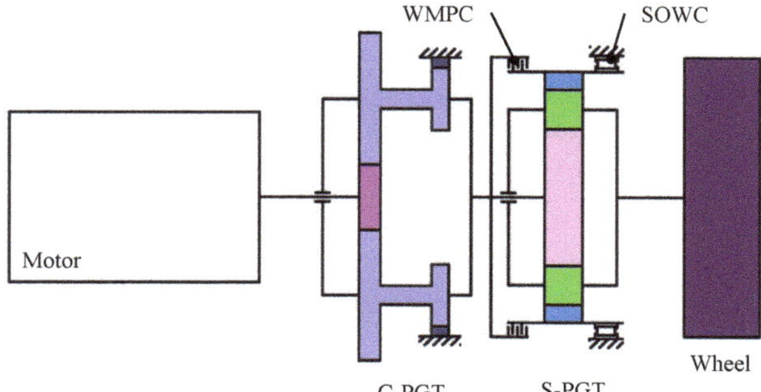

Figure 1. Block diagram of IW-AMT.

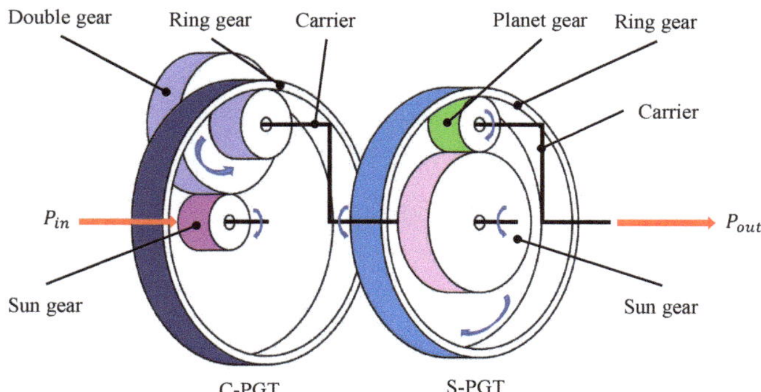

Figure 2. Block diagram of the planetary gear train.

As shown in Figure 2, we use a compound planetary gear train (C-PGT) composed of a sun gear, three double planetary gears mounted on the carrier, and ring gear to realize the first deceleration stage [18]. The sun gear meshes with the larger planet gears, and the ring gear meshes with the smaller planet gears. The driving motor power is input through the sun gear and output by the carrier of C-PGT. The C-PGT realizes a large fixed transmission ratio with less mass and packaging space.

$$v_{SP1} = \omega_{S1} r_{S1} \tag{1}$$

where v_{SP1} is the pitch linear velocity of the large planet gear and the sun gear, and r_{S1} and ω_{S1} are the radius and rotational speed of the sun gear, respectively.

Because the ring gear is fixed on the house rigidly, the contact point between the ring gear and the small planet gear has zero velocity, and the output velocity of the carrier v_{C1} can be expressed as follows:

$$v_{C1} = v_{SP1} \frac{r_{SP}}{r_{SP} + r_{LP}} \tag{2}$$

$$\omega_{C1} = \frac{v_{C1}}{r_{S1} + r_{LP}} \tag{3}$$

where r_{LP} and r_{SP} denote the radii of the large and small planets, respectively, and ω_{C1} is the rotational speed of the carrier.

The transmission ratio of C-PGT i_1 can be given as follows:

$$i_1 = \frac{\omega_{S1}}{\omega_{C1}} = \frac{r_{S1} + r_{LP}}{r_{S1}} \cdot \frac{r_{SP} + r_{LP}}{r_{SP}} \tag{4}$$

The other planetary gear set is a simple planetary gear train (S-PGT) composed of a sun gear, three planet gears, a carrier, and a ring gear. The power of the first stage is input to the sun gear and outputs to the wheel through the carrier. The S-PGT can achieve two gear ratio changes based on the action of the shift actuator. When the ring gear is fixed, the S-PGT further decelerates the drive motor. At this time, the reduction ratio can be calculated as follows:

$$i_2 = \frac{\omega_{S2}}{\omega_{C2}} = \frac{r_R + r_{S2}}{r_{S2}} > 1 \tag{5}$$

where ω_{S2} and ω_{C2} denote the rotational speeds of the sun gear and carrier, respectively, and r_R and r_{S2} denote the radii of the ring gear and sun gear.

When the sun gear and ring gear are combined, the S-PGT rotates at the same speed, and at this time,

$$i_2 = 1 \tag{6}$$

The structure of the planetary gear train is shown in Figure 3.

Figure 3. Structure of the planetary gear train.

2.2. Selectable One-Way Clutch

The SOWC was developed from OWC, and its basic operation is similar to that of OWC. By adding an independent control selection mechanism based on OWC, the selective output of power is realized. The SOWC can transmit power in two directions between the driving and the driven part, in addition to allowing for overrun in two directions. Moreover, it can effectively improve the transmission efficiency and reduce the unsprung mass, and it requires less packaging space. In the IW-AMT, the SOWC is used to reliably transmit power in both the first and reverse gears, and it is overrun in the second and neutral gears. The structure of SOWC is shown in Figure 4.

As shown in Figure 4, the SOWC is composed of an inner circle, an outer circle, reverse struts, compression springs, a selector plate, forward struts, and a worm gear [18]. The outer circle is fixed on the knuckle, struts are installed in the groove of the outer circle, and control pins on the struts are installed in the evenly arranged chute on the selector plate. The selector plate is fixedly connected to the worm gear mechanism, compression springs are installed between the struts and the outer circle, and the inner circle is fixedly connected with the ring gear of the S-PGT. As the worm gear mechanism rotates by the executive motor, the selector plate is driven to rotate, and the struts are up or down. The ring gear of the S-PGT is selectively locked in two directions or in a state of overrun. To show the working status of IW-AMT with SOWC in different gears clearly, we have developed the shift table, as shown in Table 1.

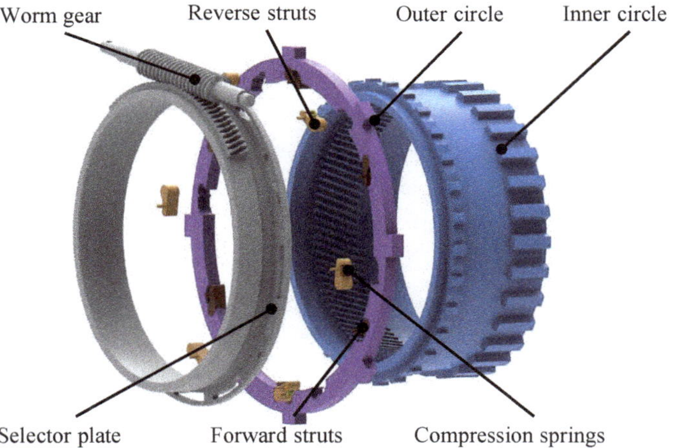

Figure 4. Structure of SOWC.

Table 1. Shift table of IW-AMT.

Gear State	Motor Rotation Direction	SOWC Forward Struts State	SOWC Reverse Struts State	WMPC State
Neutral	-	Up	Up	Disengaged
First gear	Forward	Down	Down	Disengaged
Second gear	Forward	Up	Up	Engaged
Reverse	Reverse	Down	Down	Disengaged

As shown in Figure 5, when the EV is engaged in the first gear, the selector plate rotates to drop the forward and reverse struts sequentially. The SOWC locks the ring gear of the S-PGT in the forward and reverse rotation directions and maintains the drive motor power output in the first gear ratio. When the drive motor is driving, the forward struts transmit the positive torque. When the motor instantly switches the braking state, the reverse struts transmit the reverse torque completely. There is no switching time between forward and reverse struts when the drive motor is cyclically switched between the driving and braking states. In the reverse gear, the state of the struts is the same as that of the first gear. The rotation direction and transmitted torque direction of each component are opposite.

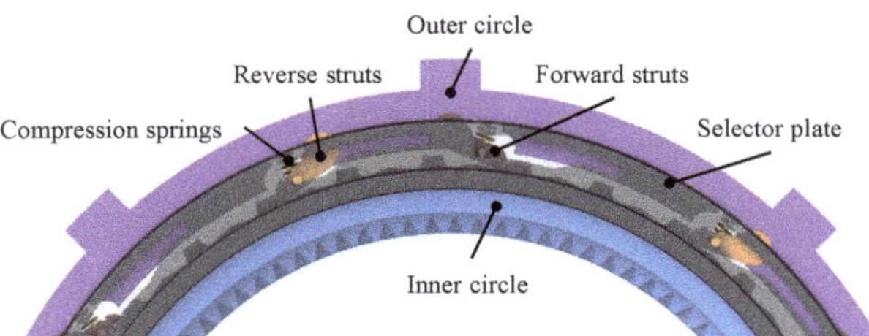

Figure 5. Schematic diagram of the first and reverse gears.

When the vehicle is in the second or the neutral gear, the SOWC is overrunning. By reasonably designing the selector plate chute, the forward and the reverse struts can be maintained in a raised state. At this time, the ring gear of the S-PGT is in a free rotation state. The corresponding schematic diagram is shown in Figure 6.

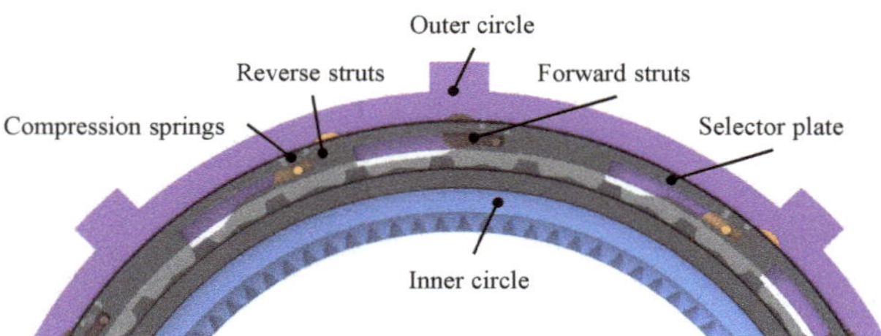

Figure 6. Schematic diagram of the second and neutral gears.

During upshifting, the selector plate rotates to sequentially raise the reverse and forward struts, so that the SOWC is overrunning. Conversely, during downshifting, the selector plate rotates to sequentially drop the forward and reverse struts, so that the SOWC locks the ring gear of the S-PGT in both directions. The schematic diagram of the intermediate state is shown in Figure 7.

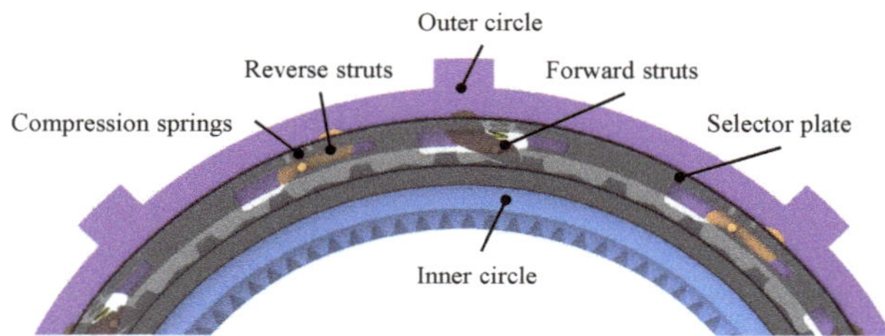

Figure 7. Schematic diagram of the intermediate state.

2.3. Overall Structure and Characteristics

Through coordinated control of the drive motor and the shift actuators, the IW-AMT can achieve a variety of power output states. The structure of the IW-AMT is depicted in Figure 8, and the main technical parameters of the IW-AMT and NSK Ltd. wheel hub motor are summarized in Table 2. NSK Ltd. (Tokyo, Japan) adopts a dual-motor design, also with a two-stage planetary gear train. In contrast, the IW-AMT has a lower unsprung mass and higher power density.

Figure 8. Structure of IW-AMT.

Table 2. Comparison of the main technical parameters.

Parameter	IW-AMT	NSK Ltd. Wheel Hub Motor [26]	Unit
Maximum output power	35	25	kW
Maximum output torque	620	850	Nm
Overall mass	15.72	32	kg
Power density	2.23	0.78	kW/kg
Torque density	39.44	26.56	Nm/kg

3. Gear Shifting Process

The IW-AMT entire gear shifting process includes the torque and the inertia phases [18]. The torque phase represents the phase of torque exchange during gear shifting. In the torque phase, the motor speed does not change suddenly, and the torque transmitted by each component is mutually exchanged. The torque phase ends when the torque synchronization is completed. The inertia phase represents the synchronization phase of the drive and driven components during the shifting process. In the inertia phase, there is no mutual exchange of torque, and the speed difference between the driving and the driven part of the clutch is gradually synchronized. The inertia phase ends when there is no speed difference in the clutch. The ideal shifting process of IW-AMT is shown in Figure 9.

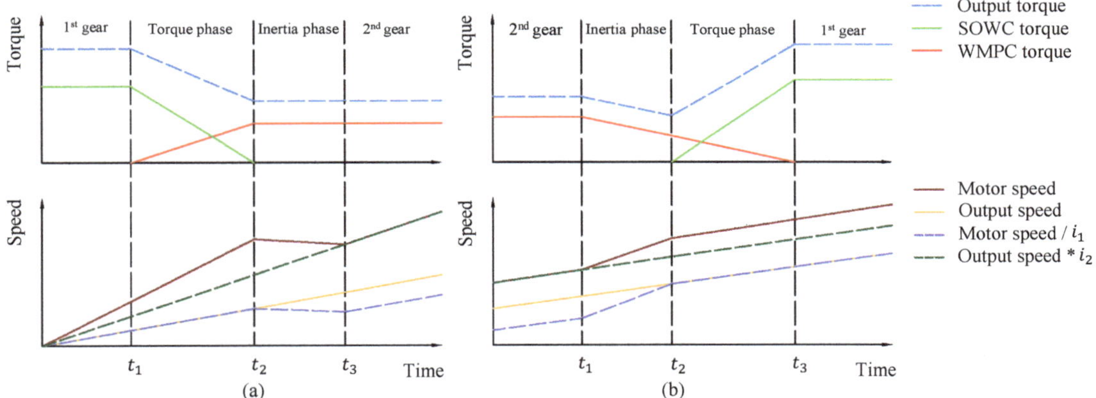

Figure 9. Ideal shifting process of IW-AMT: (**a**) Upshifting process. (**b**) Downshifting process.

During upshifting, the WMPC starts to combine, marking the start of the torque phase. With the gradual combination of the WMPC, the torque transmitted increases gradually, and the torque that the SOWC transmits decreases gradually. When the torque transmitted by the SOWC decreases to zero, the torque phase ends, and the inertia phase starts. At this time, the selector plate rotates, and the forward and reverse struts of the SOWC are raised sequentially. When the inertia phase starts, there is a speed difference between the driving part and the driven part of the WMPC. As the sliding grinding process continues, when the speed difference between the driving and the driven parts gradually decreases to zero, the inertia phase ends, and the EV starts to drive stably in the second gear.

During downshifting, the IW-AMT first enters the inertia phase, WMPC starts to separate gradually and enters a slipping state, and its transmission torque gradually decreases. As the speed difference between the driving and the driven parts of the WMPC gradually increases, the ring gear of the S-PGT decelerates gradually by the resistance torque. The inertia phase ends when the rotational speed of the ring gear is zero. At this time, the selector plate rotates and the SOWC forward and reverse struts are dropped sequentially, and the drive system enters the torque phase. As the WMPC continues to separate, its transmission torque further decreases, and the SOWC transmission torque gradually increases. The torque phase ends when the WMPC transmission torque decreases to zero, and the EV starts to drive stably in the first gear.

4. Simulation Model and Results

To clearly depict the changes in the torque and speed of each component of the IW-AMT system with the SOWC during the gear shifting process and evaluate the quality of shifting, we established a virtual prototype simulation model of the IW-AMT. The simulation model was composed of the motor and vehicle equivalent model, two-stage planetary gear train model, and SOWC model, including outer circle, inner circle, worm gear, selector plate, struts, and compression springs. The virtual prototype simulation model is shown in Figure 10, and the main parameters are summarized in Table 3.

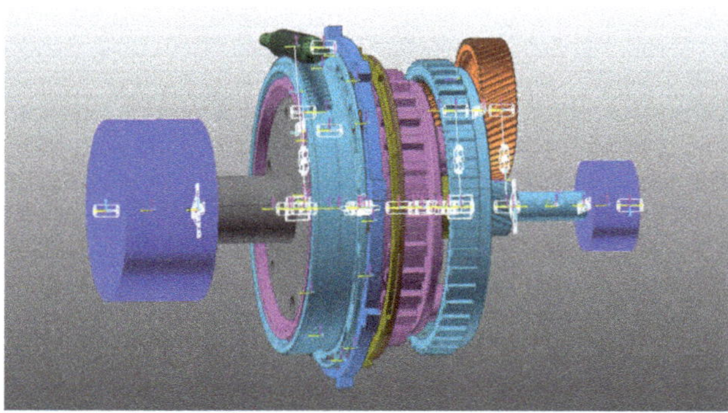

Figure 10. Virtual prototype simulation model of IW-AMT.

Table 3. Main parameters of the simulation model.

Parameter	Value	Unit
Vehicle mass	1036	kg
Tire radius	0.2979	m
Vehicle frontal area	1.4	m^2
Air resistance coefficient	0.45	-
Rolling resistance coefficient	0.011	-
Rotation mass correction coefficient	1.04	-
Max motor torque	21	Nm
Max motor speed	20,000	rpm
Motor rotational inertia	2.74	$kgcm^2$
First gear ratio	29.51	-
Second gear ratio	11.88	-
Equivalent vehicle rotational inertia	107.74	kgm^2
Worm gear ratio	35	-
Worm gear drive speed	334	rpm
SOWC strut mass	31.4	g
Compression spring stiffness coefficient	2.81	N/mm

In the virtual prototype simulation model, we modeled the motor as a rigid body whose rotational inertia is J_M. At the output end of the simulation model, we constructed a cylindrical rigid body to simulate the equivalent inertia of the vehicle. The vehicle resistance torque T_R includes the rolling resistance torque T_{Roll}, air resistance torque T_{Wind}, acceleration resistance torque T_{Acc}, and ramp resistance torque T_{Ramp}. The vehicle resistance torque was applied to the output terminal to simulate changes in the resistance torque during driving.

$$T_R = T_{Roll} + T_{Wind} + T_{Acc} + T_{Ramp} \tag{7}$$

To improve vehicle dynamics and make full use of motor power. We designed a dynamic dual-parameter shifting schedule, as shown in Figure 11, taking the intersection of the vehicle acceleration curves at different accelerator pedal openings in two adjacent gears as the shifting point. To avoid frequent shifts near the shift speed, we delayed the downshift curve by 6 km/h.

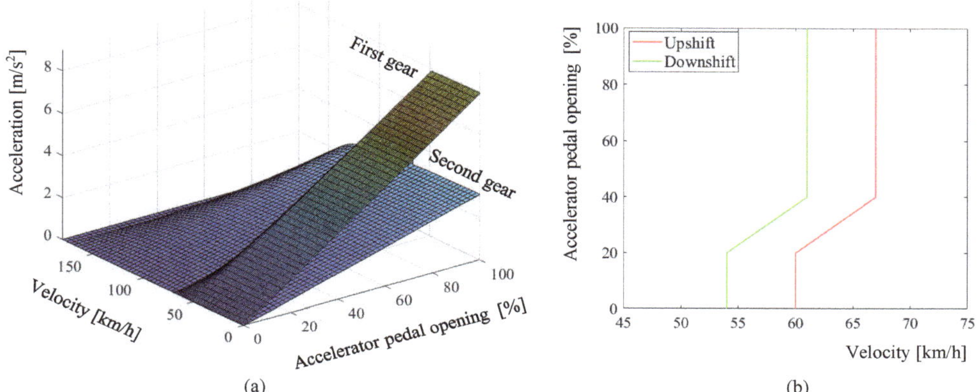

Figure 11. Dynamic dual-parameter shifting schedule: (**a**) Vehicle acceleration curves in different gears. (**b**) Upshift and downshift curves.

The jerk of the vehicle j is an important indicator for evaluating ride comfort, and the recommended value of Germany is $|j| < 10 \text{ m/s}^3$ [27]. In IW-AMT, the SOWC is used to transmit torque in the first and reverse gears, which does not have the gradual engaging process that a friction clutch does.Therefore, we select the jerk as a criterion to evaluate the quality of shifting. During the shifting process, the jerk j can be expressed as:

$$j = \frac{da}{dt} = \frac{d^2v}{dt^2} \tag{8}$$

where a and v denote the longitudinal acceleration and velocity of the vehicle.

Because we focus on the parameter changes of IW-AMT during the shifting process, the period time of the shifting process is intercepted and displayed in the simulation results.

4.1. Upshifting Process

In the initial stage of the upshifting process, the SOWC forward and reverse struts are dropped to lock the ring gear of the S-PGT. With the gradual combination of the WMPC, the upshifting process starts. The simulation results are as follows:

As shown in Figure 12, when the upshifting process starts at 1 s, as the WMPC compression force gradually increases, its transmission torque increases gradually, and the transmission torque of the SOWC forward struts decreases gradually. At approximately 2 s, the SOWC transmission torque decreases to zero, forward and reverse struts are controlled to rise sequentially. The struts can be raised smoothly when the transmission torque of the ring gear is not decreased to zero, the switching time of the struts is about 0.04 s, and the forward strut's lag behind the reverse struts is about 0.05 s. Throughout the upshifting process, the torque and the vehicle velocity change smoothly, without torque interruption, and the maximum jerk is less than 10 m/s^3.

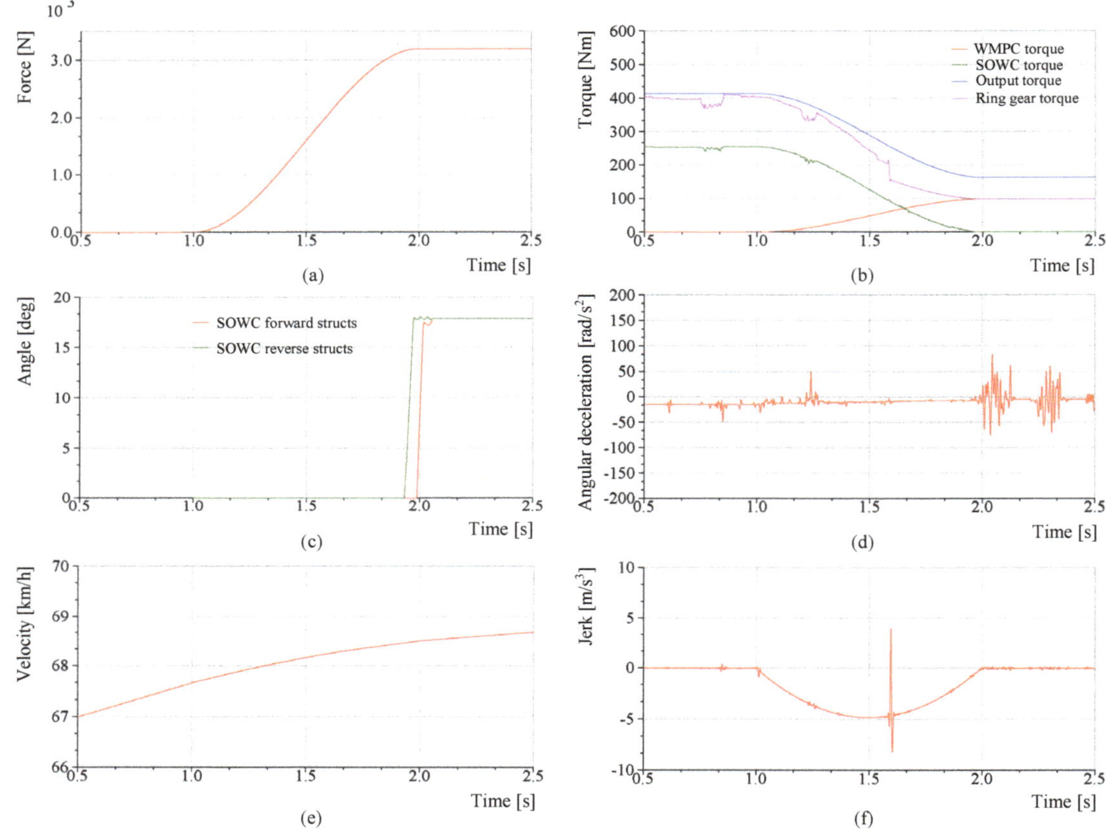

Figure 12. Simulation results of the upshifting process: (**a**) Changes of WMPC pressing force. (**b**) Changes in torque transmitted by each component. (**c**) Changes of struts rotation angle. (**d**) Changes of ring gear angular deceleration. (**e**) Changes in vehicle speed. (**f**) Changes in jerk.

4.2. Downshifting Process

The downshifting process is basically the opposite of the upshifting process. In the initial stage of the downshifting process, the SOWC is overrunning, its forward and reverse struts are raised, and the WMPC is in a stable combined state. With the gradual separation of the WMPC, the downshifting process commences. To simulate the power downshift of the vehicle, we increase the ramp resistance torque to simulate the vehicle climbing condition for fulfilling the power downshift condition. The simulation results are as follows:

As shown in Figure 13, the downshifting process starts at approximately 1.7 s, and as the WMPC compression force decreases, the transmission torque gradually decreases. This causes the S-PGT ring gear to gradually reduce its rotational speed under the action of the resistance torque, and there is a tendency for reverse rotation. When the rotational speed of the ring gear decreases to zero, the SOWC forward and reverse struts are dropped sequentially. When the forward struts come into contact with the ring gear, the rotation angle curve fluctuates; at this time, the forward struts have locked the ring gear and then the reverse struts dropped smoothly. The reverse struts dropped behind the forward struts at about 0.05 s but dropped completely about 0.05 s earlier than the forward struts. At approximately 2.7 s, the WMPC is completely separated, all of the torque is transmitted by the SOWC. The torque fluctuation during the entire downshifting process is greater than that during the upshifting process. The maximum jerk is approximately 41 m/s^3, and part of the reason may be that the ring gear has a higher deceleration when the struts dropped.

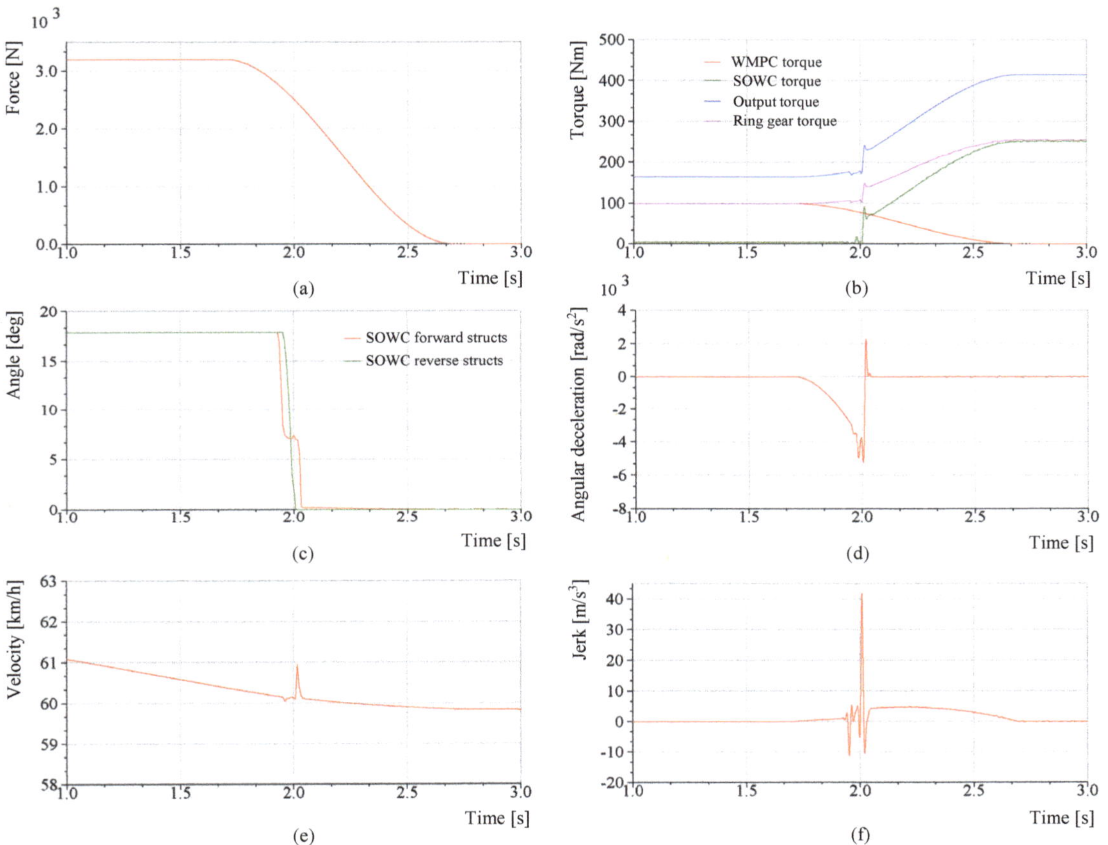

Figure 13. Simulation results of the downshifting process: (**a**) Changes of WMPC pressing force. (**b**) Changes in torque transmitted by each component. (**c**) Changes of struts rotation angle. (**d**) Changes of ring gear angular deceleration. (**e**) Changes in vehicle speed. (**f**) Changes in jerk.

4.3. Improvement of Jerk

To reduce the peak jerk of the vehicle during downshifting and improve ride comfort, we analyze the momentary state of the SOWC forward struts that are dropped during downshifting.

As can be seen from the previous simulation results, when the SOWC forward struts are dropped for approximately 2 s during the downshifting process, the vehicle torque fluctuates considerably. One of the reasons for this fluctuation is that the ring gear of the S-PGT decelerates to a greater extent when the WMPC gradually separates. When the SOWC forward struts are dropped, it resists deceleration and produces a jerk. Therefore, we start by reducing the deceleration of the ring gear to reduce the jerk when the struts are dropped. The IW-AMT shift process indicates that if the WMPC continues to transmit torque during the sliding process, the deceleration of the ring gear under the action of the resistance torque can be decreased, and in this manner, the jerk induced when the forward struts are dropped can be reduced. The improved WMPC compression force curve and downshift simulation results are as follows:

As shown in Figure 14, after the improvement of the WMPC combined curve, the jerk caused by the falling of the SOWC forward struts decreased. The torque and velocity of the vehicle during the downshifting process were relatively stable, and the peak jerk was significantly reduced. The maximum jerk was approximately 23 m/s^3. Furthermore, we started with the SOWC structure and attempted to improve it to reduce the jerk.

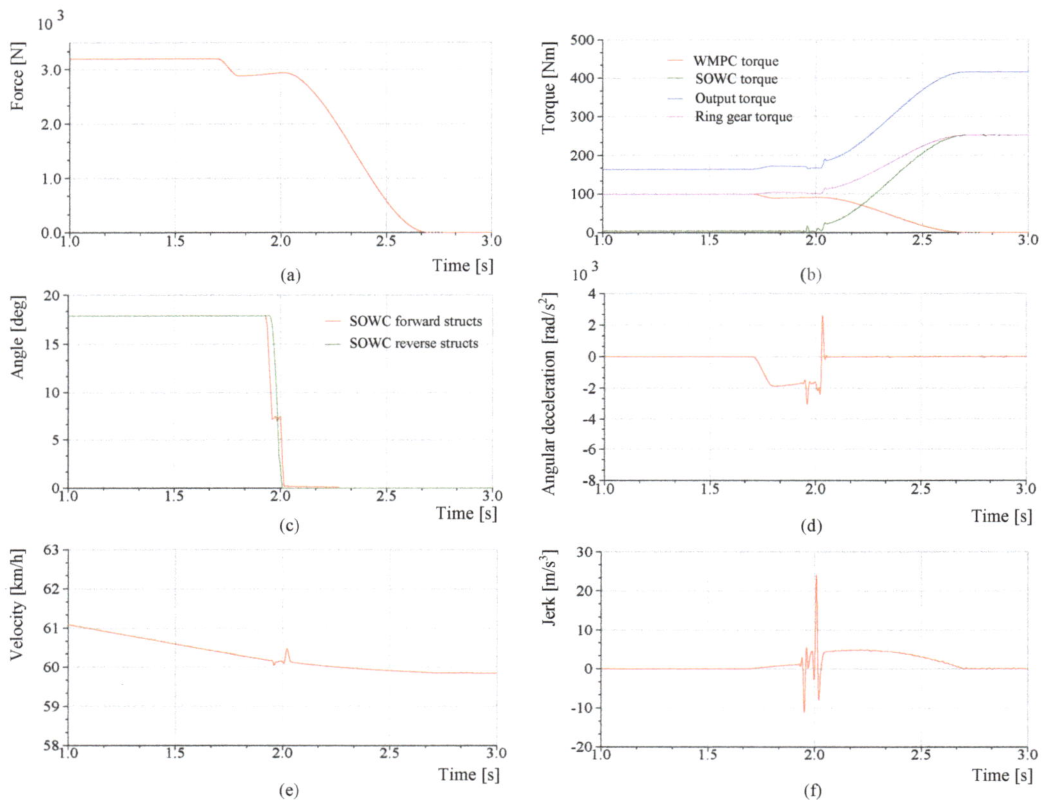

Figure 14. Simulation results of the downshifting process after the improvement of the WMPC combined curve: (**a**) Changes of WMPC pressing force. (**b**) Changes in torque transmitted by each component. (**c**) Changes of struts rotation angle. (**d**) Changes of ring gear angular deceleration. (**e**) Changes in vehicle speed. (**f**) Changes in jerk.

Many studies have indicated that reducing the freeway angle of the SOWC is beneficial for reducing the jerk when the struts are dropped [28], but owing to the size limitation of the SOWC structure, the potential for improvement of the freeway angle is small. The SOWC freeway angle designed in this paper was 2°.

$$\theta_i = \frac{360°}{z} \cdot \frac{a}{n} \tag{9}$$

where θ_i is the freeway angle, a the number of struts grouping, n the number of struts, and z the number of teeth in the inner circle.

Furthermore, we refer to the principle of engine corner cushion cushioning and try to arrange a cushion with a certain stiffness and damping on the outer circle of the SOWC to alleviate the instantaneous impact when the SOWC forward struts are dropped.

The stiffness coefficient of the cushion is 693.595 N/mm, and the damping coefficient is 0.466 N/(mm/s). The simulation results are shown in Figure 15, where the maximum jerk is approximately 13 m/s^3. The simulation results show that the placement of cushions has a significant effect on reducing impact.

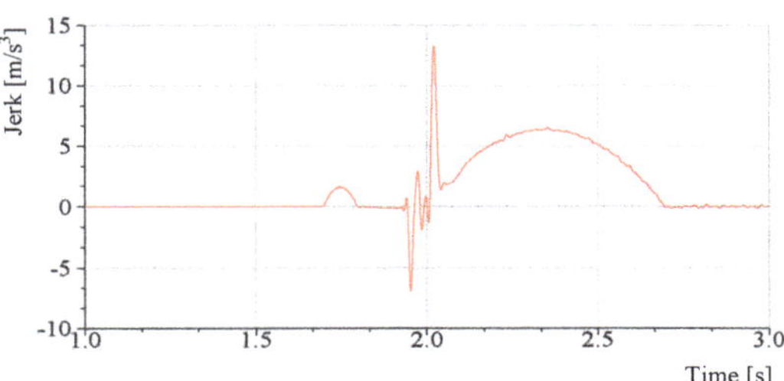

Figure 15. Jerk change during the downshifting process after the improvement of the SOWC structure.

5. Discussion

The simulation results of the upshifting process indicated that the vehicle speed and torque changed smoothly, and the jerk was less than 10 m/s^3. The was ascribed to the fact that the SOWC struts do not drop instantly during the upshifting process, and the magnitude of jerk is strongly related to the WMPC combination curve. The simulation results of the downshifting process indicated that when the WMPC separation curve was opposite to the combined curve, the impact of the jerk was approximately 41 m/s^3, and the shift impact was obvious. This is because the SOWC forward struts must drop during the downshifting process to instantly lock the ring gear of the S-PGT. Then, we start from the instantaneous state of the SOWC. After we improved the WMPC separation curve, the vehicle torque, and speed changed more smoothly, and the jerk was approximately 23 m/s^3, which represented a certain decrease but was nevertheless higher than the 10 m/s^3 recommended value of Germany. Furthermore, we improved the SOWC structure by imitating the principle of engine corner pads and evenly arranged cushion rubber pads on the SOWC outer circle. Thereafter, the simulation results indicated that the jerk due to downshifting was approximately 13 m/s^3, since the SOWC does not follow a continuous process of combination and separation as in the case of the WMPC.

6. Conclusions

In this paper, we proposed an in-wheel two-speed AMT to improve the efficiency of the drive system and vehicle performance. In addition, to develop an efficient and compact drive system, we described the use of a SOWC as a new clutch solution for the IW-AMT. The mass of the proposed IW-AMT is only 15.72 kg, and its power and torque densities can be up to 2.23 kW/kg and 39.44 Nm/kg, and it effectively reduces the unsprung mass of the distributed in-wheel drive.

In future works, we will use relevant optimization theories to track and optimize the combination and separation curves of the WMPC to achieve the best shift effect with the SOWC.

Author Contributions: Conceptualization, D.M. and B.G.; methodology, Y.W.; software, D.M.; validation, D.M., F.W. and B.G.; writing—original draft preparation, D.M.; writing—review and editing, D.M. and B.G.; visualization, D.M.; supervision, B.G. All authors have read and agreed to the published version of the manuscript.

Funding: This research was funded by the National Nature Science Foundation of China No. 61803173, in part by Jilin Provincial Science, Technology Department No. 20200301011RQ, and in part by the Jilin Provincial Science Foundation of China under Grant 20200201062JC.

Acknowledgments: We acknowledge the members of Qingdao Legee Transmission System Technology Co., Ltd. for their support.

Conflicts of Interest: The authors declare no conflict of interest.

References

1. Ding, N.; Prasad, K.; Lie, T.T. The electric vehicle: A review. *Int. J. Electr. Hybrid Veh.* **2017**, *9*, 49–66. [CrossRef]
2. Grunditz, E.A.; Thiringer, T.; Saadat, N. Acceleration, drive cycle efficiency, and cost tradeoffs for scaled electric vehicle drive system. *IEEE Trans. Ind. Appl.* **2020**, *56*, 3020–3033. [CrossRef]
3. Mo, W.; Walker, P.D.; Fang, Y.; Wu, J.; Ruan, J.; Zhang, N. A novel shift control concept for multi-speed electric vehicles. *Mech. Syst. Signal Process.* **2018**, *112*, 171–193. [CrossRef]
4. Ahssan, M.R.; Ektesabi, M.M.; Gorji, S.A. Electric vehicle with multi-speed transmission: A review on performances and complexities. *SAE Int. J. Altern. Powertrains* **2018**, *7*, 169–182. [CrossRef]
5. Rimpas, D.; Chalkiadakis, P. Electric vehicle transmission types and setups: A general review. *Int. J. Electr. Hybrid Veh.* **2021**, *13*, 38–56. [CrossRef]
6. Montazeri-Gh, M.; Pourbafarani, Z. Simultaneous design of the gear ratio and gearshift strategy for a parallel hybrid electric vehicle equipped with AMT. *Int. J. Veh. Des.* **2012**, *58*, 291–306.
7. Gao, B.; Liang, Q.; Xiang, Y.; Guo, L.; Chen, H. Gear ratio optimization and shift control of 2-speed I-AMT in electric vehicle. *Mech. Syst. Signal Process.* **2015**, *50*, 615–631. [CrossRef]
8. Cui, J.; Tan, G.; Tian, Z.; Agyeman, P. Parameter Optimization of Two-Speed AMT Electric Vehicle Transmission System. *SAE Tech. Pap.* **2020**. [CrossRef]
9. Tian, Y.; Ruan, J.; Zhang, N.; Wu, J.; Walker, P. Modelling and control of a novel two-speed transmission for electric vehicles. *Mech. Mach. Theory* **2018**, *127*, 13–32. [CrossRef]
10. Fang, S.; Song, J.; Song, H.; Tai, Y.; Li, F.; Nguyen, T.S. Design and control of a novel two-speed uninterrupted mechanical transmission for electric vehicles. *Mech. Syst. Signal Process.* **2016**, *75*, 473–493. [CrossRef]
11. Yue, H.; Zhu, C.; Gao, B. Fork-less two-speed I-AMT with overrunning clutch for light electric vehicle. *Mech. Mach. Theory* **2018**, *130*, 157–169. [CrossRef]
12. Feng, S.; Magee, C.L. Technological development of key domains in electric vehicles: Improvement rates, technology trajectories and key assignees. *Appl. Energy* **2020**, *260*, 114264. [CrossRef]
13. Kim, D.; Shin, K.; Kim, Y.; Cheon, J. Integrated design of in-wheel motor system on rear wheels for small electric vehicle. *World Electr. Veh. J.* **2010**, *4*, 597–602. [CrossRef]
14. Jian, L. Research status and development prospect of electric vehicles based on hub motor. In Proceedings of the 2018 China International Conference on Electricity Distribution (CICED), Tianjin, China, 17–19 September 2018; pp. 126–129.
15. de Carvalho Pinheiro, H.; Galanzino, E.; Messana, A.; Sisca, L.; Ferraris, A.; Airale, A.G.; Carello, M. All-Wheel Drive Electric Vehicle Modeling and Performance Optimization. *SAE Tech. Pap.* **2020**. [CrossRef]
16. Gunji, D.; Matsuda, Y.; Kimura, G. Wheel Hub Motor. U.S. Patent 8,758,178, 24 June 2014.
17. Wang, W.; Chen, X.; Wang, J. Motor/generator applications in electrified vehicle chassis—A survey. *IEEE Trans. Transp. Electrif.* **2019**, *5*, 584–601. [CrossRef]
18. Meng, D.; Tian, M.; Miao, L.; Wang, Y.; Hu, J.; Gao, B. Design and modeling of an in-wheel two-speed AMT for electric vehicles. *Mech. Mach. Theory* **2021**, *163*, 104383. [CrossRef]
19. Kim, J.; Choi, S.B. Design and modeling of a clutch actuator system with self-energizing mechanism. *IEEE/ASME Trans. Mechatron.* **2010**, *16*, 953–966. [CrossRef]
20. Kim, D.H.; Kim, J.W.; Choi, S.B. Design and modeling of energy efficient dual clutch transmission with ball-ramp self-energizing mechanism. *IEEE Trans. Veh. Technol.* **2019**, *69*, 2525–2536. [CrossRef]
21. Lee, C.J.; Samie, F.; Kao, C.K. Control of selectable one-way clutch in GM six-speed automatic transmissions. *Dyn. Syst. Control Conf.* **2009**, *48937*, 605–609.
22. Bird, N.J.; Bindra, R.; Klaser, J. Development and challenges of electrically selectable one-way clutches. *SAE Int. J. Engines* **2017**, *10*, 1338–1350. [CrossRef]
23. Xu, X.; Dong, P.; Liu, Y.; Zhang, H. Progress in automotive transmission technology. *Automot. Innov.* **2018**, *1*, 187–210. [CrossRef]
24. Fan, Y. Study on the Transmission Characteristics of the Multi-gear Multi-degree-of-freedom Hybrid Planetary Gear Automatic Transmission Based on the Line Method. *IOP Conf. Ser. Earth Environ. Sci.* **2020**, *512*, 012168. [CrossRef]
25. Beiser, C. Compact and Efficient Electric Propulsion Systems Enabled by Integrated Electric Controllable Clutches. In *CTI Symposium 2018*; Springer: Berlin/Heidelberg, Germany, 2020; pp. 20–42.
26. Yamamoto, S.; Morita, R.; Oike, M. Transmission-equipped wheel hub motor for passenger cars. *ATZ Worldw.* **2018**, *120*, 28–33. [CrossRef]
27. Yu, H.L.; Xi, J.Q.; Zhang, F.Q.; Hu, Y.H. Research on gear shifting process without disengaging clutch for a parallel hybrid electric vehicle equipped with AMT. *Math. Probl. Eng.* **2014**, *2014*, 985652. [CrossRef]
28. Samie, F.; Lee, C.J.; Pawley, B. Selectable one-way clutch in GM's RWD 6-speed automatic transmissions. *SAE Int. J. Engines* **2009**, *2*, 307–313. [CrossRef]

Article

Optimization Design for the Planetary Gear Train of an Electric Vehicle under Uncertainties

Xiang Xu [1], Jiawei Chen [2,*], Zhongyan Lin [1], Yiran Qiao [1], Xinbo Chen [1], Yong Zhang [3], Yanan Xu [1] and Yan Li [2]

[1] School of Automotive Studies, Tongji University, Shanghai 201804, China; xiangxu@tongji.edu.cn (X.X.); 1931577@tongji.edu.cn (Z.L.); qiaoyiran@tongji.edu.cn (Y.Q.); chenxinbo@tongji.edu.cn (X.C.); 1610840@tongji.edu.cn (Y.X.)
[2] School of Mechanical Engineering, Tongji University, Shanghai 201804, China; liyanliyan910@126.com
[3] College of Mechanical Engineering and Automation, Huaqiao University, Xiamen 361021, China; zhangyong@hqu.edu.cn
* Correspondence: chen_jiawei@tongji.edu.cn

Abstract: The planetary gear train is often used as the main device for decelerating and increasing the torque of the drive motor of electric vehicles. Considering the lightweight requirement and existing uncertainty in structural design, a multi-objective uncertainty optimization design (MUOD) framework is developed for the planetary gear train of the electric vehicle in this study. The volume and transmission efficiency of the planetary gear train are taken into consideration as optimization objectives. The manufacturing size, material, and load input of the planetary gear train are considered as uncertainties. An approximate direct decoupling model, based on subinterval Taylor expansion, is applied to evaluate the propagation of uncertainties. To improve the convergence ability of the multi-objective evolutionary algorithm, the improved non-dominated sorting genetic algorithm II (NSGA-II) is designed by using chaotic and adaptive strategies. The improved NSGA-II has better convergence efficiency than classical NSGA-II and multi-objective particle swarm optimization (MOPSO). In addition, the multi-criteria decision making (MCDM) method is applied to choose the most satisfactory solution in Pareto sets from the multi-objective evolutionary algorithm. Compared with the multi-objective deterministic optimization design (MDOD), the proposed MUOD framework has better reliability than MDOD under different uncertainty cases. This MUOD method enables further guidance pertaining to the uncertainty optimization design of transportation equipment, containing gear reduction mechanisms, in order to reduce the failure risk.

Keywords: optimization design; vehicle structure design; uncertainty; deceleration device

Citation: Xu, X.; Chen, J.; Lin, Z.; Qiao, Y.; Chen, X.; Zhang, Y.; Xu, Y.; Li, Y. Optimization Design for the Planetary Gear Train of an Electric Vehicle under Uncertainties. *Actuators* **2022**, *11*, 49. https://doi.org/10.3390/act11020049

Academic Editor: Richard M. Stephan

Received: 9 December 2021
Accepted: 3 February 2022
Published: 5 February 2022

Publisher's Note: MDPI stays neutral with regard to jurisdictional claims in published maps and institutional affiliations.

1. Introduction

In recent years, electric vehicle technology has developed rapidly [1,2]. The planetary gear reducer is used in electric vehicles due to its high transmission efficiency and compact structure. Due to the space limitation of electric vehicles, the design of compact planetary gear trains has become a key issue. Numerous optimization methods are involved in the gear train design. For example, Parmar et al. [3] proposed a novel multi-objective optimization method, for planetary gear trains, using NSGA-II. Miler et al. [4] chose transmission volume and power loss as design objectives, and they optimized the parameters of the planetary gear train with multi-objective optimization. Sedak et al. [5] proposed a constrained multi-objective nonlinear optimization problem for planetary gearboxes, based on a hybrid element heuristic algorithm, considering gear volume, center distance, contact ratio, and power loss as optimization objectives. Patil et al. [6] proposed a multi-objective optimization strategy to minimize the total volume and power loss of the two-stage helical gearbox and spur gearbox. Compared to the single-objective optimization method with tribological constraints, the multi-objective optimization results in less power loss. Savsani et al. [7] used the particle swarm optimization algorithm, and the simulated annealing algorithm, to carry out the optimization design of the lightweight spur gear transmission system, and, resultingly, this method is deemed to be suitable for the single-objective or multi-objective optimization design of the multi-stage spur gear transmission. Considering the above research, the main challenge of gear transmission design is in reducing weight and power loss. At present, the optimization design method of planetary gear trains mainly considers the determining system parameters and implements the conventional deterministic optimization method. However, for practical engineering structures, many uncertainties are observed in the material properties, manufacturing, and measurement [8–12]. To obtain a reliable structural design, the uncertainties of the planetary gear train of electric vehicles need to be considered.

Uncertainty optimization in engineering design has gradually attracted attention [13–15]. For example, Xian et al. [16] proposed an effective analysis framework for stochastic optimization pertaining to non-linear viscous dampers of energy dissipation structures, which was applied to the uncertainty optimization of non-linear viscous dampers of suspension bridges. Lü et al. [17] proposed an efficient approach for the optimization design of dual uncertain structures, taking into account the dual robust design and the possibility of failure, quickly estimating the dual uncertain target of fuzzy random variables, and equivalently solving the possibility constraints involving fuzzy randomness. Baek et al. [18] developed a design method of a composite microwave absorbing structure using reliability-based optimization (RBO), which considers the failure probability. Compared with the results of deterministic optimization (DO), it was found that the total thickness of the reliability design method increased slightly, but RBO significantly reduced the failure probability. Fang et al. [19] developed an effective multi-objective uncertainty optimization program in order to design car doors. The program analyzed the impact of changing the uncertainty conditions and improving the reliability level, and it provided clear design information for decision-makers. Zhang et al. [20] proposed a reliable uncertainty optimization design route for obtaining optimal energy-absorbing structures. The study found that the solution obtained, by uncertainty optimization, sacrificed certain demand performance, but it was more reliable than deterministic design. The above studies have carried out the uncertainty optimization based on the probability model, which is highly dependent on statistical data. Considering that the distribution of uncertainty requires a lot of data, it is of a high cost to obtain effective probability data from a practical engineering perspective.

To overcome the limitation, of uncertainty optimization, due to the lack of data, some interval uncertainty modes have been gradually developed and applied to engineering optimization [21,22]. The interval uncertainty model mainly focuses on the upper and lower boundaries of uncertainty values, which is easier to implement than the probabilistic uncertainty model. Inuiguchi et al. [23] proposed a linear multi-objective strategy based on maximum and minimum regret criteria to solve the problem of interval uncertainty in the objective function. Fu et al. [24] developed a multi-objective direct structural optimization method for solving interval uncertainty. This method uses the satisfaction value of the interval possibility model to deal with non-linear uncertain constraints, and it judges the feasibility and infeasibility of individual design vectors. Wu et al. [25] proposed a non-probabilistic robust topology optimization method for interval uncertain structures. The method uses the Chebyshev interval inclusion function to realize the non-invasiveness of the interval algorithm. Wang et al. [26] developed an effective interval uncertain optimization design strategy using Legendre polynomial chaotic expansion, which is more efficient than the conventional method. Hou et al. [27] carried out the uncertainty optimization, pertaining to the energy efficiency of ships in icy areas, considering the interval parameters; the optimization results provided practical guidance for the energy-saving design of ships in the case of uncertainty in the actual environment. Yu et al. [28] regarded friction coefficient, material properties, and wear element thickness as interval uncertainty factors, and proposed an uncertainty optimization method for the noise suppression of the brake system.

The above studies have developed a highly effective uncertainty optimization method based on the interval model, and they have applied it to solve practical engineering problems. The interval model has been validated as a highly applicable uncertainty optimization method. Uncertainties in the manufacturing and operation of the planetary gear train of electric vehicles are unavoidable. The process of efficiently solving multi-objective uncertainty problems for the planetary gear train of electric vehicles is still a key issue. Therefore, a multi-objective uncertainty optimization design (MUOD) framework is developed for the planetary gear train of an electric vehicle in this study. Section 2 describes the detailed methodology of MUOD. Section 3 describes the design requirement of the planetary gear train of an electric vehicle. Section 4 shows the optimization results. The main conclusions are drawn in Section 5.

2. Methodology

2.1. Multi-Objective Uncertainty Optimization Problem

In general, the multi-objective deterministic optimization design (MDOD) model can be expressed as follows [29,30]:

$$\begin{cases} \min f(x) = \{f_1(x), f_2(x), \dots, f_q(x)\} \\ \text{s.t.} \begin{cases} G_i(x) \leq 0, & i = 1, 2, \dots l \\ h_j(x) = 0, & j = 1, 2, \dots g \\ x \in \{S\} \end{cases} \end{cases} \tag{1}$$

In the formula, $\{f_1, \dots, f_q\}$ are the objective functions and q is the number of objectives. $G_i(X)$ is the inequality constraint and l is the number of its constraints; $h_j(X)$ is the equality constraint, and g is the number of its constraints; and $\{S\}$ is the design space. Different from the conventional deterministic optimization, the uncertainties of optimization variables and other relevant design parameters need to be considered during actual processing. Stochastic probability models are often used to construct uncertainty models, but the distribution information of uncertainties is unknown due to the lack of test samples. Therefore, the interval uncertainty model is employed in this study [31]. The multi-objective deterministic optimization can be transformed into the interval uncertainty problem, as follows:

$$\begin{cases} \min f\left(x^I, d^I\right) = \left\{f_1\left(x^I, d^I\right), f_2\left(x^I, d^I\right), \dots, f_q\left(x^I, d^I\right)\right\} \\ \text{s.t.} \begin{cases} G_i\left(x^I, d^I\right) \leq 0, & i = 1, 2, \dots l \\ h_j\left(x^I, d^I\right) = 0, & j = 1, 2, \dots g \\ d^{IC} - d^{IR} \leq d^{IC} \leq d^{IC} + d^{IR} \\ x \in \{S\} \end{cases} \end{cases} \tag{2}$$

In the formula, x^I and d^I are interval design variables and other relevant design parameters, respectively. The superscripts IC and IR represent the nominal value and interval radius, respectively. The interval radius of an interval value reflects its fluctuation range and can be expressed as uncertainty deviation. When the design variables and other relevant parameters are interval values, the relationship of reliability-based possibility degree P_d can be used to transform the interval uncertainty models into general non-interval models [31]. For the interval values A_1, A_2 and $A_1 \leq A_2$,

$$P_d(A_1 \leq A_2) = \begin{cases} 0, & A_1^{IL} \geq A_2^{IU} \\ 0.5 \cdot \dfrac{A_2^{IU} - A_1^{IL}}{A_1^{IU} - A_1^{IL}} \cdot \dfrac{A_2^{IU} - A_1^{IL}}{A_2^{IU} - A_2^{IL}}, & A_2^{IL} \leq A_1^{IL} < A_2^{IU} \leq A_1^{IU} \\ \dfrac{A_2^{IL} - A_1^{IL}}{A_1^{IU} - A_1^{IL}} + 0.5 \cdot \dfrac{A_2^{IU} - A_2^{IL}}{A_1^{IU} - A_1^{IL}}, & A_1^{IL} < A_2^{IL} < A_2^{IU} \leq A_1^{IU} \\ \dfrac{A_2^{IL} - A_1^{IL}}{A_1^{IU} - A_1^{IL}} + \dfrac{A_1^{IU} - A_2^{IL}}{A_1^{IU} - A_1^{IL}} \cdot \dfrac{A_2^{IU} - A_1^{IU}}{A_2^{IU} - A_2^{IL}} + 0.5 \cdot \dfrac{A_1^{IU} - A_2^{IL}}{A_1^{IU} - A_1^{IL}} \cdot \dfrac{A_1^{IU} - A_2^{IL}}{A_2^{IU} - A_2^{IL}}, & A_1^{IL} < A_2^{IL} \leq A_1^{IU} < A_2^{IU} \\ \dfrac{A_2^{IU} - A_1^{IU}}{A_2^{IU} - A_2^{IL}} + 0.5 \cdot \dfrac{A_1^{IU} - A_1^{IL}}{A_2^{IU} - A_2^{IL}}, & A_2^{IL} \leq A_1^{IL} < A_1^{IU} \leq A_2^{IU} \\ 1, & A_1^{IU} < A_2^{IL} \end{cases} \tag{3}$$

The superscripts *IL* and *IU* represent the lower and the upper values, respectively. The reliability-based possibility degree of the interval level should be given beforehand based on the actual reliable problem. Therefore, the multi-objective uncertainty optimization model can be expressed as:

$$
\begin{cases}
\min f\left(x^{IC}, d^{IC}\right) = \left\{ f_1\left(x^{IC}, d^{IC}\right), f_2\left(x^{IC}, d^{IC}\right), \ldots, f_q\left(x^{IC}, d^{IC}\right) \right\} \\
\text{s.t.}
\begin{cases}
P_{d\_i}(G_i\left(x^I, d^I\right) \leq 0) \geq \lambda_i, \quad i = 1, 2, \ldots l \\
h_j\left(x^I, d^I\right) = 0, \quad j = 1, 2, \ldots g \\
d^{IL} \leq d^{IC} \leq d^{IU} \\
x \in \{S\}
\end{cases}
\end{cases}
\tag{4}
$$

In the formula, λ_i is the requirement of reliability-based possibility degree, and it also represents the equivalent reliability with different constraints. The main optimization goals and constraints have been described in Section 3.

Nested optimization design is often used in interval uncertainty optimization, which treats the uncertainty analysis problem as an internal optimization problem. The purpose of inner optimization is to evaluate the propagation of uncertainty and feed it back to the outer optimization route. It is worth considering that adding a new optimization solver will cause low computational efficiency. Therefore, Taylor expansion, as an effective decoupling method, is applied to analyze the propagation of uncertainty in this study [32]. The constraint function $G_i\left(x^I, d^I\right)$ can be approximately constructed by first-order Taylor expansion, that is:

$$
G_i\left(x^I, d^I\right) \approx G_i\left(x^{IC}, d^{IC}\right) + \sum_{i=1}^{n} \frac{\partial G_i\left(x^{IC}, d^{IC}\right)}{\partial x_i^{IC}} x_i^{IR} + \sum_{j=1}^{m} \frac{\partial G_i\left(x^{IC}, d^{IC}\right)}{\partial d_i^{IC}} d_i^{IR}
\tag{5}
$$

Therefore, the lower and upper bounds of the constraint function can be expressed as follows:

$$
G_i^{IL}\left(x^I, d^I\right) \approx G_i\left(x^{IC}, d^{IC}\right) - \left| \sum_{i=1}^{n} \frac{\partial G_i\left(x^{IC}, d^{IC}\right)}{\partial x_i^{IC}} \right| x_i^{IR} - \left| \sum_{j=1}^{m} \frac{\partial G_i\left(x^{IC}, d^{IC}\right)}{\partial d_i^{IC}} \right| d_i^{IR}
\tag{6}
$$

$$
G_i^{IU}\left(x^I, d^I\right) \approx G_i\left(x^{IC}, d^{IC}\right) + \left| \sum_{i=1}^{n} \frac{\partial G_i\left(x^{IC}, d^{IC}\right)}{\partial x_i^{IC}} \right| x_i^{IR} + \left| \sum_{j=1}^{m} \frac{\partial G_i\left(x^{IC}, d^{IC}\right)}{\partial d_i^{IC}} \right| d_i^{IR}
\tag{7}
$$

Generally, the Taylor formula can achieve the best approximation in the case of a small interval uncertainty. Further, the calculation accuracy can be improved by establishing a subinterval to compensate for the nonlinear approximation error. For the uncertainty values U,

$$
U_s = \left[U^{IL} + \frac{2(s-1)U^{IR}}{S\_n}, U^{IL} + \frac{2sU^{IR}}{S\_n} \right], \quad s = 1, 2, \ldots, S\_n
\tag{8}
$$

In the formula, U_s and S_n are the sth subinterval and the subinterval number, respectively. The subinterval number can be determined by referring to the number of uncertain parameters. The interval range of constraint function $G_i(U)$ is expressed as follows:

$$
G_i(U) = \left[\min\left(G_i^{IL}(U_1) \ldots G_i^{IL}(U_s) \right), \max\left(G_i^{IU}(U_1) \ldots G_i^{IU}(U_s) \right) \right]
\tag{9}
$$

$$
G_i^{IL}(U) = \min\left(G_i^{IL}(U_1) \ldots G_i^{IL}(U_s) \right), \quad G_i^{IU}(U) = \max\left(G_i^{IU}(U_1) \ldots G_i^{IU}(U_s) \right)
\tag{10}
$$

Through the above interval uncertainty analysis method, the uncertain information of constraint function $G_i(U)$ can be solved by using the approximate direct decoupling method.

2.2. Improved Evolutionary Algorithm

The classical non-dominated sorting genetic algorithm (NSGA-II) generally uses the random function to generate the initial population [33], and its population uniformity is poor. The crossover probability and mutation probability of classical NSGA-II are set to a fixed value, respectively, and the optimization algorithm falls into the premature problem. Therefore, this paper adopts the improved NSGA-II designed by using chaotic and adaptive evolutionary strategies in order to obtain the multi-objective solution set.

Here, a chaotic strategy is used to generate the initial population of a multi-objective evolutionary algorithm, which can improve the diversity of the population. Tent map is one of the most commonly used mapping functions for generating chaotic sequences [34]. Here, the main steps of population chaos initialization and assignment, using the Tent mapping method, are as follows:

Step 1: Randomly generate an N-dimensional random number vector, $X_1 = (X_{11}, \ldots X_{1j}, \ldots X_{1N_v})$, $X_{1j} \in [0,1]$, where N_v is the number of optimization variables.

Step 2: The improved Tent mapping method is used to calculate the chaotic component of each optimized variable, as follows:

$$X_{(i+1,j)} = \begin{cases} T\left(X_{(i,j)}\right) + 0.1 \cdot \text{rand}(0,1), & X_{(i,j)} \in [0, 0.25, 0.5, 0.75] \text{ or } X_{(i,j)} = X_{(i-\delta,j)}, \ \delta \in [1,2,3,4] \\ T\left(X_{(i,j)}\right), & \text{else} \end{cases} \tag{11}$$

$$T\left(X_{(i,j)}\right) = \begin{cases} 2X_{(i,j)}, & 0 \leq X_{(i,j)} \leq 0.5 \\ 2\left(1 - X_{(i,j)}\right), & 0.5 < X_{(i,j)} \leq 1 \end{cases} \tag{12}$$

In the formula, $i = 1, 2, \ldots, P_s$, and P_s is the population size; $j = 1, 2, \ldots, N_v$.

Step 3: Substitute each chaotic component obtained in Step 2 into the real range of each optimization variable, as follows:

$$x_{(i,j)} = x_j^{lower} + X_{(i,j)}\left(x_j^{upper} - x_j^{lower}\right) \tag{13}$$

In the formula, x_j^{lower} and x_j^{upper} are the lower and upper bounds of the jth optimized variable respectively.

Here, the adaptive evolutionary strategy mainly improves the crossover and mutation operators. The adaptive crossover probability and mutation probability are generated according to the number of iterations, which is helpful to accelerate the convergence of optimization. In this study, the exponential function is applied to the adaptive adjustment mode of crossover probability and mutation probability. The calculation formula is described as follows:

$$\begin{cases} pc(n_i) = 1 - \dfrac{1.5e^{\left(-\frac{n_i}{n_t}\right)}}{1 + e^{\left(-\frac{n_i}{n_t}\right)}} pc(0) \\ \\ pm(n_i) = \dfrac{1.5e^{\left(-\frac{n_i}{n_t}\right)}}{1 + e^{\left(-\frac{n_i}{n_t}\right)}} pm(0) \end{cases}, \tag{14}$$

In the formula, $pc(n_i)$ and $pm(n_i)$ are the crossover probability and mutation probability at the n_i^{th} iteration; $pc(0)$ and $pm(0)$ are the initial crossover probability and mutation probability respectively; n_t is the total evolutionary generation.

2.3. Multi-Criteria Decision Making (MCDM) Method

Usually, the Pareto solution set in multi-objective optimization can provide decision-makers with numerous feasible design schemes at the early stage of design, but it cannot directly obtain the most satisfactory solution. In addition, the weight method aggregates multi-objective optimization into a single comprehensive objective to obtain the ideal optimal solution. However, although some decision-makers are full of engineering experience, it is nonetheless difficult to assign the optimal weight to each optimization objective. Therefore, as a multi-criteria decision making (MCDM) model, grey relational analysis (GRA) will be applied to select the most satisfactory scheme in Pareto sets [35–37]. Here, the GRA with entropy weight method is proposed to identify the most satisfactory solution. The normalization method can be adopted in the grey relation analysis, depending on the characteristics of the original sequence. When the target value of the original sequence is "the larger the better", the original sequence can be normalized as:

$$y_i^*(k) = \frac{y_i(k) - \min[y_i(k)]}{\max[y_i(k)] - \min[y_i(k)]} \tag{15}$$

In the formula, $y_i^*(k)$ is a new sequence after normalization; $\max[y_i(k)]$ is the maximum value of the original sequence; and $\min[y_i(k)]$ is the minimum value of the original sequence. When the target value of the original sequence is "the smaller the better", the original sequence can be normalized as:

$$y_i^*(k) = \frac{\max[y_i(k)] - y_i(k)}{\max[y_i(k)] - \min[y_i(k)]} \tag{16}$$

After normalization, the grey relational coefficient $\gamma_i(k)$, which is used to quantify the relationship between the target and actual normalized results, can be formulated as [26]:

$$\gamma_i(k) = \frac{\nabla_{min} - \rho \nabla_{max}}{\nabla_{oi}(k) + \rho \nabla_{max}} \tag{17}$$

In the formula, $\nabla_{oi}(k)$ is the deviation between reference sequence $\nabla_{oi}(k)$ and the compared sequence $x_i^*(k)$, as follows:

$$\nabla_{oi}(k) = \|y_i^*(k) - y_o(k)\| \tag{18}$$

$$\nabla_{min} = \min_{\forall j \in i} \min_{\forall k} \|y_j^*(k) - y_o(k)\| \tag{19}$$

$$\nabla_{max} = \max_{\forall j \in i} \max_{\forall k} \|y_j^*(k) - y_o(k)\| \tag{20}$$

ρ is the distinguishing coefficient, $\rho \in [0, 1]$, and $\rho = 0.5$ in this study. After obtaining the grey relational coefficient, the grey relational grade c_i is presented in a weighted sum of the grey relational coefficients, as follows:

$$c_i = \frac{1}{n} \sum_{k=1}^{n} \gamma_i(k) \tag{21}$$

For the actual engineering requirements, the effect of each criterion on the design objectives is not exactly the same; resultingly, Equation (25) can be modified to

$$\begin{cases} c_i = \sum\limits_{j=1}^{n} w_k \gamma_i(k) \\ \sum\limits_{k=1}^{n} w_k = 1 \end{cases} \tag{22}$$

In the formula, w_k is a weight of kth criterion. In this study, w_j is determined by the entropy weight method. The weight is calculated by using the entropy weight method according to the variation degree of each criterion.

Different from the analytic hierarchy process (AHP) [38], the entropy weight method can objectively obtain the weight of each criterion according to the amount of information provided by each criterion and the correlation between the criteria, which overcomes the subjectivity in determining the weight of the criterion. Assuming O_{ik} is the ith alternative value of the kth evaluation criterion, and the initial evaluation matrix is $O = (O_{ik})_{m \times n}$.

The proportion of the ith alternative value of the kth evaluation criterion is [39]:

$$P_{ik} = \frac{O_{ik}}{\sum_{i=1}^{m} O_{ik}} (i = 1, 2 \dots, m; k = 1, 2, \dots, n) \tag{23}$$

The entropy e_k of the kth criterion is:

$$e_k = -\sum_{i=1}^{m} p_{ik} \ln(p_{ik}) / \ln(m) \tag{24}$$

When p_{ik} is equal to 0, to ensure that $\ln(p_{ik})$ is meaningful, Equation (23) can be modified to:

$$P_{ik} = \frac{O_{ik} + 1}{m + \sum_{i=1}^{m} O_{ik}}. \tag{25}$$

Therefore, the entropy weight of the kth criterion can be expressed as follows:

$$w_k = \frac{1 - e_k}{\sum_{k=1}^{n}(1 - e_k)}. \tag{26}$$

2.4. Main Processes of MUOD

The main steps of the multi-objective uncertainty optimization design (MUOD) framework are as shown in Figure 1:

Step 1: Define multi-objective optimization problems, optimization variables, objectives, and constraint functions. This step is similar to conventional deterministic multi-objective optimization.

Step 2: The interval optimization problem is transformed into a deterministic optimization problem using the relationship of reliability-based possibility degree. It should be noted that the interval uncertainty transformation mainly aims at the inequality constraints in the multi-objective optimization model. The lower and upper values of constraint functions can be solved directly by using the approximate direct decoupling method.

Step 3: The execution process of the multi-objective evolutionary algorithm. An improved multi-objective evolutionary algorithm is applied to calculate the transformed mathematical model in Step 2. The improved NSGA-II is designed in Section 2.2. The initial crossover probability and mutation probability are 0.8 and 0.1.

Step 4: The execution process of the MCDM method. The MCDM method is described in Section 2.3. The GRA with entropy weight method is applied to choose the most satisfactory solution in Pareto sets. The objective weight is calculated by using the entropy weight method, and the grey relational grade is calculated by Equation (26).

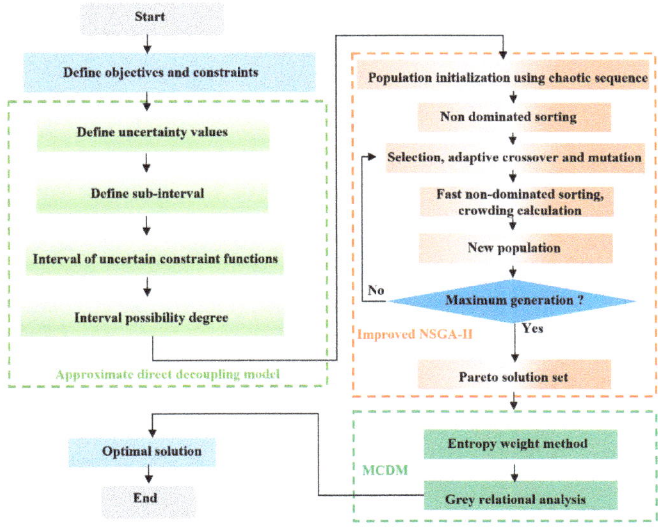

Figure 1. Main steps of MUOD framework.

3. Design Requirements of the Planetary Gear Train

3.1. Main Design Variables and Optimization Objectives

The electric drive system and its planetary gear train are shown in Figure 2. The planetary gear train is used to reduce the speed and increase the output torque of the motor. Since the helical gear has the advantages of good meshing, stable transmission, and low noise, the helical planetary gear train is designed in this study. The main parameters of an electric commercial vehicle are shown in Table 1, which are provided by a vehicle company. The transmission ratio of the gear and the final output torque can be calculated according to the vehicle parameters.

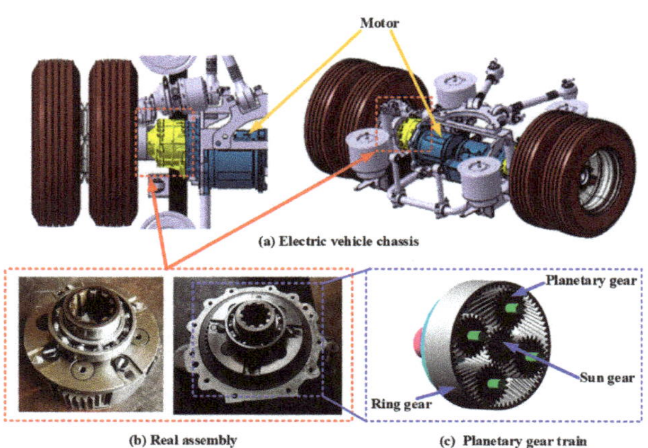

Figure 2. The electric drive system of the vehicle.

Table 1. Main parameters of full vehicle design.

Parameters	Values
Body size (length, width, height) (mm)	7232, 2240, 2820
Wheelbase (mm)	3935
Curb mass (kg)	5000
Full load mass (kg)	8500
Front/rear wheel track (mm)	1901/1630
Rolling radius (mm)	373
Maximum speed (km/h)	100
Maximum climbing degree	30%
Maximum speed in 30 min (km/h)	90

The planetary gear train should be compact; that is, the overall volume of the planetary gear train should be small enough to facilitate the arrangement of the electric drive system in the chassis of the electric vehicle. Smaller gear volume corresponds to lighter weight, which is more conducive to the improvement of energy efficiency. Therefore, the volume of the helical planetary gear train is used as the optimization objective function. To simplify the calculation, the volume v of the ring gear is chosen as the design objective, as follows:

$$v = \frac{1}{4}\pi b d_r^2, \tag{27}$$

In the formula, d_r is the pitch circle diameter of the ring gear; b is the tooth width. In general, the power loss in gear transmission mainly includes the friction loss caused by gear tooth surface meshing, the bearing loss, and the stirring loss of lubricating oil. The meshing friction power loss is the main reason for the gear transmission power loss. Therefore, the transmission efficiency, considering the meshing friction power loss, is regarded as the second design objective. The planetary gear train mainly includes external and internal meshing of the gear, as shown in Figure 3a,b.

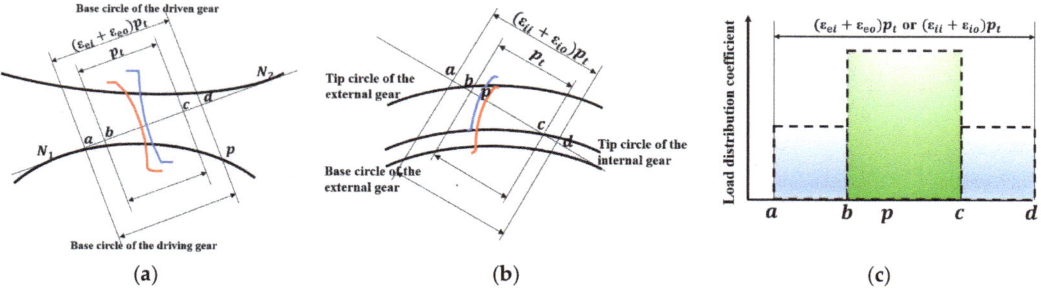

Figure 3. Schematic diagram of two kinds of gear meshing: (**a**) external meshing; (**b**) internal meshing; and (**c**) load distribution coefficient.

Here, the transmission efficiency η_{ex} of external meshing can be expressed as [40]:

$$\eta_{ex} = 1 - \frac{\mu Z_p p_t \left(\frac{1}{Z_s} + \frac{1}{Z_p}\right) f_{ex}(\varepsilon)}{R_{bp} - \mu R_p (\varepsilon_{ei} - \varepsilon_{eo}) \sin \alpha_t + \mu p_t f_{ex}(\varepsilon)} \tag{28}$$

In the formula, μ is the friction coefficient; Z_s is the number of teeth of the sun gear; Z_p is the number of teeth of the planet gear; α_t is the transverse pressure angle; p_t is the transverse circular pitch; R_{bp} and R_p are the base circle radius and pitch circle radius of planet gear respectively; and ε_{ei} and ε_{eo} are the meshing in and meshing out contact ratio of the gear external meshing, respectively, as follows:

$$\varepsilon_{ei} = \frac{Z_s (tg\alpha_{as} - tg\alpha')}{2\pi} \tag{29}$$

$$\varepsilon_{eo} = \frac{Z_p (tg\alpha_{ap} - tg\alpha')}{2\pi} \tag{30}$$

In the formula, α' is the working pressure angle; α_{as} and α_{ap} are the tooth top pressure angle of sun gear and planet gear respectively. As shown in Figure 3c, assuming that the load distribution coefficient in the regions "a-b" and "c-d" is 0.5, $f_{ex}(\varepsilon)$ can be expressed as follows:

$$f_{ex}(\varepsilon) = 0.5 \left(\varepsilon_{ei}^2 + \varepsilon_{eo}^2 - \varepsilon_{ei} - \varepsilon_{eo} + 1\right) \tag{31}$$

Similarity, the transmission efficiency η_{in} of internal meshing can be expressed as follows [40]:

$$\eta_{in} = 1 - \frac{\mu Z_P p_t \left(\frac{1}{Z_p} - \frac{1}{Z_r}\right) f_{in}(\varepsilon)}{R_{bp} - \mu R_p (\varepsilon_{ii} - \varepsilon_{io}) \sin \alpha_t + \mu p_t f_{in}(\varepsilon)} \tag{32}$$

In the formula, Z_r is the is the number of teeth of the ring gear; ε_{ii} and ε_{io} are the meshing in and meshing out contact ratio of the gear external meshing, respectively, as follows:

$$\varepsilon_{ii} = \frac{\sqrt{R_{ap}^2 - R_{bp}^2} - R_p \sin \alpha_t}{p_t} \tag{33}$$

$$\varepsilon_{io} = \frac{R_r \sin \alpha_t - \sqrt{R_{ar}^2 - R_{br}^2}}{p_t} \tag{34}$$

Same as Equation (31), assuming that the load distribution coefficient is 0.5, $f_{in}(\varepsilon)$ can be expressed as follows:

$$f_{in}(\varepsilon) = 0.5 \left(\varepsilon_{ii}^2 + \varepsilon_{io}^2 - \varepsilon_{ii} - \varepsilon_{io} + 1\right). \tag{35}$$

Therefore, the transmission efficiency η_p of the planetary reduction gear train can be expressed as follows [41]:

$$\eta_p = \frac{R_s}{2(R_r - R_p)} + \left[1 - \frac{R_s}{2(R_r - R_p)}\right]\eta_{ex}\eta_{in} \tag{36}$$

In this study, six main parameters are considered as design variables to find the minimum volume and maximum transmission efficiency. These design variables X_v include the teeth number of the sun gear Z_s, the teeth number of the planet gear Z_p, the teeth number of the ring gear Z_r, helix angle β, face width b, and normal module m_n. Z_s, Z_p, and Z_r are integers, β and b are continuous, and m_n is discrete. The alternative modulus m_n is shown in Equation (38). Table 2 shows the detailed information of all design variables.

$$X_v = \left\{Z_s, Z_p, Z_r, \beta, b, m_n\right\}, \tag{37}$$

$$m_n \in \{2,\ 2.25,\ 2.5,\ 2.75,\ 3,\ 3.5,\ 4,\ 4.5,\ 5\} \tag{38}$$

Table 2. Design variables.

Design Variables	Lower Bound	Upper Bound
Z_s	20	30
Z_p	20	30
Z_r	60	80
β	20	30
b	30	50
m_n	/	/

3.2. Main Design Constraints

The gear design should meet the specified constraints to meet the actual geometric, load, and material requirements. The main constraints of the planetary gear train in this study are as follows.

3.2.1. Equally Spaced Planets

To prevent the gear teeth from interfering with the mating gear, the gear teeth of all gears must mesh with the center gear teeth at the same time. The installation requirement needs to meet the following conditions:

$$\frac{Z_s + Z_r}{n_p} = \text{integer}, \tag{39}$$

In the formula, n_p is the number of planet gears.

3.2.2. Equally Spaced Planets

According to the actual power requirement of an electric vehicle, the transmission ratio r_i of the planetary gear train needs to meet the following conditions:

$$4.1 \leq r_i \leq 4.6, \tag{40}$$

$$r_i = 1 + \frac{Z_r}{Z_s}. \tag{41}$$

3.2.3. Tooth Width Coefficient

The size of the tooth width is related to the strength of the gear; the larger the tooth width, the higher the strength. However, it should be noted that, if the tooth width is too large, there will be a larger number of tooth contact errors as well as a more uneven load distribution in the tooth direction. Therefore, it is critical to select an applicable tooth width. Here, the tooth width coefficient Φ_d is the primary indicator of the tooth width design, which needs to meet the following constraints [42]:

$$0.7 \leq \Phi_d \leq 4, \tag{42}$$

$$\Phi_d = \frac{b}{d_s}. \tag{43}$$

3.2.4. Minimum Teeth of No-Undercut

Gear undercutting not only weakens the root of gear teeth while reducing the bending strength, but it also reduces the coincidence degree. Therefore, undercutting should be avoided in the gear design stage. The minimum number of teeth without undercutting of the helical cylindrical gear needs to meet the following constraints:

$$\{Z_s, Z_p, Z_r\} \geq 17 \cos^3 \beta. \tag{44}$$

3.2.5. Concentric Constraint

The center distance between sun gear, ring gear, and planetary gear should be equal. The concentric constraint is:

$$Z_s + Z_p = Z_r - Z_p. \tag{45}$$

3.2.6. Adjacency Constraint

To prevent the planet gears from colliding with each other, it is necessary to ensure that the planet gears have a certain clearance on their connecting lines; that is, the sum of the tooth top circle radius of two adjacent planetary gears shall be less than the center distance of two adjacent planetary gears. The adjacency constraint is:

$$d_{ap} < 2l_{sp} \sin\left(\frac{\pi}{n_p}\right), \tag{46}$$

In the formula, d_{ap} is the addendum circle diameter of planet gear; l_{sp} is the center distance between the sun gear and the planet gear.

3.2.7. Contact Stress Requirement

The planetary gear train should be able to resist material failure (deformation and fracture) during contact behavior. The real contact stress requirement σ_c needs to meet the following constraint [43]:

$$\sigma_c = 0.418\sqrt{F_n E\left(\frac{1}{\rho_a} + \frac{1}{\rho_p}\right)/b} \leq [\sigma_c] \tag{47}$$

In the formula, F_n is the normal load; E is the elastic modulus of the material; $[\sigma_c]$ is the allowable contact stress, and the gear material is 40Cr in this study; and ρ_a and ρ_p denote the radius of curvature at the nodes of the driving and driven gears, respectively. In the planetary gear system, the contact stress mainly occurs between the sun gear and the planetary gear, and between the planetary gear and the ring gear. However, considering that the dangerous position is usually present between the sun gear and the planetary gear, this study will focus on the contact stress between the sun gear and the planetary gear.

3.2.8. Bending Stress Requirement

The bending stress requirement σ_{w_s} needs to meet the following constraint [42]:

$$\sigma_{w\_s} = \frac{2T_t K_\sigma / d}{b\pi m_n Y K_\varepsilon} \leq [\sigma_{w\_s}], \tag{48}$$

In the formula, T_t is the transmitted torque; d is the pitch circle diameter; K_σ is the stress concentration factor of contact ratio, $K_\sigma = 1.5$; K_ε is the influence factor of contact ratio, $K_\varepsilon = 2.0$; and Y is the tooth profile coefficient. According to the empirical formula, the tooth profile coefficient of the sun gear can be expressed as follows:

$$Y = 0.1735 - \frac{0.717}{Z_v} - \frac{8.37}{Z_v^2} + \frac{53.84}{Z_v^3}, \tag{49}$$

In the formula, Z_v is a virtual number of teeth pertaining to the helical gear, $Z_v = Z / \cos^3 \beta$.

4. Optimization Results and Discussions

In this study, the helix angle β, face width b, elastic modulus E, and input torque T are considered uncertain. The uncertain helix angle β and face width b are regarded as the uncertainty of manufacturing size. The uncertain elastic modulus E is regarded as the uncertainty of material. The uncertain input torque T is regarded as the uncertainty of load input. This study defines three uncertainty cases with different uncertainty deviations, which correspond to different degrees of uncertainty deviations, as shown in Table 3. Therefore, the constraint functions related to the above uncertain values can be regarded as uncertainty constraints. Uncertainty constraints mainly include tooth width coefficient, minimum teeth of no-undercut, adjacency constraint, contact stress constraint, and bending stress constraint. The nominal values of elastic modulus E^{IC} and input load T_t^{IC} are 210 GPa and 3936 N·m, respectively. All requirements of reliability-based possibility degree λ are defined as 0.8.

Table 3. Three uncertainty cases.

Uncertainties	Case 1	Case 2	Case 3
β^{IR} (°)	2	3	4
b^{IR} (mm)	2	3	4
E^{IR} (GPa)	10.5	21	31.5
T_t^{IR} (N·m)	393.6	590.4	787.2

Here, the classical NSGA-II and multi-objective particle swarm optimization (MOPSO) are implemented in order to explore the feasibility of the improved NSGA-II. The initial population size is 400, the maximum number of iterations is 200, and the objective number of non-dominated solutions is 200. Figure 4 shows the iterative history of MDOD by using MOPSO, NSGA-II, and improved NSGA-II. The number of non-dominated solutions obtained by improved NSGA-II increases steadily, and improved NSGA-II can obtain non-dominated solutions more efficiently than MOPSO and NSGA-II. Therefore, the improved NSGA-II designed in this paper is effective, and it contains better optimization potential than the classical NSGA-II and MOPSO. The improved NSGA-II will be implemented for MUOD.

Figure 5 shows the optimal Pareto solution sets of MDOD and MUOD. There is an intense conflict between volume and transmission efficiency, which cannot achieve the common optimization; that is, the further improvement of one objective will inevitably worsen the other objective. The Pareto solution set of deterministic optimization design is lower than that of uncertainty optimization design, and the optimization objective of deterministic optimization design is better than uncertainty optimization design. In general, the inequality constraint of deterministic optimization is mainly concentrated near the constraint boundary, so its Pareto solution set has more loose space, and it is easier to obtain the better solution. It should be noted that, with the increase of uncertainty, the optimization results of MUOD tend to be conservative. To obtain the optimal solutions in different multi-objective optimization models, this study makes a trade-off analysis on the Pareto solution set by using the MCDM method. The optimal results of MDOD and MUOD are shown in Appendix A. Table A1 presents the optimization results of MDOD, and Tables A2–A4 present the optimization results of MUOD. The alternatives of different optimization methods are sorted according to the grey correlation degree, and the optimal solutions of all optimization methods are shown in bold. It is found that all moduli are 2, which means that the optimization space for modulus is small. Here, three uncertainty cases are substituted into all optimization results, and the obtained constraints are shown in Table 4. The optimal result of MDOD shows that the λ of the upper bound of tooth width coefficient is less than 0.8, and the bending stress is less than 0.8 when the uncertainty range is the largest (Case 3). From another perspective, the bending stress constraint and the tooth width coefficient constraint are the most prone to failure types. In the three uncertainty cases, MUOD meets the reliability requirements for all uncertainty constraints. A higher λ indicates that the farther the optimization result is from the boundary of the inequality constraint, the higher its reliability. Compared with the conventional MDOD, the MUOD proposed in this study can design a more reliable planetary gear train and reduce the risk of constraint failure. The results show that MUOD sacrifices certain performance, but it is more reliable than MDOD.

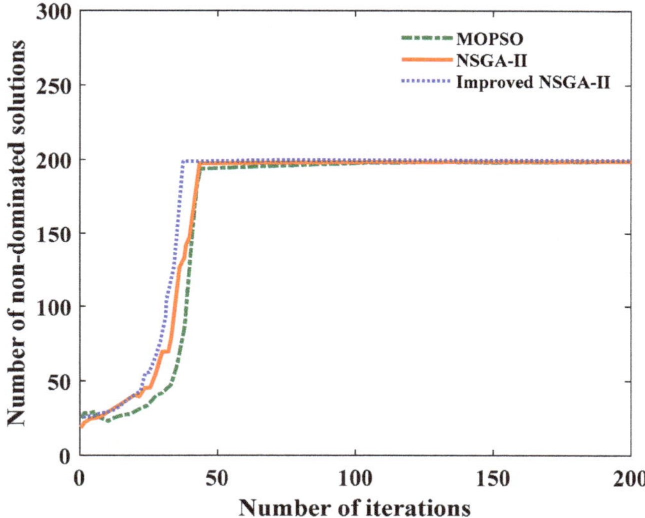

Figure 4. Iterative history of non-dominated solutions.

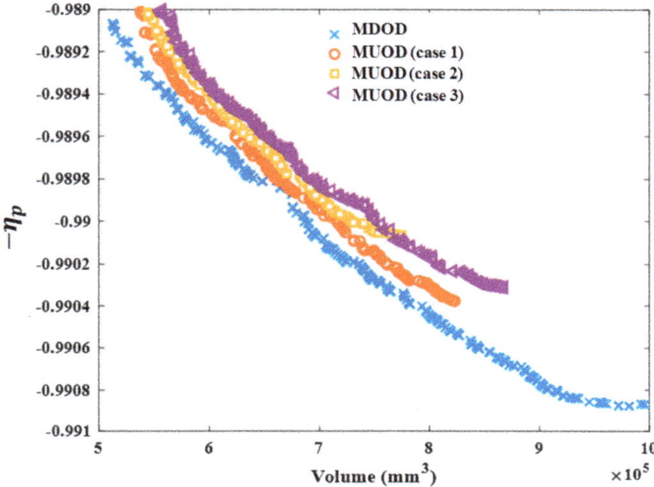

Figure 5. The Pareto front results with different uncertainties.

Table 4. Reliability-based possibility degree λ of four solution sets under three uncertainty cases.

Constraint	MDOD (Case 1)	MUOD (Case 1)	MDOD (Case 2)	MUOD (Case 2)	MDOD (Case 3)	MUOD (Case 3)
Lower bound of tooth width coefficient	1	1	1	1	1	1
Upper bound of tooth width coefficient	0.64	0.80	0.60	0.81	0.57	0.81
Minimum teeth of no-undercut for Z_s	1	1	1	1	1	1
Minimum teeth of no-undercut for Z_p	1	1	1	1	1	1
Minimum teeth of no-undercut for Z_r	1	1	1	1	1	1
Adjacency constraint	1	1	1	1	1	1
Contact stress	1	1	1	1	1	1
Bending stress	1	1	0.84	0.85	0.76	0.98

5. Conclusions

In order to design a reasonable planetary reduction gear system, matching the electric vehicle motor, this study proposes a multi-objective uncertainty optimization design (MUOD) framework for the planetary gear train of an electric vehicle. An approximate direct decoupling model, based on subinterval Taylor expansion, is applied to evaluate the propagation of uncertainties; the improved evolutionary algorithm is designed by using chaotic and adaptive evolutionary strategies. The volume and transmission efficiency of the planetary gear are optimization objectives. The optimization results of MUOD show that the Pareto front gradually moves to the upper right corner with the uncertainty increases. The most satisfactory solutions (improving lightweight and improving transmission efficiency) pertaining to different multi-objective optimization models can be obtained by the MCDM method. Compared with the conventional multi-objective deterministic optimization design (MDOD) method, the uncertainty optimization design of the planetary gear train sacrifices certain performance. When the reliability-based possibility degree λ is defined as 0.8, the optimization results of MUOD always meet this requirement, but at least one constraint violates this requirement in MDOD. As the degree of uncertainty increases, the optimization results of MUOD tend to be conservative, but MUOD is more reliable than MDOD. The uncertainty of planetary gear trains of electric vehicles is very complex in actual working conditions. The MUOD framework proposed in this study is able to continue carrying out optimization design with more complex high dimension uncertainties and objectives in order to ensure that the structure has better potential to resist the risk of failure.

Author Contributions: Writing—review & editing, X.X.; writing—original draft, X.X., J.C.; software, X.X., J.C., Z.L., Y.Q., X.C., Y.Z. and Y.X.; methodology, X.X., J.C., Z.L., Y.Q., X.C., Y.Z., Y.X. and Y.L.; investigation, X.X., Z.L., Y.Q., Y.Z., Y.X. and Y.L.; project administration, X.C. All authors have read and agreed to the published version of the manuscript.

Funding: This research was funded by "Project of Shanghai Science and Technology Committee, grant number 20511104602" and "Prospective Technology Project of Nanchang Intelligent New Energy Vehicle Research Institute, grant number 17092380013".

Conflicts of Interest: The authors declare no conflict of interest.

Appendix A. Optimization Results

Table A1. Optimization results from MDOD.

Ranking	b (mm)	β (°)	m_n (mm)	Z_p	Z_r	Z_s	c_i
1	**31.60**	**24.91**	**2**	**22**	**64**	**20**	**0.9333**
2	31.63	25.32	2	22	64	20	0.9195
3	31.63	25.46	2	22	64	20	0.9153
4	32.00	22.52	2	25	71	21	0.6758
5	32.00	22.55	2	25	71	21	0.6753
6	31.94	22.69	2	25	71	21	0.6752
7	32.00	22.56	2	25	71	21	0.6751
8	31.96	22.66	2	25	71	21	0.6751
9	31.95	22.69	2	25	71	21	0.6748
10	32.01	22.83	2	25	71	21	0.6709
⋮	⋮	⋮	⋮	⋮	⋮	⋮	⋮

Table A2. Optimization results from MUOD (Case 1).

Ranking	b (mm)	β (°)	m_n (mm)	Z_p	Z_r	Z_s	c_i
1	**34.81**	**22.38**	**2**	**24**	**70**	**22**	**0.9333**
2	34.81	22.39	2	24	70	22	0.9331
3	34.87	22.38	2	24	70	22	0.9247
4	34.86	22.51	2	24	70	22	0.9241
5	34.87	22.59	2	24	70	22	0.9201
6	34.87	22.60	2	24	70	22	0.9200
7	34.87	22.61	2	24	70	22	0.9193
8	34.95	22.67	2	24	70	22	0.9106
9	35.11	22.70	2	24	70	22	0.8968
10	35.11	22.74	2	24	70	22	0.8951
⋮	⋮	⋮	⋮	⋮	⋮	⋮	⋮

Table A3. Optimization results from MUOD (Case 2).

Ranking	b (mm)	β (°)	m_n (mm)	Z_p	Z_r	Z_s	c_i
1	**32.56**	**22.31**	**2**	**22**	**64**	**20**	**0.9333**
2	32.56	22.50	2	22	64	20	0.9292
3	32.69	22.57	2	22	64	20	0.9214
4	32.73	22.67	2	22	64	20	0.9175
5	32.69	23.01	2	22	64	20	0.9117
6	32.69	23.02	2	22	64	20	0.9114
7	32.69	23.03	2	22	64	20	0.9114
8	32.73	23.16	2	22	64	20	0.9064
9	32.73	23.21	2	22	64	20	0.9055
10	32.73	23.28	2	22	64	20	0.9039
⋮	⋮	⋮	⋮	⋮	⋮	⋮	⋮

Table A4. Optimization results from MUOD (Case 3).

Ranking	b (mm)	β (°)	m_n (mm)	Z_p	Z_r	Z_s	c_i
1	34.84	22.33	2	23	67	21	0.9333
2	34.86	22.33	2	23	67	21	0.9319
3	35.15	23.05	2	23	67	21	0.8859
4	35.15	23.14	2	23	67	21	0.8823
5	35.17	23.17	2	23	67	21	0.8798
6	35.17	23.18	2	23	67	21	0.8795
7	35.27	23.28	2	23	67	21	0.8699
8	35.27	23.42	2	23	67	21	0.8652
9	35.33	23.46	2	23	67	21	0.8592
10	35.38	23.49	2	23	67	21	0.8558
⋮	⋮	⋮	⋮	⋮	⋮	⋮	⋮

References

1. Wang, R.; Chen, Y.; Feng, D.; Huang, X.; Wang, J. Development and performance characterization of an electric ground vehicle with independently actuated in-wheel motors. *J. Power Sources* **2011**, *196*, 3962–3971. [CrossRef]
2. Hang, P.; Chen, X. Towards autonomous driving: Review and perspectives on configuration and control of four-wheel independent drive/steering electric vehicles. *Actuators* **2021**, *10*, 184. [CrossRef]
3. Parmar, A.; Ramkumar, P.; Shankar, K. Macro geometry multi-objective optimization of planetary gearbox considering scuffing constraint. *Mech. Mach. Theory* **2020**, *154*, 104045. [CrossRef]
4. Miler, D.; Žeželj, D.; Lončar, A.; Vučković, K. Multi-objective spur gear pair optimization focused on volume and efficiency. *Mech. Mach. Theory* **2018**, *125*, 185–195. [CrossRef]
5. Sedak, M.; Rosić, B. Multi-objective optimization of planetary gearbox with adaptive hybrid particle swarm differential evolution algorithm. *Appl. Sci.* **2021**, *11*, 1107. [CrossRef]
6. Patil, M.; Ramkumar, P.; Shankar, K. Multi-objective optimization of the two-stage helical gearbox with tribological constraints. *Mech. Mach. Theory* **2019**, *138*, 38–57. [CrossRef]
7. Savsani, V.; Rao, R.V.; Vakharia, D.P. Optimal weight design of a gear train using particle swarm optimization and simulated annealing algorithms. *Mech. Mach. Theory* **2010**, *45*, 531–541. [CrossRef]
8. Sun, G.; Zhang, H.; Fang, J.; Li, G.; Li, Q. A new multi-objective discrete robust optimization algorithm for engineering design. *Appl. Math. Model* **2018**, *53*, 602–621. [CrossRef]
9. Xu, X.; Chen, X.; Liu, Z.; Yang, J.; Xu, Y.; Zhang, Y.; Gao, Y. Multi-objective reliability-based design optimization for the reducer housing of electric vehicles. *Eng. Optim.* **2021**, 1–17. [CrossRef]
10. Xu, X.; Chen, X.; Liu, Z.; Xu, Y.; Zhang, Y. Reliability-based design for lightweight vehicle structures with uncertain manufacturing accuracy. *Appl. Math. Model* **2021**, *95*, 22–37. [CrossRef]
11. Fang, J.; Gao, Y.; Sun, G.; Xu, C.; Li, Q. Multiobjective robust design optimization of fatigue life for a truck cab. *Reliab. Eng. Syst. Saf.* **2015**, *135*, 1–8. [CrossRef]
12. Dawood, T.; Elwakil, E.; Novoa, H.M.; Delgado, J.F.G. Soft computing for modeling pipeline risk index under uncertainty. *Eng. Fail. Anal.* **2020**, *117*, 104949. [CrossRef]
13. Li, W.; Gao, L.; Xiao, M. Multidisciplinary robust design optimization under parameter and model uncertainties. *Eng. Optim.* **2020**, *52*, 426–445. [CrossRef]
14. Chen, Z.; Li, T.; Xue, X.; Zhou, Y.; Jing, S. Fatigue reliability analysis and optimization of vibrator baseplate based on fuzzy comprehensive evaluation method. *Eng. Fail. Anal.* **2021**, *127*, 105357. [CrossRef]
15. Yuan, R.; Tang, M.; Wang, H.; Li, H. A Reliability analysis method of accelerated performance degradation based on bayesian strategy. *Access* **2019**, *7*, 169047–169054. [CrossRef]
16. Xian, J.; Su, C. Stochastic optimization of uncertain viscous dampers for energy-dissipation structures under random seismic excitations. *Mech. Syst. Signal Process.* **2022**, *164*, 108208. [CrossRef]
17. Lü, H.; Yang, K.; Huang, X.; Yin, H.; Shangguan, W.-B.; Yu, D. An efficient approach for the design optimization of dual uncertain structures involving fuzzy random variables. *Comput. Methods Appl. Mech. Eng.* **2020**, *371*, 113331. [CrossRef]
18. Baek, S.M.; Lee, W.J. Design method for radar absorbing structures using reliability-based design optimization of the composite material properties. *Compos. Struct.* **2021**, *262*, 113559. [CrossRef]
19. Fang, J.; Gao, Y.; Sun, G.; Li, Q. Multiobjective reliability-based optimization for design of a vehicledoor. *Finite Elem. Anal. Des.* **2013**, *67*, 13–21. [CrossRef]
20. Zhang, Y.; Xu, X.; Sun, G.; Lai, X.; Li, Q. Nondeterministic optimization of tapered sandwich column for crashworthiness. *Thin Walled Struct.* **2018**, *122*, 193–207. [CrossRef]
21. Li, F.; Luo, Z.; Sun, G.; Zhang, N. An uncertain multidisciplinary design optimization method using interval convex models. *Eng. Optim.* **2013**, *45*, 697–718. [CrossRef]

22. Xu, X.; Chen, X.; Liu, Z.; Zhang, Y.; Xu, Y.; Fang, J.; Gao, Y. A feasible identification method of uncertainty responses for vehicle structures. *Struct. Multidiscip. Optim.* **2021**. [CrossRef]
23. Inuiguchi, M.; Sakawa, M. Minimax regret solution to linear programming problems with an interval objective function. *Eur. J. Oper. Res.* **1995**, *86*, 526–536. [CrossRef]
24. Fu, C.; Liu, Z.; Deng, J. A direct solution framework for structural optimization problems with interval uncertainties. *Appl. Math. Model.* **2020**, *80*, 384–393. [CrossRef]
25. Wu, J.; Gao, J.; Luo, Z.; Brown, T. Robust topology optimization for structures under interval uncertainty. *Adv. Eng. Softw.* **2016**, *99*, 36–48. [CrossRef]
26. Wang, L.; Yang, G.; Li, Z.; Xu, F. An efficient nonlinear interval uncertain optimization method using legendre polynomial chaos expansion. *Appl. Soft Comput.* **2021**, *108*, 107454. [CrossRef]
27. Hou, Y.; Xiong, Y.; Zhang, Y.; Liang, X.; Su, L. Vessel energy efficiency uncertainty optimization analysis in ice zone considering interval parameters. *Ocean Eng.* **2021**, *232*, 109114. [CrossRef]
28. Lü, H.; Yu, D. Brake squeal reduction of vehicle disc brake system with interval parameters by uncertain optimization. *J. Sound Vib.* **2014**, *333*, 7313–7325. [CrossRef]
29. Fonseca, C.M.; Fleming, P.J. Multiobjective optimization and multiple constraint handling with evolutionary algorithms I. A unified formulation. *Trans. Syst. Man Cybern. Part A Syst. Hum.* **1998**, *28*, 26–37. [CrossRef]
30. Fonseca, C.M.; Fleming, P.J. Multiobjective optimization and multiple constraint handling with evolutionary algorithms II. Application example. *Trans. Syst. Man Cybern. Part A Syst. Hum.* **1998**, *28*, 38–47. [CrossRef]
31. Jiang, C.; Han, X.; Liu, G.R.; Liu, G.P. A nonlinear interval number programming method for uncertain optimization problems. *Eur. J. Oper. Res.* **2008**, *188*, 1–13. [CrossRef]
32. Meng, D.; Hu, Z.; Guo, J.; Lv, Z.; Xie, T.; Wang, Z. An uncertainty-based structural design and optimization method with interval taylor expansion. *Structures* **2021**, *33*, 4492–4500. [CrossRef]
33. Deb, K.; Pratap, A.; Agarwal, S.; Meyarivan, T. A fast and elitist multiobjective genetic algorithm: NSGA-II. *Trans. Evol. Comput.* **2002**, *6*, 182–197. [CrossRef]
34. Paknejad, P.; Khorsand, R.; Ramezanpour, M. Chaotic improved PICEA-g-based multi-objective optimization for workflow scheduling in cloud environment. *Futur. Gener. Comput. Syst.* **2021**, *117*, 12–28. [CrossRef]
35. Ju-Long, D. Control problems of grey systems. *Syst. Control Lett.* **1982**, *1*, 288–294. [CrossRef]
36. Li, C.H.; Tsai, M.J. Multi-objective optimization of laser cutting for flash memory modules with special shapes using grey relational analysis. *Opt. Laser Technol.* **2009**, *41*, 634–642. [CrossRef]
37. Wei, G.-W. Gray relational analysis method for intuitionistic fuzzy multiple attribute decision making. *Expert Syst. Appl.* **2011**, *38*, 11671–11677. [CrossRef]
38. Emovon, I.; Oghenenyerovwho, O.S. Application of MCDM method in material selection for optimal design: A review. *Results Mater.* **2020**, *7*, 100115. [CrossRef]
39. Kumar, R.; Singh, S.; Bilga, P.S.; Singh, J.; Singh, S.; Scutaru, M.-L.; Pruncu, C.I. Revealing the benefits of entropy weights method for multi-objective optimization in machining operations: A critical review. *J. Mater. Res. Technol.* **2021**, *10*, 1471–1492. [CrossRef]
40. Li, C.; Qin, D.T.; Shi, W.K. Reference Efficiency of Planetary Gear Train. *J. Chongqing Uni.* **2006**, *29*, 11–14.
41. Chen, Y.; Ishibashi, A.; Sonoda, K.; Matubara, M. Studies on noise and vibration of planetary gear drives for automatic transmission of passenger cars. *Trans. Jpn. Soc. Mech. Eng. Ser. C* **2000**, *66*, 634–639. [CrossRef]
42. Lin, Z.; Zhang, J.; Xu, X.; Chen, J.; Chen, X. Optimization design of distributed drive vehicle reducer based on improved particle swarm optimization algorithm. In Proceedings of the 3rd World Conference on Mechanical Engineering and Intelligent Manufacturing (WCMEIM), Shanghai, China, 4–6 December 2020; IEEE: Piscataway, NJ, USA, 2020. [CrossRef]
43. Wang, J. *Automotive Design*, 4th ed.; China Machine Press: Beijing, China, 2011.

Article

On the Lightweight Truss Structure for the Trash Can-Handling Robot [†]

Jiawei Chen [1], Yan Li [1], Xiang Xu [2] and Xinbo Chen [2,*]

[1]   School of Mechanical Engineering, Tongji University, Shanghai 201804, China; Chen_jiawei@tongji.edu.cn (J.C.); liyanliyan910@126.com (Y.L.)
[2]   School of Automotive Studies, Tongji University, Shanghai 201804, China; xiangxu@tongji.edu.cn
*   Correspondence: chenxinbo@tongji.edu.cn
†   This paper is an extended version of the conference paper: Yan Li; Jiawei Chen; Xiang Xu; Zhongyan Lin and Xinbo Chen. Optimization of Garbage Dumping Mechanism of Intelligent Sanitation Vehicle Based on Particle Swarm Algorithm. In Proceedings of the Intelligent Manufacturing and Automation Technology (MEMAT 2021), Guilin, China, 23–25 April 2021.

Abstract: With the rapid development of cities, the automated and intelligent garbage transportation has become an important direction for technological innovation of sanitation vehicles. In this paper, a vehicle-mounted trash can-handling robot is proposed. In order to reduce the cost of the robot and increase the loading capacity of the intelligent sanitation vehicles, a lightweight design method is proposed for the truss structure of the robot. Firstly, the parameters of the robot that are related to the load are optimized by multi-objective parameter optimization based on particle swarm optimization. Then, the material distribution of the truss structure is optimized by topology optimization under multiple load cases. Finally, the thickness of the truss structure parts is optimized by discrete optimization under multiple load cases. The optimization results show that the mass of the truss structure is reduced by 8.72%, the inherent frequency is increased by 61.08%, and the maximum stress is reduced by 10.98%. The optimization results achieve the goal of performance optimization of the intelligent sanitation vehicle, and prove the feasibility of the proposed lightweight design method.

Keywords: intelligent sanitation vehicle; trash can-handling robot; truss structure; multi-objective parameter optimization; topology optimization; discrete optimization; multiple load cases

Citation: Chen, J.; Li, Y.; Xu, X.; Chen, X. On the Lightweight Truss Structure for the Trash Can-Handling Robot. *Actuators* **2021**, *10*, 214. https://doi.org/10.3390/act10090214

Academic Editor: Hicham Chaoui

Received: 15 July 2021
Accepted: 30 August 2021
Published: 31 August 2021

Publisher's Note: MDPI stays neutral with regard to jurisdictional claims in published maps and institutional affiliations.

1. Introduction

With the rapid development of cities, the production of municipal solid waste is increasing year by year, which has a non-negligible impact on the residents' living standard [1,2]. With the goal of efficient and environmental-friendly urban cleaning work, the automated and intelligent garbage transportation has become an important direction for technological innovation of sanitation vehicles. To this end, the authors' team has developed a vehicle-mounted trash can-handling robot. This robot has realized fully automated operations including trash can identification, trash can-handling, garbage dumping, and trash can resetting. In this paper, the composition and basic functions of the robot will be briefly introduced. On the basis of force analysis, this paper will study the lightweight design of the robot's truss structure, in order to further improve its working performance.

The developed trash can-handling robot is shown in Figure 1. The robot consists of a mechanical system, a driving system (hydraulic system), a control system and a perception system, as shown in Figure 2. Furthermore, the mechanical system is mainly composed of a manipulator, a telescopic boom and a truss structure. As the end-effector of the robot, the manipulator has a longitudinal adjustment range of ± 0.25 m and a lateral telescopic distance of 1 m, which reduces the technical requirements for drivers. The driving system is mainly composed of hydraulic components such as hydraulic motor and hydraulic cylinder. The roller chain system is used to transmit the power of the hydraulic motor to drive the manipulator to move along the guide rail. The control system is mainly composed of sensors, controllers and a human–machine interaction module. The driver can set the robot in automatic or manual mode through the touch screen or the operation panel installed in the cab. Finally, the perception system is mainly composed of two cameras and a lidar. The lighting lamp is used to ensure good lighting conditions in the working environment.

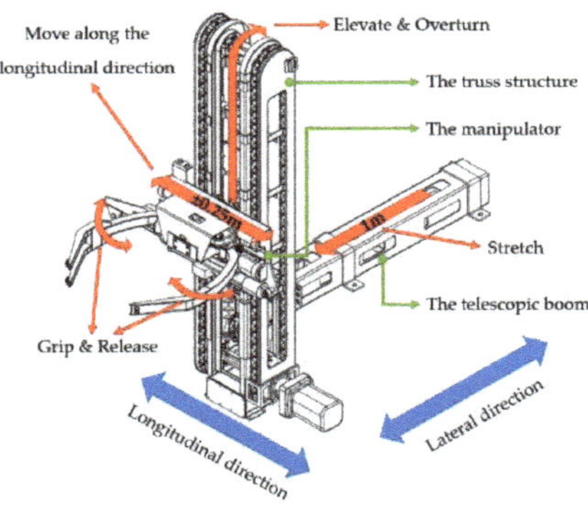

Figure 1. The trash can-handling robot.

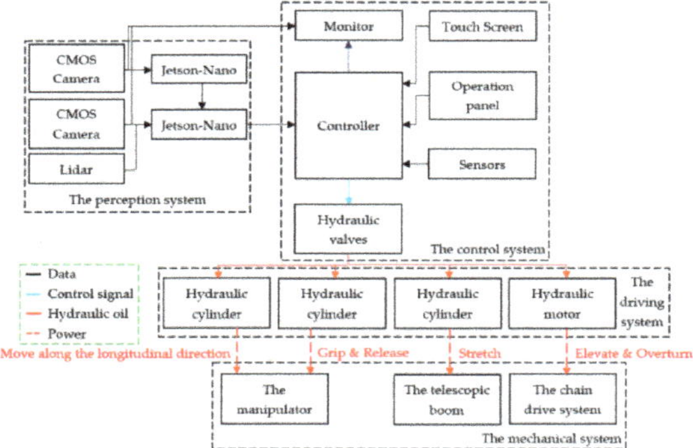

Figure 2. The composition of the trash can-handling robot.

The automatic workflow of the robot is as follows:

1. After the driver parks the sanitation vehicle next to the trash can, the perception system sequentially detects the type of the trash can, the relative position of the trash can, pedestrians and obstacles. If the position of the trash can is beyond the working range of the robot, the driver will be prompted to make adjustments;
2. The perception system converts the relative position information of the trash can into control data and then sends it to the control system;
3. According to the preset control strategy, the control system controls the manipulator through the hydraulic components to complete the garbage loading operation.

As the robot is installed on the side of the vehicle, the heavy mechanical structure will cause the vehicle to roll, which has a detrimental impact on the vehicle's handling performance and the robot's control accuracy. Otherwise, the truss structure is the key load-bearing component of the robot. Due to the complex load of the robot, the truss structure is required to have high load-bearing capacity such as rigidity and strength. Therefore, the lightweight design of the truss structure is very important to ensure the performance of the robot.

However, currently all kinds of lifting equipment mainly use multi-link mechanism. The relevant research mainly focuses on the optimization of the position of the hinge points [3–5]. So, there is little research on the optimization of the lifting equipment similar to the robot in this paper. For the truss robot with similar structure, many scholars have carried out static characteristic analysis, dynamic characteristic analysis and comprehensive analysis on the truss structure. On this basis, the structural size of the truss structure is optimized [6–9]. However, in these studies, the loads and constraints of the truss structure are quite different from those of the robot in this paper. Therefore, the reference value of these studies is limited.

For lightweight design, the main methods are structure optimization, process lightweight and material lightweight [10–12]. The structure optimization can be further divided into size optimization, shape optimization and topology optimization. At present, size optimization and shape optimization have been widely used in engineering, such as lightweight design of loading platform of flat transport vehicle, lightweight design of soybean harvester's frame, and comprehensive optimization design of column of double spindle horizontal machining center [13–15]. In addition, according to the type of design variables, size optimization can be divided into discrete size optimization and continuous size optimization [16]. In general, the results of continuous size optimization need to be rounded according to the available size parameters, so the results of discrete size optimization are more in line with the actual needs of engineering [17]. At the same time, there are more optimization variables for discrete optimization, such as cross section [18,19] and material [20–22]. In addition, the variables in the assignment problem and scheduling problem are also discrete, so discrete optimization is also applied to solve these problems. Furthermore, the discrete optimization that optimizes multiple optimization variables at the same time can obtain better optimization results [23]. However, due to the increase of the dimension of optimization variables, the solution of discrete size optimization is becoming more and more difficult. Some scholars reduce the computational cost by making discrete design variables continuous [24], while many other scholars propose their optimization methods based on different algorithms, which is a research hotspot in recent years. For example, Kaveh et al. [25] proposed an improved Shuffle Jaya algorithm for discrete size optimization of bone structure; Degertekin et al. [26] proposed an improved hybrid HS algorithm for large-scale truss structure's size optimization.

Topology optimization is mainly used in the conceptual design stage. Common topological optimization methods include: homogenization method, variable density method, evolutionary structural optimization method, level set method, etc. [27]. At present, the research on topology optimization is divided into optimization strategy and engineering application. The purpose of the research on optimization strategy is to improve the accuracy of stress prediction [28,29]. The objects of engineering application include the optimization design of car body [30,31], the mechanism design of aero-engine [32,33], the optimization design of the compliant mechanism using composite materials [34,35], and the design of parts manufactured through additive manufacturing [36,37], etc. It can be seen that topology optimization has been applied in many disciplines.

Based on the research above, it can be found that most of the current researches are focused on the optimization methods in specific design stage. Without a systematic design route, the optimization methods can only meet specific engineering needs. At the same time, the trash can-handling robot proposed in this paper also has the demand of performance optimization. Therefore, based on the load analysis and optimization, the topology optimization in the conceptual design stage and the discrete size optimization in the engineering design stage, this paper proposes a lightweight design method for the truss structure in the robot. The main research route of this paper is as follows: in the second chapter, the kinematic and dynamic equations of the manipulator is established. The multi-objective optimization of the parameters related to the robot's load is carried out through the particle swarm algorithm to reduce the load of the truss structure. In the third chapter, three typical load cases of the truss structure are set, and the topology optimization of the truss structure under multiple load cases is carried out. In the fourth chapter, the discrete size optimization of the truss structure parts' thickness under multiple load cases is carried out through the sequential quadratic programming solver. The fifth chapter summarizes the lightweight design method used in this paper.

2. Multi-Objective Optimization of Parameters Related to Robot Load

Some parameters of the robot will affect the load of its truss structure. Therefore, it is necessary to optimize these parameters first. In this chapter, this paper establishes the kinematic and dynamic equations of the manipulator. Then, the load-related parameters are optimized through the particle swarm algorithm.

2.1. Establishment of Kinematic Equation

This paper takes the movement of the manipulator after grabbing the trash can as the analysis object, and makes the following settings:

1. There is no relative displacement between the manipulator and the trash can;
2. The garbage in the trash can does not move during the whole operation, and the position of the center of mass remains unchanged;

According to the time sequence, the movement of the manipulator can be divided into three stages, as shown in Figure 3. The lifting movement and the turning movement are respectively linear movement and circular movement, which will not be analyzed here. This paragraph will mainly analyze the transition movement.

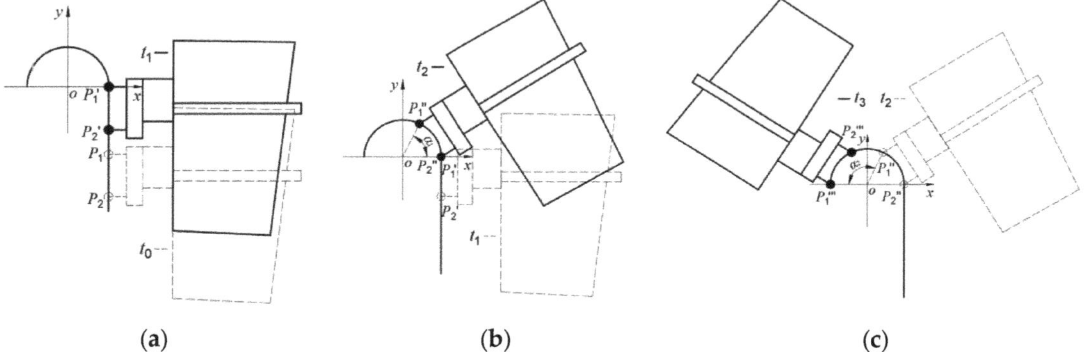

(a)   (b)   (c)

Figure 3. Three stages of the manipulator's movement: (**a**) lifting movement; (**b**) transition movement; (**c**) turning movement.

In the transition movement, the position of the manipulator is shown in Figure 4. In the figure, oxy is the world coordinate system; $o_1x_1y_1$ is the tool coordinate system; P_1' is the center point of the upper groove wheels; P_2' is the center point of the lower groove wheels; P_{cm}' is the equivalent center of mass of the manipulator and load; r_1 is the arc radius of the dumping track; l_1 is the center distance between the upper and lower groove wheels; l_2 is the distance between the point P_2' and the y_1 axis; l_3 is the distance between the point P_{cm}' and the y_1 axis; l_4 is the distance between the point P_{cm}' and the x_1 axis; α is the rotation angle of the manipulator; β is the pitch angle of the manipulator.

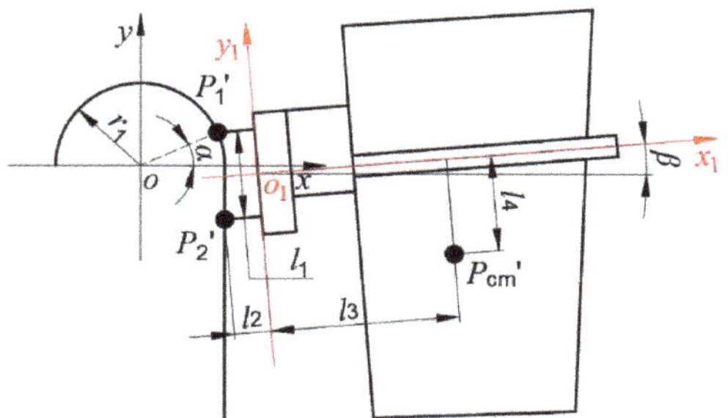

Figure 4. Schematic diagram of the manipulator's position.

For the points P_1' and P_2', the speed and acceleration can be expressed as follows:

$$\begin{cases} v_1^c = -(k_1 v_0 + k_2 x_v)\sin a + (k_1 v_0 + k_2 x_v)\cos a\, i \\ v_2^c = (k_1 x_v + k_2 v_0)i \\ a_1^c = -\dfrac{(k_1 v_0 + k_2 x_v)^2}{r_1}\cos a - k_2 x_a \sin a - \left[\dfrac{(k_1 v_0 + k_2 x_v)^2}{r_1}\sin a - k_2 x_a \cos a\right]i \\ a_2^c = k_1 x_a i \end{cases} \quad (1)$$

In the equations, the superscript 'c' means that the quantity is in complex form; k_1 and k_2 are the coefficients indicating that the driving force acts on the axis of the upper or the lower groove wheels, and there are only two cases: $\begin{cases} k_1 = 1 \\ k_2 = 0 \end{cases}$ or $\begin{cases} k_1 = 0 \\ k_2 = 1 \end{cases}$; v_0 is the linear velocity of the chain system; x_v and x_a are unknown variables, and their value can be calculated through the following equations.

$$\begin{cases} 0 = Re\left\{\dfrac{\left(P_2^c - P_1^c\right)^*\left(v_2^c - v_1^c\right)}{\left|P_2^c - P_1^c\right|^2}\right\} \\ \omega = Im\left\{\dfrac{\left(P_2^c - P_1^c\right)^*\left(v_2^c - v_1^c\right)}{\left|P_2^c - P_1^c\right|^2}\right\} \\ -\omega^2 = Re\left\{\dfrac{\left(P_2^c - P_1^c\right)^*\left(a_2^c - a_1^c\right)}{\left|P_2^c - P_1^c\right|^2}\right\} \end{cases} \quad (2)$$

In the equations, the superscript '*' means that the quantity is the conjugate complex number of itself.

Then the velocity and acceleration of the point P_{cm}' can be obtained through the complex interpolation method [38], as shown in Equation (3).

$$\begin{cases} v_{cm}^c = v_1^c + \dfrac{P_{cm}'^c - P_1'^c}{P_2'^c - P_1'^c}\left(v_2^c - v_1^c\right) \\ a_{cm}^c = a_1^c + \dfrac{P_{cm}'^c - P_1'^c}{P_2'^c - P_1'^c}\left(a_2^c - a_1^c\right) \end{cases} \quad (3)$$

2.2. Establishment of Dynamic Equation

Taking the scheme in which the driving force acts on the axis of the lower groove wheels as an example, the force analysis of the manipulator is shown in Figure 5. In the figure, C is the instantaneous center of velocity of the manipulator; F_1 is the equivalent force of the gravity of the manipulator and load; F_t is the driving force; N_1 and N_2 are the normal force; f_1 and f_2 are the friction force.

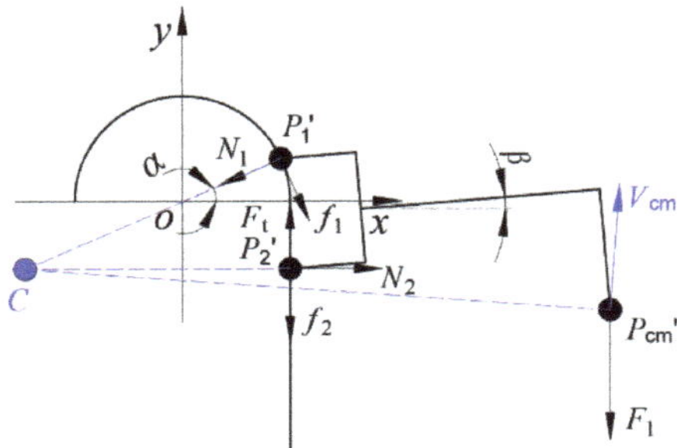

Figure 5. Force analysis diagram of the manipulator.

According to the theorem of kinetic energy and the balance relationship of forces, the dynamic equations of the manipulator in the lifting movement can be expressed as follows:

$$\{F_t, N_1, N_2\} = \begin{cases} N_1 = N_2 = \frac{l_2+l_3}{l_1}F_1 \\ F_t - f_1 - f_2 - F_1 = ma_{cm,y} \end{cases} . \tag{4}$$

The dynamic equations of the manipulator in the transition movement can be expressed as follows:

$$\{F_t, N_1, N_2\} = \begin{cases} (k_1F_t - f_1)s_1 + (k_2F_t - f_2)s_2 - F_1s_3 = \Delta E_k \\ -(k_1F_t - f_1)\sin\alpha - N_1\cos a + N_2 = ma_{cm,x} \\ (k_1F_t - f_1)\cos a - N_1\sin\alpha + k_2F_t - f_2 - F_1 = ma_{cm,y} \end{cases} . \tag{5}$$

In the equations, s_1 is the moving distance of the upper groove wheels in unit time; s_2 is the moving distance of the lower groove wheels in unit time; s_3 is the height change of the point P_{cm} in unit time; ΔE_k is the kinetic energy change of the manipulator and load in unit time.

The dynamic equations of the manipulator in the turning movement can be expressed as follows:

$$\{F_t, N_1, N_2\} = \begin{cases} (k_1F_t - f_1)s_1 + (k_2F_t - f_2)s_1 - F_1s_3 = 0 \\ -(k_1F_t - f_1)\sin\alpha - N_1\cos\alpha - (k_2F_t - f_2)\sin(\alpha - \theta) + N_2\cos(\alpha - \theta) = ma_{cm,x} \\ (k_1F_t - f_1)\cos\alpha - N_1\sin\alpha + (k_2F_t - f_2)\cos(\alpha - \theta) + N_2\sin(\alpha - \theta) - F_1 = ma_{cm,y} \end{cases} . \tag{6}$$

In the equations, θ is a fixed angle. It can be expressed as follows:

$$\theta = 2\arcsin\left(\frac{l_1}{2r_1}\right). \tag{7}$$

2.3. Mathematical Model of the Multi-Objective Optimization of the Load-Related Parameters

1. Design variables: this paper takes the arc radius r_1, the center distance l_1 and the time consumption t_4 of the dumping action as the design variables;
2. Constraints: to ensure that the manipulator can dump garbage smoothly, the pitch angle of the manipulator must be greater than 135°;
3. Optimization objective: the purpose of parameter optimization is to reduce the load on the truss structure. Therefore, the optimization objective is to minimize the maximum instantaneous power P_{max}, the average power $\overline{P}$, the maximum change of the instantaneous driving force $\Delta F_{t,max}$ and the time consumption t_4 of the dumping action.

The mathematical model of the optimization of the load-related parameters can be expressed as follows:

$$\begin{cases} X = [r_1,\ l_1,\ t_4] \\ t_4 = t_3 - t_1 \\ \min F(X) = \sum c_{1,i} \cdot f_i(X) \\ s.t. \begin{cases} r_1 \in [50, 200] \\ l_1 \in [80, 210] \\ l_1 < \sqrt{2} r_1 \\ t_4 \in [1, 4] \end{cases} \end{cases} \tag{8}$$

2.4. Results of the Multi-Objective Optimization of the Load-Related Parameters

Particle swarm optimization (PSO) was proposed by Kennedy and Eberhart in 1995 [39]. In order to improve the optimization efficiency, this paper adopts particle swarm optimization algorithm with improved weight coefficient [40]. The process of the optimization iteration is shown in Figure 6.

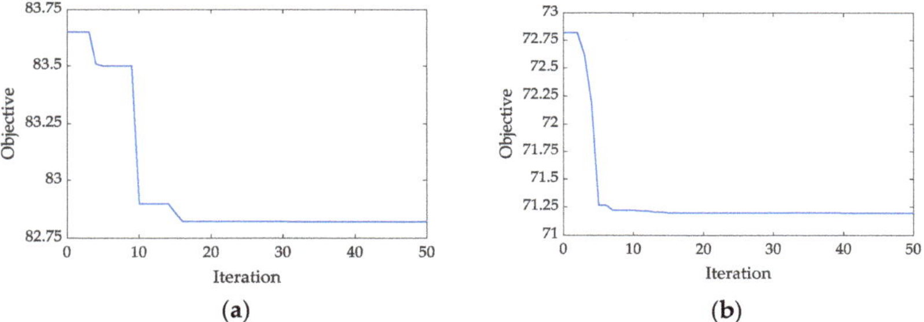

Figure 6. The process of PSO optimization iteration: (**a**) optimization scheme A; (**b**) optimization scheme B.

The optimization results are shown in Table 1.

Table 1. Results of the multi-objective optimization of the load-related parameters.

Parameter	Initial Design Scheme [1]	Optimization Scheme A [2]	Optimization Scheme B [3]
r_1 (mm)	120	130	148
l_1 (mm)	150	157	210
t_4 (s)	2	3.45	3.36
P_{max} (W)	3764.17	2028.33	1936.71
$\overline{P}$ (W)	1817.43	1028.25	1115.82
$\Delta F_{t,max}$ (N)	2442.92	722.22	349.32

[1] The initial design scheme is the design scheme of the prototype. [2] The driving force in optimization scheme A acts on the axis of the upper groove wheels. [3] The driving force in optimization scheme B acts on the axis of the lower groove wheels.

According to Table 1, both optimization scheme A and B have obvious optimization effect. The maximum instantaneous power of optimization scheme B is reduced by 1827.46 W, the average power is reduced by 701.61 W and the maximum change of the instantaneous driving force is reduced by 2093.6 N, which is more effective than that of the optimization scheme A. Therefore, optimization scheme B is the reasonable optimization scheme.

3. Topology Optimization of the Truss Structure under Multiple Load Cases

If the truss structure has a reasonable material distribution, the material can fully play its role, which is an important basis for the lightweight design [41]. In this chapter, this paper first analyzes and calculates the load on the truss structure, and then determines three typical load cases. Finally, the topology optimization of the truss structure under multiple load cases is carried out.

3.1. Analysis of the Load on the Truss Structure

The load on the truss structure mainly comes from the manipulator and the roller chain system, as shown in Figure 7. The definition of each load is shown in Table 2.

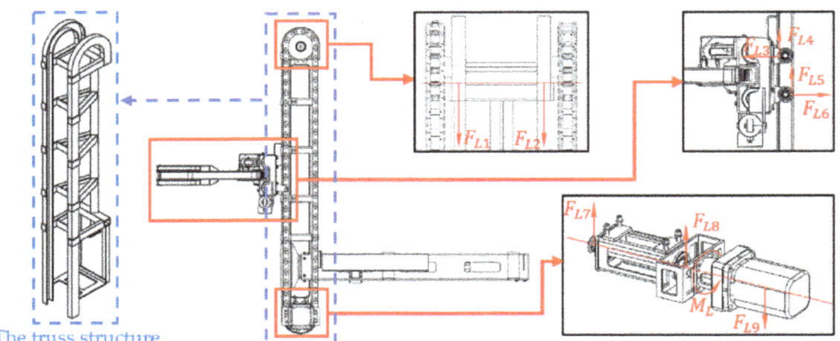

Figure 7. The load on the truss structure.

Table 2. The definition of the load.

Load	Definition
F_{L1}, F_{L2}	The force of the driven sprocket assembly acting on the truss structure.
F_{L3}, F_{L6}	The force of the groove wheels acting on the track.
F_{L4}, F_{L5}	The friction force of the groove wheels acting on the track.
F_{L7}, F_{L8}	The force of the drive sprocket assembly acting on the truss structure.
F_{L9}	The gravity of the hydraulic motor.
M_L	The torque of the hydraulic motor acting on the truss structure.

According to the optimization results above, the detailed truss structure parameters are shown in Table 3. In the table, l_5 is the length of the vertical track, and μ is the coefficient of friction.

Table 3. Parameters of the truss structure.

r_1(mm)	l_1(mm)	l_2(mm)	l_3(mm)	l_4(mm)	l_5(mm)	m(kg)	μ
148	210	68.5	383.65	308.84	1720	320.35	0.1

The control method of the robot is 'Sliding Mode Variable Structure Control' [42]. The preset linear velocity of the chain system is shown in Figure 8a. The corresponding speed and acceleration of the point P_{cm} are shown in Figure 8b,c.

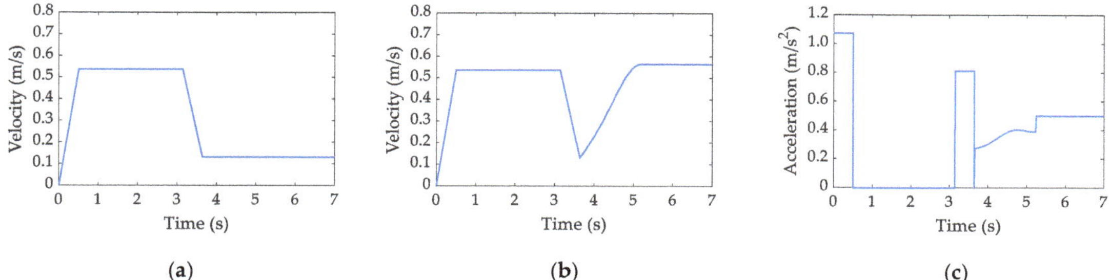

Figure 8. (**a**) Preset linear velocity of the chain system; (**b**) velocity of the point P_{cm}; (**c**) acceleration of the point P_{cm}.

According to the dynamic equations, the driving force required for the motion of the manipulator and the normal force of the manipulator acting on the track are shown in Figure 9. It can be seen that when the manipulator enters the circular arc section of the track, the driving force and the normal force increase significantly. In the lifting motion, the driving force required by the manipulator is the largest when accelerating.

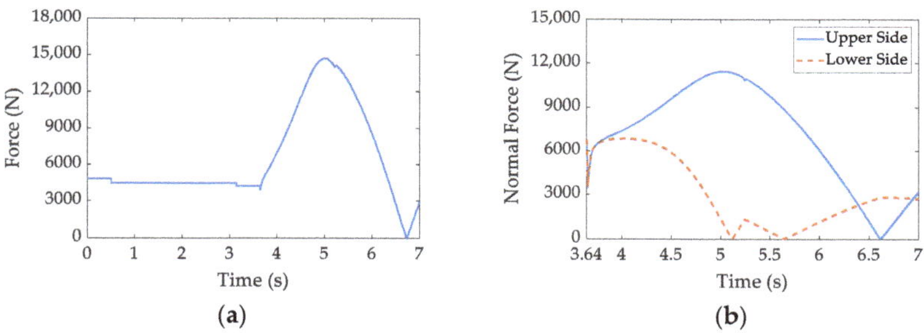

Figure 9. (**a**) Driving force required for the motion of the manipulator; (**b**) normal force of the manipulator acting on the track.

Then the calculation formula of the truss structure load defined in Table 2 can be expressed as follows:

$$\begin{cases} F_{L1} = F_{L2} = F_t + (m_c + m_w)g \\ F_{L3} = 0.5N_1 \\ F_{L4} = \mu F_{L3} \\ F_{L5} = \mu F_{L6} \\ F_{L6} = 0.5N_2 \\ F_{L7} = F_{L8} = 0.5F_t - m_w g \\ F_{L9} = m_m g \\ M_L = F_t r_1 + J_e \alpha \end{cases} \tag{9}$$

In the formula, m_c is the mass of the roller chain on one side; m_w is the mass of a sprocket; m_m is the mass of the hydraulic motor; J_e is the equivalent moment of inertia of all rotating parts.

3.2. Topology Optimization under Multiple Load Cases

3.2.1. Determination of Load Cases

Based on the analysis and calculation results of the truss structure load, the states when the manipulator is in the acceleration lifting movement, the transition movement, and on standby are regarded as three typical load cases in this paper. The schematic diagrams of the typical load cases are shown in Figure 10.

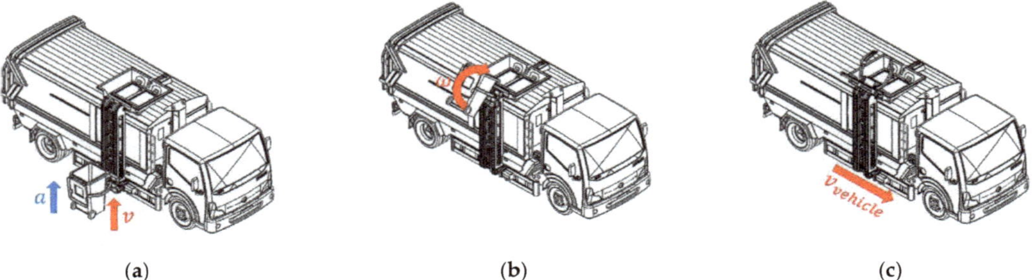

(a) (b) (c)

Figure 10. Three typical load cases: (**a**) load case A; (**b**) load case B; (**c**) load case C.

The value of the truss structure load under three load cases are shown in Table 4.

Table 4. The value of the truss structure load.

Load Case	$F_{L1}(N)$	$F_{L2}(N)$	$F_{L3}(N)$	$F_{L4}(N)$	$F_{L5}(N)$	$F_{L6}(N)$	$F_{L7}(N)$	$F_{L8}(N)$	$F_{L9}(N)$	$M_L(N{\cdot}m)$
Load case A	5160.07	5160.07	3379.75	337.97	337.97	3379.75	2335.97	2335.97	342.02	716.12
Load case B	15,041.05	15,041.05	5721.5	572.15	57.08	570.75	7276.46	7276.46	342.02	2178.05
Load case C	3620.2	3620.2	2169.76	0	0	280.14	2149.46	2149.46	342.02	660.46

3.2.2. Mathematical Model of Topology Optimization under Multiple Load Cases

In this paper, the optimization objective is to minimize the weighted strain energy of the truss structure under multiple load cases. The ratio of the optimized volume to the initial volume is the constraint. The mathematical model of the optimization can be expressed as follows:

$$
\begin{cases}
X = [x_1, \ x_2, \ x_3, \ \cdots] \\
\min T(X) = \sum c_{2,i}\Delta t_i(X) \\
s.t.\begin{cases} \dfrac{V_i(X)}{V_0} \leq z \\ 0 \leq x_j \leq 1, \ j \in N^* \end{cases}
\end{cases}
\tag{10}
$$

In the formula, $T(X)$ is the weighted strain energy; $c_{2,i}$ is the weight coefficient of the i-th load case, whose value is $1/3$; $t_i(X)$ is the strain energy of the i-th load case; $V_i(X)$ is the optimized volume; V_0 is the initial volume; z is the volume fraction; x_j is the material density of the j-th unit.

3.2.3. Results of Topology Optimization under Multiple Load Cases

The truss structure is a kind of frame parts. The typical structure of this type of parts is cubic shape and triangular prism shape. According to the connection relationship between the truss structure and other parts, the truss structure can be designed as a combination of cubic shape and triangular prism shape. The optimization model is shown in Figure 11a.

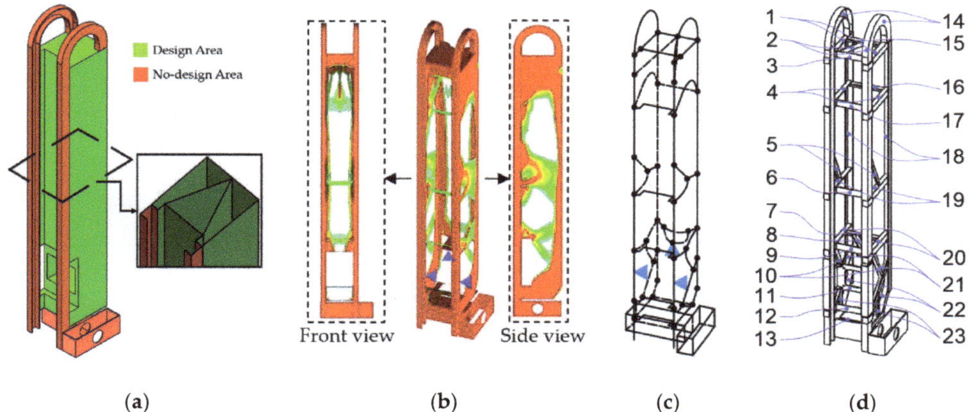

Figure 11. Technical route of the topology optimization: (**a**) optimization model; (**b**) material distribution; (**c**) force transmission route map; (**d**) conceptual configuration model.

Through the finite element optimization solver Optistruct, the material distribution of the truss structure is obtained, as shown in Figure 11b. After simplifying the material distribution, the corresponding force transmission route map is formed, as shown in Figure 11c. As the technological conditions and processing efficiency need to be considered in practical engineering, the truss structure is mainly welded by sheet metal parts and angle iron. The conceptual configuration model is shown in Figure 11d.

4. Discrete Optimization of the Truss Structure under Multiple Load Cases

Based on the conceptual configuration model, this chapter will optimize the section size of the parts. In this chapter, this paper firstly establishes the mathematical model of discrete optimization under multiple load cases. Then, the optimization is carried out based on different preference settings. Finally, this paper compares the optimization results.

4.1. Mathematical Model of Discrete Optimization under Multiple Load Cases

If the three parameters of the length, width and thickness of the part are all taken as optimization variables, the optimization will have a large feasible set. At the same time, the change of the length and width of different parts will cause the change of the connection form, which will increase the computational cost [43]. Therefore, this paper has determined the length and width of each part in the conceptual configuration model to improve the efficiency of optimization solution.

The optimization objective is to maximize the inherent frequency, and minimize the maximum stress and the mass of the truss structure under multiple load cases. The thickness of the parts is the optimization variable, and the yield strength of the material is the constraint. The mathematical model of discrete optimization can be expressed as follows:

$$\begin{cases} X = [thk_1, \ thk_2, \ \cdots, \ thk_{23}] \\ Thk = [1, \ 1.5, \ 2, \ 2.5, \ \cdots, \ 10] \\ \min S(X) = \sum c_{3,i} \left(\frac{\sigma_i(X) - \sigma_{i,min}}{\sigma_{i,max} - \sigma_{i,min}}\right) + q_1\left(\frac{m(X) - m_{min}}{m_{max} - m_{min}}\right) + q_2\left(\frac{f_{max} - f(X)}{f_{max} - f_{min}}\right) \\ s.t. \begin{cases} \sigma_i(X) \leq [\sigma] \\ thk_j \in Thk, \ j = 1, \ 2, \cdots, \ 23 \end{cases} \end{cases} \quad (11)$$

In the formula, $S(X)$ is the comprehensive optimization objective; thk_j is the thickness of the j-th part; $\sigma_i(X)$ is the maximum stress of the i-th load case; $m(X)$ is the mass of the truss structure; m_{max} and m_{min} are the maximum and minimum mass of the truss structure under the constraint; $f(X)$ is the inherent frequency of the truss structure; f_{max} and f_{min} are the maximum and minimum values in the optimization with the inherent frequency of the truss structure as the optimization objective; $[\sigma]$ is the yield strength of the material; Thk is the set of available material thickness; $c_{3,i}$ is the weight coefficient of the i-th load case, whose value is equal to $c_{2,i}$; q_1 and q_2 are the correction factors, whose value is $1/3$ as well.

4.2. Results of Discrete Optimization under Multiple Load Cases

Due to the large difference in the density of different types of garbage [44], the typical loads of the robots that perform different tasks are different. In order to make the optimization more targeted, this paper sets preference mass (optimization scheme A), preference performance (optimization scheme B) and no preference (optimization scheme C) lightweight schemes respectively. Then the sequential quadratic programming (SQP) solver is applied to solve the mathematical model. The process of optimization iteration is shown in Figure 12, and the optimization results are shown in Figure 13.

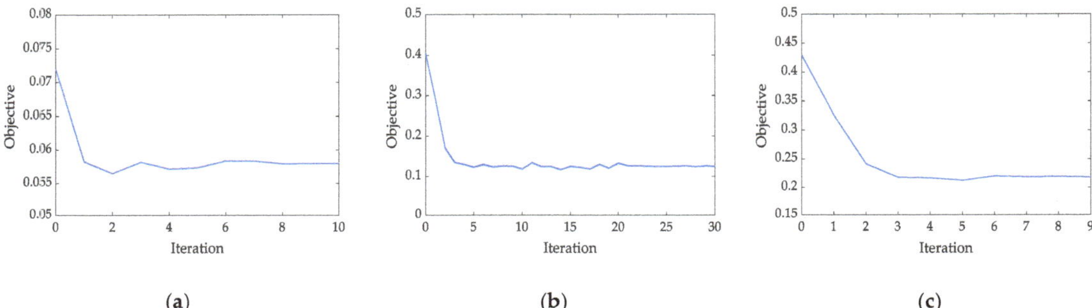

(a) (b) (c)

Figure 12. The process of optimization iteration: (**a**) optimization scheme A; (**b**) optimization scheme B; (**c**) optimization scheme C.

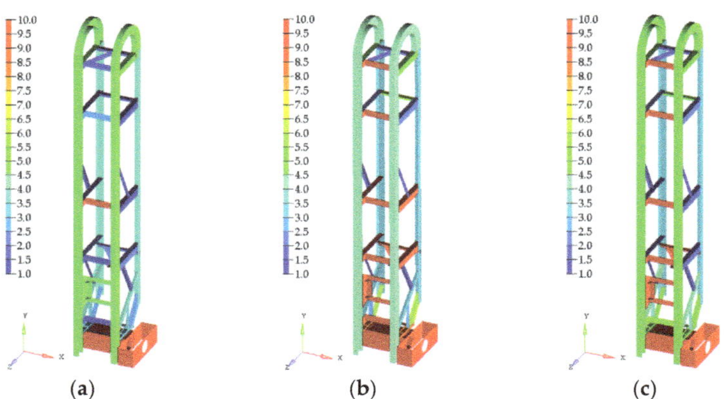

(a) (b) (c)

Figure 13. Optimization results: (**a**) optimization scheme A; (**b**) optimization scheme B; (**c**) optimization scheme C.

The thickness of the truss structure parts is shown in Table 5.

Table 5. The thickness of the truss structure parts.

Part Number	Initial Design Scheme [1] (mm)	Optimization Scheme A (mm)	Optimization Scheme B (mm)	Optimization Scheme C (mm)
1	5.0	5.0	5.0	5.0
2	5.0	1.0	1.0	1.0
3	4.0	1.5	10.0	10.0
4	5.0	1.0	2.0	2.0
5	3.0	1.0	1.0	1.0
6	4.0	10.0	10.0	10.0
7	5.0	1.5	10.0	9.0
8	3.0	2.0	10.0	10.0
9	4.0	1.0	10.0	10.0
10	3.0	1.0	1.0	1.0
11	4.0	4.5	9.5	10.0
12	4.0	1.0	10.0	5.5
13	5.0	3.0	8.0	4.5
14	5.0	4.5	4.5	4.5
15	5.0	1.0	1.0	1.0
16	5.0	1.0	6.0	5.0
17	4.0	2.5	10.0	10.0
18	5.0	3.5	3.0	3.0
19	5.0	1.0	9.5	1.5
20	5.0	2.0	9.5	1.5
21	5.0	3.5	3.0	4.0
22	5.0	2.5	6.0	5.5
23	3.0	3.0	2.5	3.5

[1] The initial design scheme is established according to the prototype. For example, if the thickness of the guide rail in the prototype is 5 mm, the thickness of the guide rail in the initial design scheme is also 5 mm.

It can be seen from Figure 13a and Table 6 that when the preference of the optimization scheme is set to mass, the mass of the truss structure is 58.37 kg, which is reduced by 18.99%. The inherent frequency, maximum stress and maximum deformation of the truss structure haven't been optimized. The maximum stress is close to the material's yield stress of 680 MPa. Therefore, this optimization scheme requires higher-strength materials. From Figure 13b and Table 6, it can be seen that when the preference of the optimization scheme is set to performance, the performance of the truss structure is significantly improved, while the mass is only reduced by 0.33 kg. The lightweight design effect is not significant. From Figure 13c and Table 6, it can be seen that when there is no preference for the optimization, the maximum stress is reduced by 70.97 MPa, the maximum deformation is increased by 0.2 mm, the inherent frequency is increased by 6.23 Hz, and the mass is reduced by 6.28 kg. The performance and mass of the truss structure have all been optimized. Therefore, optimization scheme C is the reasonable optimization scheme.

Table 6. Performance comparison of optimization schemes.

Performance	Initial Design Scheme	Optimization Scheme A	Optimization Scheme B	Optimization Scheme C
m (kg)	72.05	58.37	71.72	65.77
f (Hz)	10.20	10.38	16.92	16.43
$d_{1,max}$ (mm)	1.10	1.75	0.93	1.08
$d_{2,max}$ (mm)	2.19	3.41	1.98	2.39
$d_{3,max}$ (mm)	0.77	1.51	0.64	0.77
$\sigma_{1,max}$ (MPa)	218.16	273.30	216.47	222.75
$\sigma_{2,max}$ (MPa)	646.09	666.74	573.67	575.12
$\sigma_{3,max}$ (MPa)	202.37	322.62	128.29	226.16

5. Lightweight Design Method of the Robot Truss Structure

The lightweight design method used in this paper are summarized as follows:

1. This paper first established the kinematic and dynamic equations of the manipulator (load). Then the variables that are related to the load were optimized through the particle swarm algorithm to reduce the load of the truss structure;

2. This paper then determined the typical load cases of the truss structure. The topology optimization under multiple load cases was carried out to optimize the material distribution of the truss structure. The conceptual configuration model was established through model reconstruction method;

3. Based on the conceptual configuration model, this paper finally reduced the dimensions of the optimization variables according to the technological conditions and processing efficiency. The sequential quadratic programming solver was applied to optimize the thickness of the truss structure parts under multiple load cases.

The flow chart of this method is shown in Figure 14.

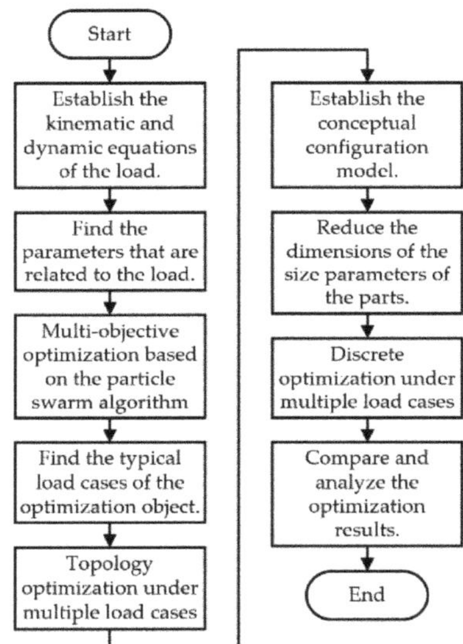

Figure 14. Flow chart of the method.

6. Conclusions

Aiming at the performance optimization requirement of the trash can-handling robot, this paper optimizes its truss structure and proposes a systematic lightweight design method. The main research conclusions are as follows:

1. In this paper, the kinematic and dynamic equations of the manipulator was established through the complex interpolation method and the theorem of kinetic energy. The particle swarm algorithm was used to optimize the load-related parameters. This provides a new method for the optimization of the equipment moving along the guide rail in the future. After the optimization, the maximum instantaneous power required by the robot for dumping garbage is reduced by 48.55%, the average power is reduced by 38.60%, and the maximum change of the instantaneous driving force is reduced by 85.70%;

2. By analyzing the load of the truss structure during the operation, the states when the manipulator is in the acceleration lifting movement, the transition movement, and on standby are regarded as three typical load cases in this paper. Combined with practical engineering experience, due to the significant increase of the driving force,

the transition movement needs special attention in the design and optimization of the equipment with similar structure;

3   In this paper, three kinds of discrete optimization of the truss structure with different preference were carried out. According to the optimization results, the optimization scheme with no preference best meets the actual needs of the project. In this optimization scheme, the mass of the truss structure is reduced by 8.72%, the inherent frequency is increased by 61.08%, and the maximum stress is reduced by 10.98%;

4   The lightweight design method proposed in this paper is a new optimization method as it includes load optimization. The results show that the method is effective for the optimization of the robot's truss structure. This method can also be applied to the forward design or lightweight design of the actuators with similar structure, such as the column of vertical drilling machine. So, this method gives a reference value for actual projects.

Author Contributions: Conceptualization, J.C., Y.L., X.C. and X.X.; methodology, J.C. and X.X.; software, J.C. and X.X.; validation, J.C. and X.C.; investigation, J.C., Y.L., X.C. and X.X.; data curation, J.C. and Y.L.; writing—original draft preparation, J.C.; writing—review and editing, J.C., Y.L., X.C. and X.X.; visualization, J.C.; supervision, Y.L. and X.C.; project administration, X.C. All authors have read and agreed to the published version of the manuscript.

Funding: This research was funded by "Transverse Project of Tongji University- Research on the Key Technology of New Intelligent Self-loading and Unloading Compression Sanitation Vehicle and Its Engineering Project Two-Research on Intelligent Garbage Transfer Equipment" (grant number KH0170920191853) and "Project of Shanghai Science and Technology Commission" (grant number 20511104602).

Acknowledgments: The authors are thankful for the support of the IIV (Institute of Intelligent Vehicle).

Conflicts of Interest: The authors declare no conflict of interest.

References

1.   Chen, W.; Dong, Z. Analysis of Influencing Factors and Scale Prediction of Garbage Production in Shanghai - Research Based on Grey System Theory. Available online: https://oversea.cnki.net/KCMS/detail/detail.aspx?dbcode=CJFD&dbname=CJFDLAST2020&filename=ZSZY202005005&v=EBMK4Zr%25mmd2Fj8q0NJrqqoPvvXols3NuLzWluZqj8%25mmd2F%25mmd2FbI4rnr3%25mmd2FfcufjH4WFSksEXEB2 (accessed on 30 August 2021).

2.   Chu, X. Evaluation Research on the Implementation Effect of the Current Municipal Household Waste Classification Policy in Shanghai. Master's Thesis, Harbin Engineering University, Harbin, China, June 2020.

3.   Wei, M. Performance Study and Improved Design of the Leakage-Free Lifting Mechanism for Garbage Truck. Master's Thesis, Yangzhou University, Yangzhou, China, June 2020.

4.   Yin, C. Simulation Analysis and Key Components Lightweight of Food Garbage Truck Lifting Mechanism. Master's Thesis, Yangzhou University, Yangzhou, China, June 2021.

5.   Yang, Q.; Xue, B. Kinematics and Dynamics Analysis of the Lifting Mechanism of the New-Type Sugarcane Harvester. Available online: https://oversea.cnki.net/KCMS/detail/detail.aspx?dbcode=CJFD&dbname=CJFDAUTO&filename=GXJX202105001&v=oOprQCOQxw%25mmd2F0om%25mmd2BakEi8%25mmd2B1vKqhiUVfZfmaW4Qlc6wJn4L8fA%25mmd2BhDhI16AKb7EK7qH (accessed on 30 August 2021).

6.   Wang, B. Coupling Characteristics Analysis and Structure Optimization of Truss Robot. Master's Thesis, Hefei University of Technology, Hefei, China, April 2017.

7.   Wang, J.; Wang, X.; Fu, J.; Lu, G.; Jin, C.; Chen, Y. Static and dynamic characteristic analysis and structure optimization for crossbeam structure of heavy-duty truss robot. *J. Zhejiang Univ. Eng. Sci.* **2021**, *55*, 124–134.

8.   Li, Z.; Li, D.; Song, Y. Finite Element Analysis of Law of Thermo Mechanical Coupling Deformation and Dynamic Characteristics of Rail Beam of Truss Pipe Pulling Robot. Available online: https://oversea.cnki.net/KCMS/detail/detail.aspx?dbcode=CJFD&dbname=CJFDLAST2021&filename=JXQD202103029&v=Ldfdhg%25mmd2F21Z8X%25mmd2BOmhufPrWhMdseEIy8%25mmd2B40uPDb2asqtIjdi%25mmd2Fn57dALYr4myuf%25mmd2BD0m (accessed on 30 August 2021).

9.   Yuan, B. Analysis and Optimization of Truss Structure Based on Finite Element Method. Master's Thesis, Yangzhou University, Yangzhou, China, December 2018.

10.   Shi, G.; Chen, Y.; Yang, Y.; Jiang, X.; Song, Z. BIW architecture multidisciplinary light weight optimization design. *J. Mech. Eng.* **2012**, *48*, 110–114. [CrossRef]

11.   Xu, X.; Chen, X.; Liu, Z.; Xu, Y.; Zhang, Y. Reliability-based design for lightweight vehicle structures with uncertain manufacturing accuracy. *Appl. Math. Model.* **2021**, *95*, 22–37. [CrossRef]

12. Xu, X.; Chen, X.; Liu, Z.; Yang, J.; Xu, Y.; Zhang, Y.; Zhang, Y. Multi-Objective Reliability-Based Design Optimization for the Reducer Housing of Electric Vehicles. Available online: https://www.tandfonline.com/doi/abs/10.1080/0305215X.2021.1923704 (accessed on 30 August 2021).
13. Wang, J.; Zhang, P.; Wang, C.; Han, S. Lightweight Optimization Design of the Flat Transport Vehicle's Loading Platform Based on the high-Strength Steel. Available online: https://www.webofscience.com/wos/alldb/full-record/CSCD:6967249 (accessed on 30 August 2021).
14. Li, Y.; Xu, Z.; Zhang, L.; Du, J.; Jiang, Y. Light Weight Design of Chassis Frame for a Soybean Harvester. Available online: https://www.webofscience.com/wos/alldb/full-record/CSCD:6600797 (accessed on 30 August 2021).
15. Huang, W.; Huang, L.; Peng, T.; Liang, Q.; Luo, L. Comprehensive Optimal Design of the Column of Double Spindle Horizontal Machining Center. Available online: https://www.webofscience.com/wos/alldb/full-record/CSCD:6841618 (accessed on 30 August 2021).
16. Gao, Y.; Ma, C.; Liu, Z.; Duan, Y.; Tian, L. Optimization of Multi-Material and Beam Cross-Sectional Shape and Dimension of Skeleton-Type Body. Available online: https://oversea.cnki.net/KCMS/detail/detail.aspx?dbcode=CAPJ&dbname=CAPJLAST&filename=JLGY20201216006&v=yiDx1rS1Ygw5mkOLat4bs%25mmd2BGYRDj9LAxCGX6WtMshjMK9bIGGSEc5LzRR634jCfOf (accessed on 17 December 2020).
17. Cheng, M.; Prayogo, D.; Wu, Y.; Lukito, M. A hybrid harmony search algorithm for discrete sizing optimization of truss structure. *Autom. Constr.* **2016**, *69*, 21–33. [CrossRef]
18. Yoshimura, M.; Nishiwaki, S.; Izui, K. A multiple cross-sectional shape optimization method for automotive body frames. *J. Mech. Des.* **2005**, *127*, 49–57. [CrossRef]
19. Hou, W.; Wang, Z.; Zhang, W.; Zhang, H. Optimization design for auto-body beam section based on complex engineering constraints. *J. Mech. Eng.* **2014**, *50*, 127–133. [CrossRef]
20. Cui, X.; Zhang, H.; Wang, S.; Zhang, L.; Ko, J. Design of lightweight multi-material automotive bodies using new material performance indices of thin-walled beams for the material selection with crashworthiness consideration. *Mater. Des.* **2011**, *32*, 815–821. [CrossRef]
21. Cao, Y.; Wang, Z.; Zhou, C.; Wang, W. Optimization of circular-axis flexure hinge by considering material selection and geometrical configuration simultaneously. *J. Mech. Eng.* **2017**, *53*, 46–57. [CrossRef]
22. Cui, A.; Zhang, H.; Yu, D.; Zhang, S.; Yang, Y. A study on material selection for multi-material autobody components based on PSI method. *Automot. Eng.* **2018**, *40*, 239–244+233.
23. Zhao, Y.; Li, Y.; Chen, D.; Hou, W. Optimization Design Method of Multi-Material Car Body Structure Based on Modified Graphic Decomposition. Available online: https://www.webofscience.com/wos/alldb/full-record/CSCD:6721999 (accessed on 30 August 2021).
24. Hou, Y.; Zhang, Y.; Liu, T. Optimization of standard cross-section type selection in steel frame structures based on gradient methods. *Eng. Mech.* **2013**, *30*, 454–462.
25. Kaveh, A.; Hosseini, S.M.; Zaerreza, A. Improved shuffled Jaya algorithm for sizing optimization of skeletal structures with discrete variables. *Structures* **2021**, *29*, 107–128. [CrossRef]
26. Degertekin, S.O.; Minooei, M.; Santoro, L.; Trentadue, B.; Lamberti, L. Large-scale truss-sizing optimization with enhanced hybrid HS algorithm. *Appl. Sci.* **2021**, *11*, 3270. [CrossRef]
27. Xing, G.; Sun, Z.; Cui, X.; Wu, W.; Jia, P.; Jia, Z. Topological Optimization Design of Aero-Engine Support Structure under Multiple Loading Conditions. Available online: https://www.webofscience.com/wos/alldb/full-record/CSCD:6884473 (accessed on 30 August 2021).
28. Ding, M.; Geng, D.; Zhou, M.; Lai, X. Topology optimization strategy of structural strength based on variable density method. *J. Shanghai Jiaotong Univ.* **2021**, *55*, 764–773.
29. Picelli, R.; Townsend, S.; Brampton, C.; Norato, J.; Kim, H. Stress-based shape and topology optimization with the level set method. *Comput. Meth. Appl. Mech. Eng.* **2018**, *329*, 1–23. [CrossRef]
30. Wu, Y.; Liu, X.; Wang, F. Topology optimization design of the bracket of a special vehicle based on FEA method. In Proceedings of the 2010 Second International Conference on Computational Intelligence and Natural Computing (CINC 2010), Wuhan, China, 13–14 September 2010; pp. 352–355.
31. Chen, X.; Zhao, K.; Shen, C.; Zheng, K.; Lü, W. Topological Analysis and a Parametric Optimization Method for Aluminum Alloy Frame Body of Electric Vehicle. Available online: https://www.webofscience.com/wos/alldb/full-record/CSCD:6824257 (accessed on 30 August 2021).
32. Kober, M.; Kühhorn, A.; Rademann, J.; Mück, B. Nonlinear topology optimization of centrifugally loaded aero-engine part with newly developed optimality-criteria based algorithm. *Aerosp. Sci. Technol.* **2014**, *39*, 705–711. [CrossRef]
33. Yang, J.G.; Zhang, W.H.; Zhu, J. Optimal design of aero-engine stator structure with combined shape and topology optimization method. *Mater. Sci. Forum.* **2012**, *1441*, 623–626. [CrossRef]
34. Tong, X.; Ge, W.; Gao, X.; Li, Y. Optimization of combining fiber orientation and topology for constant-stiffness composite laminated plates. *J. Optim. Theory Appl.* **2019**, *181*, 653–670. [CrossRef]

35. Zhan, J.; Qin, Y.; Liu, M. Topological Design of Orthotropic Material Compliant Mechanisms Considering Variable Fiber Angle. Available online: https://oversea.cnki.net/KCMS/detail/detail.aspx?dbcode=CAPJ&dbname=CAPJLAST&filename=JXKX2 021061000H&v=7ll%25mmd2FrpOgPQe%25mmd2FKnRykBkh6qlngqWHUPsifvaU%25mmd2F6hoRHnoSON5jHQ3OqbSY% 25mmd2Bd8QFdF (accessed on 30 August 2021).
36. Chen, F.; Zhang, W.; Ye, D. Typical additive manufacturing joint design in civil aircrafts. Available online: https://oversea.cnki.net/KCMS/detail/detail.aspx?dbcode=CJFD&dbname=CJFDLAST2021&filename=YYLX202103028&v= rZj4GQjzvwxqTa%25mmd2BL4M4l9ZW0Xl%25mmd2Fi3K6bYFFr5QEPPM4D3aWKWJn6DLYj3Wi5ZAwc (accessed on 30 August 2021).
37. Yan, M.; Tian, X.; Peng, G.; Li, D.; Yao, R.; Zhang, W.; Bai, R.; Meng, W. Performance Study of Lightweight Composites Equipment Section Support Fabricated by Selective Laser Sintering. Available online: http://www.cjmenet.com.cn/CN/Y2019/V55/I13/144 (accessed on 30 August 2021).
38. Zhang, X. The Complex Interpolation Approach to Analysis of the Planar Motion of Rigid Bodies. Available online: https://oversea.cnki.net/KCMS/detail/detail.aspx?dbcode=CJFD&dbname=CJFDLAST2019&filename=DXWL201901002 &v=oJ4RN9i2Lsp399gicG4Y20zdGmxcL4SLG%25mmd2B5zFM81vwlOaVKM%25mmd2FTal3kN3CZh6Rl9L (accessed on 30 August 2021).
39. Kennedy, J.; Eberhart, R. Particle swarm optimization. In Proceedings of the ICNN'95—International Conference on Neural Networks, Perth, WA, Australia, 27 November–1 December 1995; pp. 1942–1948.
40. Shi, Y.; Eberhart, R.C. Fuzzy adaptive particle swarm optimization. In Proceedings of the 2001 Congress on Evolutionary Computation, Seoul, Korea, 27–30 May 2001; pp. 101–106.
41. Wang, D.; Shen, H.; Sun, Y.; Dong, H.; Ma, Y. A novel approach for conceptual structural design of complex machine elements. *J. Mech. Eng.* **2016**, *52*, 152–163. [CrossRef]
42. Zhang, J.; Wang, J.; Lü, X. Control and Simulation of Robotic Arm System of Intelligent Sanitation Vehicle. Available online: https://oversea.cnki.net/KCMS/detail/detail.aspx?dbcode=CPFD&dbname=CPFDLAST2021&filename=ZGZN20201100104 8&v=Q7FS4XFR%25mmd2BG4Tb7oQsGIvtVN8qh5towQwpfrJytHAoT4%25mmd2BgRst1wWVbxKOAKiDukwEHtdAwf%25 mmd2FRmMU%3d (accessed on 30 August 2021).
43. Zhang, Z.; Li, H. The method of truss structure layout discrete variable optimization based on ant colony optimization. *Chin. J. Comput. Mech.* **2013**, *30*, 336–342.
44. Liu, D.; Wang, Y.; Tong, Y.; Tong, Y.; Luo, Y.; Chen, W.; Tan, Z. Properties and disposal countermeasures of sorting waste. *Environ. Sanitation Eng.* **2013**, *21*, 27–30.

Article

Performance Assessment of an Electric Power Steering System for Driverless Formula Student Vehicles

Raffaele Manca [1], Salvatore Circosta [1,*], Irfan Khan [1], Stefano Feraco [1], Sara Luciani [1], Nicola Amati [1], Angelo Bonfitto [1] and Renato Galluzzi [2]

[1] Department of Mechanical and Aerospace Engineering, Politecnico di Torino, 10129 Turin, Italy; raffaele.manca@polito.it (R.M.); irfan.khan@polito.it (I.K.); stefano.feraco@polito.it (S.F.); sara.luciani@polito.it (S.L.); nicola.amati@polito.it (N.A.); angelo.bonfitto@polito.it (A.B.)

[2] School of Engineering and Sciences, Tecnologico de Monterrey, Calle del Puente 222, Mexico City 14380, Mexico; renato.galluzzi@tec.mx

* Correspondence: salvatore.circosta@polito.it

Abstract: In the context of automated driving, Electric Power Steering (EPS) systems represent an enabling technology. They introduce the ergonomic function of reducing the physical effort required by the driver during the steering maneuver. Furthermore, EPS gives the possibility of high precision control of the steering system, thus paving the way to autonomous driving capability. In this context, the present work presents a performance assessment of an EPS system designed for a full-electric all-wheel-drive electric prototype racing in Formula Student Driverless (FSD) competitions. Specifically, the system is based on the linear actuation of the steering rack by using a ball screw. The screw nut is rotated through a belt transmission driven by a brushless DC motor. Modeling and motion control techniques for this system are presented. Moreover, the numerical model is tuned through a grey-box identification approach. Finally, the performance of the proposed EPS system is tested experimentally on the vehicle through both sine-sweep profiles and co-simulated driverless sessions. The system performance is assessed in terms of reference tracking capability, thus showing favorable results for the proposed actuation solution.

Keywords: electric power steering; autonomous driving; steering actuator; driverless racing vehicles; control

Citation: Manca, R.; Circosta, S.; Khan, I.; Feraco, S.; Luciani, S.; Amati, N.; Bonfitto, A.; Galluzzi, R. Performance Assessment of an Electric Power Steering System for Driverless Formula Student Vehicles. *Actuators* **2021**, *10*, 165. https://doi.org/10.3390/act10070165

Academic Editors: Peng Hang, Xin Xia and Xinbo Chen

Received: 22 June 2021
Accepted: 14 July 2021
Published: 18 July 2021

Publisher's Note: MDPI stays neutral with regard to jurisdictional claims in published maps and institutional affiliations.

1. Introduction

The automotive industry is currently facing a substantial shift towards new mobility trends: electrification, spreading of car sharing services and autonomous driving [1]. Features like autonomous emergency braking and lane keeping assistance are referred to as Advanced Driver Assistance Systems (ADAS). These features expand from premium to mass-offering markets and represent a key-point in the progressive transition towards autonomous driving. In this context, Waymo-Google launched a self-driving car project in 2009. Later, Tesla Motor Company rolled out autopilot software on their Model S in October 2015. Ongoing efforts by most carmakers like Volvo head towards the design of autonomous vehicles [2]. A disruptive SAE Level 4 autonomy is expected to be available between 2020 and 2022 [3], while full autonomy with Level 5 technology is supposed to arrive by 2030 [2].

In this challenging scenario, Electric Power Steering (EPS) is a key technology for highly automated driving. Currently, such systems are employed to help the driver in the steering maneuvers by reducing the physical effort required, especially at low speed. Furthermore, EPS systems are used to provide the forces acting on the steering wheel as feedback to the user to preserve the driving sensation. In fact, EPS systems act on the rack-pinion steering box through an electric motor that provides the required torque to the steering column or directly to the rack. Another steering assistance technology is represented by hydraulic power systems. They exploit the pressure generated through a motor- or engine-driven pump to assist the turning of the steering wheel. Nevertheless, such technology is less efficient than the electric one, as witnessed in [4,5]. Therefore, EPS is regarded as a key technology for implementing autonomous driving features in the context of assisted and efficient mobility [6].

Intense research efforts have been dedicated to the control of EPS systems in the last decades. Research works presented by Mehrabi et al. in 2011 [4], Liao & Du [7] in 2003, Frankem & Müller in 2014 [8] and Chen & Chen in 2006 [9] focus on the control of such systems for ADAS applications. Specifically, torque control is addressed with the goal of comfort and safety optimization, as well as disturbance rejection. As stated by Groll et al. in 2006 [10], the most relevant frequencies of the driver input are below 4 Hz. Therefore, the control system must operate to reject the disturbances efficiently to avoid high-frequency oscillatory behavior [11]. For industrial applications, the tendency is to adopt PID control as a cost-effective and easy-to-tune solution. However, this control scheme may present important drawbacks when dealing with the multi-order nature of the steering system. Therein, induced resonant behavior can affect the stability or tracking performance of an EPS angle control, as demonstrated in [12]. The work presented by Govender et al. [11] with a PID for front steering angle control gives an insight into the above-mentioned robustness drawbacks. It suggests the adoption of filters, anti-windup and non-zero structures enhance the stability of the control system. Liao & Du [13] proposed a solution for the modeling and co-simulation of the EPS system and vehicle dynamics. Other efforts deal with active front steering control for automated driving applications through model predictive control (MPC) [14]. Further works enhance the MPC solution with path tracking and trajectory planning methods [15]. In [16], Daimler AG presents an EPS model and the design of a controller to ensure accurate, robust and smooth tracking of the desired trajectories of the front steering angle. The developed steering model shows significant nonlinear behavior due to the elastic elements, friction and gear ratios. Thereby, different control techniques are proposed and compared in [16,17].

Despite these comprehensive efforts, minor attention has been paid to the performance assessment of EPS systems applied to a driverless application from an actuation standpoint. The design and control target for such applications pose important challenges. During manual driving, the driver can constantly adjust the steering wheel input to compensate for disturbances from the road. Conversely, the EPS in autonomous vehicles is no longer used as an amplifier for the driver's torque since it must actuate the entire steering system [11]. Therefore, the extension of EPS systems to driverless applications is not straightforward and has not been addressed properly by the available literature.

In this context, the main contributions provided by the present work are: (i) the performance assessment of the proposed EPS actuator through a dedicated on-vehicle experimental campaign aimed at identifying the dynamic capabilities of the system for a driverless application; (ii) the validation of the discussed actuator in a driverless maneuver generated by the complete vehicle MPC controller.

The integration of the steering control system in the complete autonomous vehicle model needs the characterization of the actuator to define its dynamic behavior and the system capability for the specific automated driving scenario. Furthermore, a precise mathematical model of the controlled plant is essential to properly design a control strategy of the complete vehicle. In the case of MPC, the compensator is aimed at minimizing the vehicle lateral deviation and relative yaw angle with respect to the reference trajectory, as described in [18]. For this purpose, correct system identification of the controlled plant is fundamental. The work presented in [19] addresses the estimation of the parameters of an EPS by using different algorithms to improve the robustness of the designed controller. Nevertheless, the estimation was performed in a test bench scenario thus implying model mismatch when the system is mounted on a vehicle.

The performance assessment of an EPS system for the autonomous driving application, through in-vehicle experimental tests, is discussed in [20]. However, this paper focuses only on the tracking performance of the proposed controller, without identifying and discussing the dynamic capabilities of the actuator for the considered application.

In this perspective, this paper presents a custom EPS system for a driverless racing vehicle participating at Formula Student Driverless (FSD) events. In detail, the presented work is focused on the implementation and performance assessment of the actuator and system identification using a grey-box model. The work addresses the dynamic performance of the steering actuator integrated with the complete driverless vehicle. To this end, an in-vehicle experimental campaign is carried out to validate the identified model that will be integrated with the MPC controller of the complete driverless vehicle.

The proposed actuator layout consists of a ball screw assembly acting in parallel to the steering rack. The screw nut is actuated by using a toothed belt transmission driven by a brushless DC electric motor. Hence, the screw translates thus actuating the steering rack. This work examines the characteristics of the chosen layout and deals with its integration in the reference racing vehicle. Specifically, the system has been optimized for the integration with the considered vehicle to be compliant with the guidelines of FSD competitions [21], while also accomplishing the autonomous driving function. Electric motor, mechanical system and vehicle dynamic models have been developed and implemented in a MATLAB/Simulink environment, by using a linear grey-box model for the estimation of the unknown parameters. The EPS system is controlled through a PID with a feedforward position control loop, whereas a classic PI is used for the electric motor current control. A pole placement technique is used for the controller tuning. To validate the system and assess performance, in-vehicle experimental tests were conducted. Then, the validity of the proposed layout is demonstrated by testing the racing vehicle equipped with the developed EPS system in a driverless scenario. In particular, the steering profile computed by a vehicle dynamics MPC controller is used as a realistic reference for this verification [18].

This paper is organized as follows. Section 2 describes the considered system layout along with the selected design choices. Section 3 illustrates the system modeling and the implemented control techniques. Finally, Section 4 focuses on the system identification and on the discussion of the experimental results obtained during different maneuvers.

2. System Layout

The EPS layout for the FSD application is chosen and designed to be compliant with the regulations and guidelines provided by Formula SAE [21] and to fit into the already existing reference vehicle. The driverless class was introduced in FSD competitions in 2018. It consists of static and dynamic events aimed at evaluating the autonomous vehicle's ability to adapt to the tested driving scenario. The academic activities for developing a driverless single-seated race car provide a platform to develop and validate new technologies under challenging conditions. Self-driving racecars represent a unique opportunity to design and test software required in autonomous transport, such as redundant perception, failure detection and control in challenging conditions [22].

The steering system of the reference vehicle is a mechanical rack and pinion with a herringbone-like gear profile. The steering rack is a custom solution starting from the zRack provided by Zedaro. It presents a rack length of 264 mm to fit the limited available space in the front cross-sectional area of the monocoque, as shown in Figure 1. The C-Factor and the steering ratio are 85.5 mm and 4.3:1, respectively, with a steering wheel working angle ranging from −90 to +90 degrees. The steering column is equipped with a rotary encoder, which provides the feedback signal for the autonomous steering system control loop.

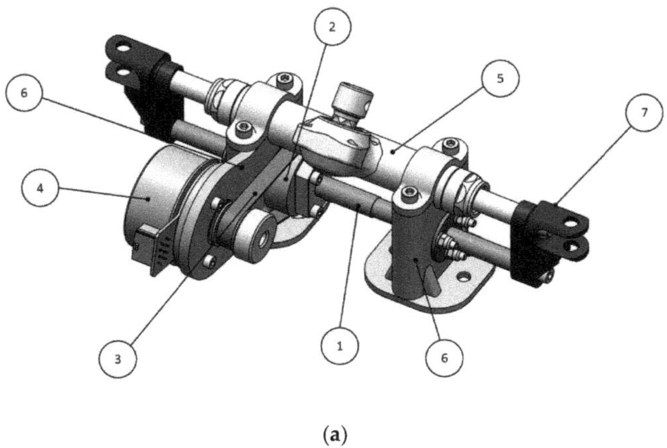

(**a**)

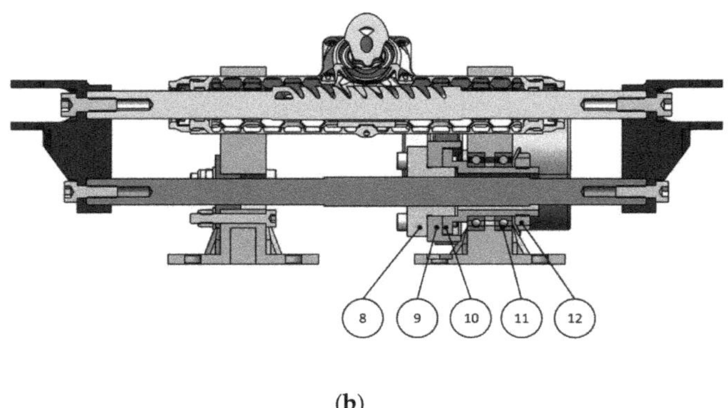

(**b**)

Figure 1. (**a**) Autonomous Steering System actuator view: 1—Ball screw; 2—Ball screw nut and bearing assembly; 3—HTD belt; 4—BLDC motor; 5—Steering rack; 6—Support structure; 7—Clevises. (**b**) Autonomous Steering System cross-section to highlight the ball screw nut and bearing assembly: 8—Ball screw nut; 9—HTD pulley; 10—Nut bearing support; 11—Ball bearings; 12—Lock nut washer.

The proposed solution design is subject to specific constraints, which can be summarized as follows:

- Small rack assembly with a rack length equal to 264 mm.
- Required total rack travel equal to 45 mm, by considering a steering wheel working angle from −90 to +90 degrees.
- Required actuation speed $v_{BS} = 45$ mm/s, by taking as target the capability of the driver to actuate the steering wheel from full left to full right in 1 s.
- Reversibility between driverless and with-driver modes must be guaranteed without mechanically dismounting any physical part, according to the competition rules [21].

Specifically, the reversibility constraint implies an integration of the EPS solution with the existing steering rack, thus reducing the available space inside the cockpit.

In this scenario, the chosen solution (Figure 1) consists of a rack that is linearly actuated by using a ball screw (1). The power unit is a brushless DC (BLDC) motor provided by

Maxon Motor AG^{TM} (Sachseln, Switzerland) (4). The nut (8) is rotated through a belt-drive transmission (3), while its translation is constrained. By converse, the screw is free to translate, thus providing a linear actuation of the rack (5).

The required rack force for the considered application is $F_{rack} = 1000$ N. It is due to the required torque needed to move the steering rack at standstill, as a worst-case scenario [22]. Then, the required ball screw torque $T_{BS} = 0.397$ Nm and the motor speed $n = 1350$ rpm and power $P_{mot} = 57$ W are computed. A Bosch Rexroth ball screw with a diameter of 12 mm and a lead of 2 mm is chosen for the present application. It features an efficiency $\eta_{BS} = 0.8$.

According to the calculated power request and due to the constraints on the available space, the Maxon^{TM} EC 60 Flat 150 W motor is selected for its compact size. It offers a nominal speed and torque of 3480 rpm and 0.401 Nm, respectively. The ball screw is parallel to the vehicle steering rack, as shown in Figure 1. The rack supports have been modified to mount the motor and accommodate the ball screw within a single assembly. The screw and the rack ends are rigidly connected by means of two clevises. The latter are coupled with the steering tie rods.

The ball screw drive is a reversible mechanism that allows the vehicle to be driven in both driverless and with-driver modes, without mechanically disconnecting the autonomous actuator. The selected belt and pulley system have a standard HTD profile with a pitch of 3 mm and a width of 9 mm. The belt drive system was sized to comply with the power transmission for this application. The bearing assembly uses two SKF four-point contact ball bearings in the "O" arrangement (11).

The BLDC motor is controlled and driven by a Maxon^{TM} EPOS 4 electronic control unit that communicates via CAN with a dSPACE^{TM} (Paterborn, Germany) MicroAutoBox. The latter is the onboard CPU of the vehicle. The reference steering angle for the control loop is computed by the trajectory planning and control algorithm implemented on the dSPACE^{TM} unit, as described in [18,23]. The algorithm receives the signals from the stereo camera, LiDAR and inertial measurement unit (IMU) sensors. The control also provides the steering command to follow the desired path. Then, the reference steering angle is compared with the feedback provided by the steering encoder. The EPOS ECU closes the position feedback loop and provides the voltage command to the electric motor to drive the autonomous steering actuator, as subsequently discussed in Section 3.3.

Figure 2a shows the positioning of the steering actuator assembly in the vehicle CAD model. In Figure 2b, we show the installation of the real prototype, together with its components.

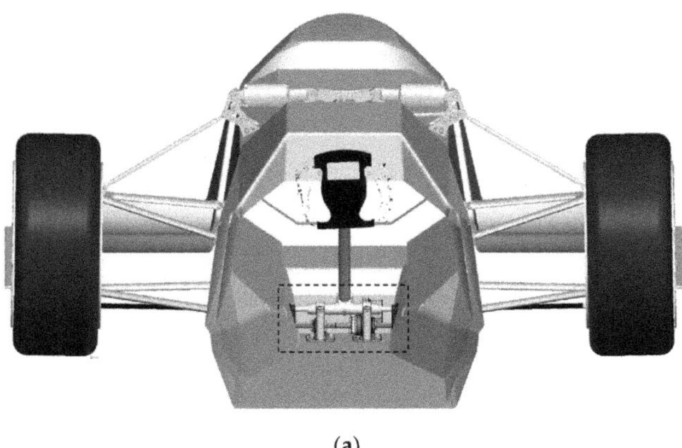

(**a**)

Figure 2. *Cont.*

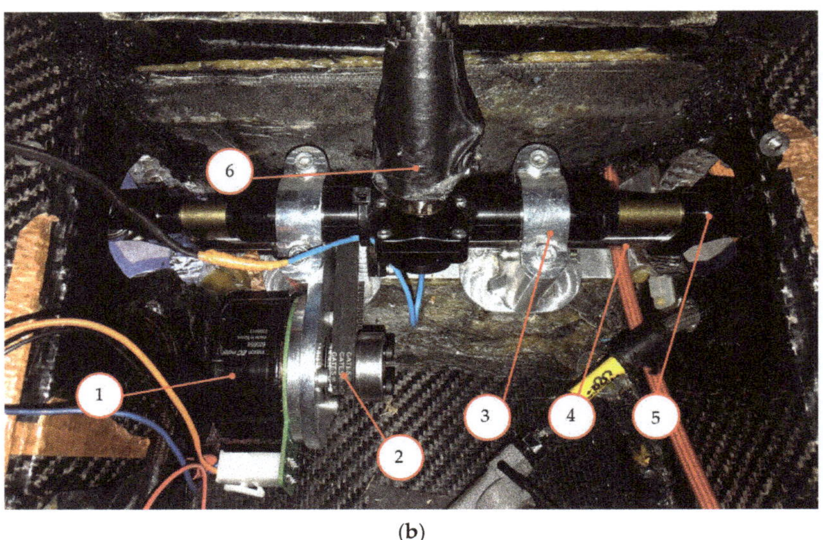

(b)

Figure 2. (**a**) Autonomous EPS system positioning in the vehicle CAD model. (**b**) EPS positioning in the actual vehicle: 1—BLDC motor; 2—HTD belt; 3—Support structure; 4—Ball screw; 5—Clevises; 6—Steering column.

3. System Modelling

For performance assessment and control tuning purposes, the system is modeled by following a linear lumped-parameter approach. The developed model is schematized in Figure 3. The model includes the motor and its controller along with the dynamic model of the autonomous steering system.

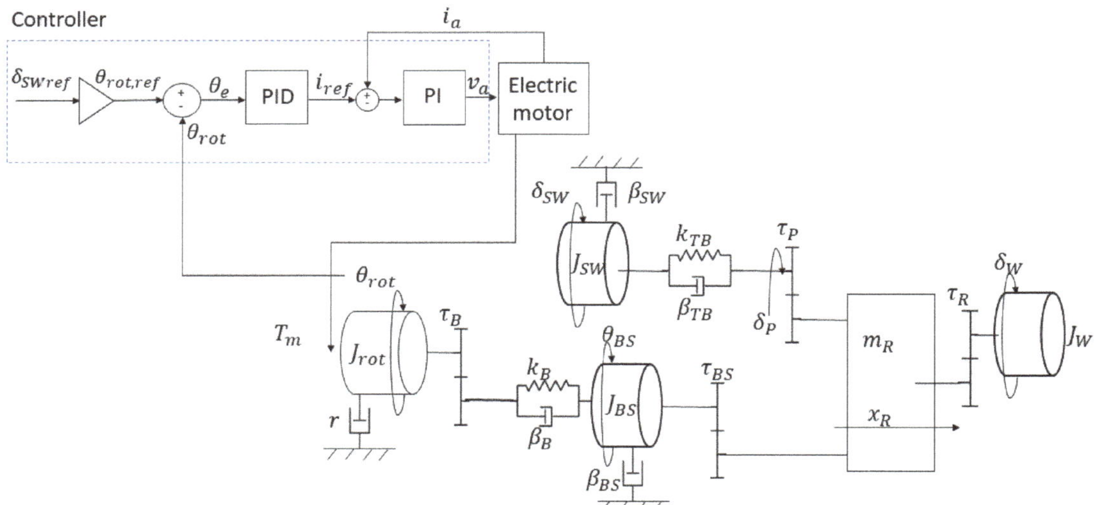

Figure 3. Dynamic model of the studied system.

3.1. Electric Motor

For simplicity, the BLDC motor can be represented as a standard DC motor. Neglecting field-weakening operation, the electrical dynamic model of the BLDC motor can be described as

$$v_a(t) = R_a i_a(t) + L_a \frac{d i_a(t)}{dt} + e(t) \tag{1}$$

where v_a is the motor terminal voltage, i_a is the motor armature current, L_a is the phase-to-phase terminal inductance, R_a is the phase-to-phase terminal resistance and e is the motor back-EMF, given by

$$e(t) = k_e \dot{\theta}_{rot} \tag{2}$$

and the electromagnetic motor torque provided by the motor is

$$T_m(t) = k_t i_a(t) \tag{3}$$

being k_e, k_t the machine characteristic constants and θ_{rot} the angular position of the motor.

The damping coefficient r of the electric motor representing the mechanical losses is computed as the ratio between the no load torque and the no load speed.

$$r = \frac{k_t I_0}{n_0} \tag{4}$$

Relevant parameters from the motor manufacturer are reported in Table 1.

Table 1. Electric motor model parameters.

Parameter	Description	Value	Unit
R_a	Resistance	0.293	(Ω)
L_a	Inductance	0.279	(mH)
k_t	Torque constant	52.5	(mNm/A)
k_e	Speed constant	52.5	(mV/(rad/s))
r	Torque/speed gradient	5.78×10^{-5}	(Nm/(rad/s))
I_0	No load current	497	(mA)
n_0	No load speed	4300	(rpm)

3.2. Mechanical System

The mechanical dynamics of the steering actuator are described through a 3-DOF system, in which friction is modeled as viscous damping and constant gear ratios are used. For the ball screw, a viscous damper with coefficient β_{BS} is introduced. The ball screw transmission ratio is τ_{BS}. The same approach is used to model the belt drive connecting the motor shaft with the ball screw nut. Damping coefficient β_B and transmission ratio τ_B are used. Furthermore, the belt stiffness is accounted for through the coefficient k_B. The steering column is modeled by considering it as a torsion bar with torsional stiffness k_{TB} and damping coefficient β_{TB}. The rack and pinion mechanism of the steering system of the vehicle is represented by the constant gear ratio τ_P that is equal to the C-factor of the rack. The ratio τ_R considers the ratio between the wheel steering angle δ_W and the linear displacement of the rack x_R. It was experimentally determined based on the acquisitions from the on-vehicle sensor data. The term β_{SW} models the friction of the steering wheel.

The equations of motion are obtained through the Lagrangian approach with the motor angle θ_{rot}, the ball-screw angle θ_{BS} and the steering angle δ_{SW} as generalized coordinates. The system differential equations are

$$J_{rot}\ddot{\theta}_{rot} + \left(r + \beta_B \tau_B^2\right)\dot{\theta}_{rot} - \beta_B \tau_B \dot{\theta}_{BS} + k_B \tau_B^2 \theta_{rot} - k_B \tau_B \theta_{BS} = T_m \tag{5}$$

$$J^*_{BS}\ddot{\theta}_{BS} + (\beta_{BS} + \beta_B + \beta_{TB}(\tau_P/\tau_{BS})^2)\dot{\theta}_{BS} - \tau_B\beta_B\dot{\theta}_{rot} - \beta_{TB}(\tau_{BS}/\tau_P)\dot{\delta}_{SW} + \\ (k_B + k_{TB}(\tau_P/\tau_{BS})^2)\theta_{BS} - k_B\tau_B\theta_{rot} - k_{TB}(\tau_{BS}/\tau_P)\delta_{SW} = 0 \tag{6}$$

$$J_{SW}\ddot{\delta}_{SW} + (\beta_{SW} + \beta_{TB})\dot{\delta}_{SW} - \beta_{TB}(\tau_{BS}/\tau_P)\dot{\theta}_{BS} + K_{TB}\delta_{SW} - K_{TB}(\tau_{BS}/\tau_P)\theta_{BS} = 0 \tag{7}$$

The equivalent inertia at the ball screw is computed using an energetic approach as

$$J^*_{BS} = J_{BS} + (m_R + m_{BS})\tau^2_{BS} + J_W\tau^2_{BS}\tau^2_R \tag{8}$$

The system is rearranged in the state-space form $\dot{x} = Ax + Bu$. The system state vector is $x = \begin{bmatrix} \theta_{rot} & \dot{\theta}_{rot} & \theta_{BS} & \dot{\theta}_{BS} & \delta_{sw} & \dot{\delta}_{sw} \end{bmatrix}^T$. The input to the system is the driving torque provided to the electric motor: $u = T_m$. The outputs of the system are the angle at the motor used to perform the PID position control and the angle at the steering column pinion end, compared with the measurement of the steering encoder sensor: $y = \begin{bmatrix} \theta_{rot} & \delta_P \end{bmatrix}^T$. The angle at the pinion is computed considering the rigid transmission ratio of the ball screw and the rack C-factor as: $\delta_P = \theta_{BS} \cdot (\tau_{BS}/\tau_P)$.

A full description of the model transmission ratios and known masses and inertia are provided in Table 2. The missing parameters, i.e., the ball screw inertia J_{BS}, the steering wheel inertia J_{SW}, the wheel inertia J_W, the belt damping β_B, the belt stiffness k_B, the damping of the ball screw β_{BS}, the damping of the steering wheel β_{SW}, the damping of the steering column β_{TB} and the stiffness of the steering column k_{TB}, are identified through the grey-box model parameters estimation optimization process described in Section 4.3.

Table 2. Dynamic model parameters.

Parameter	Description	Value	Unit
J_{rot}	Rotor inertia	0.81×10^{-4}	(kg·m^2)
m_{BS}	Ball screw mass	0.171	(kg)
m_R	Rack mass	6.5×10^{-2}	(kg)
τ_B	Belt to ball screw ratio	2	(-)
τ_{BS}	Ball screw lead	3.183×10^{-4}	(m/rad)
τ_P	Pinion C-factor	1.36×10^{-2}	(m/rad)
τ_R	Rack displacement to wheel toe angle ratio	19.4	(rad/m)
b	Longitudinal distance tie rod-kingpin axis	0.049	(m)

3.3. Electric Motor Control

The adopted control strategy for the BLDC motor has been designed and modeled based on the EPOS 4 controller architecture [24]. The most suitable operating mode for the autonomous steering actuator is the cyclic synchronous position mode provided by the EPOS 4 ECU. As presented in Figure 4, the reference steering angle ($\delta_{SW,ref}$) is computed by the trajectory planning algorithm implemented in dSPACE^{TM}. The reference is compared with the signal acquired by the steering wheel encoder sensor. Then, it is converted to a motor target position that is passed to the EPOS motor controller. Thus, the EPOS performs the position control loop and provides the current set to the inner (cascade) current control loop, which in turn will yield motion on the motor shaft.

The position control of the electric motor is implemented through a Proportional-Integral-Derivative (PID) algorithm and a feedforward control to improve the motion system setpoint. The velocity feedforward (FF_ω) serves for compensation of speed-proportional friction and its coefficient is computed as

$$FF_\omega = \frac{r}{k_t} \tag{9}$$

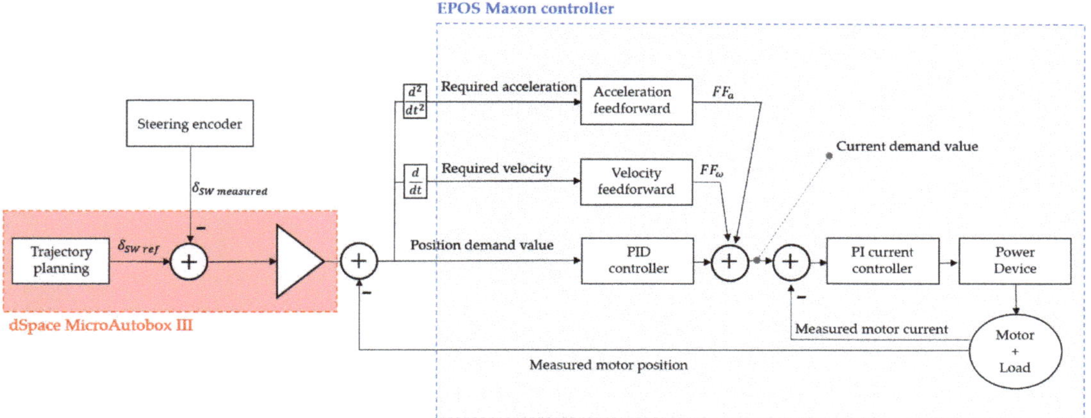

Figure 4. Electric motor control block diagram.

The acceleration feedforward (FF_a) accounts for the inertia of the system and its coefficient is determined by

$$FF_a = \frac{J^*_{mot}}{k_t} \tag{10}$$

The anti-windup method is used to prevent saturation of the command signals. The limit in current is set to 20 A according to the motor capabilities. The internal current control loop computes the voltage set to the motor. It consists of a Proportional-Integral (PI) control with an anti-windup algorithm to avoid saturation of the voltage command (24 V).

The block scheme of the controller architecture is depicted in Figure 4.

The tuning of the PI current controller is performed using the pole placement technique. Since the system can be analyzed as a linear canonical first-order system, an algebraic method can be adopted to define precise relationships between poles and the shape of the response. The tuning of the position PID controller has been performed by using the Tuning App (based on transfer functions) in the Simulink block PID controller of the Control Design MATLAB toolbox.

The parameters used for implementing the electric motor control are reported in Table 3.

Table 3. Electric motor controller parameters.

Parameter	Description	Value	Unit
$K_{P\,current}$	Current controller P gain	2000	(mV/A)
$K_{I\,current}$	Current controller I gain	2100	(mV/(A·ms))
$K_{P\,position}$	Position controller P gain	4390	(mA/rad)
$K_{I\,position}$	Position controller I gain	7352	(mA/(rad·s))
$K_{D\,position}$	Position controller D gain	159.87	((mA·s)/rad)
FF_ω	Velocity feedforward	12.59	(mA·s/rad)
FF_a	Acceleration feedforward	1.783	(mA·s^2/rad)

4. Results and Discussion

4.1. Testbed Setup

The presented system has been installed on the chassis of the considered vehicle and the experimental campaign is carried out. The electric motor of the steering actuator and its embedded Hall sensor are connected to the EPOS ECU presented in the previous section via a dedicated wiring system. The ECU is fixed on the monocoque of the vehicle. An additional microcontroller (Texas Instrument™ Launch XL-F28379d, Dallas, TX, USA) is installed on the vehicle to acquire the steering angle sensor data from the digital encoder sensor mounted on the steering rack and to provide the reference profile to the actuator ECU in real-time at a sampling frequency of 100 Hz. All the systems are connected via CAN to the PC to both provide the enabling and configurations messages to the system and to acquire the data of encoder sensor steering angle and steering profile from the TI microcontroller and the values of actual current and the actual position of the electric motor provided from the EPOS controller. Figure 5 schematizes the overall layout of the actuator testbed installed in the actual vehicle.

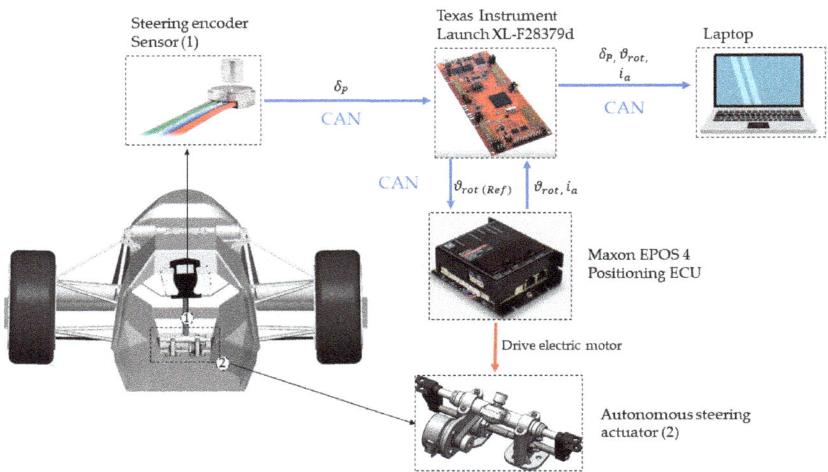

Figure 5. Autonomous steering actuator testbed setup block scheme.

All the systems are powered by the Low Voltage (LV) battery of the car i.e., a 24 V lithium-ion battery.

The performed tests are sine-sweep steering maneuvers and steering maneuvers generated by the MPC strategy that controls the driverless vehicle dynamics. The sine-sweep maneuvers are performed with both suspended and on-ground vehicles aimed at the system identification and dynamic performance assessment. Furthermore, the system tracking performance is evaluated when a reference profile generated by the MPC for the autonomous mission is provided.

4.2. Sine-Sweep Test

Different sine-sweep steering maneuvers are conducted first. The test is performed for a frequency range from 0 Hz to 10 Hz, in a time interval of 10 s with an amplitude at the steering wheel of 10 degrees. Data are acquired for two different conditions of the vehicle, namely suspended and stationary vehicles on the ground. The response to the reference steering wheel position is measured by the encoder mounted on the steering column. Results obtained in the case of the suspended vehicle are reported in Figure 6a. Similarly, Figure 6b shows the motor current and position measured by the embedded sensors.

The current measured by the motor approaches the limit of 20 A provided by the motor manufacturer. By analyzing the trend of the current, we can see sections without steering angle variations for a short period despite the high motor current and torque output. This behavior is caused by the dead band in the actuator due to the backlash and the compliance of the elastic elements present in the system, for this reason, a dead band compensator should be implemented in the controller, as described in [15].

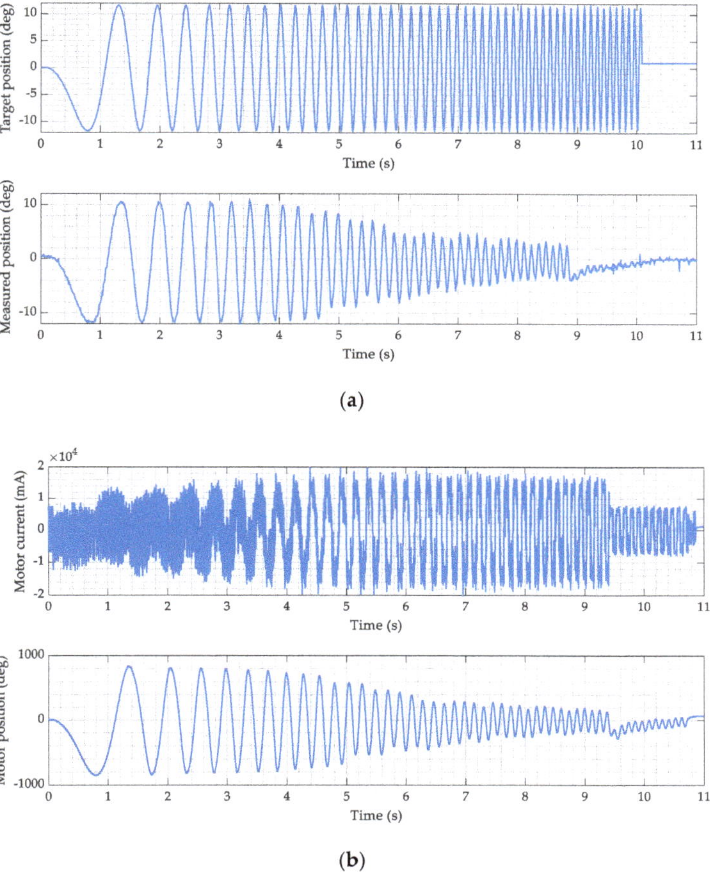

Figure 6. Sine-sweep steering maneuver—suspended vehicle, 0–10 Hz, amplitude 10 degrees. (**a**) Reference steering wheel position and measured response by the steering encoder. (**b**) Measured motor current and position.

An equivalent sine-sweep test is repeated with the vehicle on the ground (dry asphalt). In Figure 7 the data collected by the steering encoder sensor and by the motor sensors are reported.

The performed sine-sweep maneuvers in both the test conditions are processed to extract the frequency response function of the system. Figure 8 shows the Bode plot where the output and input to the system are the measured positions by the steering encoder sensor and the reference steering sine-sweep profile, respectively. The cut-off frequency for the suspended vehicle condition is at 5 Hz. The frequency response function of the test performed with the vehicle on the ground shows a cut-off frequency at 4.8 Hz.

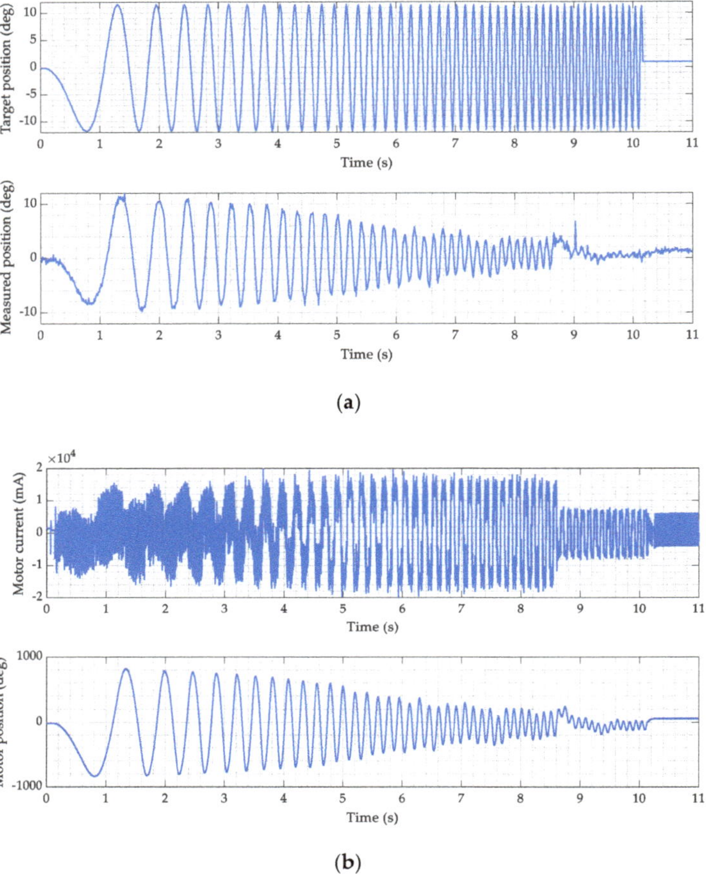

Figure 7. Sine-sweep steering maneuver—stationary vehicle on the ground, dry asphalt, 0–10 Hz, amplitude 10 degrees. (**a**) Reference steering wheel position and measured response by the steering encoder. (**b**) Measured motor current and position.

From the analysis of the Bode plot for the vehicle on the ground case, reported in Figure 8, the stick slip effect is preponderant at low frequency in the case of the vehicle on the ground. Due to friction among tires and ground, the bandwidth of the system on the ground is slightly lower than in the suspended vehicle case. For both the test conditions, the dynamic performance of the system is consistent with the considered application. As the typical steering frequencies of the driver input are below 4 Hz [10], the designed actuator shows good tracking performance below 4 Hz for both the considered tests.

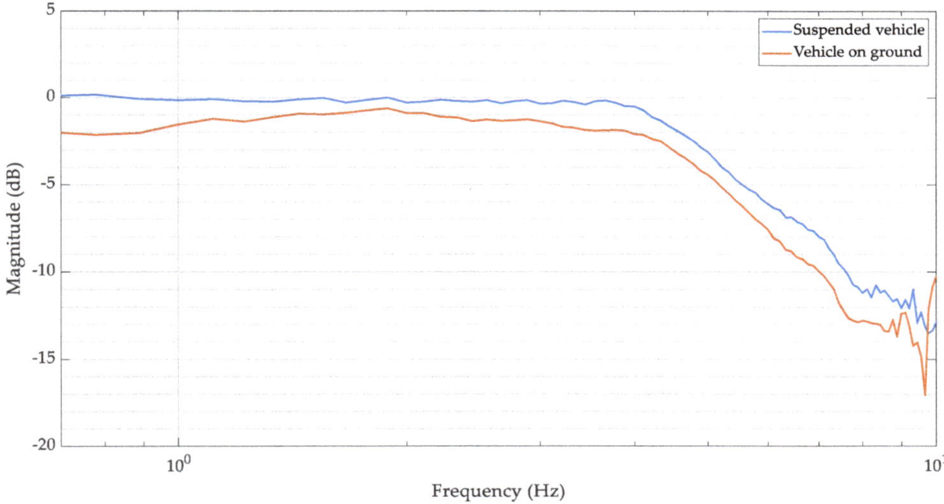

Figure 8. Experimental frequency response function of the closed loop system: suspended vehicle (blue) and vehicle on the ground (orange).

4.3. System Identification

The grey-box model estimation approach is used to identify damping and stiffness coefficients along with inertia terms related to the ball screw, steering wheel, steering column, and wheels. To this end, the experimental data reported in Section 4.2 are used. The friction losses of the system as the moments of inertia of the ball screw, of the steering wheel and column, and of the wheels were determined starting from the experimental acquisitions by using a grey-box model estimation approach. The first analytical esteem of the parameters is performed to determine the starting nominal values of the parameters.

The estimated system has nine design variables: the ball screw inertia J_{BS}, the steering wheel inertia J_{SW}, the wheel inertia J_W, the belt damping β_B, the belt stiffness k_B, the damping of the ball screw β_{BS}, the damping of the steering wheel β_{SW}, the damping of the steering column β_{TB} and the stiffness of the steering column k_{TB}.

The software tunes the model parameters to obtain a simulated response (y_{sim}) that tracks the measured response or reference signal (y_{ref}). The y_{ref} provided to the parameter estimation algorithm is the suspended vehicle sine sweep test response reported in Figure 6. The optimization method for the estimation of the parameters is the Nonlinear Least Squares aimed at minimizing the squares of the residuals in the system response. In Table 4, the identified parameters are reported.

Table 4. Identified parameters.

Parameter	Description	Value	Unit
J_{BS}	Ball screw inertia	6.5×10^{-6}	(kg·m^2)
J_{SW}	Steering wheel inertia	3.3×10^{-3}	(kg·m^2)
J_W	Wheel inertia	2.65	(kg·m^2)
k_B	Belt stiffness	0.29	(Nm/rad)
β_B	Belt damping	1.7×10^{-3}	(Nm/(rad/s))
β_{BS}	Ball screw damping	1.34×10^{-2}	(Nm/(rad/s))
k_{TB}	Torsion bar stiffness	164.75	(Nm/rad)
β_{TB}	Torsion bar damping	0	(Nm/(rad/s))
β_{SW}	Steering wheel damping	1.59	(Nm/(rad/s))

The identified values are consistent with the considered application. The response of the identified numerical model (y_{sim}) is reported in Figure 9, compared with the signal acquired by the encoder sensor during an on-vehicle experimental testing campaign (y_{ref}).

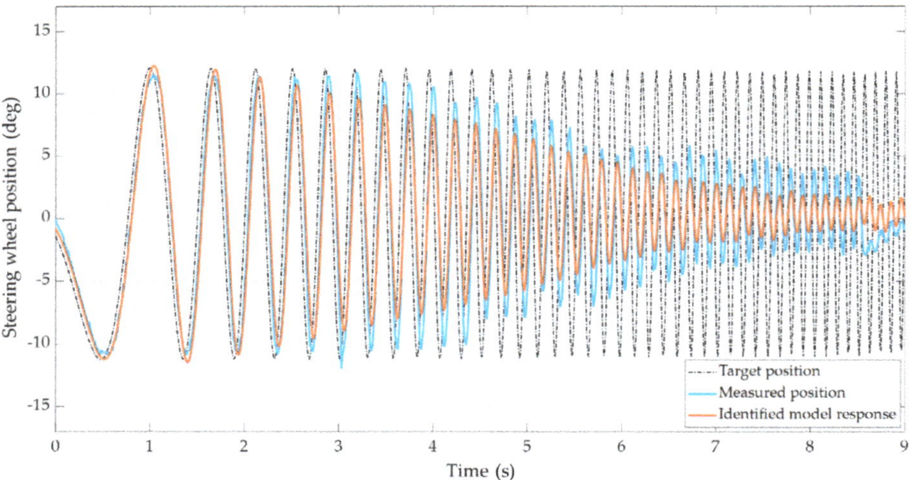

Figure 9. Steering wheel position behavior during the performed sine-sweep maneuver for validating the system identification: target position (black, dash-dotted) vs. measured position (blue, solid) vs. identified model response (orange, solid).

The identified system features one pole at the origin and a real stable pole at 334 rad/s (53.2 Hz) that are the rigid body modes. Then, two complex conjugate sets of poles appear at 30.9 rad/s (4.9 Hz) and 131 rad/s (20.9 Hz). The former low-frequency pole represents the motor inertia mode vibrating due to the belt drive compliance. The latter is the ball screw inertia vibrating due to the torsional stiffness of the steering wheel and column. The target frequency range for the present application (1 Hz) is below the frequencies of the identified mechanical system.

The goodness-of-fit between the identified model and the actual actuator has been quantified as a function of the Normalized Root Mean Square Error (NRMSE) cost function, as follows:

$$GoF = 1 - NRMSE = 1 - \frac{\|y_{ref} - y_{sim}\|_2}{\|y_{ref} - \overline{y}_{ref}\|_2} \tag{11}$$

where $\|\,\|$ indicates the 2-norm of a vector, y_{ref} is the measured position and y_{sim} is the identified position. The proposed identification has a *GoF* equal to 75%. The identified numerical response matches the experimental one.

4.4. Autonomous Driving Test

The performance of the proposed EPS system has been evaluated with a steering profile coming from the co-simulation of an autonomous driving mission. The vehicle is driven at a constant longitudinal speed equal to 5 km/h while the steering angle reference profile is properly given to the actuator for a time length of about 120 s. Figure 10a compares the steering wheel position from the encoder with the reference signal generated by the MPC strategy devoted to autonomous driving [18]. Accordingly, Figure 10b shows the followed track that yields the tracked position profile. From these results, an angular position RMSE of 3 degrees is computed for the whole track. This value is consistent with the required performance for the specific application.

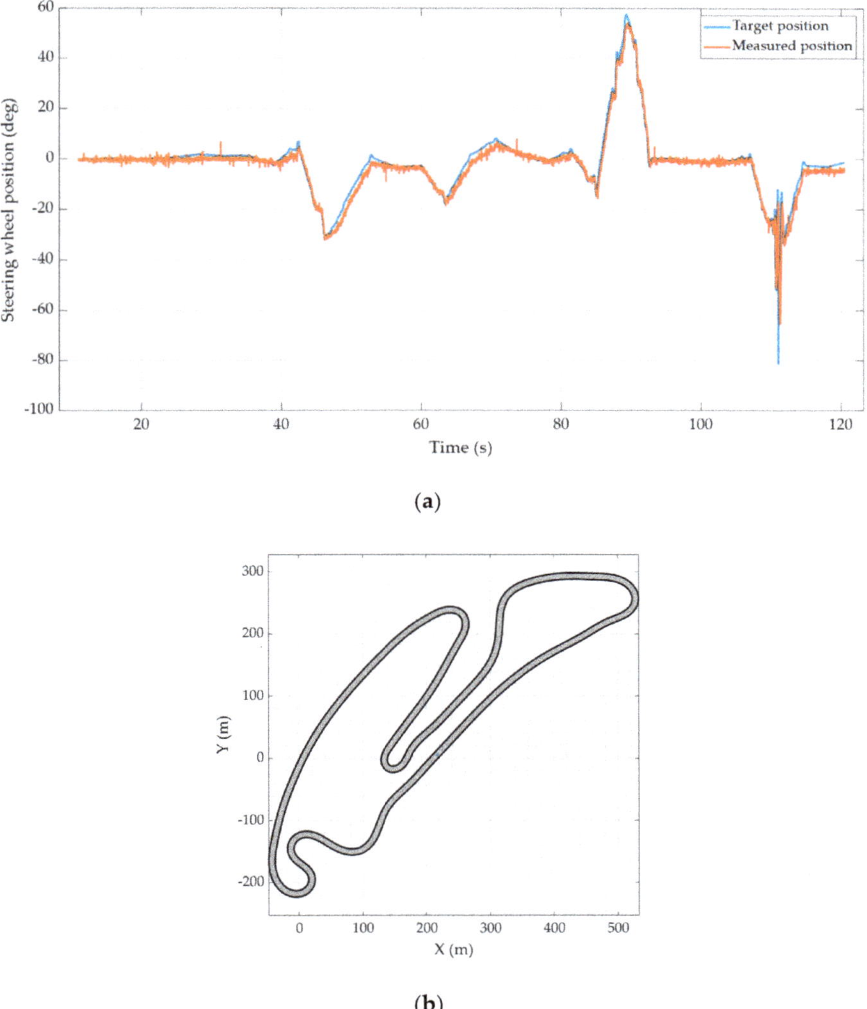

(a)

(b)

Figure 10. Autonomous driving test. (**a**) Steering wheel position measured by the encoder sensor (orange, solid) vs. target position generated by the MPC strategy devoted to autonomous driving (blue, solid). (**b**) Spatial coordinates (X,Y) of the race track related to the performed autonomous driving test.

Furthermore, the applied realistic reference highlights important features of the control. The maximum overshoot is found at 89.4 s, reaching up to 4.5 degrees of error for a very narrow time interval. It is worth noting that the position activity around 111 s presents high dynamic content and thus puts in evidence the intrinsic limitations of the actuation system above 4 Hz (max. error up to 15.7 degrees). However, this behavior is beyond the bandwidth of interest.

5. Conclusions

This paper presented the design, integration, and performance assessment of an EPS system for a driverless vehicle. In detail, an FSD racing vehicle was considered and the proposed EPS was studied for system identification and performance assessment during different maneuvers. To this end, the actuator was designed to fulfill the competition rules and both design and packaging constraints. The system model was developed by considering both the electric motor and the existing mechanical steering subsystem. Furthermore, the applied control strategy was discussed, along with the performed system identification procedure that has been conducted exploiting a grey-box model. Experimental tests in a proper laboratory environment were carried out to assess the system performance while validating the described model and performing system identification. The performance of the proposed EPS system was tested on the racetrack, during both sine-sweep maneuvers and co-simulated driverless sessions.

Sine-sweep test results highlighted the actuator capabilities in terms of bandwidth. Also, favorable matching was found between the plant model and the prototype. Finally, the system performance was assessed in terms of position reference tracking. The investigated EPS system was able to satisfy the design requirement by showing favorable tracking metrics.

As future work, an extensive validation stage on-track could be required to test the proposed system under several demanding handling maneuvers in different road conditions.

Author Contributions: Conceptualization, R.M., S.C., I.K., S.F., S.L., N.A. and R.G.; Investigation, R.M., S.C. and I.K.; Methodology, R.M., S.C. and I.K.; Supervision, N.A., A.B. and R.G.; Validation, R.M., S.C., S.F. and I.K.; Writing—original draft, R.M.; Writing—review & editing, S.C., I.K., S.F., S.L. and R.G. All authors have read and agreed to the published version of the manuscript.

Funding: This research received no external funding.

Institutional Review Board Statement: Not applicable.

Informed Consent Statement: Not applicable.

Data Availability Statement: Data available on request due to restrictions.

Acknowledgments: This work was developed in the framework of the activities of the Interdepartmental Center for Automotive Research and Sustainable Mobility (CARS) at Politecnico di Torino (www.cars.polito.it, accessed on 16 July 2021).

Conflicts of Interest: The authors declare no conflict of interest.

References

1. PwC. Available online: https://www.pwc.com/en/industries/automotive/Publications/eascy.html (accessed on 15 January 2021).
2. Mc Kinsey Center for Future Mobility. Available online: https://www.mckinsey.com/features/mckinsey-center-for-future-mobility/overview/autonomous-driving (accessed on 15 January 2021).
3. SAE J3016—Levels of Driving Automation. Available online: https://www.sae.org/news/2019/01/sae-updatesj3016automateddrivinggraphic#:~{}:text=The%20J3016%20standard%20defines%20six,graphic%20first%20deployed%20in%202016 (accessed on 15 January 2021).
4. Mehrabi, N.; Azad, N.L.; McPhee, J. Optimal disturbance rejection control design for electric power steering systems. In Proceedings of the 2011 50th IEEE Conference on Decision and Control and European Control Conference, Orlando, FL, USA, 12–15 December 2011; pp. 6584–6589.
5. Baharom, M.B.; Hussain, K.; Day, A.J. Design of full electric power steering with enhanced performance over that of hydraulic power-assisted steering. *Proc. Inst. Mech. Eng. Part D J. Automob. Eng.* **2013**, *227*, 390–399. [CrossRef]
6. Würges, M. New electrical power steering systems. *Encycl. Automot. Eng.* **2014**, 1–17. [CrossRef]
7. Liao, Y.G.; Du, H.I. Modelling and analysis of electric power steering system and its effect on vehicle dynamic behaviour. *Int. J. Veh. Auton. Syst.* **2003**, *1*, 153–166. [CrossRef]
8. Fankem, S.; Müller, S. A new model to compute the desired steering torque for steer-by-wire vehicles and driving simulators. *Veh. Syst. Dyn.* **2014**, *52* (Suppl. 1), 251–271. [CrossRef]
9. Chen, X.; Chen, X. *Control-Oriented Model for Electric Power Steering System (No. 2006-01-0938)*; SAE Technical Paper; SAE International: Warrendale, PA, USA, 2006.
10. Groll, M.V.; Mueller, S.; Meister, T.; Tracht, R. Disturbance compensation with a torque controllable steering system. *Veh. Syst. Dyn.* **2006**, *44*, 327–338. [CrossRef]
11. Govender, V.; Khazardi, G.; Weiskircher, T.; Keppler, D.; Müller, S. A PID and state space approach for the position control of an electric power steering. In *16 Internationales Stuttgarter Symposium*; Springer: Wiesbaden, Germany, 2016; pp. 755–769.
12. Carriere, S.; Caux, S.; Fadel, M. Optimal lqi synthesis for speed control of synchronous actuator under load inertia variations. *IFAC Proc. Vol.* **2008**, *41*, 5831–5836. [CrossRef]

13. Liao, Y.G.; Du, H.I. Cosimulation of multi-body-based vehicle dynamics and an electric power steering control system. *Proc. Inst. Mech. Eng. Part K J. Multi-Body Dyn.* **2001**, *215*, 141–151. [CrossRef]
14. Falcone, P.; Borrelli, F.; Asgari, J.; Tseng, H.E.; Hrovat, D. Predictive active steering control for autonomous vehicle systems. *IEEE Trans. Control. Syst. Technol.* **2007**, *15*, 566–580. [CrossRef]
15. Park, M.; Lee, S.; Han, W. Development of Steering Control System for Autonomous Vehicle Using Geometry-Based Path Tracking Algorithm. *Etri J.* **2015**, *37*, 617–625. [CrossRef]
16. Govender, V.; Müller, S. Modelling and position control of an electric power steering system. *IFAC PapersOnLine* **2016**, *49*, 312–318. [CrossRef]
17. Zakaria, M.I.; Husain, A.R.; Mohamed, Z.; Shah, M.B.N. Steering control of a steer-by-wire system vehicle with time delay and actuator saturation via anti-windup controller. *Eng. Appl. Sci. Res.* **2019**, *46*, 72–78.
18. Khan, I.; Feraco, S.; Bonfitto, A.; Amati, N. A model predictive control strategy for lateral and longitudinal dynamics in autonomous driving. In *Volume 4: 22nd International Conference on Advanced Vehicle Technologies (AVT), Proceedings of the ASME 2020 International Design Engineering Technical Conferences and Computers and Information in Engineering Conference, Virtual, Online, 17–19 August 2020*; ASME: New York, NY, USA, 2020; p. V004T04A004.
19. Nguyen, V.G.; Guo, X.; Zhang, C.; Tran, X.K. Parameter Estimation, Robust Controller Design and Performance Analysis for an Electric Power Steering System. *Algorithms* **2019**, *12*, 57. [CrossRef]
20. Naranjo, J.E.; González, C.; García, R.; de Pedro, T. Electric power steering automation for autonomous driving. In Proceedings of the International Conference on Computer Aided Systems Theory, Las Palmas de Gran Canaria, Spain, 7–11 February 2005; Springer: Berlin/Heidelberg, Germany, 2005; pp. 519–524.
21. Formula Student 2020 Rulebook. Available online: https://www.formulastudent.de/all/2020/rules/FSRules-2020-V1 (accessed on 7 December 2020).
22. Kabzan, J.; de la Iglesia Valls, M.; Reijgwart, V.; Hendrikx, H.F.C.; Ehmke, C.; Prajapat, M.; Bühler, A.; Gosala, N.; Gupta, M.; Sivanesan, R.; et al. Amz driverless: The full autonomous racing system. *J. Field Robot.* **2020**, *37*, 1267–1294. [CrossRef]
23. Feraco, S.; Luciani, S.; Bonfitto, A.; Amati, N.; Tonoli, A. A local trajectory planning and control method for autonomous vehicles based on the RRT algorithm. In Proceedings of the 2020 AEIT International Conference of Electrical and Electronic Technologies for Automotive (AEIT AUTOMOTIVE), Torino, Italy, 18–20 November 2020; pp. 1–6. [CrossRef]
24. Maxon Group. *Epos 4 Positioning Controller Application Notes*; Edition 2019-11; Maxon Group: Sachseln, Switzerland, 2019.

Article

Segment Drift Control with a Supervision Mechanism for Autonomous Vehicles

Ming Liu [1,2], **Bo Leng** [1,2,3,*], **Lu Xiong** [1,2], **Yize Yu** [1,2] and **Xing Yang** [1,2]

1 School of Automotive Studies, Tongji University, Shanghai 201804, China; 2110215@tongji.edu.cn (M.L.); xiong_lu@tongji.edu.cn (L.X.); yuyize@163.com (Y.Y.); yang_xing@tongji.edu.cn (X.Y.)
2 Clean Energy Automotive Engineering Center, Tongji University, Shanghai 201804, China
3 Postdoctoral Station of Mechanical Engineering, Tongji University, Shanghai 201804, China
* Correspondence: harrisonleng@gmail.com

Abstract: Stable maneuverability is extremely important for the overall safety and robustness of autonomous vehicles under extreme conditions, and automated drift is able to ensure the widest possible range of maneuverability. However, due to the strong nonlinearity and fast vehicle dynamics occurring during the drift process, drift control is challenging. In view of the drift parking scenario, this paper proposes a segmented drift parking method to improve the handling ability of vehicles under extreme conditions. The whole process is divided into two parts: the location approach part and the drift part. The model predictive control (MPC) method was used in the approach to achieve consistency between the actual state and the expected state. For drift, the open-loop control law was designed on the basis of drift trajectories obtained by professional drivers. The drift monitoring strategy aims to monitor the whole drift process and improve the success rate of the drift. A simulation and an actual vehicle test platform were built, and the test results show that the proposed algorithm can be used to achieve accurate vehicle drift to the parking position.

Keywords: autonomous vehicles; drift parking; open-loop control; supervision mechanism

Citation: Liu, M.; Leng, B.; Xiong, L.; Yu, Y.; Yang, X. Segment Drift Control with a Supervision Mechanism for Autonomous Vehicles. *Actuators* **2021**, *10*, 219. https://doi.org/10.3390/act10090219

Academic Editor: Hai Wang

Received: 18 July 2021
Accepted: 18 August 2021
Published: 1 September 2021

Publisher's Note: MDPI stays neutral with regard to jurisdictional claims in published maps and institutional affiliations.

1. Introduction

The stability of the vehicle chassis has always been a matter of concern. Chassis design can be divided into different classifications for different groups of people [1]. For professional drivers, the chassis usually exhibits a reduced margin of stability in the system when completing specific driving actions. This usually causes tire adhesion to reach saturation, also referred to as the limit condition. By studying the dynamic characteristics of the vehicle under extreme states, it is possible to better adapt the dynamic control boundaries of the vehicle. When the vehicle is driving on a low-adhesion road, with a low friction coefficient, it is easy for turning to cause the rear wheels to reach the adhesion limit ahead of the other wheels, and the tail of the vehicle will swing out, that is, the drift phenomenon will occur. When the vehicle drifts, it causes the vehicle's heading angle, mass center sideslip angle, and other states to change with time, accelerating, and the vehicle will be in an unstable state. Goh et al., performed experiments on the full-scale MARTY test vehicle to confirm the effectiveness of the controller on a trajectory with a curvature varying from 1/7 to 1/20 m. The vehicle speed was varied from 25 to 45 km/h [2]. Driverless vehicles are able to perform correct decision making by sensing the surrounding environmental conditions, and accurately tracking their trajectory. In addition, driverless vehicles are able to ignore driver factors such that nonprofessional drivers are able to experience the fun of the drift.

The trajectory tracking control of intelligent vehicles has developed rapidly in the last ten years. Due to the strong nonlinearity, internal dynamic instability, and under-drive of the vehicle system, achieving trajectory tracking control with high precision and high robustness remains a difficult problem. Therefore, various control methods are constantly emerging.

The linear quadratic regulator (LQR) is one of the most commonly used optimal control methods for trajectory tracking and has a small real-time calculation burden and a simple structure. Alcala et al. [3] used the Lyapunov-based control method to reconstruct the closed-loop system in the form of linear variable parameters and used the linear quadratic regulator–linear matrix inequalities (LQR-LMI) to adjust the parameters of the Lyapunov controller. The sliding mode control (SMC) method has good robustness, and still possesses a good control effect in systems with high model uncertainty. Tagne et al. [4] introduced a high-order sliding mode controller to control the steering wheel angle of autonomous vehicles in response to the current lateral displacement error. Hu et al. [5] adopted nonlinear feedback (integral sliding mode–composite nonlinear feedback) based on sliding mode control to weaken the chattering of the system in consideration of the stability of the system under tire saturation conditions. Funke et al. [6] comprehensively considered trajectory tracking, vehicle stability, and collision avoidance as the three control objectives by adjusting the weight coefficient in the MPC method, with priority being given to avoiding obstacles and maintaining vehicle stability. Liu et al. [7] used the MPC method to realize lane changing control in unmanned vehicles at high speed, while assessing vehicle stability on the basis of the phase diagram, and developed a stability envelope constraint on this basis to ensure the stability of the vehicle under high lateral conditions. Guo et al. [8] realized trajectory tracking control of four-wheel distributed-drive electric vehicles through hierarchical control. The upper layer calculates the expected front wheel angle and the direct yaw moment through the MPC method, while the lower controller assigns the direct yaw moment to each wheel motor. Kim et al. [9] considered the dynamic characteristics of the steering system in a control model and added actuator characteristic constraints to the MPC controller.

In recent years, scholars at home and abroad have performed a lot of research on vehicle drift control. Velenis et al. [10] studied the drift stability of rear-wheel-drive vehicles and demonstrated that a vehicle can only maintain an unstable balance if the vehicle's throttle and steering are controlled simultaneously. A set of backstepping controllers was designed, and these were combined with the driver's input commands to achieve control of the stability of vehicle drift along a steady circle in the simulation environment. In line with the preview control theory, Nakano et al. [11] designed a full-state feedback controller based on the linearization of a nonlinear system and tracked a steady-state circular trajectory with a drift attitude. Goh et al. [12] studied lateral displacement control, calculating the lateral force of the front and rear wheels while simultaneously controlling the stability of the sideslip angle of the mass center and directly solving the longitudinal force on the basis of the saturation of the rear tires, thus allowing a vehicle to drive along a steady circle in a state of drift balance. The control scheme proposed by Jelavic et al., switches between nonlinear model predictive control and linear feedforward feedback strategy to achieve drift [13]. An RC vehicle with a ratio of 1/10 was used for test verification. Kolter et al. [14,15] designed a set of open-loop and closed-loop fusion control algorithms. Firstly, a closed-loop controller based on the LQR algorithm was designed according to the vehicle dynamics model, which can realize the trajectory tracking control under normal working conditions. Secondly, an open-loop controller was designed according to the analysis of a data library of the motion control actions of professional drivers during drifts. At each moment of the control process, the two are switched independently according to the control effect of the controller.

In traditional research, when the vehicle is in e extreme emergency conditions on the low-adhesion roads, it becomes more likely to experience problems with poor control accuracy, which makes the vehicle lose stability or even abruptly sideslip. These algorithms have poor self-adaptability in complex environments, so it is difficult to ensure overall control stability [16,17]. Currently, research on trajectory tracking control under extreme conditions in terms of driverless vehicle motion control remains immature, and drift control is rarely studied. Based on research on the control of driverless vehicles drifting into a storage warehouse, this paper aims to achieve limited controllability of rear wheel brake lock, a state which is regarded as unstable in traditional motion tracking control The present study thus plays a role in technical exploration within the field.

This paper further explores the control of unmanned drift into a storage warehouse. The main contributions are as follows:

(1) A segment drift control strategy is designed. The depot approach section ensures that the vehicle enters the drift state when it reaches the drift trigger point. Drift in depot ensures that vehicles can complete the drift in operation with high precision.

(2) The drift monitoring strategy is proposed, including a path planning monitoring strategy, a drift-triggered state monitoring strategy, and a drift process monitoring strategy. Since the drift results are greatly affected by external disturbance factors, the proposed monitoring strategy can increase the success rate of drifts and ensure the safety and integrity of the test.

(3) Based on Simulink and CarSim, simulation experiments are carried out to verify the drift parking and monitoring strategy. Actual vehicle verification is carried out to provide the basis for the research under typical limit conditions.

The schematic diagram of the segmented drift control designed in this paper is shown in Figure 1. The OD segment is the location approaching segment, D is the drift trigger point, and the DP segment is the drift parking segment.

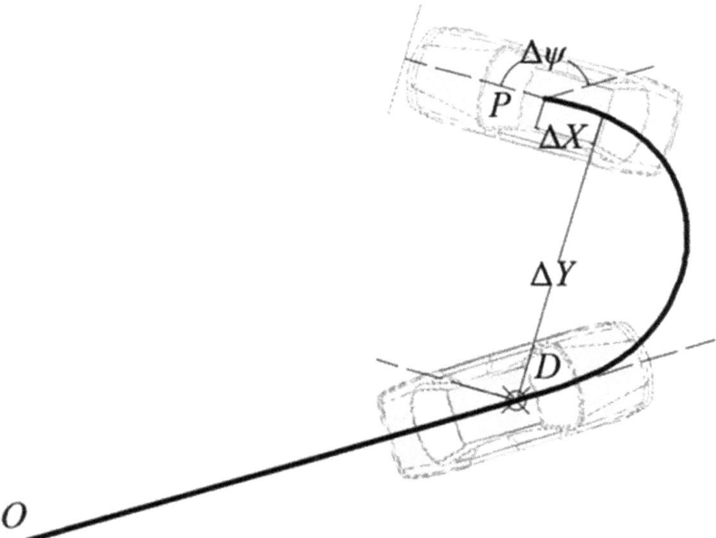

Figure 1. Schematic diagram of drift parking.

The difficulties of segment drift parking control include the following: (1) The triggering drift state of the vehicle should be consistent with the expected vehicle state. (2) The entire control system is greatly affected by external disturbance. (3) As a complex control system, the vehicle has strong parameter uncertainty and nonlinearity. Efforts to control costs impose limits on the type and quantity of onboard sensors available, so it is difficult to obtain vehicle dynamics parameters accurately and in real-time.

The rest of the article is arranged as follows: Section 2 covers the location approaching process; Section 3 covers the drift parking process; Section 4 covers the drift control supervision strategy; and Sections 5 and 6 cover both the simulation test and actual vehicle test, including a summary.

2. The Location Approaching Process

2.1. Path Planning

A Bezier curve, with as little curvature change as possible, is made between the current position of the vehicle and the drift trigger point, to serve as the travel path. The Bessel curve can be expressed as [18]:

$$q(\tau_i) = \sum_{k=0}^{m} \binom{m}{k} P_k (1 - \tau_i)^{m-k} \tau_i^{k}, \tau_i \in [0,1] \tag{1}$$

where m is the order of the Bezier curve, $q(\tau_i)$ is the interpolation point at the parameter, and τ_i, P_k is the control point with k sequence on the trajectory. By taking the value of the parameter τ_i, any interpolation point can be generated in the first control point and the last control point. A cubic Bezier curve is commonly used, where $m = 3$, and the cubic Bezier curve can be expressed as:

$$q(\tau_i) = (1 - \tau_i)^3 P_0 + 3\tau_i(1 - \tau_i)^2 P_1 + 3\tau_i^2(1 - \tau_i)P_2 + \tau_i^3 P_3 \tag{2}$$

The least-squares method is selected to fit the middle point of each reference path of the cubic Bezier curve. The sum of the squares of the fitting residuals can be expressed as:

$$S = \sum_{i=1}^{n} [p_i - q(\tau_i)]^2 \tag{3}$$

where n is the number of discrete path points contained in the cubic Bezier curve, and p_i are the discrete path points given by the cubic Bezier curve. According to the least squares method, by solving $\frac{\partial S}{\partial P_1} = 0$, $\frac{\partial S}{\partial P_2} = 0$, the two control points P_1 and P_2 in the middle of the cubic Bezier curve can be obtained. The equation can be expressed as:

$$P_1 = \frac{A_2 C_1 - A_{12} C_2}{A_1 A_2 - A_{12}^2}, \quad P_2 = \frac{A_1 C_2 - A_{12} C_1}{A_1 A_2 - A_{12}^2} \tag{4}$$

where: $A_1 = 9\sum_{i=1}^{n} \tau_i^2(1 - \tau_i)^4$, $A_2 = 9\sum_{i=1}^{n} \tau_i^4(1 - \tau_i)^2$, $A_{12} = 9\sum_{i=1}^{n} \tau_i^3(1 - \tau_i)^3$,

$C_1 = \sum_{i=1}^{n} 3\tau_i(1 - \tau_i)^2[p_i - (1 - \tau_i)^3 P_0 - \tau_i^3 P_3]$, $C_2 = \sum_{i=1}^{n} 3\tau_i^2(1 - \tau_i)[p_i - (1 - \tau_i)^3 P_0 - \tau_i^3 P_3]$.

In the initial stage of path fitting, the vehicle starting point and the drift trigger point are regarded as the first and last control points of the cubic Bezier curve, respectively. In each iteration of curve fitting, the position of the middle control point is solved according to Equation (4), and then the interpolation point corresponding to the original path point can be obtained according to Equation (2).

2.2. Trajectory Tracking

The vehicle kinematics model is shown in Figure 2. The definitions of the main terms appearing in the following equation are shown in Table 1.

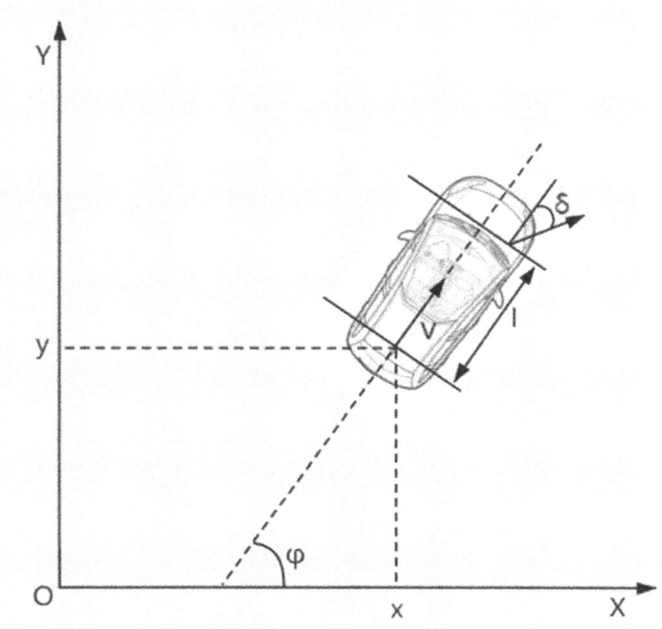

Figure 2. Kinematic model.

Table 1. Symbols and Definitions.

Symbols	Definitions
φ	Vehicle heading angle
δ	Vehicle front wheel angle
v	Vehicle speed
l	Wheel base
u_r	System reference input
χ_r	Vehicle reference status
T	Sampling time
Q	Weight matrix
R	Weight matrix
N_p	Prediction time domain
N_c	Control time domain
ρ	Weight coefficient
ε	Relaxation factor
η	Dimension of state quantity
$u_{\min}$	System reference input minimum
$u_{\max}$	System reference input maximum
(X_D, Y_D)	Drift trigger point position
ψ_D	Heading angle of drift trigger point
d_{thres}	Drift trigger distance threshold
$\Delta\psi_{thres}$	Drift trigger heading angle threshold
Δv_{thres}	Drift trigger speed threshold
l_f	Distance from centroid to front axle
l_r	Distance from centroid to rear axle

In the ground fixed coordinate system, the vehicle kinematics equation can be expressed as:

$$\begin{bmatrix} \dot{x} \\ \dot{y} \\ \dot{\varphi} \end{bmatrix} = \begin{bmatrix} \cos\varphi \\ \sin\varphi \\ \frac{\tan\delta}{l} \end{bmatrix} v \tag{5}$$

where (x, y) is the coordinate of the center of the rear axle of the vehicle, φ is the vehicle heading angle, δ is the front wheel angle, v is the longitudinal speed of the vehicle, and l is the wheelbase of the vehicle.

Defining $u(v, \delta)$ as the system input, the state variable is (x, y, φ). The system can be expressed as:

$$\dot{\chi} = f(\chi, u) \tag{6}$$

Each point on the cubic Bessel curve obtained by planning satisfies the above kinematic equation. The reference value is r. The reference trajectory can be expressed as:

$$\dot{\chi}_r = f(\chi_r, u_r) \tag{7}$$

The system equation is expanded by the Taylor series at the reference trajectory point:

$$\dot{\chi}_r = f(\chi_r, u_r) + \frac{\partial f(\chi, u)}{\partial x}\bigg| x - \chi_r + \frac{\partial f(\chi, u)}{\partial u}\bigg| (u - u_r) \tag{8}$$

where $\chi_r = (x_r, y_r, \varphi_r)$, and $u_r = (v_r, \delta_r)$. The error of the vehicle tracking model can be expressed as:

$$\dot{\tilde{\chi}} = \begin{bmatrix} \dot{x} - \dot{x}_r \\ \dot{y} - \dot{y}_r \\ \dot{\varphi} - \dot{\varphi}_r \end{bmatrix} = \begin{bmatrix} 0 & 0 & -v_r\sin\varphi_r \\ 0 & 0 & v_r\cos\varphi_r \\ 0 & 0 & 0 \end{bmatrix}\begin{bmatrix} x - x_r \\ y - y_r \\ \varphi - \varphi_r \end{bmatrix} + \begin{bmatrix} \cos\varphi_r & 0 \\ \sin\varphi_r & 0 \\ \frac{\tan\delta_r}{l} & \frac{v_r}{l\cos^2\delta_r} \end{bmatrix}\begin{bmatrix} v - v_r \\ \delta - \delta_r \end{bmatrix} \tag{9}$$

We can discretize this equation as:

$$\tilde{\chi}(k+1) = A_{k,t}\tilde{\chi}(k) + B_{k,t}\tilde{u}(k) \tag{10}$$

where $A_{k,t} = \begin{bmatrix} 1 & 0 & -v_r\sin\varphi_r T \\ 0 & 1 & v_r\cos\varphi_r T \\ 0 & 0 & 1 \end{bmatrix}$, $B_{k,t} = \begin{bmatrix} \cos\varphi_r T & 0 \\ \sin\varphi_r T & 0 \\ \frac{\tan\delta_r}{l} & \frac{v_r T}{l\cos^2\delta_r} \end{bmatrix}$, T is the sampling time.

In order to ensure that the vehicle can track the cubic Bezier curve quickly and smoothly, the objective function is designed in the following form:

$$J(k) = \sum_{i=1}^{N_p} \|\eta(k+i|t) - \eta_{ref}(k+i|t)\|_Q^2 + \sum_{i=1}^{N_c-1} \|\Delta U(k+i|t)\|_R^2 + \rho\varepsilon^2 \tag{11}$$

where Q and R are weight matrices, N_p is the prediction time domain, N_c is the control time domain, ρ is the weight coefficient, and ε is the relaxation factor. The vehicle linear error model is transformed as follows:

$$\xi(k|t) = \begin{bmatrix} \tilde{x}(k|t) \\ \tilde{u}(k-1|t) \end{bmatrix} \tag{12}$$

State-space expressions can be expressed as:

$$\xi(k+1|t) = \tilde{A}_{k,t}\xi(k|t) + \tilde{B}_{k,t}\Delta U(k|t) \tag{13}$$

$$\eta(k|t) = \tilde{C}_{k,t}\xi(k|t) \tag{14}$$

where $\widetilde{A}_{k,t} = \begin{bmatrix} A_{k,t} & B_{k,t} \\ 0_{m \times n} & I_m \end{bmatrix}, \widetilde{B}_{k,t} = \begin{bmatrix} B_{k,t} \\ I_m \end{bmatrix}$. n is the state quantity dimension and m is the control quantity dimension. The output expression of system prediction can be expressed as:

$$Y(t) = \psi_t \xi(t|t) + \Theta_t \Delta U(t) \tag{15}$$

$$\text{where}\quad Y_t = \begin{bmatrix} \eta(t+1|t) \\ \eta(t+2|t) \\ \eta(t+3|t) \\ \cdots\cdots \\ \eta(t+N_p|t) \end{bmatrix}, \quad \psi_t = \begin{bmatrix} \widetilde{C}_{t,t}\widetilde{A}_{t,t} \\ \widetilde{C}_{t,t}\widetilde{A}_{t,t}^2 \\ \widetilde{C}_{t,t}\widetilde{A}_{t,t}^3 \\ \cdots\cdots \\ \widetilde{C}_{t,t}\widetilde{A}_{t,t}^{N_p} \end{bmatrix}, \quad \Delta U_t = \begin{bmatrix} \Delta u(t|t) \\ \Delta u(t+1|t) \\ \Delta u(t+2|t) \\ \cdots\cdots \\ \Delta u(t+N_c|t) \end{bmatrix},$$

$$\Theta_t = \begin{bmatrix} \widetilde{C}_{t,t}\widetilde{B}_{t,t} & 0 & 0 & 0 \\ \widetilde{C}_{t,t}\widetilde{A}_{t,t}\widetilde{B}_{t,t} & \widetilde{C}_{t,t}\widetilde{B}_{t,t} & 0 & 0 \\ \cdots\cdots & \cdots\cdots & \cdots\cdots & \cdots\cdots \\ \widetilde{C}_{t,t}\widetilde{A}_{t,t}^{N_c}\widetilde{B}_{t,t} & \widetilde{C}_{t,t}\widetilde{A}_{t,t}^{N_c-1}\widetilde{B}_{t,t} & \cdots\cdots & \widetilde{C}_{t,t}\widetilde{A}_{t,t}\widetilde{B}_{t,t} \\ \cdots\cdots & \cdots\cdots & \cdots\cdots & \cdots\cdots \\ \widetilde{C}_{t,t}\widetilde{A}_{t,t}^{N_p-1}\widetilde{B}_{t,t} & \widetilde{C}_{t,t}\widetilde{A}_{t,t}^{N_p-2}\widetilde{B}_{t,t} & \cdots\cdots & \widetilde{C}_{t,t}\widetilde{A}_{t,t}^{N_p-N_c-1}\widetilde{B}_{t,t} \end{bmatrix}.$$

The constraint conditions of both the control quantity and increment are specified. The control quantity includes the wheel angle and the longitudinal speed of the vehicle. The control quantity constraint can be expressed as:

$$u_{\min}(t+k) \leq u(t+k) \leq u_{\max}(t+k), k = 0,1,2 \ldots N_c - 1 \tag{16}$$

The control increment constraint can be expressed as:

$$\Delta u_{\min}(t+k) \leq \Delta u(t+k) \leq \Delta u_{\max}(t+k), k = 0,1,2 \ldots N_c - 1 \tag{17}$$

The constraint equation for control quantity is transformed and the corresponding transformation matrix is obtained:

$$u(t+k) = u(t+k-1) + \Delta u(t+k) \tag{18}$$

$$U_t = 1_{N_c} \otimes u(k-1) \tag{19}$$

$$A = M_{N_c \times N_c} \otimes I_m \tag{20}$$

where 1_{N_c} is a column vector with N_c rows, $M_{N_c \times N_c}$ is the unit lower triangular matrix with dimension N_c, I_m is the identity matrix of dimension m, $\otimes$ is the Kronecker product, and $u(k-1)$ is the actual control quantity of the previous time. In combination with Equations (18)–(20), the constraint condition of the control quantity can be rewritten as:

$$U_{\min} \leq A \times \Delta U_t + U_t \leq U_{\max} \tag{21}$$

where $U_{\min}$ is the minimum set of control variables in the control time domain and $U_{\min}$ is the maximum set of control variables in the control time domain. The objective function is then transformed into a standard quadratic form:

$$J(\xi(t), u(t-1), \Delta U(t)) = \left[\Delta U(t)^T, \varepsilon\right]^T H_t \left[\Delta U(t)^T, \varepsilon\right] + G_t \left[\Delta U(t)^T, \varepsilon\right] \tag{22}$$

$$s.t. \quad \begin{aligned} \Delta U_{\min} \leq \Delta U_t \leq \Delta U_{\max} \\ U_{\min} \leq A\Delta U_t + U_t \leq U_{\max} \end{aligned} \tag{23}$$

where $H_t = \begin{bmatrix} \Theta_t^T Q \Theta_t + R & 0 \\ 0 & \rho \end{bmatrix}$, $G_t = \begin{bmatrix} 2e_t^T Q \Theta_t & 0 \end{bmatrix}$, e_t is the tracking error in the prediction time domain. In each control cycle, the control input increment can be expressed as:

$$\Delta U_t^* = \left[\Delta u_t^*, \Delta u_{t+1}^*, \Delta u_{t+2}^*, \dots, \Delta u_{t+N_c-1}^* \right]^T \tag{24}$$

The first element of the optimal sequence control is applied to the control system as the optimal control increment in this cycle, until the next period solves the new optimal control quantity according to the real-time system state. The vehicle finally reaches the drift trigger point, and the vehicle will begin to drift when the drift-triggering condition is met.

3. The Process of Drift Parking

3.1. Drift Open-Loop Control

Vehicle drift is triggered based on the longitudinal coupling characteristics of the tire. According to the tire force ellipse shown in Figure 3, when the rear wheel applies enough braking force to lock the wheel, the longitudinal force reaches the road adhesion limit, and the lateral force provided by the rear wheel is close to zero. At this time, the front wheel turns at a specific angle to produce a lateral force, and the rear wheel cannot provide a balanced lateral force. The lateral force of the front wheel produces a yaw moment on the body, which makes the rear axle sideslip, triggering the drift [19].

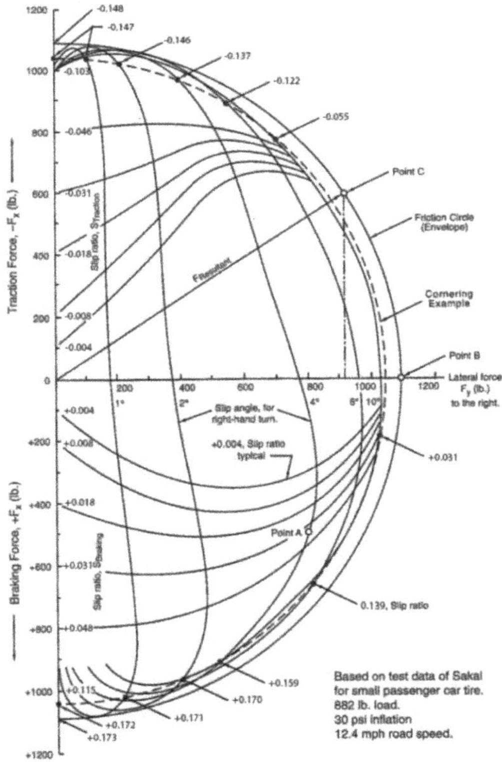

Figure 3. Tire force friction ellipse.

To create a sample of drifting instances to study, it is not necessary to float the vehicle into the warehouse during sampling, but only to carry out repeated tail-flick tests of the rear-wheel brake locking in the same field under the same vehicle conditions [20,21]. The vehicle starts in a static state, begins to accelerate, and then applies a drift after reaching a specific speed. Changes in the vehicle state and the action sequence from the initial time of the drift starting to the vehicle coming to a complete stop and achieving stability are recorded, such that:

$$
\begin{aligned}
S(k) &= s_1, s_2, \ldots s_k \\
A(k) &= a_1, a_2, \ldots, a_k
\end{aligned}
\tag{25}
$$

The recorded action A includes the desired steering wheel angle and the desired brake fluid pressure of the vehicle. The recorded vehicle state S includes the X, Y direction coordinates and heading angle. The purpose of state sequence $S(k)$ is to record the position change at the end of the drift process $(\Delta X, \Delta Y)$, and the change in the heading angle, $\Delta \psi$. The absolute coordinates (X_D, Y_D) and heading angle, ψ_D, of drift trigger point D can be calculated by using Equation (26) according to the coordinates (X_P, Y_P) and heading angle (i.e., ψ_P) of the target location during drift test:

$$
\begin{aligned}
X_D &= X_P + (\Delta X \cos \psi_P - \Delta Y \sin \psi_P) \\
Y_D &= Y_P + (\Delta X \sin \psi_P + \Delta Y \cos \psi_P) \\
\psi_D &= \psi_P - \Delta \psi
\end{aligned}
\tag{26}
$$

The vehicle trajectory and heading sequence are used as the reference sequence in order to monitor whether the vehicle drifts according to the expected trajectory. The drift process failure monitoring strategy outlined in Section 4 was designed based on this premise.

3.2. Design of Drift Trigger Conditions

Directed by the motion tracking controller, the vehicle travels along the planned route and gradually accelerates to the desired speed. When the vehicle is running, the state of the vehicle is monitored in real-time to determine whether it is consistent with the expected drift trigger state. The judgment conditions are as follows:

(1) Begin by calculating the distance between the current vehicle position coordinates (X, Y) and the drift trigger point (X_D, Y_D), $d(k) = \sqrt{(X - X_D)^2 + (Y - Y_D)^2}$. Compare this with the distance obtained previously, to determine if $d(k) - d(k-1) \leq 0$, and $d(k) < d_{thres}$. The result indicates whether the vehicle meets the position condition triggered by drift.

(2) Calculate whether the difference between the actual speed and the expected speed, Δv is less than Δv_{thres}. If $\Delta v < \Delta v_{thres}$, the vehicle meets the speed condition triggered by a drift.

(3) Calculate whether the difference between the actual heading angle and the expected heading angle, $\Delta \psi$ is less than $\Delta \psi_{thres}$. If $\Delta \psi < \Delta \psi_{thres}$, the vehicle meets the heading angle condition triggered by drift.

(4) Judge whether the current steering wheel angle exceeds the limit value. If $|\delta| < \delta_{thres}$, it means that the vehicle meets the yaw motion condition triggered by drift.

Since the vehicle drifting into the warehouse is simulated by the recurrence of the tail-flick action, the motion state cannot be feedback-controlled during drifting with our methodology, so the consistency between the vehicle state at the drift trigger time and the expected vehicle state must be high. If the motion planner or motion tracking controller fails during the depot approach and the vehicle triggers a drift in the wrong state, it will not drift to the depot. It may collide with the pile barrels or other obstacles near the depot, and aggravate the wear of the rear tires. Therefore, it was necessary to design a failure monitoring strategy for the drift entry action. When the vehicle motion state meets the specific conditions and cannot drift into the warehouse successfully, some measures should be taken to stop the drift entry action.

4. Monitoring Strategy

4.1. Supervision Strategy of the Path Planning Algorithm

In this paper, a cubic Bezier curve is used to connect the vehicle starting point and the drift trigger point and serves as the vehicle's approach path. In the process of generating the path, the algorithm is used to ensure the path has a minimum change in the curvature. In addition to the geometric constraints outlined in the planning stage, the path should also meet the following constraints:

(1) Maximum curvature constraint: The minimum radius of the path should be greater than the minimum turning radius of the vehicle while satisfying the corner constraint of the path-following controller.
(2) Maximum attachment constraint: The tire force should be less than the maximum tire force provided by the road surface at the maximum curvature of the constrained path, and also less than the tire force provided at the larger curvature when running at the maximum speed, so as to avoid wheel slip.
(3) Longitudinal velocity constraint: The planned path should be long enough to allow the vehicle to accelerate at the maximum acceleration and reach the desired drift longitudinal speed at the drift trigger point.

The curvilinear path is treated as being connected by several small circles, and the problem of a vehicle driving along the curvilinear path is simplified as a steady-state circular problem. The relationship between the curvature of the path and the steering angle of the vehicle can now be obtained. If the vehicle maintains a constant speed while moving in a circular motion with a certain radius, R_0, then the radius and the front wheel angle of δ will demonstrate the following relationship:

$$\delta = \left(1 + K \cdot v_x^2\right) \frac{l}{R_0} \tag{27}$$

where v_x is the vehicle speed; R_0 is the turning radius; l is the vehicle wheelbase; and K is the stability factor, $K = \frac{m}{l^2}\left(l_f C_{\alpha f} - l_r C_{\alpha r}\right)$. According to Equation (27), when the front wheel angle $\delta_{\max}$ corresponds to the maximum steering wheel angle in the controller constraint, the upper bound of the path curvature constraint is reached when the vehicle reaches maximum speed. In other words, the expression of the maximum curvature constraint is as follows:

$$\kappa \leq \frac{k_\delta \delta_{\max}}{(1 + K \cdot v_{\max}^2)l} \tag{28}$$

where k_δ is the safety factor, and the value range is [0, 1]. The maximum expected speed on the path is $v_{\max}$, which is equal to the drift trigger speed v_{target}. The vehicle motion is simplified to the steady-state circular driving problem, and the maximum attachment constraint of the path is deduced. The front axle does not slip during steering, provided the following criterion is met:

$$F_{yf} \leq \mu F_{zf} \tag{29}$$

where μ is the road adhesion coefficient, F_{yf} is front axle lateral force, F_{zf} is front axle vertical force, and v_y is the lateral speed. The vehicle is simplified as a linear model with two degrees of freedom, and the front axle lateral force can be expressed as:

$$F_{yf} = C_{\alpha f}\left(\delta - \frac{v_y + l_f \omega}{v_x}\right) \tag{30}$$

Due to the steady circular motion of the vehicle:

$$R_0 = \frac{v_x}{\omega} \tag{31}$$

We then substitute Equations (27), (30), and (31) into Equation (29):

$$\kappa \leq \frac{\frac{v_y}{v_x} + \frac{\mu F_{zf}}{C_{\alpha f}}}{l \cdot (1 + K v_x^2) - l_f} \tag{32}$$

Ignoring the lateral acceleration of the vehicle makes the inequality constraint stricter, and the maximum path attachment constraint is obtained:

$$\kappa \leq \frac{\mu F_{zf}}{C_{\alpha f}\left[l \cdot (1 + K v_x^2) - l_f\right]} \tag{33}$$

When determining the longitudinal speed of the vehicle at a certain point on the path, it is assumed that the vehicle meets the road adhesion and road surface constraints, and thus accelerates with the maximum longitudinal acceleration. The longitudinal speed can then be determined by the distance from the point to the starting point:

$$v_x(s) = \min\left(v_{\text{target}}, \sqrt{2 a_{\max} s}\right) \tag{34}$$

$$a_{\max} = \min\left(\mu g, \frac{T_{\max} i}{m r}\right) \tag{35}$$

where v_{target} is the target speed at the drift trigger point, s is the path length from the starting point to a certain point, $T_{\max}$ is the peak torque of the driving motor, i is the transmission ratio of the reduction mechanism, m is the mass of the whole vehicle, and r is the wheel radius.

In addition, it is necessary to verify the distance from the starting point to the drift trigger point in order to ensure that the vehicle can achieve maximum acceleration towards the drift trigger speed before reaching the drift trigger point. The longitudinal speed constraint equation is expressed as follows:

$$s \geq \frac{v_{\text{target}}^2}{2 a_{\max}} \tag{36}$$

where $a_{\max}$ is determined by Equation (35). After planning a path connecting the starting point and the drift trigger point, Equations (28), (33) and (36) can be utilized to check whether the constraint conditions are met, so as to judge the feasibility of the proposed path. If the conditions are not met, it means that the path planning fails. The approach path to the depot must be re-planned by adjusting the initial vehicle position and the initial heading angle.

4.2. Drift Process Monitoring Strategy

The vehicle drift process may be influenced by changes in the vehicle road system, which cause the vehicle system to produce different responses under the same control input. When the same site and the same vehicle conditions are tested, the vehicle road system may change due to factors including:

(1) Tire characteristics: Tire wear occurs in the process of lock slip, which leads to the change of tire characteristics. The change of tire force will directly affect the corresponding relationship between the steering wheel angle input and the drift trajectory output, so that the actual drift trajectory does not match the expected trajectory.

(2) Road conditions: Due to the influence of temperature and humidity, the adhesion condition of the road surface may differ between the tail-flick test and the drift test, which changes the tire force under the same load and slip rate.

(3) Vehicle status: Changes in the vehicle load size and distribution lead to changes in vehicle mass and the centroid position, which affects the tire force.

Through the sampling of vehicle states in the tail-flick test, the expected trajectory and the expected heading angle sequence of the vehicle drift process were obtained. The vehicle state is (X_t, Y_t, ψ_t) at a certain time during the drift. The expected state closest to the current state in the expected sequence is calculated by Equation (37):

$$k = \underset{k}{\mathrm{argmin}}\left(\begin{bmatrix} \|X_t - X_{ref}(k)\| \\ \|Y_t - Y_{ref}(k)\| \\ \|\psi_t - \psi_{ref}(k)\| \end{bmatrix}^{\mathrm{T}} \times \begin{bmatrix} \omega_X & & \\ & \omega_Y & \\ & & \omega_\psi \end{bmatrix} \times \begin{bmatrix} \|X_t - X_{ref}(k)\| \\ \|Y_t - Y_{ref}(k)\| \\ \|\psi_t - \psi_{ref}(k)\| \end{bmatrix}\right) \tag{37}$$

where ω_X, ω_Y, and ω_ψ are the weight coefficient, which is used to balance the influence of different distance and angle dimensions. After obtaining the expected state at the current moment, the weighted error vector between the actual vehicle state and the expected state at the current moment is calculated:

$$e_t = [\omega_X, \omega_Y, \omega_\psi]^{\mathrm{T}} \cdot \begin{bmatrix} \|X_t - X_{ref}(k)\| \\ \|Y_t - Y_{ref}(k)\| \\ \|\psi_t - \psi_{ref}(k)\| \end{bmatrix} \tag{38}$$

The error vector e_t is compared with the error threshold vector $e_{thres} = \left[e_{thres}^X, e_{thres}^Y, e_{thres}^\psi\right]^{\mathrm{T}}$. When any component of e_t is greater than y, it is considered that the control open-loop is invalid and the vehicle cannot accurately stop in the storage position.

5. Simulation and Ground Test

A CarSim-Simulink simulation platform was built to verify the effectiveness of the drift parking algorithm, and the drift parking action in the simulation environment was realized. Key parameters of the vehicle model are shown in Table 2.

Table 2. CarSim key parameters of simulation vehicle model.

Parameter	Unit	Value	Parameter	Unit	Value
Vehicle mass	kg	1412	Vehicle length	m	4.025
Yaw moment of inertia	kg m^2	1536.7	Vehicle width	m	1.916
Wheel radius	m	0.325	Steering ratio	-	2.91
Centroid height	m	0.54	Wheelbase	m	2.91

5.1. Simulation Test of Drift Whole Process Control

A simulation experiment of the whole-process open-loop algorithm was carried out to verify the effectiveness of the algorithm.

(1) Working condition setting:

The location coordinates of the vehicle's starting point are $(-100, -50)$, and the location coordinates of the depot are $(0, 0)$. The initial vehicle heading angle is $0°$. The initial speed is 0 km/h. The target location's orientation is $180°$. The road adhesion coefficient is 1. According to the results of the tail-flick test, the preset coordinates of the drift trigger point are $(-10.69, -6.13)$, and the heading angle of the drift point is $7.50°$. The longitudinal speed triggered by the drift is 39.96 km/h, and the step angle of the drift steering wheel is $140°$.

(2) Parameter setting:

The threshold settings of the drift trigger point are $\Delta v_{thres} = 0.5$ km/h, $d_{thres} = 0.3$ m, $\psi_{thres} = 5$, and $\delta_{thres} = 5$.

(3) Simulation test results:

The simulation results are shown in Figure 4. At 11.72 s, when the vehicle reaches (−10.97, −6.34), the vehicle is 0.278 m away from the drift trigger point, and the steering wheel angle is −2.7°. The longitudinal speed is 40.1 km/h, satisfying the drift triggering condition. The change in the vehicle trajectory in the global coordinate system is shown in Figure 4. Finally, the vehicle parks at (−1.196, 0.075). The distance from the center of the warehouse's error is 0.196 m. The final heading angle is 180.2°, and the error of the storage location orientation angle is 0.2°. The location is 5.2 × 2.5 m in size. The simulation test demonstrates that the car body stops completely within the storage position range and does not interfere with the storage position line. An animation of the entire output process made in CarSim is shown in Figure 5. The figures demonstrate that the vehicles are correctly parked in the warehouse and surrounded by the four pile barrels.

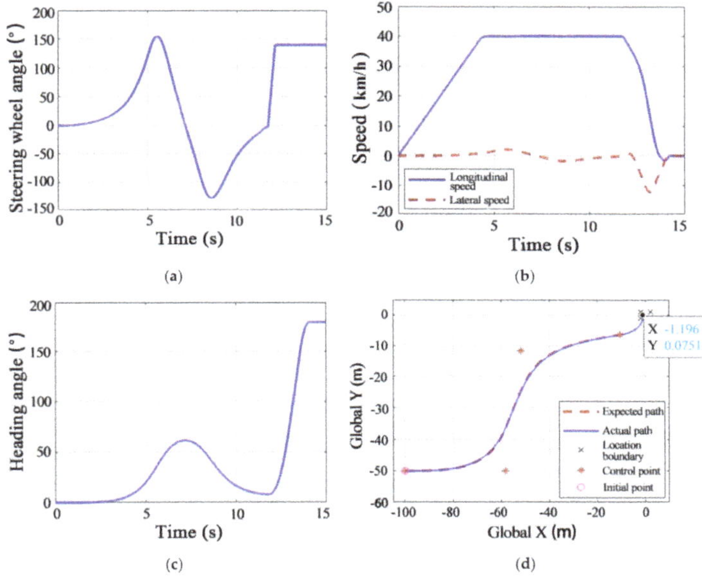

Figure 4. Simulation test results: (**a**) Steering wheel angle; (**b**) Vehicle speed; (**c**) Heading angle; and (**d**) Drift process.

Figure 5. Drift parking process.

5.2. Function Verification of Monitoring Strategy

First, the initial positions and heading angles of different vehicles are set to verify the effectiveness of the failure monitoring strategy of the path planner. The simulation results are shown in Figure 6. The coordinates of the target drift trigger point are (0, 0), and the heading angle of the drift trigger point is $0°$. The three flag bits correspond to the maximum curvature constraint, the maximum adhesion constraint, and the longitudinal speed constraint, respectively. The path planning and feasibility judgment were completed before the simulation test.

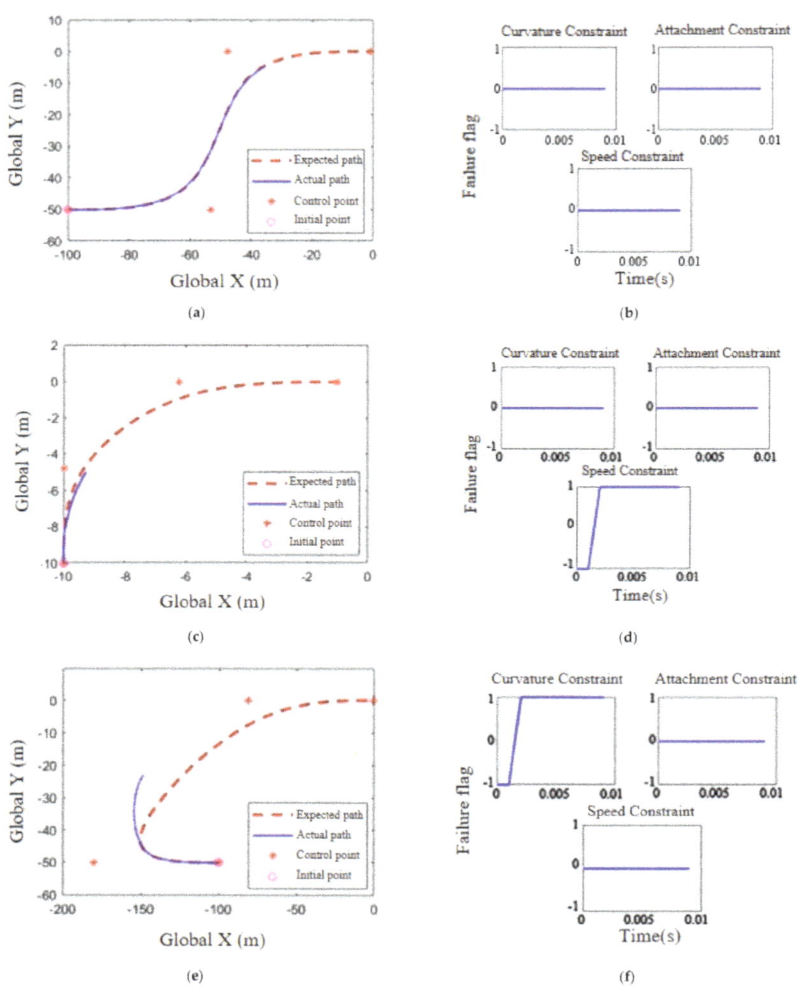

Figure 6. Simulation test of path planning failure monitoring strategy: (**a**) Vehicle path; (**b**) Failure flag; (**c**) Vehicle path; (**d**) Failure flag; (**e**) Vehicle path; and (**f**) Failure flag.

As shown in Figure 6a,b, the vehicle initial point coordinates are $(-100, -50)$, and the initial heading angle is $0°$. The simulation results show that the vehicle can complete the approach action, and the failure flag is 0.

As shown in Figure 6c,d, the vehicle initial point coordinates are $(-10, -10)$, and the initial heading angle is $90°$. the path is too short for the vehicle to accelerate to the desired drift trigger speed by the time it reaches the desired point, and thus does not meet the drift trigger conditions. The path does not meet the longitudinal speed constraint, and the corresponding flag bit is 1.

As shown in Figure 6e,f, the vehicle initial point coordinates are $(-100, -50)$, and the initial heading angle is $180°$. Due to the path's large curvature, the vehicle cannot track the path with the maximum steering wheel angle, which leads to path tracking failure. The path does not satisfy the maximum curvature constraint, and the corresponding flag bit is 1.

Next, the drift trajectory tracking and monitoring strategy and drift stop strategy are simulated and verified. Let the weight coefficients in Equation (38) be $\omega_X = 1$, $\omega_Y = 1$, $\omega_\psi = 2$. The target location is $(0, 0)$, and the heading angle of the target location is $180°$. The vehicle starting point coordinates are $(-100, -50)$, and the starting heading angle is $0°$. If the road adhesion coefficient is set to 0.5, the vehicle can complete the approaching movement on the road surface attached to the center, but it cannot drift into the warehouse according to the open-loop control law obtained from the tail-flick test when the adhesion coefficient is 1. The simulation results are shown in Figure 7. As can be seen from Figure 7b, at 13.84 s, the vehicle meets the drift trigger condition and enters the drift state. At 14.71 s, the controller detects that the vehicle deviates from the expected drift trajectory, and the drift failure flag is 1. At this time, the steering wheel angle returns to $0°$. The front axle is put under greater pressure and the pressure on the rear axle is reduced, as shown in Figure 7c. Figure 7a shows that the drift stop action makes the vehicle stop faster and greatly reduces the yaw motion. Compared with a vehicle completing the entire drift the final heading angle changed from $193°$ to $124°$, the total drift time decreased from 5.86 s to 3.56 s, and the rear wheel slip distance decreased from 28.87 m to 23.33 m. The simulation results verify the effectiveness of the strategy.

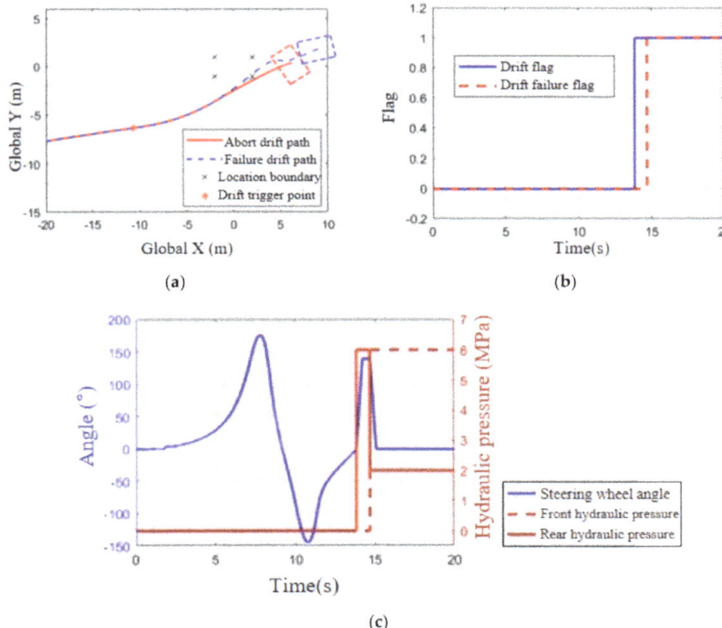

Figure 7. Simulation test of drift process failure monitoring strategy: (**a**) Vehicle path; (**b**) Drift flag; and (**c**) Actuator output.

5.3. Ground Test

The open-loop control drift algorithm was verified using the actual vehicle. First, a 170° steering wheel angle was applied to record the change in the vehicle motion state from the beginning to the end of the drift. The initial position of the vehicle is $(0, 0)$. At the end of the drift, the x-direction displacement changes by 12.27 m, the y-direction displacement changes by 11.28 m, and the heading angle changes by 75.4 degrees. In the real vehicle test, the vehicle conditions and road conditions must be consistent to achieve high-precision drift control. The data begin recording when the drift state is triggered. The change in vehicle motion state parameters across the entire drift process is shown in Figure 8.

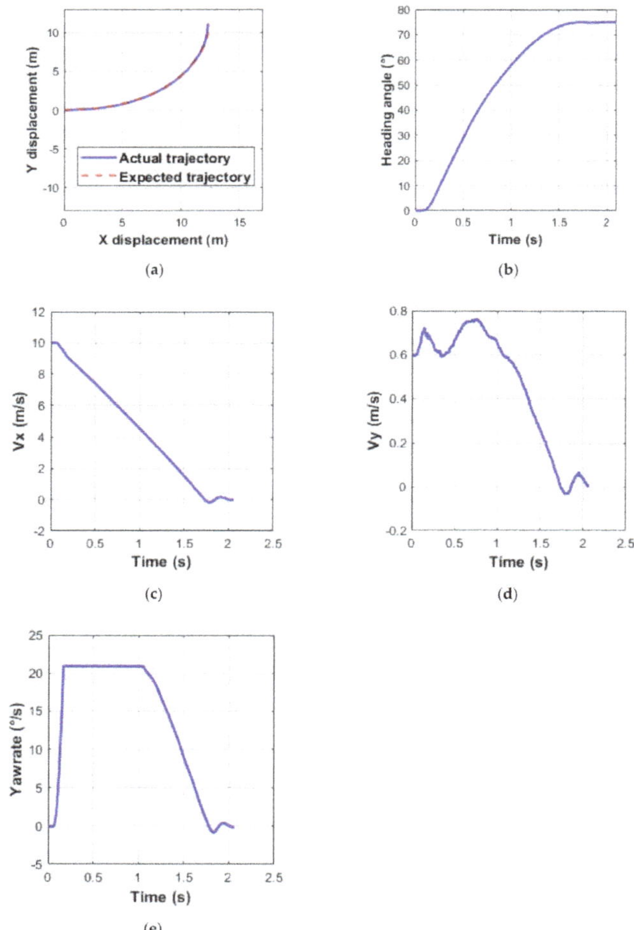

Figure 8. Ground test: (**a**) Vehicle path; (**b**) Heading angle; (**c**) Vx; (**d**) Vy; and (**e**) Yawrate.

The drift trigger point is set to $(0, 0)$, the distance threshold of the drift trigger point is set to 0.3 m, and the heading angle error threshold is set to 3 degrees. The vehicle meets the drift trigger condition and begins to drift. During the entire process of the vehicle drifting and entering the warehouse, the heading angle changes by 75.1°, the x-direction displacement changes by 12.32 m, and the y-direction displacement changes by 11.05 m. In contrast to the collected data, the heading angle deviation is 0.3°, the x-direction displacement deviation is 0.05 m, the y-direction displacement deviation is 0.23 m, and the vehicle completes the drift.

6. Discussion and Conclusions

The actual vehicle test is compared with the simulation experiment. In the simulation experiment, the distance between the drifting vehicle and the center of the parking location is 0.196 m. In the actual vehicle test, when the vehicle completes its drift, the distance between the vehicle and the center of the parking location is 0.235 m. This indicates a 3.9% accuracy difference between the two. The accuracy of the heading angle deviation is 1%. These differences stem from fluctuations in the drift trigger point and the state of the vehicle road system during the actual vehicle test.

In this paper, a segmented drift algorithm is designed to extend the handling ability beyond the limit of vehicle stability. By tracking the planned path, the vehicle can reach the drift trigger point and apply the open-loop control rate. In the simulation test, the vehicle drifts into the parking location from 0.196 m away, with a heading angle deviation of 0.2 degrees. In the ground test, the deviation between the final position of the vehicle and the center position of the parking location is 0.235 m, and the deviation of the heading angle is 0.3°. A strategy for monitoring the drift triggering condition, path planning, and vehicle state was designed. The simulation results show that the monitoring method can accurately monitor the real-time state of the vehicle and completion of the drift. The simulation and real vehicle test results show that the segmented drift control method can achieve high-precision drift parking.

The research of segmented drift control has the following significance:

(1) Through the combination of path planning, path tracking, and an open-loop control algorithm, it can realize the action of the driverless vehicle drifting into the warehouse, which demonstrates the potential of further research on driverless vehicle under extreme conditions.

(2) The segmented drift control strategy is designed to make the vehicle complete a drift during its approach of the warehouse. In order to ensure that there are no major changes to the vehicle road system, the open-loop control rate can effectively complete the drift.

(3) The realization of the whole drift process requires the initial state of the vehicle and the vehicle path system to be consistent with the acquisition path, which leads to the low success rate of drift parking. Constraints on the planned path and drift trigger state can significantly improve the success rate of drift storage. The monitoring strategy of the drift process can also ensure the integrity and safety of the test.

In subsequent research based on this paper, the tire inflation state should also be fully considered as part of the road system. The tire characteristics and road adhesion coefficient could be used as input for improving the robustness of the system. Future research could try to employ reinforcement learning methods in drift control experiments.

Author Contributions: Conceptualization, M.L., Y.Y. and B.L.; methodology, M.L. and Y.Y.; software, M.L.; validation, M.L., Y.Y. and B.L.; formal analysis, M.L.; investigation, M.L.; resources, M.L. and Y.Y.; data curation, M.L. and Y.Y.; writing—original draft preparation, M.L. and X.Y.; writing—review and editing, M.L. and X.Y.; visualization, M.L.; supervision, L.X. and B.L.; project administration, B.L.; funding acquisition, M.L. and B.L. All authors have read and agreed to the published version of the manuscript.

Funding: This research was funded by the National Natural Science Foundation of China (Grant No. 52002284).

Institutional Review Board Statement: Not applicable as our studies did not involve humans or animals.

Informed Consent Statement: Not applicable as our studies did not involve humans or animals.

Data Availability Statement: Detailed data are contained within the article. More data that support the findings of this study are available from the author M.L. upon reasonable request.

Acknowledgments: The authors thank wish to express their thanks for the assistance of the School of Automotive Studies, Tongji University and the Clean Energy Automotive Engineering Center, Tongji University.

Conflicts of Interest: The authors declare no conflict of interest. The funders had no role in the design of the study; in the collection, analyses, or interpretation of data; in the writing of the manuscript; or in the decision to publish the results.

References

1. Zhang, F. Trajectory Planning and Motion Control for Extreme Maneuvers of Autonomous Vehicles. Ph.D. Thesis, Tsinghua University, Beijing, China, 2018.
2. Goh, J.Y.; Goel, T.; Gerdes, J.C. A controller for automated drifting along complex trajectories. In Proceedings of the 14th International Symposium on Advanced Vehicle Control (AVEC 2018), Beijing, China, 14–18 September 2018; pp. 1–6.
3. Alcala, E.; Puig, V.; Quevedo, J.; Escobet, T.; Comasolivas, R. Autonomous vehicle control using a kinematic Lyapunov-based technique with LQR-LMI tuning. *Control Eng. Pract.* **2018**, *73*, 1–12. [CrossRef]
4. Tagne, G.; Talj, R.; Charara, A. Higher-order sliding mode control for lateral dynamics of autonomous vehicles, with experimental validation. In Proceedings of the IEEE Intelligent Vehicles Symposium, Gold Coast, QLD, Australia, 23–26 June 2013; pp. 678–683.
5. Hu, C.; Wang, R.; Yan, F. Integral Sliding Mode-Based Composite Nonlinear Feedback Control for Path Following of Four-Wheel Independently Actuated Autonomous Vehicles. *IEEE Trans. Transp. Electrif.* **2016**, *2*, 221–230. [CrossRef]
6. Funke, J.; Brown, M.; Erlien, S.M.; Gerdes, J.C. Collision Avoidance and Stabilization for Autonomous Vehicles in Emergency Scenarios. *IEEE Trans. Control Syst. Technol.* **2017**, *25*, 1204–1216. [CrossRef]
7. Liu, K.; Gong, J.; Kurt, A.; Chen, H.; Ozguner, U. Dynamic Modeling and Control of High-Speed Automated Vehicles for Lane Change Maneuver. *IEEE Trans. Intell. Veh.* **2018**, *3*, 329–339. [CrossRef]
8. Guo, J.; Luo, Y.; Li, K.; Dai, Y. Coordinated path-following and direct yaw-moment control of autonomous electric vehicles with sideslip angle estimation. *Mech. Syst. Signal Process.* **2018**, *105*, 183–199. [CrossRef]
9. Kim, E.; Kim, J.; Sunwoo, M. Model predictive control strategy for smooth path tracking of autonomous vehicles with steering actuator dynamics. *Int. J. Automot. Technol.* **2014**, *15*, 1155–1164. [CrossRef]
10. Velenis, E.; Katzourakis, D.; Frazzoli, E.; Tsiotras, P.; Happee, R. Steady-state drifting stabilization of RWD vehicles. *Control Eng. Pract.* **2011**, *19*, 1363–1376. [CrossRef]
11. Nakano, H.; Kinugawa, J.; Kosuge, K. Control of a four-wheel independently driven electric vehicle with a large sideslip angle. In Proceedings of the 2014 IEEE International Conference on Robotics and Biomimetics (ROBIO 2014), Bali, Indonesia, 5–10 December 2014; pp. 265–270.
12. Goh, J.Y.; Gerdes, J.C. Simultaneous stabilization and tracking of basic automobile drifting trajectories. In Proceedings of the IEEE Intelligent Vehicles Symposium, Gothenburg, Sweden, 19–22 June 2016; pp. 597–602.
13. Jelavic, E.; Gonzales, J.; Borrelli, F. Autonomous Drift Parking using a Switched Control Strategy with Onboard Sensors. *IFAC-PapersOnLine* **2017**, *50*, 3714–3719. [CrossRef]
14. Kolter, J.Z.; Plagemann, C.; Jackson, D.T.; Ng, A.; Thrun, S. A probabilistic approach to mixed open-loop and closed-loop control, with application to extreme autonomous driving. In Proceedings of the IEEE International Conference on Robotics Automation, Anchorage, AK, USA, 3–7 May 2010; pp. 839–845.
15. Kolter, J.Z. *Learning and Control with Inaccurate Models*; Stanford University: San Francisco, CA, USA, 2010.
16. Sakurama, K.; Nakano, K. Path tracking control of Lagrange systems with obstacle avoidance. *Int. J. Control Autom. Syst.* **2012**, *10*, 50–60. [CrossRef]
17. Rosolia, U.; De Bruyne, S.; Alleyne, A.G. Autonomous vehicle control: A nonconvex approach for obstacle avoidance. *IEEE Trans. Control Syst. Technol.* **2016**, *25*, 469–484. [CrossRef]
18. Gong, J.; Liu, K.; Qi, J. *Model Predictive Control for Self-Driving Vehicles*, 2nd ed.; Beijing Institute of Technology Press: Beijing, China, 2019; pp. 89–90.
19. Zhang, F.; Gonzales, J.; Li, S.; Borrelli, F.; Li, K. Drift control for cornering maneuver of autonomous vehicles. *Mechatronics* **2018**, *54*, 167–174. [CrossRef]
20. Gonzales, J.M. Planning and Control of Drift Maneuvers with the Berkeley Autonomous Race Car. Ph.D. Thesis, University of California, Berkeley/San Francisco, CA, USA, 2018.
21. Lau, T.K. Learning autonomous drift parking from one demonstration. In Proceedings of the 2011 IEEE International Conference on Robotics Biomimetics, Karon Beach, Thailand, 7–11 December 2011; pp. 1456–1461.

Article

Research on an Intelligent Driving Algorithm Based on the Double Super-Resolution Network

Taoyang Hang [1], Bo Li [2,*] , Qixian Zhao [3], Shaoyi Bei [2], Xiao Han [4], Dan Zhou [2] and Xinye Zhou [2]

1   College of Mechanical Engineering, Jiangsu University of Technology, Changzhou 213001, China; hty1360302750@163.com
2   College of Automobile and Traffic Engineering, Jiangsu University of Technology, Changzhou 213001, China; bsy1968@126.com (S.B.); zd13952018530@163.com (D.Z.); e2090391825@163.com (X.Z.)
3   Department of Computer Science and Engineering, University of Minnesota Twin Cities, Minneapolis, MN 55455, USA; zxq1360302750@163.com
4   Suzhou Automotive Research Institute, Tsinghua University, Beijing 215000, China; hanx20@mails.tsinghua.edu.cn
*   Correspondence: jslgfly@jsut.edu.cn

Abstract: Semantic segmentation plays a very important role in image processing, and has been widely used in intelligent driving, medicine, and other fields. With the development of semantic segmentation, the model has become more and more complex and the resolution of training pictures is higher and higher, so the requirements for required hardware facilities have become higher and higher. Many high-precision networks are difficult to apply in intelligent driving vehicles with limited hardware conditions, and will bring delay to recognition, which is not allowed in practical application. Based on the Dual Super-Resolution Learning (DSRL) network, this paper proposes a network model for training high-resolution pictures, adding a high-resolution convolution module which improves segmentation accuracy and speed while reducing computation. In a CamVid dataset, taking the road category as an example, IOU is 95.23%, which is 4% higher than DSRL, the real-time segmentation time of the same video is reduced by 46% from 120 s to 65 s, and the segmentation effect is better and faster, which greatly alleviates the recognition delay caused by high-resolution input.

Keywords: semantic segmentation; high-resolution atlas training; super-resolution

Citation: Hang, T.; Li, B.; Zhao, Q.; Bei, S.; Han, X.; Zhou, D.; Zhou, X. Research on an Intelligent Driving Algorithm Based on the Double Super-Resolution Network. *Actuators* **2022**, *11*, 69. https://doi.org/10.3390/act11030069

Academic Editors: Peng Hang, Xin Xia and Xinbo Chen

Received: 16 December 2021
Accepted: 18 February 2022
Published: 23 February 2022

Publisher's Note: MDPI stays neutral with regard to jurisdictional claims in published maps and institutional affiliations.

1. Introduction

Semantic segmentation is a basic computer vision task. Its purpose is to classify each pixel in the picture. It is widely used in the fields of intelligent driving, medical imaging, and pose analysis. According to research [1], when traditional cars are replaced by private autonomous vehicles, the number of cars owned by each family can be reduced, the maintenance cost will be less than traditional cars, and the mileage of family vehicles will increase by 57%. According to a survey, consumers are willing to pay the premium related to the purchase of vehicles equipped with automatic equipment. Research [2] shows that cumulative energy and greenhouse gas can be reduced by 60% in the basic case after a series of strategic deployments, and can be further reduced by 87% through accelerated grid decarburization, dynamic performance sharing, vehicle life extension, the improved efficiency of computer systems, the improved fuel efficiency of new vehicles, etc. Therefore, intelligent driving vehicles will be widely used. However, in the field of intelligent driving, semantic segmentation needs to maintain real-time detection while maintaining high accuracy. However, in an application with limited hardware facilities, a high-precision network cannot be put into use, and the recognition delay is also very large. The following are some classic networks for semantic segmentation: UNet [3], Deeplabs [4–6], PSPNet [7], SegNet [8], etc. These semantic segmentation networks usually need to use high-resolution atlas training to achieve high accuracy. High-resolution pictures can effectively transfer the features in pictures and facilitate network learning. Therefore, high-resolution features are very important in high-precision networks. At present, there are two main ways to maintain high-resolution performance. One is to use void convolution to maintain high-resolution features, and the other is to combine top-down paths and horizontal connections, such as with UNet. Both methods can effectively prevent feature disappearance due to too much convolution, but these methods themselves consume very many computing resources. On this basis, taking high-resolution images as input will further increase the amount of network computing and image segmentation time. In order to reduce the cost of automatic driving, some studies [9] have improved the hardware by using a fisheye camera instead of a vision and LiDAR odometer system. In recent years, the compressed network used in devices with limited hardware resources has attracted people's attention, but there is still a certain gap between the prediction accuracy of the current network and the network model trained by high-resolution atlas. In order to reduce the gap between the two networks above, some compressed networks also choose high-resolution pictures as input (for example, 1024×2048 or 512×1024). In order to reduce the burden on the network when high-resolution pictures are used as input, ESPNets [10,11] have been proposed to accelerate convolution calculation by using split merge or reducing the expand principle. Others use efficient classification networks (such as MobileNet [12] and ShuffleNet [13]) or some compression technologies (such as pruning [14] and vector quantization) to accelerate segmentation, but the effect is not ideal. The existing convolution kernel has two main disadvantages: one is that the receptive field is small and difficult to capture in long-distance dependence; the other is that the information between channels is redundant. On this basis, D Li [15] et al. proposed involution; that is, the convolution kernel is multiplexed in space and independent in the channel, which can be used to accelerate the speed of convolution. Li Wang [16] et al. proposed a dual super-resolution learning network (DSRL): a compressed network for high-resolution atlas training that has a certain improvement compared with the previous methods, but the DSRL network is still poor at detecting the details of objects. Therefore, in this paper, a new network framework is designed based on DSRL to alleviate this problem. More specifically, the network in this paper consists of two parts: one part is the super-resolution network, and the other is the high-resolution picture convolution network. The internal convolution is used to replace the partial convolution, which not only reduces the network parameters, but also improves the segmentation accuracy.

2. Materials and Methods

2.1. Dual Super-Resolution Learning

The Dual Super-Resolution Learning (DSRL) network is a dual super-resolution learning network based on image super-resolution in order to maintain a high-resolution display. The DSRL network aims to reconstruct high-resolution images with low-resolution input. The network model has two main modules: one is Semantic Segmentation Super-Resolution (SSSR) and the other is Single Image Super-Resolution (SISR). In addition, there is a Feature Affinity (FA) module. SSSR integrates the idea of super-resolution into the existing semantic segmentation, and the fine-grained structure based on the FA module further enhances the high-resolution features of SSSR streams. In addition, the two streams share the same feature extractor and optimize SISR branches during training.

The structure of DSRL is shown in Figure 1. The decoding module of DSRL consists of two parts. One is the SSSR module and the other is SISR, which shares the same feature extraction module. SSSR is the process of generating the final segmentation result only through upsampling; SISR is the process of image recovery from low resolution to high resolution.

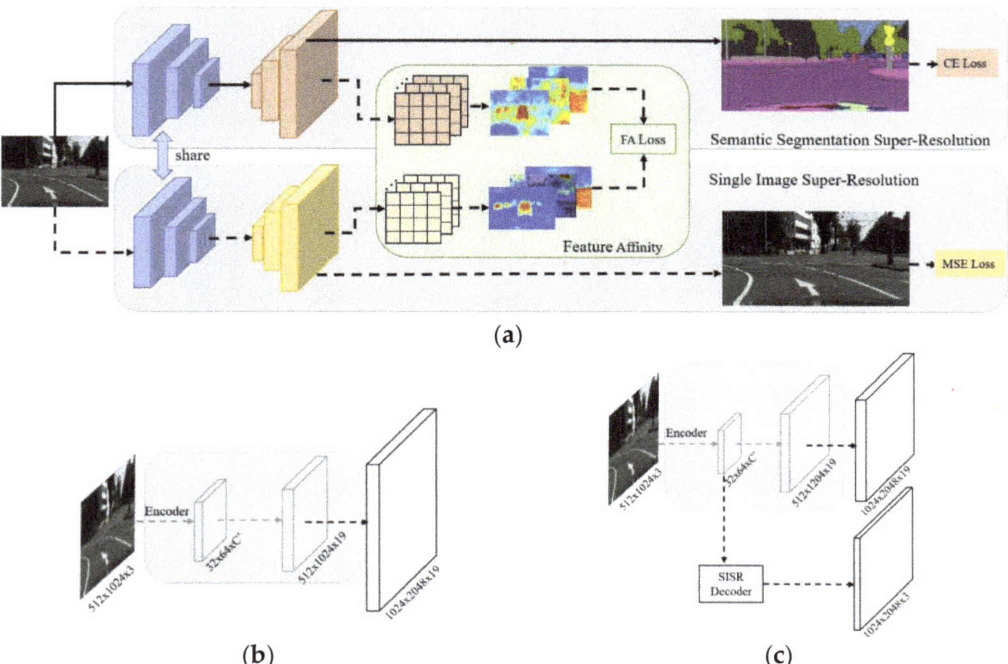

(a)

(b)   **(c)**

Figure 1. Dual Super-Resolution Learning (DSRL) network structure: (**a**) DSRL network structure; (**b**) Semantic Segmentation Super-Resolution (SSSR) realizes image segmentation only by upsampling; (**c**) SSSR + Single Image Super-Resolution (SISR) restore from low-resolution feature layer to high resolution of original image.

2.2. You Only Look One-Level Feature

The Feature Pyramid Network [17] (FPN) is a basic component in the recognition system used to detect objects with different scales. The FPN framework is shown in Figure 2. The main core benefits of FPN are two: on the one hand, FPN can fuse multi-scale feature maps to obtain better representation; on the other hand, it is a divide-and-conquer strategy, which detects targets on different levels of feature maps according to different scales of targets. Qian Chen [18] et al. proposed You Only Look One-level Feature. This paper studies the influence of two gain fittings of FPN on a single-stage detector. In this paper, FPN is regarded as a Multiple-in-Multiple-out (MiMo) encoder. Four types of encoders are studied: Multiple-in-Multiple-out (MiMo), Multiple-in-Single-out (MiSo), Single-in-Multiple-out (SiMo), and Single-in-Single-out (SiSo). It is found that the SiMo encoder has only one input feature, and the C5 feature layer can achieve the same performance as the MiMo encoder without feature fusion. The results are shown in Figure 3. These phenomena illustrate two facts:

(1) C5 feature provides sufficient semantic information for object detection at different scales, which enables the SiMo encoder to achieve the same results as the MiMo encoder;

(2) The benefit of multi-scale feature fusion is far less important than the divide-and-conquer strategy, so multi-scale feature fusion may not be the most significant benefit of FPN.

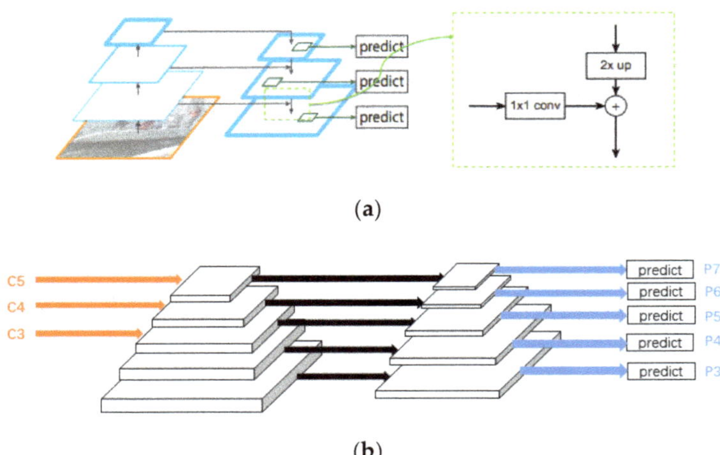

(**a**)

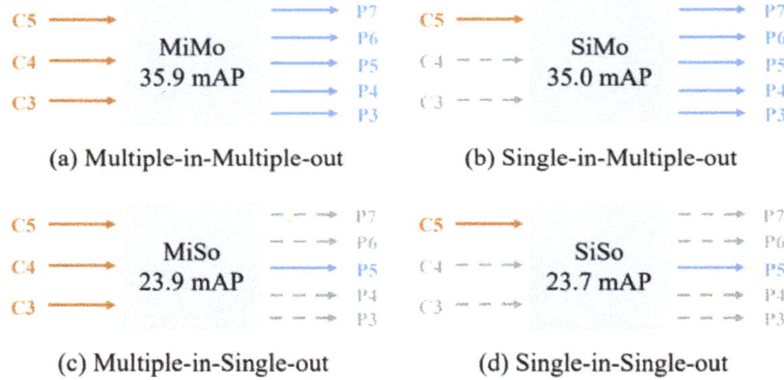

(**b**)

Figure 2. Feature Pyramid Networks (FPN) network structure: (**a**) FPN overall network structure; (**b**) The last three layers of the feature extraction module are C3~C5, respectively, and the prediction modules are P3~P7, respectively.

<table>
<tr><td>

C5 →
C4 → MiMo 35.9 mAP
C3 →
→ P7
→ P6
→ P5
→ P4
→ P3

(a) Multiple-in-Multiple-out

</td><td>

C5 →
C4 - - → SiMo 35.0 mAP
C3 - - →
→ P7
→ P6
→ P5
→ P4
→ P3

(b) Single-in-Multiple-out

</td></tr>
<tr><td>

C5 →
C4 → MiSo 23.9 mAP
C3 →
- - → P7
- - → P6
→ P5
- - → P4
- - → P3

(c) Multiple-in-Single-out

</td><td>

C5 →
C4 - - → SiSo 23.7 mAP
C3 - - →
- - → P7
- - → P6
→ P5
- - → P4
- - → P3

(d) Single-in-Single-out

</td></tr>
</table>

Figure 3. Results of four input and output combinations of FPN. Using C3~C5 level feature layers of the backbone and the feature layers of P3~P7 as the final output, compare the mAP (mean Average Precision) indicators of the four decoders: (**a**) MiMo; (**b**) SiMo; (**c**) MiSo; (**d**) SiSo.

2.3. Involution

Ordinary convolution has the following two characteristics: the spatial invariance of convolution, and channel specificity. It also has two defects: one is that the receptive field is small and difficult to capture in long-distance dependence, and the other is the redundancy of information between channels. On this basis, D Li et al. proposed the concept of involution. The involution is structurally opposed to ordinary convolution. The convolution kernel is shared in the channel dimension, and the special convolution kernel in the spatial dimension can make the modeling more flexible. The structure of involution is shown in Figure 4.

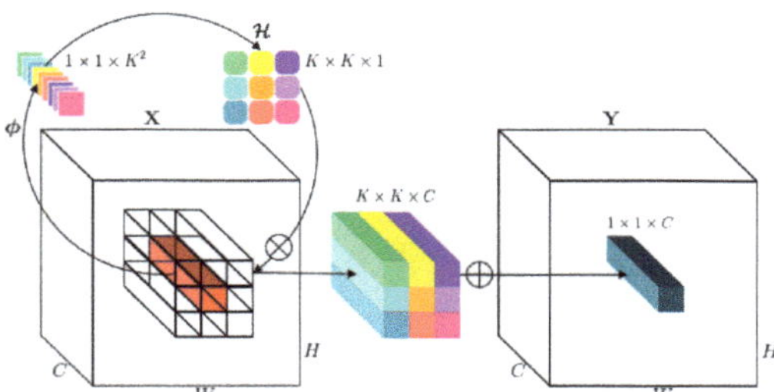

Figure 4. Involution structure (the involution kernel $\mathcal{H}_{i,j} \in \mathbb{R}^{K \times K \times 1}$ ($G = 1$ in this example for ease of demonstration) is yielded from the function ϕ conditioned on a single pixel at (i, j), followed by a channel-to-space rearrangement. The multiply–add operation of involution is decomposed into two steps, with $\otimes$ indicating multiplication broadcast across C channels and $\oplus$ indicating summation aggregated within the $K \times K$ spatial neighborhood).

The convolution kernel size of involution is $H \times W \times K \times K \times G$, among $G << C$. This means that all channels share convolution kernels. In the involution, the fixed weight matrix is not used as in the ordinary convolution, but the corresponding involution kernel is generated according to the characteristic graph. Spatial specificity makes the convolution kernel have the ability to capture multiple feature representations at different spatial locations, and improves the problem of long-distance pixel dependence. The channel invariance performance reduces the redundant information between channels to a certain extent and improves the computing efficiency of the network. In essence, this design from ordinary convolution to internal convolution redistributes the computing power at the top level, and the essence of network design is the distribution of computing power, in order to adjust the limited computing power to the position where it can give full play to its performance. This involution module is easy to implement and can be easily combined with various network models. It can easily replace conventional convolution to realize an excellent backbone network structure.

2.4. Network Structure

In the network model of Dual Super-Resolution Learning (DSRL), in order to reduce the impact of high-resolution pictures as input on the increase of network computing, firstly, sub-sampling the high-resolution image of 960×720 to 480×360, and the picture size becomes half of the original. For the low-resolution feature layer, simple upsampling is carried out through Semantic Segmentation Super-Resolution (SSSR) and Single Image Super-Resolution (SISR) to restore to the original image size. This article compares the color pictures of the original size, 1/2 downsampling, and 1/2 downsampling + 2x upsampling; the pictures are not visually different, and we use the operator of [-1 -1 -1; -1 8 -1; -1 -1 -1] to extract the edges of the above three graphs. It can be found that the edge features extracted from the original image have more noise, but the image details are also well preserved. The edge feature noise extracted after 1/2 downsampling is reduced, but the details of the object also become rough; the edge feature noise and object details extracted after 1/2 downsampling + 2x upsampling are greatly reduced. In the following experiment, parts of these three images are used as input and the segmentation effects are compared. The experimental results show that although downsampling will reduce the noise, the missing details are more important, and the amount of noise has little effect on accuracy. Images and their respective extracted edge features as shown in Figure 5.

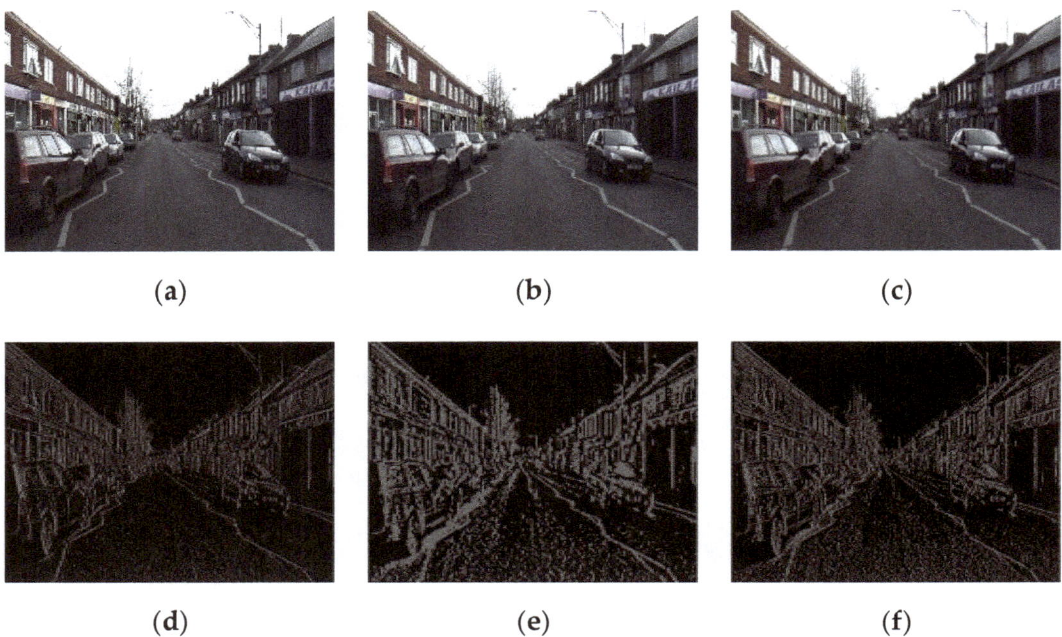

Figure 5. Picture features: (**a**) Original RGB picture; (**b**) 1/2 downsampling RGB picture; (**c**) 1/2 downsampling + 2x upsampling RGB picture; (**d**) Original RGB picture's edge features; (**e**) 1/2 downsampling RGB picture's edge features; (**f**) 1/2 downsampling + 2x upsampling RGB picture's edge features.

Therefore, this paper proposes a new network model based on the Dual Super-Resolution Learning (DSRL) network model to improve the above problems. The network is divided into two modules. One is the low-resolution image convolution module based on the super-resolution theory; the other is the convolution module of high-resolution pictures. In this paper, only the C5-level feature layer is extracted with reference to You Only Look One-level Feature (YOLOF). The C5-level feature layer has sufficient semantic information, so the low-resolution convolution module does not carry out feature fusion, expands the receptive field range through expansion convolution, and then recovers to high resolution through upsampling. However, since the image is downsampled twice at the beginning, resulting in the loss of features of the original image, a convolution module of the high-resolution image is added to the network to make up for the loss of features caused by the reduction of resolution. In order to avoid the proliferation of network parameters caused by the convolution of high-resolution images, this module only performs a small amount of convolution, and partial convolution is replaced by internal convolution to reduce the amount of calculation. The network structure is shown in Figure 6, maintaining two branches during training and two branches during testing. Pruning occurred during testing to remove Mean Square Error (MSE) loss branches and to reduce the amount of calculation.

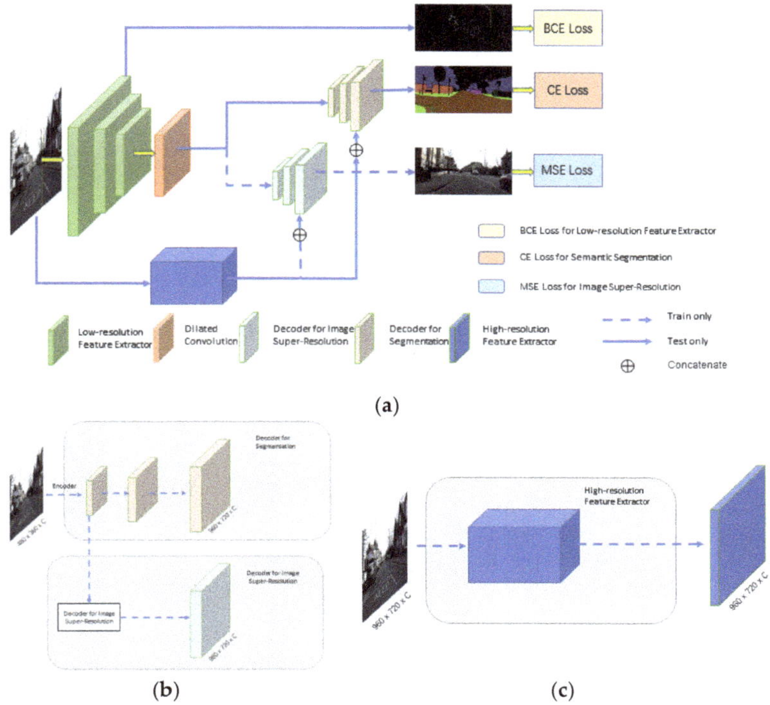

Figure 6. Network structure: (**a**) MY network structure; (**b**) Low-resolution convolution module; (**c**) High-resolution module convolution module.

2.5. Loss Function

The network loss function consists of three parts: one is the cross-entropy loss function composed of the network output and the actual segmentation graph, and the other is the binary-cross-entropy loss function composed of the network low-dimensional feature layer and the feature graph sampled under the actual segmentation graph to the corresponding size. The last part consists of the Mean Square Error (MSE) between the network output and the actual picture. The real segmentation's edge features are shown in Figure 7 (edge extraction from ground truth). The Cross-Entropy (CE) loss function is shown in Formula (1). y_i and p_i refer to the segmentation predicted probability and the corresponding category for pixel i. The Binary Cross-Entropy (BCE) loss function is shown in Formula (2). y_i and x_i refer to the target value and the value of model output. The Mean Square Error is shown in Formula (3). x_i and y_i refer to the target value and the value of model output. The whole loss function is shown in Formula (4). w_1 and w_2 are set as 0.2 and 0.4.

$$L_{CE} = \frac{1}{N} \sum_{i=1}^{N} -y_i \log(p_i) \tag{1}$$

$$L_{BCE} = -\frac{1}{N} \sum_{i=1}^{N} [y_i \log x_i + (1 - y_i) \log(1 - x_i)] \tag{2}$$

$$L_{MSE} = \frac{1}{N} \sum_{i=1}^{N} \|x_i - y_i\| \tag{3}$$

$$L = w_1 L_{MSE} + w_2 L_{BCE} + L_{CE} \tag{4}$$

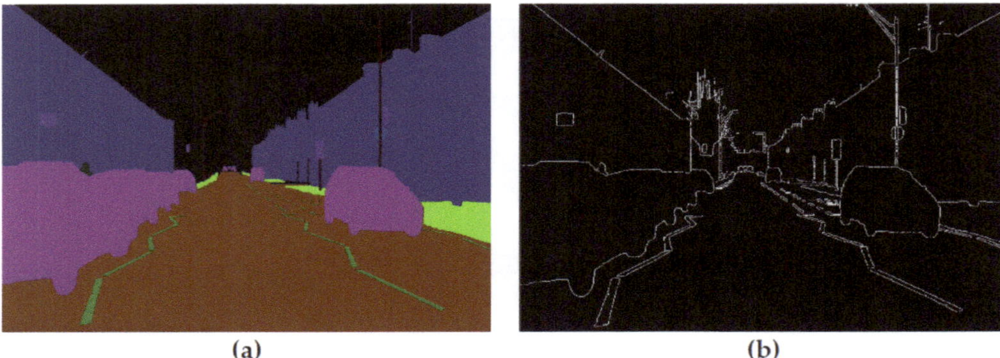
(a)   (b)

Figure 7. (**a**) Ground truth segmentation; (**b**) Edge features of real segmentation.

3. Results

3.1. Construction of Dataset

In this paper, a CamVid (Cambridge-driving Labeled Video Database) dataset was selected, which was composed of 960×720 high-resolution pictures intercepted by videos taken during the real driving process of vehicles. It was divided into 32 categories, such as bicycles, roads, cars, and so on. This paper divided the training set, verification set, and test set according to the proportion of 7:2:1. In order to enhance the generalization ability of the model, data enhancement methods such as flipping and clipping were used for the training set data.

3.2. Network Model Evaluation Index

Assuming that there are k classes (including $k-1$ target classes and one background class), $k-1$ represents the total number of pixels belonging to the i class predicted as j class, and specifically, p_{ii} represents TP (true positive); p_{ij} indicates FP (false positive); and p_{ji} indicates FN (false negatives). The evaluation indicators included the following categories:

(1) PA (Pixel Accuracy): The ratio between the number of pixels correctly classified and all pixel points is shown in Formula (5).

$$\text{PA} = \frac{\sum_{i=0}^{k} p_{ii}}{\sum_{i=0}^{k} \sum_{j=0}^{k} p_{ij}} \tag{5}$$

The larger the value of the evaluation index, the more accurate the predicted pixel classification is.

(2) MPA (Mean Pixel Accuracy) calculated the average value based on the proportion of correctly classified pixel points to all pixel points, and the formula is shown in (6).

$$\text{MPA} = \frac{1}{k+1} \frac{\sum_{i=0}^{k} p_{ii}}{\sum_{i=0}^{k} \sum_{j=0}^{k} p_{ij}} \tag{6}$$

(3) MIOU (Mean Intersection over Union): The ratio between the intersection between the real value and the predicted value and the union between the real value and the predicted value is averaged, and the formula is shown in (7).

$$\text{MIOU} = \frac{1}{k+1} \sum_{i=0}^{k} \frac{p_{ii}}{\sum_{j=0}^{k} p_{ij} + \sum_{j=0}^{k} p_{ji} - p_{ii}} \tag{7}$$

(4) DICE: The ratio of the intersection of 2 times the predicted result and the real result to the predicted result plus the real result is shown in Formula (8), where X represents the real value, Y represents the predicted value.

$$\text{DICE} = \frac{2|X \cap Y|}{|X| + |Y|} \tag{8}$$

The larger the value of the evaluation index, the more accurate the predicted pixel classification is.

3.3. Analysis of Training Results

The framework of the neural network built in this paper was PyTorch. The model of the graphics card used was RTX2060 8G. The size of DSRL and MY network parameters in this paper are shown in Table 1.

Table 1. Network model parameters.

Model	Estimated Total Size	Params Size
DSRL	8091.60 (MB)	231.03 (MB)
MY	5438.59 (MB)	40.88 (MB)

We compared the road classes with the largest proportion in the CamVid dataset, and the results are shown in Tables 2 and 3.

Table 2. Input of high-resolution network is original image.

Evaluating Indicator	DSRL	MY
IOU	91.17%	95.23%
PA	94.42%	98.99%
DICE	56.59%	60.49%

Table 3. Input of high-resolution network is original image and 1/2 downsampling + 2x upsampling.

Evaluating Indicator	DSRL	MY
IOU	95.23%	92.25%
PA	98.99%	97.86%
DICE	60.49%	60.25%

The experimental results show that the total network parameters in this paper were reduced from 8091 MB to 5438 MB. Compared with the DSRL network, the network structure in this paper improved the values of IOU, PA, and DICE:

(1) The IOU value increased from 91.17% to 95.23%;
(2) PA value increased from 94.42% to 98.99%;
(3) DICE increased from 56.59% to 60.49%.

The road segmentation diagram is shown in Figure 8 (the red part is the result of road segmentation by the network, and the gray part is the standard value). The segmentation results of the DSRL network were not good for the segmentation of small objects similar to small lane lines. However, after adding the high-resolution image convolution module in this paper, the segmentation effect of small objects was improved, which shows that the high-resolution convolution module added in this model can effectively make up for the loss of the input image due to 1/2 downsampling. Although the noise will be reduced after downsampling, the priority is not as good as it is for the object details.

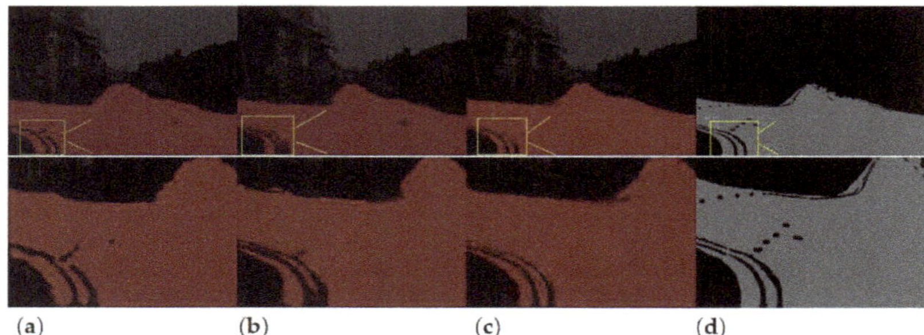

(a)　　　　　　(b)　　　　　　(c)　　　　　　(d)

Figure 8. Road segmentation picture: (**a**) DSRL (**b**) MY (1/2 downsampling + 2x upsampling) (**c**) MY (original picture) (**d**) Ground Truth.

VGG16, ResNet101, ResNet50, and CSPdarkNet53 were used as backbone networks to compare the total network parameters, parameter size, and PA, IOU, and DICE. The results are shown in Tables 4 and 5.

Table 4. Network evaluation parameters and parameter sizes of various backbone networks.

Backbone	Estimated Total Size	Params Size
VGG16	3751.10 (MB)	76.87 (MB)
ResNet50	6948.62 (MB)	113.41 (MB)
ResNet101	6517.05 (MB)	185.86 (MB)
CSPDarkNet53	5438.59 (MB)	40.88 (MB)

Table 5. Comparison of evaluation indexes of various backbone networks.

Backbone	IOU	PA	DICE
VGG16	94.38%	96.51%	60.21%
ResNet50	92.25%	97.65%	60.25%
ResNet101	91.44%	96.55%	60.22%
CSPDarkNet53	95.23%	96.55%	60.49%

It can be seen from Tables 4 and 5 that the network model with VGG16 as the backbone network could reach IOU, PA, and DICE similarly to the network model with ResNet50 and ResNet101 as the backbone network with less parameters. Taking the original image as the high-resolution network input, the comparison of various backbone network segmentation images is shown in Figure 9 (the red part is the result of the segmentation of the road class by the network).

As can be seen from various backbone network segmentation pictures in Figure 9 (the red part is the result of the segmentation of the road class by the network):

(1)　The network with VGG16 as the backbone can be achieved with half as few parameters than ResNet50 and ResNet101 with a similar effect. In terms of the segmentation accuracy of the lane line part of the road, the accuracy of VGG16 and ResNet50 is similar. Both lane lines can be clearly segmented, which is better than ResNet101. In terms of the segmentation accuracy of the tire shape at the bottom of the car, the segmentation accuracy of VGG16 is slightly better than ResNet50 and ResNet101, which can better fit the tire shape.

(2)　The tire shape segmentation accuracy of the network with CSPdarkNet53 as the backbone is better than VGG16, ResNet50, and ResNet101 on the lane line and the bottom of the vehicle, and fits better with the lane line and tire shape.

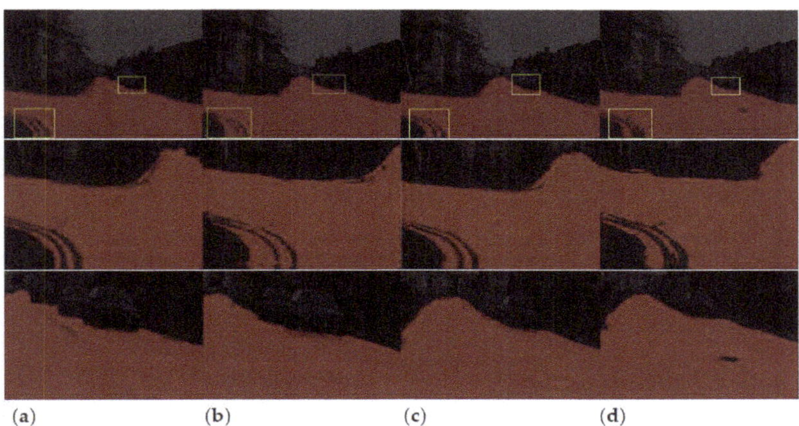

Figure 9. Comparison of various backbone network segmentation pictures: (**a**) CSPDarknet53; (**b**) VGG16; (**c**) ResNet50; (**d**) ResNet101.

Comparing ordinary convolution, ResNet, and CSPdarknet (the above three convolution structures are shown in Figure 10), it can be found that CSPdarknet cuts the input feature map to the channel, and only uses half of the original feature map to input into the residual network for processing. In forward propagation, the other half is directly spliced by the channel with the output of the residual network at the end. The advantages of doing this are as follows:

(1) Only half of the input is involved in the calculation, which can greatly reduce the amount of calculation and memory consumption;

(2) In the process of back propagation, a completely independent gradient propagation path is added, which can prevent feature loss caused by excessive convolution, and there is no reuse of gradient information.

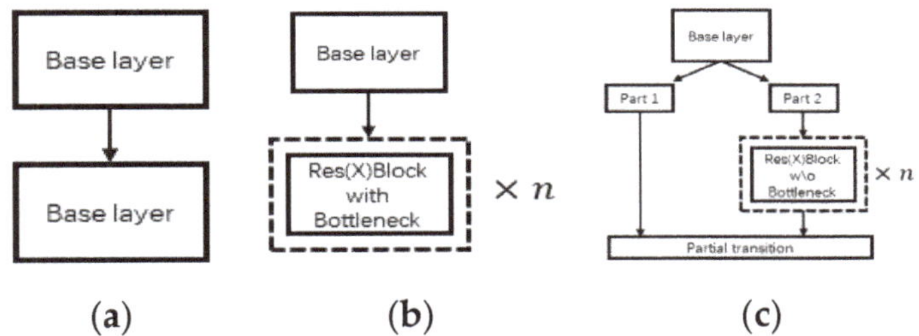

Figure 10. Convolutional structure: (**a**) Ordinary convolution; (**b**) ResNet; (**c**) CSPdarknet.

Take a video shot while driving using a single RTX2060 8G graphics card as an example: the video FPS is 25 frames, and the video resolution is 1920 × 1080, for a total of 12 s. The DSRL network takes 120 s; our network takes 65 s, a 46% reduction in time. The comparison of the segmentation results between the DSRL network and our network (the red part is the actual segmentation result) is shown in Figure 11:

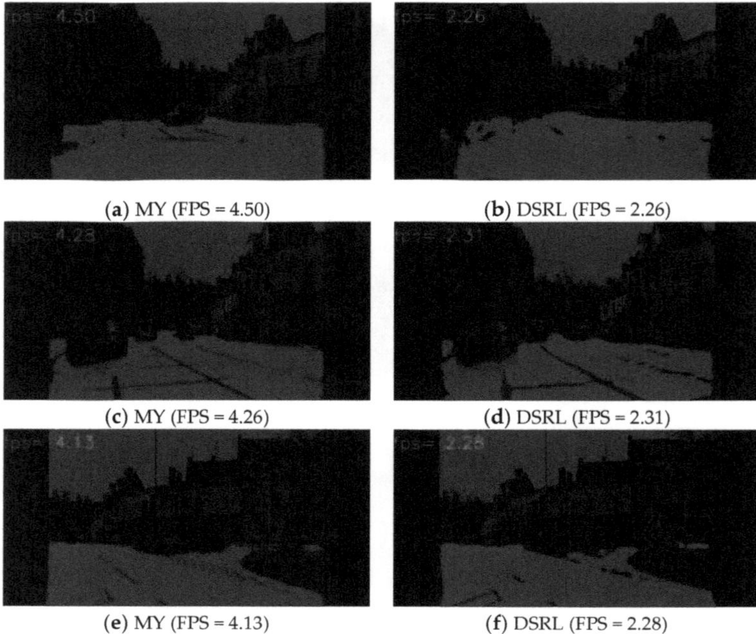

(**a**) MY (FPS = 4.50)   (**b**) DSRL (FPS = 2.26)

(**c**) MY (FPS = 4.26)   (**d**) DSRL (FPS = 2.31)

(**e**) MY (FPS = 4.13)   (**f**) DSRL (FPS = 2.28)

Figure 11. Comparison of the segmentation results of the DSRL network and our own network (the red part is the actual segmentation result). (**a**,**c**,**e**) are the segmentation results of our network on the video; (**b**,**d**,**f**) are the segmentation results of DSRL network at the same time point of the same video.

It can be seen from the above two sets of comparison charts that the fps of the DSRL network can only reach about 2 frames (up to 2.31 frames) in the actual driving video, whereas our network can achieve about 4 frames (up to 4.5 frames). The segmentation is smoother. From the above pictures, we can see that our network segmentation is faster and more accurate, and the segmentation effect is better for detailed parts such as lane lines.

Taking a single image with a resolution of 960 × 720 as input, a speed comparison between DSRL and our network segmentation is shown in Table 6. From the comparison in Table 6, we can see that the time used by our network is reduced compared with the DSRL network.

Table 6. The speed comparison between DSRL and our network segmentation.

Network	DSRL	MY
Picture1	1.36(s)	1.11(s)
Picture2	2.05(s)	1.70(s)
Picture3	2.25(s)	1.71(s)
Picture4	2.50(s)	1.72(s)
Picture5	2.16(s)	1.73(s)

4. Conclusions

In view of the high demand for hardware equipment for training and using high-resolution atlases, this paper proposes a new network model based on Dual Super-Resolution Learning (DSRL), an added high-resolution convolution module, and a discarded Feature Pyramid Network (FPN), which can effectively compensate for the downsampling of high-resolution images while reducing the amount of computation. Features are missing, and the study found that downsampling reduces noise as a lower priority than details in the picture. Our network model can segment small features better than the DSRL network, and has lower hardware requirements and faster processing speed. In terms of the actual driving video segmentation time, time is reduced by 46%, from 120 s to 65 s, which can be used in actual driving. The recognition is smoother and more accurate during driving, which greatly reduces the delay caused by high-resolution input during actual driving, thus proving the effectiveness of our method. However, the delay still exists, the detailed segmentation of objects is still lacking, and the network structure can continue being improved.

Author Contributions: Conceptualization, T.H.; methodology, T.H.; software, T.H.; validation, T.H.; formal analysis, B.L., Q.Z. and S.B.; investigation, T.H.; resources, T.H., B.L. and S.B.; data curation, T.H.; writing—original draft preparation, T.H., D.Z. and X.Z.; writing—review and editing, T.H. and B.L.; visualization, T.H.; supervision, X.H., B.L. and S.B.; project administration, B.L. and S.B.; funding acquisition, B.L. and S.B. All authors have read and agreed to the published version of the manuscript.

Funding: This research was funded by The Natural Science Foundation of the Jiangsu Higher Education of China under grant number 21KJA580001, and The National Natural Science Foundation of China under grant number 5217120589. The APC was funded by 5217120589.

Institutional Review Board Statement: Not applicable.

Informed Consent Statement: Not applicable.

Data Availability Statement: Not applicable.

Conflicts of Interest: The authors declare no conflict of interest.

References

1. Saleh, M.; Hatzopoulou, M. Greenhouse gas emissions attributed to empty kilometers in automated vehicles. *Transp. Res. Part D Transp. Environ.* **2020**, *88*, 102567. [CrossRef]
2. Gawron, J.H.; Keoleian, G.A.; De Kleine, R.D.; Wallington, T.J.; Kim, H.C. Deep decarbonization from electrified autonomous taxi fleets: Life cycle assessment and case study in Austin, TX. *Transp. Res. Part D Transp. Environ.* **2019**, *73*, 130–141. [CrossRef]
3. Ronneberger, O.; Fischer, P.; Brox, T. U-net: Convolutional networks for biomedical image segmentation. In Proceedings of the International Conference on Medical Image Computing and Computer-Assisted Intervention, Munich, Germany, 5–9 October 2015; Springer: Cham, Switzerland, 2015; pp. 234–241.
4. Chen, L.C.; Papandreou, G.; Kokkinos, I.; Murphy, K.; Yuille, A.L. Semantic image segmentation with deep convolutional nets and fully connected crfs. *arXiv* **2014**, arXiv:1412.7062.
5. Chen, L.C.; Papandreou, G.; Kokkinos, I.; Murphy, K.; Yuille, A.L. Deeplab: Semantic image segmentation with deep convolutional nets, atrous convolution, and fully connected crfs. *IEEE Trans. Pattern Anal. Mach. Intell.* **2017**, *40*, 834–848. [CrossRef] [PubMed]
6. Chen, L.C.; Zhu, Y.; Papandreou, G.; Adam, H.; Schroff, F. Encoder-decoder with atrous separable convolution for semantic image segmentation. In Proceedings of the European Conference on Computer Vision (ECCV), Munich, Germany, 8–14 September 2018; pp. 801–818.
7. Zhao, H.; Shi, J.; Qi, X.; Wang, X.; Jia, J. Pyramid scene parsing network. In Proceedings of the IEEE Conference on Computer Vision and Pattern Recognition, Honolulu, HI, USA, 21–26 June 2017; pp. 2881–2890.
8. Badrinarayanan, V.; Kendall, A.; Cipolla, R. Segnet: A deep convolutional encoder-decoder architecture for image segmentation. *IEEE Trans. Pattern Anal. Mach. Intell.* **2017**, *39*, 2481–2495. [CrossRef] [PubMed]
9. Yu, J.; Yu, Z. Mono-Vision Based Lateral Localization System of Low-Cost Autonomous Vehicles Using Deep Learning Curb Detection. *Actuators* **2021**, *10*, 57. [CrossRef]
10. Mehta, S.; Rastegari, M.; Caspi, A.; Shapiro, L.; Hajishirzi, H. Espnet: Efficient spatial pyramid of dilated convolutions for semantic segmentation. In Proceedings of the European Conference on Computer Vision (ECCV), Munich, Germany, 8–14 September 2018; pp. 552–568.
11. Mehta, S.; Rastegari, M.; Shapiro, L.; Hajishirzi, H. Espnetv2: A light-weight, power efficient, and general purpose convolutional neural network. In Proceedings of the IEEE/CVF Conference on Computer Vision and Pattern Recognition, Long Beach, CA, USA, 15–20 June 2019; pp. 9190–9200.
12. Sandler, M.; Howard, A.; Zhu, M.; Zhmoginov, A.; Chen, L.C. Mobilenetv2: Inverted residuals and linear bottlenecks. In Proceedings of the IEEE Conference on Computer Vision and Pattern Recognition, Salt Lake City, UT, USA, 18–23 June 2018; pp. 4510–4520.
13. Ma, N.; Zhang, X.; Zheng, H.T.; Sun, J. Shufflenet v2: Practical guidelines for efficient cnn architecture design. In Proceedings of the European Conference on Computer Vision (ECCV), Munich, Germany, 8–14 September 2018; pp. 116–131.
14. Han, S.; Mao, H.; Dally, W.J. Deep compression: Compressing deep neural networks with pruning, trained quantization and huffman coding. *arXiv* **2015**, arXiv:1510.00149.

15. Li, D.; Hu, J.; Wang, C.; Zhu, L.; Zhang, T.; Chen, Q. Involution: Inverting the inherence of convolution for visual recognition. In Proceedings of the IEEE/CVF Conference on Computer Vision and Pattern Recognition, Nashville, TN, USA, 20–25 June 2021; pp. 12321–12330.
16. Wang, L.; Li, D.; Zhu, Y.; Shan, Y.; Tian, L. Dual super-resolution learning for semantic segmentation. In Proceedings of the IEEE/CVF Conference on Computer Vision and Pattern Recognition, Seattle, WA, USA, 13–19 June 2020; pp. 3774–3783.
17. Kim, S.W.; Kook, H.K.; Sun, J.Y.; Kang, M.C.; Ko, S.J. Parallel feature pyramid network for object detection. In Proceedings of the European Conference on Computer Vision (ECCV), Munich, Germany, 8–14 September 2018; pp. 234–250.
18. Chen, Q.; Wang, Y.; Yang, T.; Zhang, X.; Cheng, J.; Sun, J. You only look one-level feature. In Proceedings of the IEEE/CVF Conference on Computer Vision and Pattern Recognition, Nashville, TN, USA, 20–25 June 2021; pp. 13039–13048.

Communication

Optimal Control Method of Path Tracking for Four-Wheel Steering Vehicles

Xiaojun Tan [1,2], Deliang Liu [1] and Huiyuan Xiong [1,*]

[1] School of Intelligent Systems Engineering, Sun Yat-sen University, Guangzhou 510006, China; tanxj@mail.sysu.edu.cn (X.T.); liudliang5@mail2.sysu.edu.cn (D.L.)
[2] Southern Marine Science and Engineering Guangdong Laboratory (Zhuhai), Zhuhai 519082, China
* Correspondence: xionghy@mail.sysu.edu.cn

Abstract: Path tracking is a key technique for intelligent electric vehicles, while four-wheel steering (4WS) technology is of great significance to improve its accuracy and flexibility. However, the control methods commonly used in path tracking for a 4WS vehicle cannot take full advantage of the additional steering freedom of the 4WS vehicle, because of restricting the relationship between the front and rear wheels steering angle. To address this issue, we derive a kinematic model without the restriction based on the small-angle assumption. Then, the objective function and constraints of system control quantity optimization are designed based on the tracking error model. After the optimization problem is solved in the form of quadratic programming with constraints, the control sequence with the smallest performance index is obtained through rolling optimization. The proposed method is tested on a high-fidelity Carsim/Simulink co-simulation platform and an experimental vehicle. The results show that the standard deviation of the lateral error and the yaw angle error of the algorithm is less than 0.1 m and 3.0°, respectively. Compared with the other two algorithms, the control of the front and rear wheels angle of this method is more flexible and the tracking accuracy is higher.

Keywords: four-wheel steering; model predictive control; path tracking

Citation: Tan, X.; Liu, D.; Xiong, H. Optimal Control Method of Path Tracking for Four-Wheel Steering Vehicles. *Actuators* **2022**, *11*, 61. https://doi.org/10.3390/act11020061

Academic Editors: Peng Hang, Xin Xia and Xinbo Chen

Received: 18 January 2022
Accepted: 16 February 2022
Published: 18 February 2022

Publisher's Note: MDPI stays neutral with regard to jurisdictional claims in published maps and institutional affiliations.

1. Introduction

In recent years, unmanned vehicles have become a research hotspot due to the increase of various traffic problems such as traffic congestion and traffic accidents [1]. The key technologies mainly include environmental perception, precise localization, planning and decision-making, and motion control. Path tracking is one of the key problems of motion control for autonomous vehicles, which is denoted as tracking a predetermined path by controlling the lateral and yaw movement of the vehicle [2]. Thus, it can be defined as minimizing the lateral offset and heading errors [3].

Path tracking control methods can be mainly divided into two categories. One is geometry-based, which mainly includes pure pursuit (PP) [4] and Stanley [5], etc. The other is model-based, represented by synovial membrane control [6], linear quadratic regulator (LQR) [7,8] and model predictive control (MPC) [9,10], etc. Geometry-based control methods are often used in low-speed scenarios, with good interpretability and fast calculation speed. Model-based methods mainly based on dynamic models are often used for stability control of high-speed vehicles [11], whose disadvantages include poor real-time performance and the difficulty to obtain kinetic parameters accurately [12]. However, the above studies are mostly based on front-wheel steering (FWS) vehicles. The only control input for lateral tracking control is the front-wheel steering angle, which limits the ability of path tracking control.

To improve the flexibility and stability of vehicles, the concept of the 4WS vehicle was proposed in the late 1980s [13]. At low speed, the steering modes of a 4WS vehicle are more diverse than FWS vehicles [14,15]. The front and rear wheels can be turned in reverse phase to reduce the turning radius and improve maneuverability. At high speed, a 4WS vehicle can improve handling stability by steering the front and rear wheels in phase to ensure zero slip angle and ideal yaw rate [16]. Making full use of the additional degrees of freedom of the 4WS vehicle can independently control the path and attitude of the vehicle, reduce the yaw motion required by the body, and improve the responsiveness of the vehicle heading change [2]. At the same time, the vehicle has better path tracking performance due to the improvement of flexibility [17].

Aiming at the path tracking problem of the 4WS vehicle, Ye et al. [18] designed a strategy to switch steering modes include active front and rear steering (AFRS), Ackermann steering, and crab steering for achieving accurate path-following of the vehicle. Hiraoka et al. [6] proposed a 4WS vehicle path tracking controller based on the sliding mode control theory, which uses front and rear control points for tracking. However, the above methods restrict the steering freedom of the 4WS vehicle and reduce flexibility. Wu et al. [19] developed a novel rear-steering-based decentralized control (RDC) algorithm for the 4WS vehicle. Yin et al. [20] carried out a new distribution controller to allocate driving torques to four-wheel motors, which can use each tire to generate yaw moment and achieve a quicker yaw response. Fnadi et al. [21] synthesized a new controller for dynamic path tracking by using constrained model predictive control (MPC) for double steering off-road vehicles, which takes into account steering and sliding constraints to ensure safety and lateral stability. However, these methods only use rear-wheel steering within a small turning angle range and are not suitable for flexible control of 4WS vehicle at low speed.

Aiming at these challenges, a tracking error model unrestrained on the front and rear wheels steering (UFRWS) relationship of the 4WS vehicle is established in this paper. As shown in Figure 1, a predictive controller based on this model is proposed, which performs lateral motion control, and forms a trajectory tracking controller with the PI controller that performs longitudinal control. The advantage of this controller is that it can fully utilize the steering freedom of the 4WS vehicle, improving tracking accuracy and flexibility.

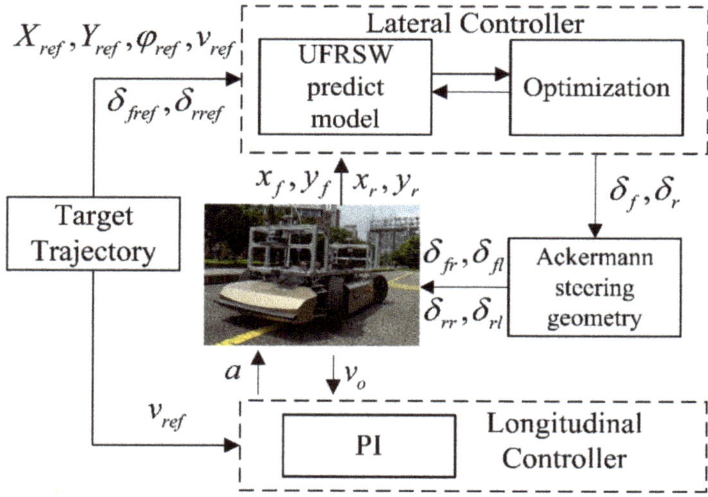

Figure 1. Trajectory tracking controller framework diagram.

2. Kinematics Model of 4WS Vehicle

The kinematic model is the basis of trajectory planning and tracking control. To reduce the complexity of the controller design, the 4WS vehicle kinematics model can be simplified to a single-track model with the assumption of pure rolling and small steering angle as shown in Figure 2. The points $F(x_f, y_f)$ and $B(x_r, y_r)$ are the center of the front and the rear axle of the vehicle, respectively. The point $M(x, y)$ is the geometric center of the vehicle, and the point C is the center of rotation of the vehicle. R denotes the radius of rotation of the vehicle. The wheelbase L is the distance between the front and rear axles, and the wheel track W refers to the distance between the left and right wheels. The heading angle φ refers to the angle between the body direction and the X axis in the global coordinate system XOY. The center of mass slip angle β is the angle between the speed v_m at the point M and the direction of the body.

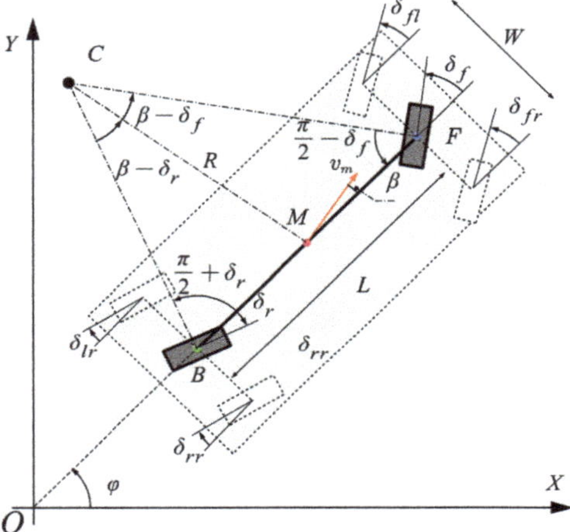

Figure 2. Schematic diagram of relevant variables of 4WS vehicle kinematics model.

The 4WS vehicle bicycle model is gray as shown in Figure 2. Its front steering angle δ_f and rear steering angle δ_r should satisfy the Ackerman steering geometric relationship. So, the steering angle of each wheel $\delta_i (i = fr, fl, rr, rl)$ satisfies the Equation (1).

$$\begin{cases} \tan \delta_{fl} = \dfrac{\tan \delta_f}{1 - \frac{W}{2L}(\tan \delta_f - \tan \delta_r)} \\ \tan \delta_{fr} = \dfrac{\tan \delta_f}{1 + \frac{W}{2L}(\tan \delta_f - \tan \delta_r)} \\ \tan \delta_{rl} = \dfrac{\tan \delta_r}{1 - \frac{W}{2L}(\tan \delta_f - \tan \delta_r)} \\ \tan \delta_{rr} = \dfrac{\tan \delta_r}{1 + \frac{W}{2L}(\tan \delta_f - \tan \delta_r)} \end{cases} \tag{1}$$

We take M as the control point. Then, the nonlinear kinematics equations of the 4WS vehicle bicycle model in the global coordinate system can be expressed as

$$\begin{cases} \dot{X} = v_m \cos(\varphi + \beta) \\ \dot{Y} = v_m \sin(\varphi + \beta) \\ \dot{\varphi} = \dfrac{v_m \cos(\beta)}{L}\left(\tan\left(\delta_f\right) - \tan(\delta_r)\right) \\ \beta = \arctan\left(\dfrac{\tan \delta_r + \tan \delta_f}{2}\right) \end{cases} \tag{2}$$

Most of the existing path tracking lateral control methods are designed for the application of front-wheel steering vehicles. Therefore, to apply PP and MPC methods to four-wheel steered vehicles, this article regards 4WS as FWS vehicles with the wheelbase halved by restricting the steering angles of the front and the rear to be equal and out of phase as shown in Figure 3. Then, Equation (2) can be simplified to Equation (3), which is the symmetrical front and the rear wheels steering (SFRWS) model of the 4WS vehicle. The

PP and MPC methods based on the SFRWS model can be used as the comparison algorithm in this article for the method based UFRWS model.

$$\begin{cases} \dot{X} = v_m \cos\varphi \\ \dot{Y} = v_m \sin\varphi \\ \dot{\varphi} = 2v_m \tan\delta_f / L \end{cases} \tag{3}$$

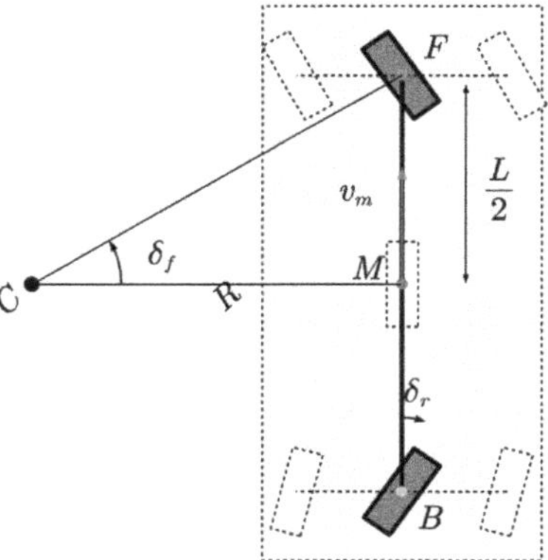

Figure 3. Schematic diagram of the SFRWS kinematics model of the 4WS vehicle.

Obviously, due to the constraint between the front and rear wheel angle relationship, the SFRWS model limits the steering freedom of 4WS vehicle, which reduces flexibility. For this reason, this paper proposes a predictive control method based on the unconstrained steering model of 4WS.

From the trigonometric function operation, we can get the Equation (4).

$$\tan\delta_r + \tan\delta_f = \tan(\delta_f + \delta_r)(1 - \tan\delta_r \tan\delta_f) \tag{4}$$

We can further simplify the kinematics model because the vehicle turning angle is less than $30°$.

$$\tan\delta_r + \tan\delta_f \approx (\delta_f + \delta_r) \tag{5}$$

Combining Equations (1) and (5), we can get a simplified non-linear 4WS kinematics model unrestrained on the front and the rear wheel steering relationship (UFRWS). The model is as follows:

$$\begin{cases} \dot{X} = V\cos(\psi + \left(\frac{\delta_r + \delta_f}{2}\right)) \\ \dot{Y} = V\sin(\psi + \left(\frac{\delta_r + \delta_f}{2}\right)) \\ \dot{\psi} = \frac{V\cos\left(\frac{\delta_r + \delta_f}{2}\right)}{\ell_f + \ell_r}\left(\delta_f - \delta_r\right) \end{cases} \tag{6}$$

3. Optimal Predictive Control Based Different Model

In this section, the objective functions and constraints of system control quantity optimization are designed based on the UFRWS model and SFRWS model. The optimization problem is solved in the form of a constrained quadratic programming and rolling optimization is performed.

3.1. Linear Discrete Tracking Error Model Based UFRWS Model

We define the state vector $\chi = [\begin{array}{ccc} e_x & e_y & e_\varphi \end{array}]^T$, the control input $u = [\begin{array}{cc} \delta_f & \delta_r \end{array}]^T$, where the error e_x is the difference between the actual position of the vehicle and the reference position in the X direction, the error e_y is in the Y direction, and the heading error

e_φ is the difference between the vehicle heading angle and the reference heading angle. Then, the tracking error model based on UFRWS model can be obtained.

$$\dot{\chi} = \begin{bmatrix} \dot{e}_x \\ \dot{e}_y \\ \dot{e}_\varphi \end{bmatrix} = \begin{bmatrix} \dot{X} - \dot{X}_{ref} \\ \dot{Y} - \dot{Y}_{ref} \\ \dot{\varphi} - \dot{\varphi}_{ref} \end{bmatrix} = f(\chi, u) \tag{7}$$

Since the reference points are all on the reference trajectory, Equation (7) can be expanded by the first-order Taylor expansion at the reference state quantity $\chi_{ref} = [\ 0 \quad 0 \quad 0\ ]^T$, and we can get:

$$\dot{\chi} = \left.\frac{\partial f(\chi, u)}{\partial \chi}\right|_{\substack{\chi = \chi_{ref}, \\ u = u_{ref}}} \left(\chi - \chi_{ref}\right) + \left.\frac{\partial f(\chi, u)}{\partial u}\right|_{\substack{\chi = \chi_{ref}, \\ u = u_{ref}}} \left(u - u_{ref}\right) \tag{8}$$

Based on the Jacobi matrix, the state space form of the linear tracking error model can be developed as follows:

$$\begin{cases} \dot{\chi} = A\chi + Bu + W \\ \eta = C\chi \end{cases}, \tag{9}$$

where η is the state transition matrix, C is the 3×3 identity matrix.

Discretizing the continuous system Equation (9) by using the forward Euler method can obtain a linear discrete tracking error model Equation (10).

$$\chi(k+1) = A_d\chi(k) + B_du(k) + W_d, \tag{10}$$

$$\text{Where} \quad A_d \quad = \quad I + AT \quad = \begin{bmatrix} 1 & 0 & -Tv_{ref}\sin\left(\varphi_{ref} + \beta_{ref}\right) \\ 0 & 1 & Tv_{ref}\cos\left(\varphi_{ref} + \beta_{ref}\right) \\ 0 & 0 & 1 \end{bmatrix},$$

$$B_d \quad = \quad BT \quad = \begin{bmatrix} -\frac{1}{2}Tv_{ref}\sin\left(\varphi_{ref} + \beta_{ref}\right) & -\frac{1}{2}Tv_{ref}\sin\left(\varphi_{ref} + \beta_{ref}\right) \\ \frac{1}{2}Tv_{ref}\cos\left(\varphi_{ref} + \beta_{ref}\right) & \frac{1}{2}Tv_{ref}\cos\left(\varphi_{ref} + \beta_{ref}\right) \\ -Tv_{ref}\frac{\sin\beta_{ref}(\delta_f - \delta_r) - \cos\beta_{ref}}{2L} & -Tv_{ref}\frac{\sin\beta_{ref}(\delta_f - \delta_r) + \cos\beta_{ref}}{2L} \end{bmatrix},$$

$$W_d = WT = \begin{bmatrix} \frac{1}{2}v_{ref}\sin\left(\varphi_{ref} + \beta_{ref}\right)(\delta_f + \delta_r)T \\ -\frac{1}{2}v_{ref}\cos\left(\varphi_{ref} + \beta_{ref}\right)(\delta_f + \delta_r)T \\ \frac{1}{2}\delta_f v_{ref}(\sin\beta_{ref}(\delta_f - \delta_r) - \cos\beta_{ref})T/L + \frac{1}{2}\delta_r v_{ref}(\sin\beta_{ref}(\delta_f - \delta_r) + \cos\beta_{ref})T/L \end{bmatrix}.$$

3.2. Linear Discrete Tracking Error Model Based SFRWS Model

Different from the UFRWS-based model, the control input of the SFRWS-based system is only the front wheel angle, that is $u = \delta_f$. In the same way, the discretization model based SFRWS model can be obtained as follows:

$$\chi(k+1) = A_d\chi(k) + B_du(k) + W_d, \tag{11}$$

$$\text{where } A_d = \begin{bmatrix} 1 & 0 & -Tv_{ref}\sin\left(\varphi_{ref}\right) \\ 0 & 1 & Tv_{ref}\cos\left(\varphi_{ref}\right) \\ 0 & 0 & 1 \end{bmatrix}, B_d = \begin{bmatrix} 0 \\ 0 \\ \frac{Tv_{ref}}{L\cos^2\delta_f} \end{bmatrix}, W_d = WT = \begin{bmatrix} 0 \\ 0 \\ -\frac{Tv_{ref}\delta_f}{L\cos^2\delta_f} \end{bmatrix}.$$

3.3. State Prediction Model

According to Equation (11), the state quantity at each moment can be predicted.

$$
\begin{aligned}
\chi(k+1|k) &= A_d\chi(k) + B_d u(k) + W_d \\
\chi(k+2|k) &= A_d{}^2\chi(k) + A_d B_d u(k) + B_d u(k+1) + A_d W_d + W_d \\
\chi(k+3|k) &= A_d{}^3\chi(k) + A_d^2 B_d u(k) + A_d B_d u(k+1) + B_d u(k+2) + A_d^2 W_d + A_d W_d + W_d
\end{aligned}
\tag{12}
$$

$$
\vdots
$$

$$
\chi(k+N_p|k) = A_d{}^{N_p}\chi(k) + A_d^{N_p-1} B_d u(k) + \ldots + A_d^{N_p-N_c-1} B_d u(k+N_c) + A_d^{N_p-1} W_d + \ldots + A_d^{N_p-N_p} W_d
$$

Then, the state prediction equation is:

$$
Y(k) = \Psi\chi(k) + \Theta U(k) + W_e,
\tag{13}
$$

where
$$
Y(k) = \begin{bmatrix} \eta(k+1|k) \\ \eta(k+2|k) \\ \eta(k+3|k) \\ \cdots \\ \eta(k+N_p|k) \end{bmatrix}, \quad
U(k) = \begin{bmatrix} u(k|k) \\ u(k+1|k) \\ u(k+2|k) \\ \cdots \\ u(k+N_c|k) \end{bmatrix}, \quad
\Psi = \begin{bmatrix} CA_d \\ CA_d{}^2 \\ CA_d{}^3 \\ \cdots \\ CA_d{}^{N_p} \end{bmatrix},
$$

$$
\Theta = \begin{bmatrix}
CB_d & & & \\
CA_d B_d & CB_d & & \\
CA_d{}^2 B_d & CA_d B_d & CB_d & \\
\cdots & & & \\
CA_d{}^{N_p-1} B_d & CA_d{}^{N_p-2} B_d & \cdots & CA_d{}^{N_p-N_c-1} B_d
\end{bmatrix}
\quad W_e =
$$

$$
\begin{bmatrix}
CW_d \\
CA_d W_d + CW_d \\
CA_d^2 W_d + CA_d W_d + CW_d \\
\vdots \\
CA_d^{k-1} W_d + CA_d^{k-2} W_d + \cdots + CW_d
\end{bmatrix}.
$$

3.4. Scrolling Optimization

The control goal of the system is to make the 4WS vehicle track the target trajectory quickly and stably. Therefore, it is necessary to optimize the state quantity of the system, the control quantity, and the amount of change in the control quantity.

At the k moment, the amount of change in the control quantity is defined as:

$$
\Delta u(k) = u(k) - u(k-1).
\tag{14}
$$

Then:

$$
\Delta U(k) = \begin{bmatrix} \Delta u(k) \\ \Delta u(k+1) \\ \cdots \\ \Delta u(k+N_c) \end{bmatrix} = \begin{bmatrix} 1 & -1 & & \\ -1 & 1 & & \\ & & \cdots & \\ & & -1 & 1 \end{bmatrix} U = DU.
\tag{15}
$$

The design objective function is as follows:

$$
J(k) = \sum_{i=1}^{N_p} \|\chi(k+i)\|_Q^2 + \sum_{i=1}^{N_c-1} \left\| U(k+i) - U_{ref}(k+i) \right\|_{R_1}^2 + \sum_{i=1}^{N_c-1} \|\Delta U(k+i)\|_{\Delta R}^2,
\tag{16}
$$

where the first part is to make the current state error that is the lateral error and heading error close to 0. The importance of each state quantity can be adjusted by changing the weight value in the matrix R. The second part is to minimize the error between the control quantity and the reference value. The weight of each control variable can be set by adjusting the weight value in the matrix R_1. The third part is to make the change of the control variable as small as possible to reduce the output angular velocity value. The parameters and weights can be set by adjusting the matrix ΔR.

This paper mainly considers the control quantity limit constraint and control increment constraint in the control process. The expression form of the control quantity is as follows:

$$u_{\min}(k) \leqslant u(k) \leqslant u_{\min}(k), k = 0, 1, \cdots, N_c - 1, \tag{17}$$

The expression for the control increment is as follows:

$$\Delta u_{\min}(k) \leqslant \Delta u(k) \leqslant \Delta u_{\min}(k), k = 0, 1, \cdots, N_c - 1, \tag{18}$$

Define $E = \Phi X_0 + W, R_2 = D^T \Delta R D$. After simplifying, the cost function is transformed into a standard format for quadratic objective functions with linear constraints.

$$\begin{aligned}
\min J &= \tfrac{1}{2} U^T H U + f^T U \\
st. & \\
\begin{bmatrix} U_{\min} \\ \Delta U_{\min} \end{bmatrix} &< \begin{bmatrix} I \\ D \end{bmatrix} U < \begin{bmatrix} U_{\max} \\ \Delta U_{\max} \end{bmatrix}
\end{aligned} \tag{19}$$

where $H = \left(\Theta^T Q\Theta + R_1 + R_2\right), f = \left(E^T Q\Theta - U_r^T R_1\right)^T$, $U_{\min}$ is the lower limit sequence of the angle value $u_{\min}$, $U_{\max}$ is the upper limit sequence of the angle value $u_{\max}$, $\Delta U_{\min}$ is the lower limit sequence of the angle rate value $\Delta u_{\min}$, and $\Delta U_{\max}$ is the upper limit sequence of the angle rate value $\Delta u_{\max}$.

In each control cycle, the effective set method is used to solve the Equation (16), then the optimal control sequence in the control time domain is obtained.

$$U^* = \begin{bmatrix} u_k^* & u_{k+1}^* & \cdots & u_{k+N_c-1}^* \end{bmatrix} \tag{20}$$

The first element in the control sequence is used as the actual control input to act on the system. After entering the next cycle, the optimal control sequence is recalculated and the first control increment acts on the control system. So, scrolling realizes the optimal control of vehicle trajectory tracking.

4. Experimental Results and Discussion

This article establishes a high-fidelity dynamic model in Carsim based on four-wheel steer-by-wire vehicle, which forms a joint platform with Simulink for simulation experiments. After that, real vehicle trajectory tracking experiments were carried out. The Pure Pursuit based on the SFRWS (PP-SFRWS) model tracking method, the Model Predict Control based on the SFRWS (MPC-SFRWS) model, and the Model Predict Control based on the model unconstrained the front and the rear wheels steering (MPC-UFRWS) are compared and verified.

4.1. 4WS Vehicle Experiment Platform

The electrical system for the four-wheel steer-by-wire chassis used in the experiment is as shown in Figure 4. The system has dual motors and two steer-by-wire modules to control the rotation and steering of front and rear wheels respectively. So, the 4WS vehicle has more freedom degrees used for attitude control. A combined positioning system uses Global Positioning System (GPS) and Inertial Navigation System (INS). The upper computer is used to monitor and collect data from the controller area network (CAN). All control algorithms code is downloaded to the ECU and run in the ECU.

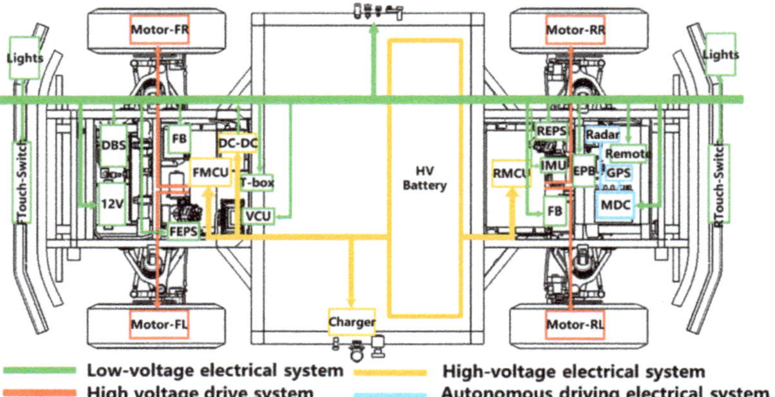

Figure 4. Steer-by-wire electrical system for Chassis of 4WS vehicle.

The main structural parameters of 4WS AGV are shown in Table 1.

Table 1. Vehicle parameters.

Parameters	Symbol	Value	Unit
The quality of the whole vehicle	m	700	kg
Wheelbase	L	1.9	m
Wheel track	W	1.2	m
Maximum steering angle	δ_{max}	30	°/s
Maximum steering angle rate	$\Delta\delta_{max}$	20	°/s

The vehicle drive control topology can be divided into the chassis domain and the autonomous driving domain as shown in Figure 5. The vehicle control unit (VCU) communicate with the motor control unit (MCU), the steering by wire (SBW), the electrical hydraulic brake (EHB), the electric park brake (EPB), the battery management system (BMS), and instrument electronic control unit (I-ECU) through the CAN bus to obtain the information of the remote control, the automatic driving domain computer (Industrial Personal Computer, IPC), and the parallel driving controller.

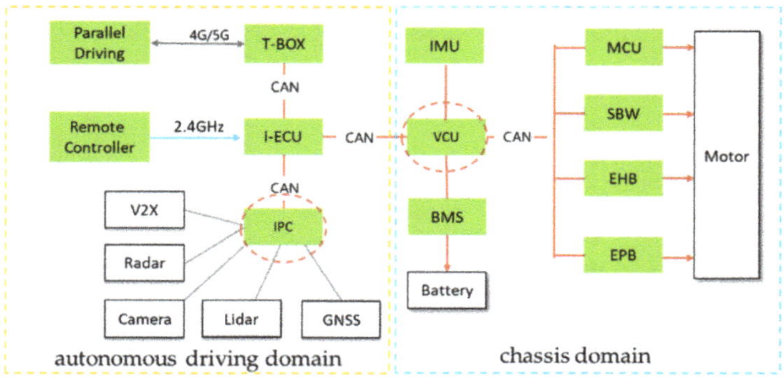

Figure 5. The software architecture of the chassis platform.

In the trajectory tracking control system established in this paper, lidar is mainly used to establish point cloud map and positioning. As shown in Figure 6, the acquisition of vehicle pose in this paper mainly relies on the fusion positioning system composed of lidar and GNSS-RTK.

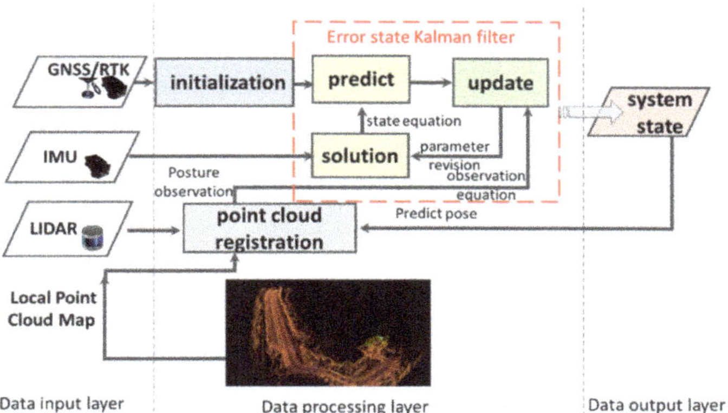

Figure 6. The location system framework.

When the vehicle starts outdoors, the navigation system is initialized with the GNSS-RTK information at the starting point. When the IMU data are received, the state variables of navigation system (position, speed, attitude, etc.) are updated recursively, and the recursive prediction is calculated to represent the uncertainty of the error state. When the LIDAR point cloud is received, the local map is used for registration, and the pose information of the vehicle relative to the local map is obtained. Taking the pose information as the observation, the new error state quantity and the filter gain are calculated. The parameters of the navigation module are modified according to the filter gain to realize the data fusion between IMU and LIDAR. As a result, vehicle pose is generated accurately and output to the control module.

4.2. Simulation Platform Construction

Based on the vehicle parameters of the experimental platform, a vehicle dynamics simulation model is established in Carsim, where the road adhesion coefficient is set to 0.8, and the rolling resistance coefficient is set to 0.8. The trajectory tracking controller is built by the S-function module in Simulink. A co-simulation platform is established with Carsim as shown in Figure 7.

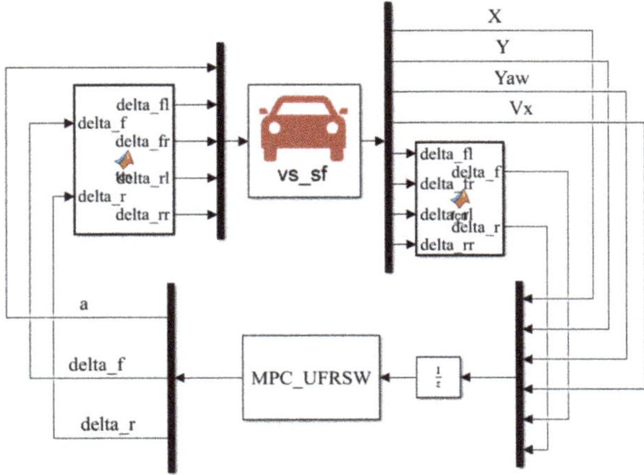

Figure 7. Carsim/Simulink co-simulation platform.

In the vehicle trajectory tracking simulation, the double lane change maneuver is a commonly used reference trajectory in the trajectory tracking test. In this paper, the function equation of the double lane change trajectory used in the simulation is as follows

$$\begin{cases} Y_{ref}(X) = \frac{d_{y1}}{2}(1 + \tanh(z_1)) - \frac{d_{y2}}{2}(1 + \tanh(z_2)) \\ \varphi_{ref}(X) = \tan^{-1}\left(d_{y1}\left(\frac{1}{\cosh(z_1)}\right)^2\left(\frac{1.2}{d_{x1}}\right) - d_{y2}\left(\frac{1}{\cosh(z_2)}\right)^2\left(\frac{1.2}{d_{x2}}\right)\right) \end{cases} \tag{21}$$

where $z_1 = \frac{2.4}{25}(X - 27.19) - 1.2$, $z_2 = \frac{2.4}{21.95}(X - 56.46) - 1.2$, $d_{x1} = 25$, $d_{x2} = 21.95$, $d_{y1} = 4.05$, $d_{y2} = 5.7$.

4.3. Analysis of Simulation Results

The simulation and real vehicle verification results are represented based on the trajectory of the vehicle's geometric center point. As shown in Figure 8a,b, the MPC-UFRSS method has better tracking performance for double lane change maneuver. At a speed of 5 m/s, the maximum lateral tracking error of the MPC-UFRWS method does not exceed 0.01 m, with 0.03 m for the SFRSW method and 0.1 m for the PP-SFRSW method.

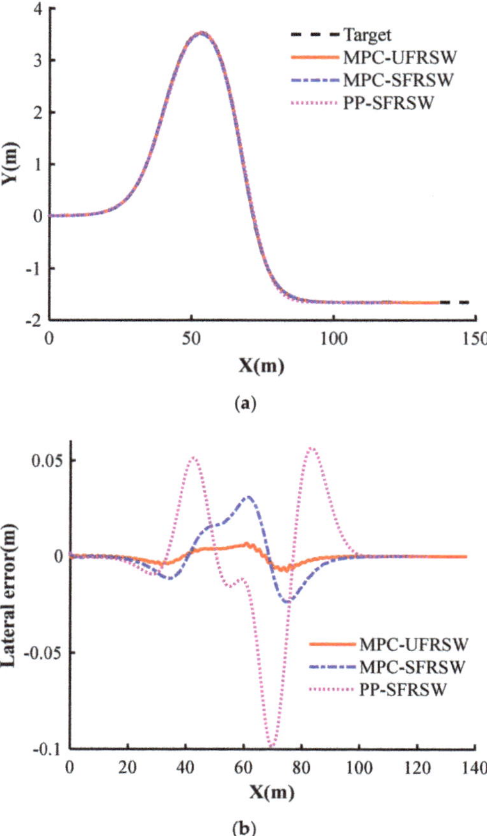

Figure 8. *Cont.*

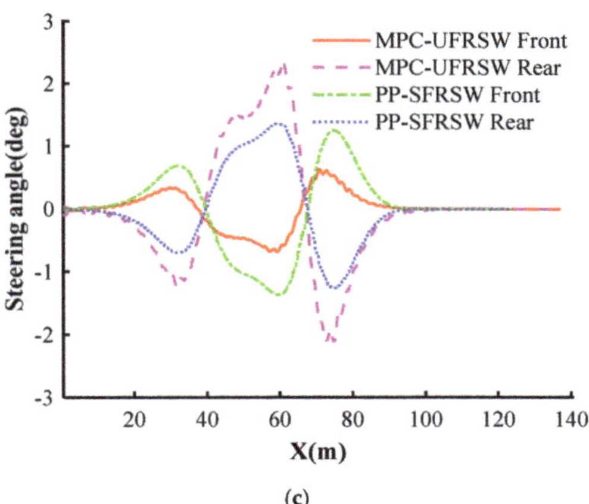

(c)

Figure 8. Simulation results of different algorithms. (**a**) Trajectory tracking; (**b**) Lateral error; (**c**) Steering angle.

As shown in Figure 8c, the PP-SFRSW method has the hard constraint between the front and rear wheels that the equal steering angles and opposite phase, while the MPC-UFRWS does not have this limitation. Therefore, the latter steering control changes are more flexible in corners, and the front and rear wheels steering forms are more diverse. As shown in Figure 8, when the vehicle enters the curve, the MPC-UFRWS method is relatively gentle, while the lateral tracking error of the PP-SFRSW increases sharply. The front and rear wheels angle adjustment of the MPC-UFRWS is larger, which can respond to the change of tracking error faster and track reference trajectory more flexibly.

4.4. Simulation Comparison at Different Speeds

The trajectory tracking of the MPC-UFRWS method comparison at different speeds is as shown in Figure 9. The greater speed, the greater the control error and the greater the overshoot. Similar to other methods based on the kinematics model, the proposed method is suitable for low-speed conditions. When the speed is less than 5 m/s, the lateral tracking error is less than 0.01 m. When the speed is 10 m/s, the lateral tracking error is greater than 0.06 m. However, when the speed is 15 m/s, the vehicle path deviates far from the reference path.

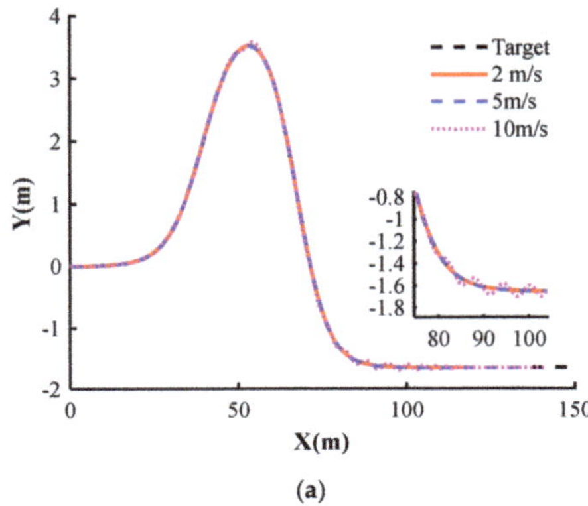

(a)

Figure 9. *Cont.*

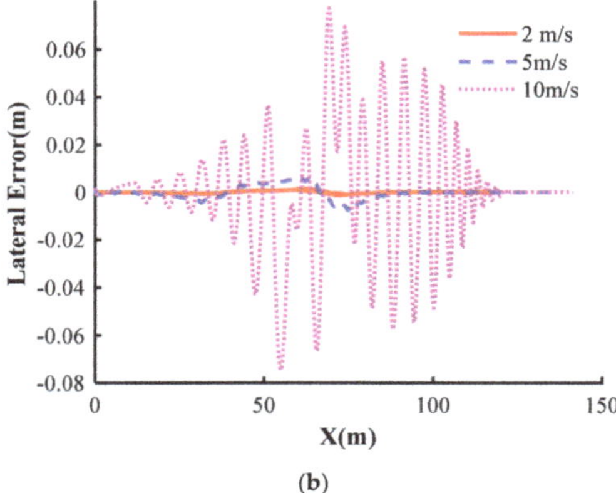

(b)

Figure 9. Tracking comparison at a different speed. (**a**) Trajectory tracking; (**b**) Lateral error.

4.5. Analysis of Real Vehicle Verification Results

To ensure the safety of personnel and vehicles, the vehicle experiment was carried out in an open space of Sun Yat-sen University. To verify the performance of the vehicle tracking straight and curved lines at the same time, we choose a recorded B-like trajectory as the reference trajectory considering the site constraints. The test site and vehicle are as shown in Figure 10.

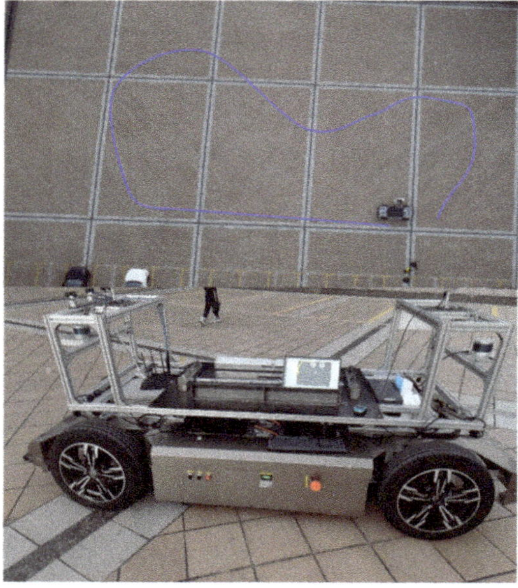

Figure 10. Experimental site and vehicle.

The real vehicle verification results of MPC-UFRWS, PP-SFRWS, and MPC-SFRWS are as shown in Figure 11 and Table 2, and analysis are as follows:

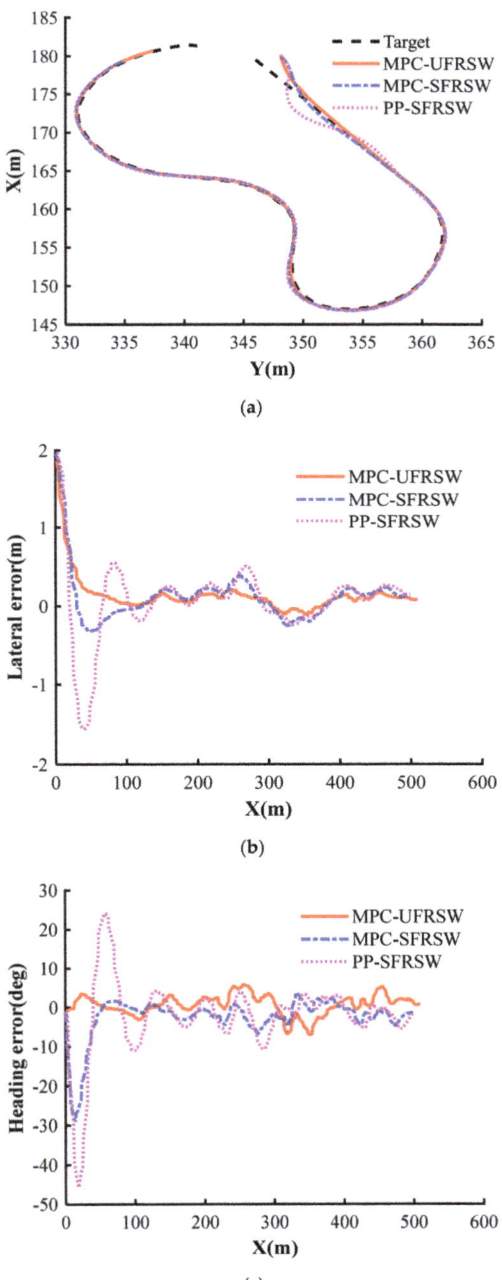

Figure 11. *Cont.*

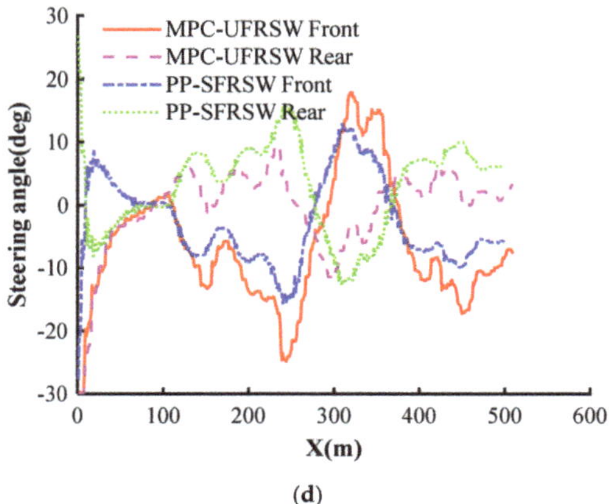

(**d**)

Figure 11. Results of different methods in vehicle experiments. (**a**) Trajectory tracking; (**b**) Lateral error; (**c**) Heading error; (**d**) Steering angle.

Table 2. Result data of vehicle verification.

Method	Max LatErr	MaxYawErr	stdPLatErr	stdPYawErr
PP-SFRWS	0.7663	12.2829	0.3168	3.4911
MPC-SFRWS	0.3488	7.2850	0.1202	1.7169
MPC-UFRWS	0.1998	6.7354	0.0678	2.6064

1. Compared with the PP-SFRWS method, the trajectory tracking accuracy of the MPC-UFRSW is higher. At the speed of 2 m/s, the maximum lateral error and yaw angle error are reduced by 0.567 m and 5.55°, respectively. This is because the Pure Pursuit method only determines the control input based on the deviation of the current measured values from the reference value, while the MPC method uses the prediction model to estimate the future deviation value and determine the current values in a rolling optimization manner.
2. The MPC-UFRWS method has higher tracking accuracy than the MPC-SRFRS. At a speed of 2 m/s, the maximum lateral error and yaw angle error are reduced by 0.15 m and 0.5°, respectively. This is because MPC-UFRWS does not constrain the relationship between the front and rear wheels angle, which can give full play to the more flexible characteristics of 4WS vehicle and avoid oversteering or understeering.
3. The proposed method enables better steering flexibility of 4WS vehicle. At the starting point, the lateral error between the vehicle and the reference trajectory is about 2.5 m. When the vehicle starts to drive, the MPC-URFSW method makes the vehicle merge into the trajectory in an approximate crab-walking manner, while the front and rear wheels turn in phase or out of phase in the corners. Therefore, the proposed method can realize the switching of various steering modes such as out-of-phase steering and crab steering without adding additional judgment.
4. Compared with the simulation results, the real vehicle tracking error increases several times. The reason may be that the curvature change rate of the B-like rail is larger than the double lane change maneuver. At the same time, external factors such as actuators, sensors, and road surfaces may also cause large tracking errors during vehicle verification.

5. Conclusions

In this paper, a path tracking controller with unconstrained front and rear wheels steering is established for the trajectory tracking control of 4WS vehicle. The controller relies on the proposed optimization function considering the tracking accuracy and control flexibility. Then, the simulation and vehicle verification are carried out to prove the effectiveness of the controller. In the Carsim/Simulink platform, MPC-UFRSW has higher tracking accuracy than PP-SFRSW and MPC-SFRSW when tracking double lane change trajectory at a speed not exceeding 10 m/s. In the real vehicle experiment with B-like curve, the steering angle of the front and rear wheels of the proposed controller changes more flexibly. The maximum lateral error and yaw angle error are reduced by 60% and 9%, respectively. The simulation and real vehicle verification results show that the proposed method has more flexible steering modes and higher tracking accuracy. The next work will further analyze the stability of the trajectory tracking of the 4WS vehicle at high speed from the dynamic point of view.

Author Contributions: Conceptualization, X.T. and D.L.; methodology, H.X.; validation, D.L., X.T. and H.X.; investigation, D.L.; data curation, D.L.; writing—original draft preparation, X.T. and D.L.; writing—review and editing, H.X.; visualization, D.L.; project administration, H.X. All authors have read and agreed to the published version of the manuscript.

Funding: This research was funded by the Natural Science and Technology Special Projects under Grant 2019-1496, Southern Marine Science and Engineering Guangdong Laboratory (Zhuhai) under Grant SML2020SP011, and the Key-Area Research and Development Program of Guangdong Province under Grant 2020B090921003.

Institutional Review Board Statement: Not applicable.

Informed Consent Statement: Not applicable.

Data Availability Statement: Detailed data are contained within the article. More data that support the findings of this study are available from the author D.L. upon reasonable request.

Acknowledgments: The authors are thankful for the support of the School of Intelligent Systems Engineering, Sun Yat-sen University, China Nuclear Power Engineering Co., Ltd., and Southern Marine Science and Engineering Guangdong Laboratory. At the same time, we thank Guohui Wu and Yuelong Pan for their help in the experiment and writing.

Conflicts of Interest: The authors declare no conflict of interest.

References

1. Paden, B.; Cap, M.; Yong, S.Z.; Yershov, D.; Frazzoli, E. A Survey of Motion Planning and Control Techniques for Self-Driving Urban Vehicles. *IEEE Trans. Intell. Veh.* **2016**, *1*, 33–55. [CrossRef]
2. Hang, P.; Chen, X. Path tracking control of 4-wheel-steering autonomous ground vehicles based on linear parameter-varying system with experimental verification. *Proc. Inst. Mech. Eng. Part I J. Syst. Control. Eng.* **2021**, *235*, 411–423. [CrossRef]
3. Cui, Q.; Ding, R.; Wei, C.; Zhou, B. Path-tracking and lateral stabilisation for autonomous vehicles by using the steering angle envelope. *Veh. Syst. Dyn.* **2020**, *59*, 1672–1696. [CrossRef]
4. Coulter, R. Implementation of the Pure Pursuit Path Tracking Algorithm. Carnegie-Mellon UNIV Pittsburgh PA Robotics INST, 1992; p. CM U-RI-TR-92-01.
5. Thrun, S.; Montemerlo, M.; Dahlkamp, H.; Stavens, D.; Aron, A.; Diebel, J.; Fong, P.; Gale, J.; Halpenny, M.; Hoffmann, G.; et al. Stanley: The robot that won the DARPA Grand Challenge. *J. Field Robot.* **2006**, *23*, 661–692. [CrossRef]
6. Hiraoka, T.; Nishihara, O.; Kumamoto, H. Automatic path-tracking controller of a four-wheel steering vehicle. *Veh. Syst. Dyn.* **2009**, *47*, 1205–1227. [CrossRef]
7. Jiang, Z.; Xiao, B. LQR optimal control research for four-wheel steering forklift based-on state feedback. *J. Mech. Sci. Technol.* **2018**, *32*, 2789–2801. [CrossRef]
8. Park, M.; Kang, Y. Experimental Verification of a Drift Controller for Autonomous Vehicle Tracking: A Circular Trajectory Using LQR Method. *Int. J. Control Autom. Syst.* **2021**, *19*, 404–416. [CrossRef]
9. Falcone, P.; Borrelli, F.; Asgari, J.; Tseng, H.E.; Hrovat, D. Predictive Active Steering Control for Autonomous Vehicle Systems. *IEEE Trans. Contr. Syst. Technol.* **2007**, *15*, 566–580. [CrossRef]
10. Kabzan, J.; Hewing, L.; Liniger, A.; Zeilinger, M.N. Learning-based Model Predictive Control for Autonomous Racing. *IEEE Robot. Autom. Lett.* **2019**, *4*, 3363–3370. [CrossRef]
11. Geng, K.; Chulin, N.A.; Wang, Z. Fault-Tolerant Model Predictive Control Algorithm for Path Tracking of Autonomous Vehicle. *Sensors* **2020**, *20*, 4245. [CrossRef] [PubMed]
12. Jiang, J.; Astolfi, A. Lateral Control of an Autonomous Vehicle. *IEEE Trans. Intell. Veh.* **2018**, *3*, 228–237. [CrossRef]

13. Furukawa, Y.; Yuhara, N.; Sano, S.; Takeda, H.; Matsushita, Y. A Review of Four-Wheel Steering Studies from the Viewpoint of Vehicle Dynamics and Control. *Veh. Syst. Dyn.* **1989**, *18*, 151–186. [CrossRef]
14. Oksanen, T.; Backman, J. Guidance system for agricultural tractor with four wheel steering. *IFAC Proc. Vol.* **2013**, *46*, 124–129. [CrossRef]
15. Yan, Y.; Geng, K.; Liu, S.; Ren, Y. A New Path Tracking Algorithm for Four-Wheel Differential Steering Vehicle. In Proceedings of the 2019 Chinese Control and Decision Conference (CCDC), Nanchang, China, 3–5 June 2019; pp. 1527–1532.
16. Liang, Y.; Li, Y.; Zheng, L.; Yu, Y.; Ren, Y. Yaw rate tracking-based path-following control for four-wheel independent driving and four-wheel independent steering autonomous vehicles considering the coordination with dynamics stability. *Proc. Inst. Mech. Eng. Part D J. Automob. Eng.* **2021**, *235*, 260–272. [CrossRef]
17. Hang, P.; Chen, X. Towards Autonomous Driving: Review and Perspectives on Configuration and Control of Four-Wheel Independent Drive/Steering Electric Vehicles. *Actuators* **2021**, *10*, 184. [CrossRef]
18. Ye, Y.; He, L.; Zhang, Q. Steering Control Strategies for a Four-Wheel-Independent-Steering Bin Managing Robot. *IFAC-PapersOnLine* **2016**, *49*, 39–44. [CrossRef]
19. Wu, Y.; Li, B.; Zhang, N.; Du, H.; Zhang, B. Rear-Steering Based Decentralized Control of Four-Wheel Steering Vehicle. *IEEE Trans. Veh. Technol.* **2020**, *69*, 10899–10913. [CrossRef]
20. Yin, D.; Wang, J.; Du, J.; Chen, G.; Hu, J.-S. A New Torque Distribution Control for Four-Wheel Independent-Drive Electric Vehicles. *Actuators* **2021**, *10*, 122. [CrossRef]
21. Fnadi, M.; Plumet, F.; Benamar, F. Model Predictive Control based Dynamic Path Tracking of a Four-Wheel Steering Mobile Robot. In Proceedings of the IEEE/RSJ International Conference on Intelligent Robots and Systems (IROS), Macau, China, 3–8 November 2019.

 *actuators*

MDPI

Article

Adaptive Cruise Control System Evaluation According to Human Driving Behavior Characteristics

Lin Liu [1], Qiang Zhang [2], Rui Liu [3], Xichan Zhu [1,*] and Zhixiong Ma [1]

[1] School of Automotive Studies, Tongji University, Shanghai 201804, China; linliu0301@163.com (L.L.); mzx1978@tongji.edu.cn (Z.M.)
[2] State Key Laboratory of Vehicle NVH and Safety Technology, Chongqing 401122, China; zhangqiang2@caeri.com.cn
[3] School of Automobile, Chang'an University, Xi'an 710064, China; liuruiaza@163.com
* Correspondence: zhuxichan@tongji.edu.cn

Abstract: With the rapid and wide implementation of adaptive cruise control system (ACC), the testing and evaluation method becomes an important question. Based on the human driver behavior characteristics extracted from naturalistic driving studies (NDS), this paper proposed the testing and evaluation method for ACC systems, which considers safety and human-like at the same time. Firstly, usage scenarios of ACC systems are defined and test scenarios are extracted and categorized as safety test scenarios and human-like test scenarios according to the collision likelihood. Then, the characteristic of human driving behavior is analyzed in terms of time to collision and acceleration distribution extracted from NDS. According to the dynamic parameters distribution probability, the driving behavior is divided into safe, critical, and dangerous behavior regarding safety and aggressive and normal behavior regarding human-like according to different quantiles. Then, the baselines for evaluation are designed and the weights of different scenarios are determined according to exposure frequency, resulting in a comprehensive evaluation method. Finally, an ACC system is tested in the selected test scenarios and evaluated with the proposed method. The tested vehicle finally got a safety score of 0.9496 (full score: 1) and a human-like score as fail. The results revealed the tested vehicle has a remarkably different driving pattern to human drivers, which may lead to uncomfortable ride experience and user-distrust of the system.

Keywords: ACC; safety evaluation; human-like evaluation; naturalistic driving study; driving behavior characteristic

Citation: Liu, L.; Zhang, Q.; Liu, R.; Zhu, X.; Ma, Z. Adaptive Cruise Control System Evaluation According to Human Driving Behavior Characteristics. *Actuators* **2021**, *10*, 90. https://doi.org/10.3390/act10050090

Academic Editor: Hai Wang

Received: 10 April 2021
Accepted: 23 April 2021
Published: 27 April 2021

Publisher's Note: MDPI stays neutral with regard to jurisdictional claims in published maps and institutional affiliations.

1. Introduction

Advanced Driver Assistance Systems (ADAS) are drawing increasing attention due to their potential in enhancing traffic safety, reducing driving workload and improving traffic efficiency. With wide studies on the control strategies of ADAS like adaptive cruise control system (ACC) [1,2], lane-keeping system (LKS) [3], automated emergency braking system (AEB) [4], etc., the functionalities of such systems are well studied and qualified. It follows that improving the anthropomorphism should also be incorporated into the development of these systems [5]. A human-like driving behavior pattern could enhance riding comfort and user trust and therefore improve user acceptance and increase usage frequency [6–8]. In the meantime, the surrounding drivers could better understand the vehicles adapting human-like driving patterns and make a natural interaction in the human-robot mixed traffic environment [9,10]. With the development of human-like ADAS, there raises the need for a testing and evaluation method considering human-like behavior.

Considering the motion state of the leading vehicle and the host vehicle, and the human driver's commands, an adaptive cruise control (ACC) system automatically controls the longitudinal motion of the host vehicle and provides the driver with driving risk tips to reduce the driving task strength and guarantee driving safety [11–13]. At present, ACC systems are widely used. In 2018, about 11.8% of all car models were equipped with ACC systems as the standard configuration in America [14].

As a mature function already on the market, there are several testing standards or regulations for ACC systems come from organizations such as ISO [11], SAE [13], GB [12], FMCSA [15] and so on. ISO 15622 and GB/T 20608 share three similar basic testing procedures: target acquisition range test, target discrimination test, and curve capability test, involving three test scenarios. Besides, the scenario for the deceleration ability test of the system is supplemented in SAE J2399. All these related standards focus on the test of functionality, that is, whether the functions such as longitudinal ranging and speed controlling could be realized.

On the other hand, studies on ACC systems or ADAS testing and evaluation methods mainly focus on the safety issue. Li et al. [16] evaluated impacts of ACC parameters on reducing collision risks on congested freeways. Qiu et al. [17] proposed a model for assessing the probability of accidents of ADAS systems, i.e., from the safety perspective. Focusing on collision avoidance, Stark et al. [18] carried out a simulation to estimate the performance of the state-of-art ADAS.

Although these standards and studies on testing and evaluation methods of ACC systems have produced great achievements on assessment of the system, none of them take the human-like behavior into consideration. Therefore, in this paper, we proposed a testing and evaluation method for ACC systems involving both safety and human-like performance. This method has the following two advantages: 1. this ready-to-use method provides testing scenarios generated from real driving data to ensure consistency to the real implementation environment; 2. the result is quantitatively evaluated from both the safety and human-like perspectives. This work offers an improvement for existing testing and evaluation methods in terms of a more real and efficient testing scenario set and more a comprehensive evaluation index, which is of great significance for further improvement of ACC systems.

The rest of the paper is arranged as follows: in Section 2 the real usage scenarios of ACC systems are defined based on the naturalistic driving study (NDS) and test scenarios are extracted according to the collision likelihood. In Section 3, the driving behavior characteristics of human drivers are obtained from NDS by statistical analysis. Evaluation indexes for both safety and human-like are designed separately in Section 4. Finally, an ACC system is tested and evaluated in Section 5 with the proposed method, following with the conclusion as ending in Section 6.

2. Testing Scenarios Extraction

In order to evaluate the safety and human-like of ACC systems, it is necessary to clarify the operating domains of ACC systems and then extract test scenarios accordingly. In this section, the operating domain is firstly defined according to the function design of ACC systems and then classified into usage scenarios according to the vehicle's motion state. Then, the test scenarios were classified into safety and human-like testing based on collision likelihood. Finally, a set of testing scenarios is generated and summarized in a table.

2.1. Usage Scenarios Definition

An ACC system performs longitudinal motion control of the vehicle according to the motion state of the host vehicle and the leading vehicle and the command from drivers. Under the premise of meeting the functional requirements, ACC systems can accurately track the following target among multiple leading vehicles and accurately measure the distance between them [12]. The system input is the motion state of the leading vehicle and the host vehicle and the command from drivers, and the output is the longitudinal motion control of the host vehicle. Therefore, the implementation scenario of ACC systems can be simplified to a two-vehicle scenario consisting of only the host vehicle and the vehicle in front (if any). What needs to be emphasized is that ACC systems only control the longitudinal movement of the host vehicle, making the movement of the vehicle limited to a single lane. In summary, the operating domain of ACC systems can be defined as a car-following (or free cruise if there is no leading vehicle) scenario in a single lane, which is referred to as a car following scenario in the following text.

During a car following process, the host vehicle keeps in a single lane while the leading car may change lanes, drive far away, etc., resulting in a scenario transition. In order to describe this scenario transition, the car-following process is decomposed into stable driving states (S) and events (A). A stable driving state S refers to the car-following process during which the leading vehicle target does not change. The stable driving state can be further classified into two types according to the presence of the leading target: (1) the host vehicle follows a fixed leading target in the lane (car following), denoted as S_{cf}, (2) the host vehicle travels in the lane with the prescribed speed without any leading target (free cruise), denoted as S_{fc}. An event A refers to the process that the movement of other traffic participants causes a change of the stable driving state of the host vehicle, including the appearance, disappearance, and change of the leading target. With an event, the host vehicle changes from one stable driving state to another stable driving state. The events that may occur during the car-following process include: cut-in A_{ci}, cut-out A_{co}, vehicle-approaching A_{va}, vehicle-distancing A_{vd}, etc.

Take the car following process in Figure 1 as an example. At t_0, the host vehicle changes lanes into a new lane and the following process starts: firstly, the host vehicle follows the leading vehicle 1, which is a stable driving state S_{cf}. Then at time t_1, event A_{ci} occurs, i.e., a new vehicle 2 drives into the front of the host vehicle and works as the new leading vehicle. The host vehicle enters the second stable driving state S_{cf}. At t_2, the leading vehicle travels far beyond the ACC system recognition range. Since there is no leading target in front of the host vehicle, it enters the cruise control state S_{fc}, and so on. The process can be described as: $S_{cf} \xrightarrow{A_{ci}} S_{cf} \xrightarrow{A_{vd}} S_{fc} \dots$

In a stable driving state, the following target and driving lane are fixed. Therefore, an ACC system is only required with the basic function, i.e., keeping a reasonable following range to the leading vehicle. However, when an event occurs, the changing following target and following state will put forward higher requirements on the performance of ACC systems. In existing standards (GB/T 20608 and ISO 15622), the three basic performance test scenarios, i.e., target acquisition range test, target discrimination test, and curve capability test, are all in S_{cf} processes, making the testing scenarios less challenging than real operation scenarios. To comprehensively evaluate the system performance, the S_{fc} process and the various events should also be included in the testing scenario set.

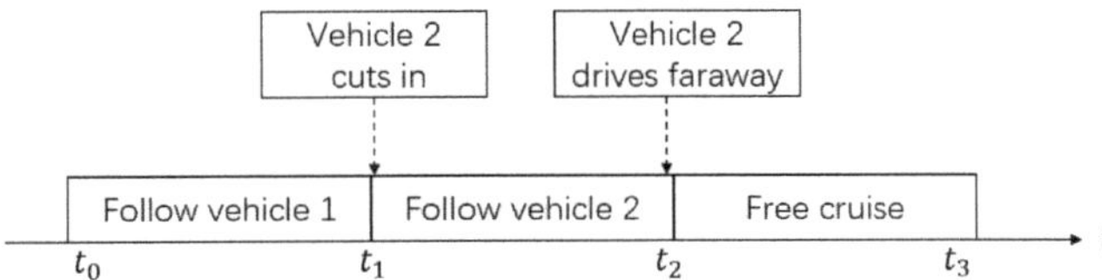

Figure 1. Car following process example.

2.2. Testing Scenarios

2.2.1. Human-Like Testing Scenario

Human-like indicates that the driving behavior of the vehicle controlled by ACC systems should be similar to that of a human driver, avoiding abnormal driving behaviors which may affect the surrounding traffic flow or the ride comfort. Human-like testing scenarios consist of S_{cf} and S_{fc} due to the following two reason: 1. the short occurrence of events making S_{cf} and S_{fc} processes occupy most of the driving distance and driving time during the car following process. Therefore, the behavior of the vehicle during S_{cf} and S_{fc} impacts the comfort experience of the driver for a longer time than events. 2. In S_{cf} and

S_{fc} processes, the scenarios are safe and then the driver focuses on driving experiences rather than the safety issue, making the human-like question significant.

2.2.2. Safety Testing Scenario

The most common danger happens in a car following process is a rear-end event caused by a too-small follow-up distance. To find proper scenarios for safety testing, events are analyzed to confirm the collision likelihood. When the leading vehicle cuts in A_{ci} or the leading vehicle approaching the leading vehicle A_{va} due to the speed difference, the following distance will reduce gradually or even suddenly. The host vehicle needs to brake to ensure safety, so A_{ci} and A_{va} are included in safety testing scenarios. When the leading vehicle cuts out A_{co} or drives far-away A_{vd}, the following distance increases, which does not involve any safety issues. Hence A_{co} and A_{vd} are excluded in the safety testing scenarios. Stop-and-go scenario refers to the car following process in which the leading vehicle decelerates to a full stop and then accelerates again. This process is common in traffic jams. At this time, timely and sufficient brake control is required to ensure a safe distance. The stop-and-go scenario is included in the testing scenario and recorded as A_{sg}.

The testing scenarios above only offer brief descriptions of the behaviors of the two vehicles during testing. In Table 1 scenarios are further detailed with the speed settings of vehicles.

Table 1. List of testing scenarios.

Type	Scenarios	Code	Cases and Description
human-like test	car-following	S_{cf1}	the leading vehicle speeds up from 30 km/h to 50/70/90/120 km/h
		S_{cf2}	the leading vehicle slows down from 50/70/90/120 km/h to 30 km/h
	cruising	S_{fc1}	the host vehicle speeds up from 30 km/h to 50/70/90/120 km/h
		S_{fc2}	the host vehicle slows down from 50/70/90/120 km/h to 30 km/h
safety test	cut-in	A_{ci}	the host vehicle is at 40 km/h and the leading vehicle cuts in with speed of 40 km/h and the range is 50 m
	vehicle appears	A_{va}	the host vehicle approaches the 40 km/h leading vehicle with speed of 50/70/110 km/h
	stop-go	A_{sg}	the leading vehicle slows down from 60 km/h to a full stop and then accelerates to 60 km/h

3. Human Driving Characteristic

The behavior characteristics of human drivers are the baseline for quantitative evaluation. Therefore, firstly, naturalistic driving data and critical driving data are used to analyze the real human driving pattern. Then the boundary among safe, critical, and dangerous driving behavior domains and among normal, aggressive, and critical driving behavior domains are extracted for safe and critical evaluation respectively. Finally, the scenario frequency parameters are integrated, and this chapter obtains an evaluation method that can be used for comprehensive quantitative evaluation of the system.

3.1. Human Driving Data

3.1.1. Naturalistic Driving Data

The naturalistic driving study refers to the driving data collection with the usage of unobtrusive observation methods. Since driver behaviors are collected from real traffic environment without disturbing the driver, the naturalistic driving study can collect massive amounts of traffic environment data, driving behavior data, and vehicle dynamical data, which can reflect the real driving needs and driving characteristic of human drivers [19,20]. Therefore, NDS is suitable as the resource date for vehicle development, testing, and verification. At present, various NDS projects were carried out all over the world, including the

100-Car Naturalistic Driving Study project [19] and the SHRP2 project [20] in the United States, the PROLOGUE project in Europe [21] and so on. The data used in this paper comes from a large-scale naturalistic driving study carried out in Shanghai, China. This project lasted for 18 months with 8 vehicles and 32 drivers. Each driver drove a vehicle for six months. The experiment vehicle is equipped with 4 cameras, which record the road environment in front of the vehicle with 2 different viewing angles, the driver's hand and pedal operations separately as shown in Figure 2. The vehicle's motion was also collected from the CAN-bus and an accelerometer. A total of 7402 were collected, which lasts for 3594 h and travels 129,935 km.

Figure 2. Naturalistic driving study.

3.1.2. Critical Driving Data—'500-Cases'

In NDS, critical or dangerous scenarios are very limited due to the extremely low frequency of danger. Most scenarios are safe and therefore only reflect human driving behavior under safe scenarios. Since the driving characteristic in critical scenarios varies a lot from that in safe scenarios, it is necessary to obtain critical driving data as a supplement. Therefore, the data set '500-Cases' is introduced. The '500-Cases' is generated from a critical scenarios collecting project carried out in Shanghai. Dashcams were installed on taxis, police cars, and some private cars to collect the critical scenarios with longitudinal deceleration greater than 0.4 g or lateral acceleration greater than 0.4 g. The cam will record the driving states 15 s before and 5 s after the time that the trigger value is reached. The sampling frequency is 2 Hz for speed and 30 Hz for acceleration and the frequency of video information is 30 Hz. A total of 4000 cases were collected during the 4 years test. Finally, a total of about 500 critical scenarios and 8 collisions were obtained and formed the 500-Cases data set.

3.2. Driving Behavior Characteristic

3.2.1. Joint Distribution of Speed-$\frac{1}{TTC}$

Time to collision TTC is a parameter commonly used to describe the degree of criticality of a car-following scenario. It was first proposed by Hayward as the time that two vehicles will collide if both of them maintain the current motion state, which is equal to the relative distance between the two vehicles divided by the relative speed [22]. In general, the larger the TTC is, the lower the risk level is. Usually, TTC is distributed in (0,+). However, when the speed of the two vehicles are similar, the value of TTC is very large, which brings inconvenience to the calculation and visualization of TTC distribution. Therefore, the value of $\frac{1}{TTC}$ at the braking time is introduced as the objective risk-level indicator.

Figure 3 shows the joint distribution of speed and $\frac{1}{TTC}$ of 78 dangerous car-following scenarios from '500-Cases'. After the regression coefficient test, the significance level $p < 0.001$, i.e., there is a significant regression relationship between $\frac{1}{TTC}$ at the start of braking and the speed of the vehicle. The linear fitting equation is: $TTC = -0.0717v + 1.2145$. Therefore, the influence of speed should be considered when $\frac{1}{TTC}$ is used to divide the safe, critical, and dangerous driving behavior domain. As shown as the green line in Figure 3, the 5% percentile of $\frac{1}{TTC}$ works as the boundary between the safe and the critical driving behavior domains. The linear quantile-regression equation is: $TTC = -0.0937v + 2.103$. As shown as the red line in Figure 3, the 95% percentile of $\frac{1}{TTC}$ works as the boundary between the critical and dangerous driving behavior domains. The linear quantile regression equation is: $TTC = -0.0057v + 0.1684$.

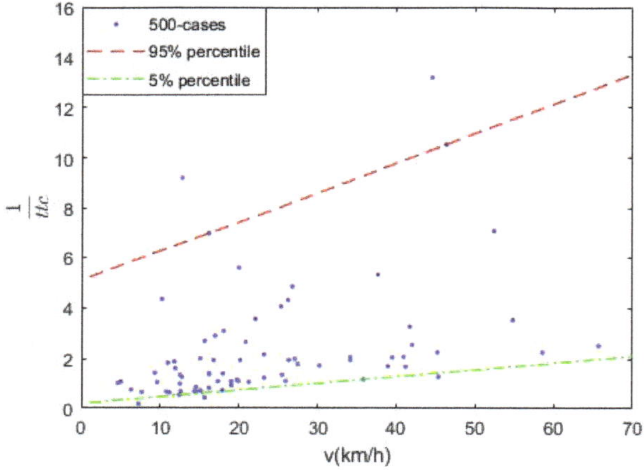

Figure 3. Joint-distribution of speed-$\frac{1}{TTC}$.

3.2.2. Joint Distribution of Velocity-Acceleration

Acceleration can directly reflect the driver's intention to control the vehicle. Therefore, the acceleration distribution obtained from NDS indicates the probability distribution of the driver's operation in the scenario. So, the system acceleration falls in the interval of higher probability indicates that it is similar to the human driver's operation, and the opposite means that the operation is poorly human-like. Acceleration can be therefore used as a characterization of the human-like of system. Besides, the magnitude of acceleration also indicates the driver's understanding of the current scenario state from safety perspective. For example, in a dangerous situation, the driver often applies a large deceleration to avoid collisions. Therefore, a very large deceleration tends to characterize the driver's subjective understanding of the current scenario as a high level of danger. Therefore, acceleration can be used as a subjective safety characterization of the system. In the following, NDS are used to obtain the joint-distribution of speed and acceleration of human drivers and critical, aggressive, and normal driving behavior domains are divided.

From NDS, 1000 journeys were randomly selected. The longitudinal velocity (km/h, hereafter referred to as velocity) and longitudinal acceleration (m/s², hereafter referred to as acceleration) are rolling averaged with the time window of 1s. The joint velocity-acceleration distribution is shown in Figure 4. As the speed increases, the range of acceleration expands and then narrows, indicating that the driver's acceleration and deceleration behavior becomes more violent in the low and medium speed intervals (0–15 m/s²) and becomes more cautious in the high speed interval. The driver's control of acceleration is clearly related to the speed.

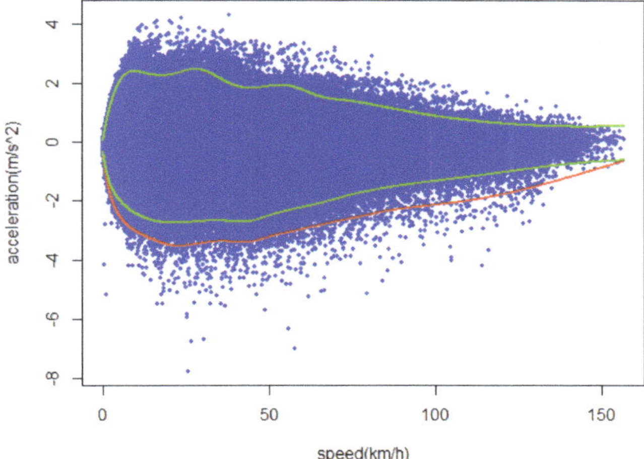

Figure 4. Joint-distribution of speed-acceleration.

Figure 5 shows the probability density function of acceleration at speeds $v = \{10 \text{ km/h},$ $20 \text{ km/h}, \ldots, 120 \text{ km/h}\}$. According to Liu Rui, the empirical distribution of acceleration in different speed intervals basically conforms to the Pareto distribution [23]. That is, under the same speed interval, the acceleration near 0 m/s^2 has the largest proportion, and the probability of extreme acceleration and deceleration is low. In normal scenarios, drivers generally control the distance to the leading vehicle by adding or subtracting speed gently; in case of danger, drivers tend to take emergency braking measures to avoid collisions, which results in a large deceleration.

As shown in Figure 4, the 95% quantile-regression line (green) is selected as the boundary between the normal and aggressive driving behavior domains; the lower 99% quantile-regression line (red) is selected as the boundary between the aggressive and critical driving behavior domains.

3.3. Testing Scenarios Frequency Weights

Since the frequency of different scenarios in the driving process is different, their weighting in evaluation should also varies, thus introducing the testing scenario frequency weight W_i. Data of one weekday are randomly selected from NDS for statistical analyze. Five of the eight test vehicles produced data with a cumulative driving time of 5 h, 22 min, and 24 s, and a cumulative mileage of 202 km. Among them, the car-following process totaled 162 km. In the proposed human-like testing scenarios, the proportion of mileage accounted for by the scenario is used as the scenario frequency weight, that is:

$$\begin{aligned} W_{S_{cf}} = P(S_{cf}) = 0.73, \\ W_{S_{fc}} = P(S_{fc}) = 0.27. \end{aligned} \tag{1}$$

For safety testing scenarios, the frequency ratio of the occurrence of each event in NDS was used as the scenario frequency weight as shown in Table 2.

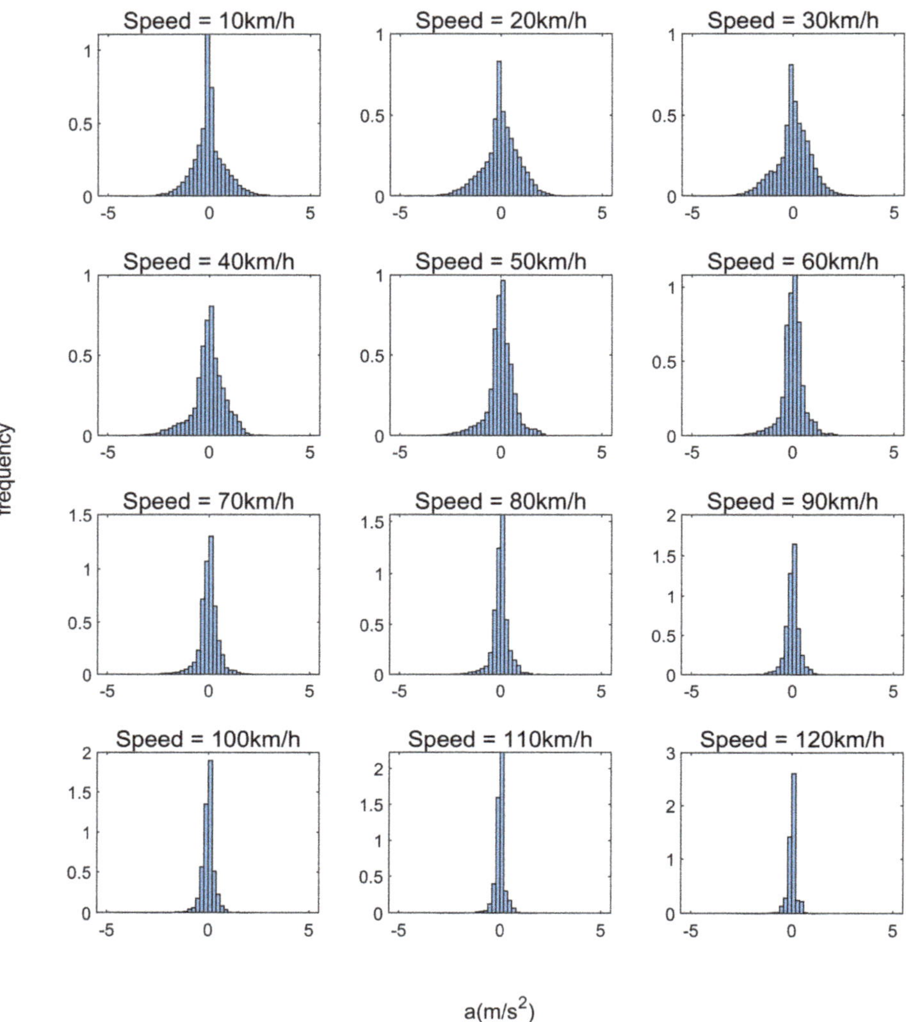

Figure 5. Frequency of acceleration with different speed.

Table 2. Event statistic.

Event	Count	Frequency	Weight
Cut-in	113	48.29%	$W_{A_{ci}} = 0.4829$
Stop-go	76	32.48%	$W_{A_{sg}} = 0.3248$
Vehicle-appear	45	19.23%	$W_{A_{va}} = 0.1923$
Sum	234	100.00%	

4. Evaluation Method for Safety and Human-Like

4.1. Scoring Method for Safety

The safety evaluation involves two indicators: $\frac{1}{TTC}$ and a. For $\frac{1}{TTC}$, the boundaries among safe, critical, and dangerous driving behavior domains are all natural baselines for objective safety, denoted as $L_{TTC,l}$ and $L_{TTC,h}$ respectively. For a, the boundary between critical and aggressive driving behavior domains works as the baseline for subjective safety and denoted as L_a hereafter.

The ACC system is scored after completing all scenarios of the safety testing. If a collision occurs during the test then the safety level is 'fail' and $P_{os} = 0$. In the absence of a collision, the objective safety of the system is calculated withing the following equation.

$$P_{os,i} = \frac{1}{V} \sum_{v=0}^{V} P_{os,i}^{v} P_{os,i}^{v} = \begin{cases} 0 & if \frac{1}{TTC^v} > \frac{1}{TTC_h^v} \\ 1 - \dfrac{\sqrt{(\frac{1}{TTC^v} - \frac{1}{TTC_l^v})^2}}{\sqrt{\frac{1}{TTC_h^v} - \frac{1}{TTC_l^v}}} & if \frac{1}{TTC_l^v} \le \frac{1}{TTC^v} \le \frac{1}{TTC_h^v} \\ 1 & if \frac{1}{TTC^v} < \frac{1}{TTC_l^v} \end{cases} \tag{2}$$

where i is the scenario number, and $P_{os,i}$ is the objective safety level of the vehicle in scenario i. v is the speed of the host vehicle, $P_{os,i}^{v}$ is the objective safety level of the vehicle in scenario i when the speed is v, V is the maximum speed of the vehicle in the test scenario, and the speed resolution is $0.1m/s$; $\frac{1}{TTC_h^v}$ and $\frac{1}{TTC_l^v}$ are the values of the baselines $L_{TTC,h}$ and $L_{TTC,l}$ when the speed is v respectively .

The subjective safety of the ACC system with acceleration as indicator is calculated using the following equation.

$$P_{ss,i} = \frac{1}{V} \sum_{v=0}^{V} P_{ss,i}^{v} P_{ss,i}^{v} = \begin{cases} 1 & if\, a^v \ge a_s^v \\ 1 - \dfrac{\sqrt{(a_a^v - a^v)^2}}{a_a^v} & if\, a^v < a_a^v \end{cases} \tag{3}$$

where $P_{ss,i}$ is the subjective safety level of the vehicle in scenario i. $P_{ss,i}^{v}$ is the subjective safety level of the vehicle when the vehicle speed is v in scenario i. a_a^v is the value of the baseline L_a when the speed is v. a^v is the acceleration value of the host vehicle when the speed of the vehicle is v in the test. In the same scenario, the comprehensive safety of the system is recorded as the mean value of the subjective and objective safety, and after completing all safety testing scenarios, the safety level of the system is obtained as follows.

$$P_s = \frac{1}{2} \sum_{i=1}^{N} W_i \times (P_{os,i} + P_{ss,i}) \tag{4}$$

where, N is the total number of human-like test scenarios, and W_i is the scenario frequency weight of the i scenario. The value range of the safety level P_s is $[0, 1]$. The closer the P_s is to 1, the higher the safety level is, indicating that the vehicle is less likely to enter a dangerous state during the driving process.

4.2. Scoring Method for Human-Like

Acceleration is the direct reflection of drivers intention in longitudinal direction. When there is no critical issue, drivers seldom apply violent acceleration or hard deceleration. Therefore, as shown in Figure 6 the 95% quantile-regression line (green dash line) is selected as the reference for human-like evaluation. Acceleration out of this range may cause discomfort or unsafe feeling of drivers.

Besides, the requirement for acceleration during the operation of the ACC system in the international standard ISO15622 [11] is also applied and denoted as the red dotted line. As shown in the figure, the area enclosed by the green dash line is much smaller and narrower than the area enclosed by the red dotted line, indicating a higher requirement proposed from human driving characteristics than from the ISO standard.

Therefore, green dash line is defined as the full-score line L_f. The acceleration within these lines could be regarded as a human-like behavior. Furthermore, the red dotted line is defined as the passing line. The behavior which exceeds these lines will fail the test.

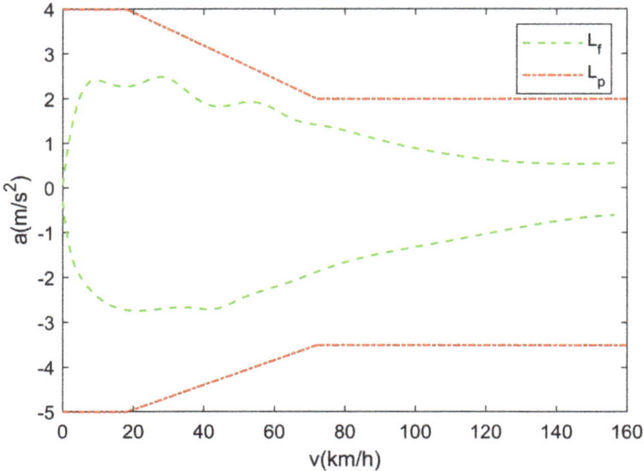

Figure 6. Full-score and passing line of acceleration.

The system is scored after completing all scenarios of the human-like test. If the acceleration exceeds the passing line L_p in any scenario during the test, it can be considered that the ACC system does not conform to the international standard. Then the system is judged as 'fail' without subsequent scoring, and the human-like level P_h is recorded as 0. On the premise of meeting the passing line, the calculation method of the human-like score in a single test scenario is:

$$P_{h,i} = \frac{1}{V} \sum_{v=0}^{V} P_{h,i}^v, \; P_{h,i}^v = \begin{cases} 1 - \frac{\sqrt{(a^v - a_h^v)^2}}{a_h^v} & if\, a^v > a_h^v \\ 1 & if\, a_l^v \le a^v \le a_h^v \\ 1 - \frac{\sqrt{(a_l^v - a^v)^2}}{|a_l^v|} & if\, a^v < a_l^v \end{cases} \tag{5}$$

where $P_{h,i}$ is the human-like level of the vehicle in scene i; v is the speed of the vehicle, $P(h,i)^v$ is the human-like level of the vehicle when the speed is v in scene i, V is the maximum speed of the vehicle in this test scenario, and the resolution of the speed is 0.1 m/s; a^v is the acceleration of the vehicle when the vehicle speed is v , a_h^v is the acceleration value of $L_{f,h}$ when the speed is v, and a_l^v is the acceleration value of $L_{f,l}$ when the velocity is v.

After completing all human-like testing scenarios, the human-like level of the ACC system is calculated as:

$$P_h = \sum_{i=1}^{N} W_i \times P_{h,i} \tag{6}$$

where, i is the scenario number, N is the total number of human-like testing scenarios, and W_i is the scenario frequency weight of the i scenario. The value range of the vehicle human-like level P_h is $[0, 1]$. The closer the P_h is to 1, the higher the human-like level is, indicating that the vehicle motion control conforms to the driving habits of human drivers.

5. Test and Results

5.1. Test Vehicle and Data Processing

A vehicle equipped with the ACC system was selected for the test. On a flat and straight road, the leading vehicle is controlled by an experienced driver according to the scenario description and the host vehicle turns on the ACC system for motion control. The following data were recorded throughout the field experiment: speed and acceleration of the host vehicle and the leading vehicle (if any), distance between them and TTC. The recording frequency is 100 Hz. The data are smoothed with a sliding time window of 0.1 s, and then safety and human-like evaluations are performed separately.

5.2. Results and Discussion

5.2.1. Results of Safety Evaluation

Figures 7 and 8 describe the safety testing results in terms of $\frac{1}{TTC}$ and a respectively. Since a negative $\frac{1}{TTC}$ indicates the host vehicle is driving away from the leading vehicle, indicating a safe scenarios, the Figure 7 only demonstrates the positive $\frac{1}{TTC}$ with a collision potential.

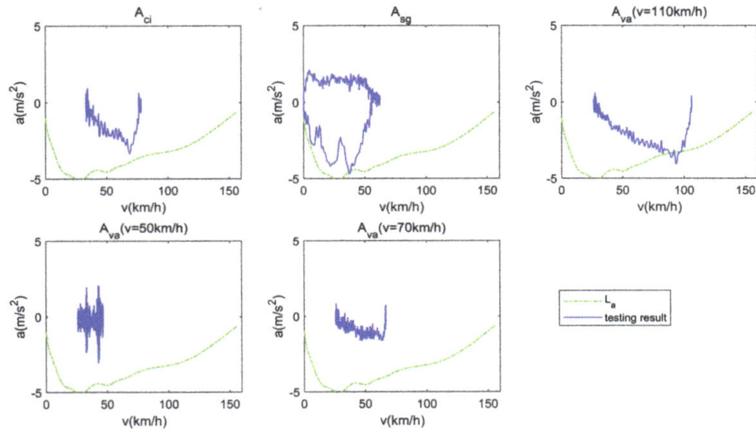

Figure 7. Safety evaluation results-$\frac{1}{TTC}$.

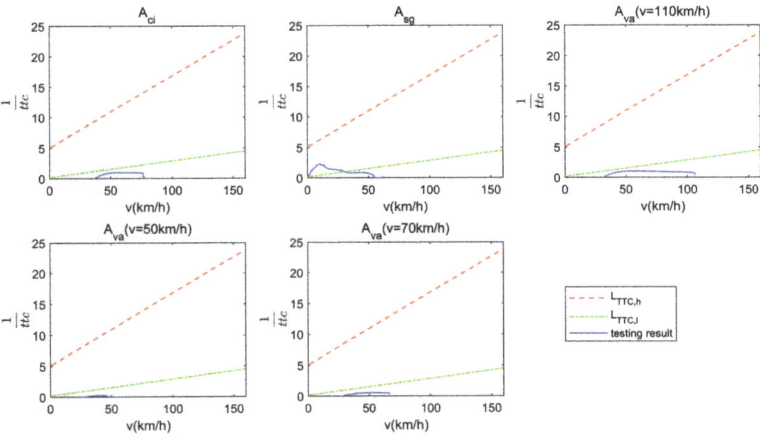

Figure 8. Safety evaluation results-a.

As shown in Figure 7, in all of the five testing scenarios, the testing vehicle did not exceed the baseline $L_{TTC,h}$. Only in scenario A_{sg} (i.e., stop and go scenario), the vehicle reached the baseline $L_{TTC,l}$ and got a score of $P_{os,2} = 0.6901$. The rest of the four scenarios all got the $P_{os,i} = 1$, $where\ i = 1,3,4,5$.

Figure 8 demonstrates the safety testing results in terms of a. In scenario $A_{va,110}$ (the testing vehicle approaching the 40 km/h leading vehicle with speed of 110 km/h),

a exceeded the baseline L_a when the speed was around 100 km/h and got a score of $P_{ss,3} = 0.9995$. The rest of the four scenarios all got the $P_{ss,i} = 1$, where $i = 1, 2, 4, 5$. The final score of safety evaluation was $P_S = 0.9496$.

5.2.2. Results of Human-Like Evaluation

Figure 9 demonstrates the human-like testing results. The upper two rows of the subfigures demonstrate the car-following scenarios with a speeding up front leading vehicle and with a slowing down front leading vehicle separately. v_f here is the speed change of the front vehicle. The lower two rows demonstrate the free-cruising scenarios separately. The dotted lines stand for the passing line from ISO 15622 while the dash lines are the full-score line from the human driving characteristics. The blue line demonstrates the real acceleration profile generated from field tests.

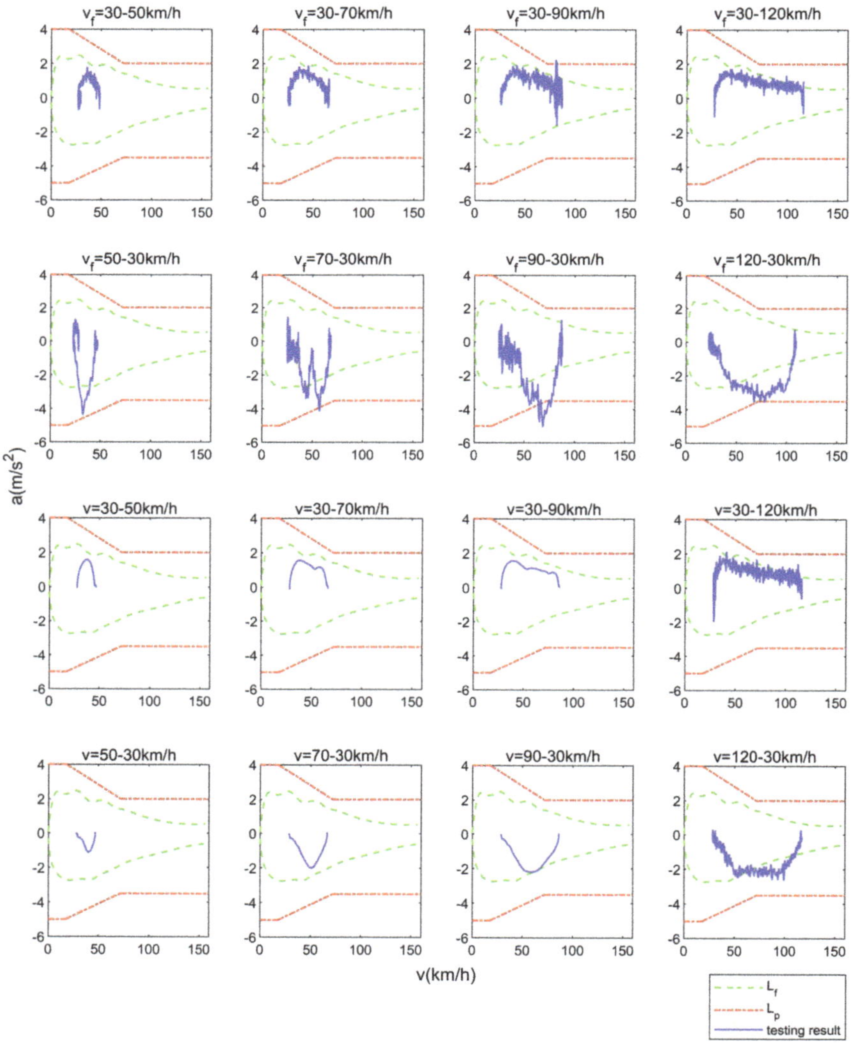

Figure 9. Human-like evaluation results under car-following scenarios with an accelerating leading vehicle.

Among all 16 testing scenarios, the testing vehicle's behavior runs out of the passing line in 4 scenarios. Therefore, the final score of the human-like evaluation was $P_h = Fail$.

All of the four failed scenarios are car-following scenarios. Compared with free cruising scenarios, the car-following scenarios are more challenging, as the longitudinal behavior decision should be restricted with a reasonable distance to the leading vehicle. In these four failed scenarios, three of them are with a decelerating leading vehicle as the decreasing following range may cause a safety issue, forcing the system to apply a relative hard brake to avoid entering a critical state. One possible improvement of the system that might be undertaken is to bring forward the timing of braking to flatten the decelerating curve. Another failed scenario is with an accelerating vehicle: the excessive acceleration may give the drivers an aggressive impression.

6. Conclusions

This paper proposed an ACC system testing and evaluation method based on human driver characteristics generated from naturalistic driving data, including testing scenarios and testing result evaluation method. The usage scenarios of the ACC system are defined and testing scenarios are then designed based on collision likelihood. The statistical analysis of real human driving data was conducted to obtain the speed-$\frac{1}{TTC}$ and speed-acceleration distributions to describe human drivers' perception of safety and driving habits. Quantiles of $\frac{1}{TTC}$ and acceleration are calculated to represent the majority behaviors.

Within the speed-$\frac{1}{TTC}$ distribution, the safe, critical, and dangerous driving behavior domains were divided by 5% and 95% quantiles, and the two boundaries were used as objective safety evaluation baseline. The normal, aggressive, and critical driving behavior domains were divided within the velocity-acceleration distribution by 5%, 95%, and 1% quantiles separately, and the boundaries were used as the baseline for human-like and subjective safety evaluation. Then the result evaluation method is accordingly designed.

An ACC system from the market is tested and evaluated. The system passed the safety tests with a score of 0.9496 (full score = 1) while failed the human-like tests. The results show the system has a more aggressive acceleration strategy and a delay on brake timing compared with human drivers.

The proposed testing and evaluation method has the following improvements compared with the existing testing protocols.

(1) The testing scenarios are derived from naturalistic driving data, improving the consistency with real driving scenarios. By introducing the scenario frequency coefficients, the final test results can reflect the real performance of the ACC system in real usage.

(2) The evaluation method is a supplement to the existing functional and safety evaluation of ACC systems for the human-like evaluation, which can help to improve the riding-comfort, user-trust, and user-acceptance of the system.

This study has the limitation that the proposed method is only applied to one ACC system due to time and budget. The comparison of the testing and evaluation results could further validate the method. Future works could apply the conception of human-like evaluation on other ADAS systems like LKS and also automated driving systems.

Author Contributions: Conceptualization, L.L., Q.Z., R.L., X.Z. and Z.M.; methodology, L.L., Q.Z. and X.Z.; validation, L.L.; investigation, L.L. and R.L.; data curation, Q.Z., X.Z. and Z.M.; writing—original draft preparation, L.L.; writing—review and editing, L.L. and X.Z.; visualization, L.L.; supervision, X.Z. All authors have read and agreed to the published version of the manuscript.

Funding: This research was funded by Shenzhen Future Intelligent Traffic Innovation Center(SFITIC), grant number TJ2020108591.

Institutional Review Board Statement: Not applicable.

Informed Consent Statement: Not applicable.

Data Availability Statement: The data presented in this study are available on request from the corresponding author. The data are not publicly available due to privacy.

Conflicts of Interest: The authors declare no conflict of interest.

References

1. Manolis, D.; Spiliopoulou, A.; Vandorou, F.; Papageorgiou, M. Real time adaptive cruise control strategy for motorways. *Transp. Res. Part C Emerg. Technol.* **2020**, *115*, 102617. [CrossRef]
2. Gong, X.; Ge, W.; Yan, J.; Zhang, Y.; Gongye, X. Review on the Development, Control Method and Application Prospect of Brake-by-Wire Actuator. *Actuators* **2020**, *9*, 15. [CrossRef]
3. Du, H.; Man, Z.; Zheng, J.; Cricenti, A.; Zhao, Y.; Xu, Z.; Wang, H. Robust control for vehicle lane-keeping with sliding mode. In Proceedings of the IEEE 2017 11th Asian Control Conference (ASCC), Gold Coast, Australia, 17–20 December 2017; pp. 84–89.
4. Han, I.C.; Luan, B.C.; Hsieh, F.C. Development of autonomous emergency braking control system based on road friction. In Proceedings of the 2014 IEEE International Conference on Automation Science and Engineering (CASE), New Taipei, Taiwan, 18–22 August 2014; pp. 933–937.
5. De Beaucorps, P.; Streubel, T.; Verroust-Blondet, A.; Nashashibi, F.; Bradai, B.; Resende, P. Decision-making for automated vehicles at intersections adapting human-like behavior. In Proceedings of the 2017 IEEE Intelligent Vehicles Symposium (IV), Los Angeles, CA, USA, 11–14 June 2017; pp. 212–217.
6. Häuslschmid, R.; von Buelow, M.; Pfleging, B.; Butz, A. SupportingTrust in autonomous driving. In Proceedings of the 22nd International Conference on Intelligent User Interfaces, Limassol, Cyprus, 13–16 March 2017; pp. 319–329.
7. Waytz, A.; Heafner, J.; Epley, N. The mind in the machine: Anthropomorphism increases trust in an autonomous vehicle. *J. Exp. Soc. Psychol.* **2014**, *52*, 113–117. [CrossRef]
8. Zihsler, J.; Hock, P.; Walch, M.; Dzuba, K.; Schwager, D.; Szauer, P.; Rukzio, E. Carvatar: Increasing trust in highly-automated driving through social cues. In Proceedings of the 8th International Conference on Automotive User Interfaces and Interactive Vehicular Applications, Ann Arbor, MI, USA, 24–26 October 2016; pp. 9–14.
9. Wei, J.; Dolan, J.M.; Litkouhi, B. A learning-based autonomous driver: Emulate human driver's intelligence in low-speed car following. In *Unattended Ground, Sea, and Air Sensor Technologies and Applications XII*; International Society for Optics and Photonics: Orlando, FL, USA. 2010; Volume 7693, p. 76930L.
10. Wang, X.; Zhu, M.; Chen, M.; Tremont, P. Drivers' rear end collision avoidance behaviors under different levels of situational urgency. *Transp. Res. Part C Emerg. Technol.* **2016**, *71*, 419–433. [CrossRef]
11. ISO 15622:2018. *Intelligent Transport Systems—Adaptive Cruise Control Systems—Performance Requirements and Test Procedures*; International Organization for Standardization, Vernier: Geneva, Switzerland, 2018.
12. China Academy of Transportation Science, W.U. *GB/T 20608-2006. Intelligent Transportation Information Systems—Adaptive Cruise Control Systems—Performance Requirements and Test Procedures*; Standards Press of China: Beijing, China, 2006.
13. Sayer, J. *Adaptive Cruise Control (ACC) Operating Characteristics and User Interface: Standard J2399*; Society of Automotive Engineers: Warrendale, PA, USA, 2003.
14. American Automobile Association. Advanced Driver Assistance Technology Names: AAA's Recommendation for Common Naming of Advanced Safety Systems. *AAA News Room*, 25 January 2019.
15. Amy Houser, John Pierowicz, R.M. *Concept of Operations and Voluntary Operational Requirements for Forward Collision Warning Systems (CWS) and Adaptive Cruise Control (ACC) Systems on Board Commercial Motor Vehicles*; Technical Report; US Department of Transportation, Federal Motor Carrier Safety Administration: Washington, DC, USA, 2005.
16. Li, Y.; Li, Z.; Wang, H.; Wang, W.; Xing, L. Evaluating the safety impact of adaptive cruise control in traffic oscillations on freeways. *Accid. Anal. Prev.* **2017**, *104*, 137–145. [CrossRef] [PubMed]
17. Qiu, S.; Rachedi, N.; Sallak, M.; Vanderhaegen, F. A quantitative model for the risk evaluation of driver-ADAS systems under uncertainty. *Reliab. Eng. Syst. Saf.* **2017**, *167*, 184–191. [CrossRef]
18. Stark, L.; Düring, M.; Schoenawa, S.; Maschke, J.E.; Do, C.M. Quantifying Vision Zero: Crash avoidance in rural and motorway accident scenarios by combination of ACC, AEB, and LKS projected to German accident occurrence. *Traffic Inj. Prev.* **2019**, *20*, S126–S132. [CrossRef] [PubMed]
19. Dingus, T.A.; Klauer, S.G.; Neale, V.L.; Petersen, A.; Lee, S.E.; Sudweeks, J.; Perez, M.A.; Hankey, J.; Ramsey, D.; Gupta, S.; et al. *The 100-Car Naturalistic Driving Study, Phase II-Results of the 100-Car Field Experiment*; Technical Report; Department of Transportation, National Highway Traffic Safety Administration, Springfield: Fairfax County, VA, USA. 2006.
20. Antin, J.F. *Design of the in-Vehicle Driving Behavior and Crash Risk Study: In Support of the SHRP 2 Naturalistic Driving Study*; Transportation Research Board: Washington, DC, USA, 2011.
21. Sagberg, F.; Eenink, R.; Hoedemaeker, M.; Lotan, T.; van Nes, N.; Smokers, R.; Welsh, R.; Winkelbauer, M. *Recommendations for a Large-Scale European Naturalistic Driving Observation Study*; PROLOGUE Deliverable D4.1; Institute of Transport Economics: Oslo, Norway, 2011.
22. Hayward, J.C. *Near Miss Determination through Use of a Scale of Danger*; Transportation Research Board: Washington, DC, USA, 1972.
23. Liu, R.; Zhu, X. Statistical Characteristics of Driver Accelerating Behavior and Its Probability Model. *arXiv* **2019**, arXiv:1907.01747.

 MDPI

Article

Research on the Identification of Tyre-Road Peak Friction Coefficient under Full Slip Rate Range Based on Normalized Tyre Model

Yinfeng Han [1,2], Yongjie Lu [1,2,*], Na Chen [1,3] and Hongwei Wang [2]

1. State Key Laboratory of Mechanical Behavior in Traffic Engineering Structure and System Safety, Shijiazhuang 050043, China; 13073782851@163.com (Y.H.); chennayu@163.com (N.C.)
2. Department of Mechanical Engineering, Shijiazhuang Tiedao University, Shijiazhuang 050043, China; whw18731230933@163.com
3. School of Information Science and Technology, Shijiazhuang Tiedao University, Shijiazhuang 050043, China
* Correspondence: luyongjie@stdu.edu.cn

Abstract: The accurate estimation of the tyre-road peak friction coefficient is the key basis for the normal operation of the vehicle active safety control system. The estimation algorithm needs to be able to adapt to various conditions encountered in the actual driving process of the vehicle and obtain the estimation results timely and accurately. Therefore, a new normalized strategy is proposed in this paper. The core is the equal ratio between the peak friction coefficient and the utilization friction coefficient between adjacent typical roads. This strategy can establish the direct connection (normalization) between tyre force and tyre-road peak friction coefficient through most tyre models in the field of vehicle dynamics and accomplish estimation by combining with the filtering algorithm. In addition, most of the vehicle dynamic estimation algorithms are limited by road excitation, and it is difficult to obtain satisfactory estimation results. This strategy can greatly reduce the system error caused by insufficient road excitation (slip rate is not 0.15–0.20) and improve the applicability of the estimation algorithm to the actual driving process of the vehicle. Finally, the magic formula (MF) tyre model is selected to describe the tyre characteristics after treatment of the normalized strategy; the tyre-road peak friction coefficient is estimated by combining the extended Kalman filter and vehicle dynamics model. Satisfactory estimation results are obtained in both simulation and real vehicle tests, which verifies the effectiveness of the proposed normalized strategy.

Keywords: tyre-road peak friction coefficient estimation; tyre model; normalization; incentive sensitivity

Citation: Han, Y.; Lu, Y.; Chen, N.; Wang, H. Research on the Identification of Tyre-Road Peak Friction Coefficient under Full Slip Rate Range Based on Normalized Tyre Model. *Actuators* **2022**, *11*, 59. https://doi.org/10.3390/act11020059

Academic Editor: Ioan Ursu

Received: 6 January 2022
Accepted: 12 February 2022
Published: 17 February 2022

1. Introduction

The tyre-road friction coefficient can describe the friction between the tyre and the road, which is very important for vehicle active safety control technology. The accurate estimation of the tyre-road friction coefficient helps to control vehicle driving performance, reduce slippage, and improve vehicle-handling stability. A large number of studies on vehicle stability have clearly put forward the use of the tyre-road friction coefficient to promote the improvement of vehicle safety control systems [1–3]. Therefore, the real-time and accurate estimation of the tyre-road friction coefficient is of great significance to improve the performance of vehicle control systems, such as the anti-lock braking system (ABS), electronic stability control (ESC), and active yaw control system (AYC) [2,4–6].

In recent years, scholars have conducted extensive research on the estimation method of the tyre-road friction coefficient and basically formed two kinds of estimation methods [7–9]: experiment-based and model-based.

The experiment-based method mainly measures the relevant signals (such as road surface morphology, tyre deformation, and noise) directly by sensors and establishes the corresponding relationships to obtain the tyre-road friction coefficient [7]. Among them, Leng B. et al. [10] accomplished road recognition and classification by extracting road color and texture features; Hong S. et al. [11] placed piezoelectric sensors inside the tyre to measure the lateral deflection of the tyre section and to estimate the tyre-road friction coefficient; J. Alonso et al. [12] used acoustic sensors to select and extract tyre noise to accomplish road recognition. The advantage of this kind of method is that it has a wide range of identification and a predictive effect [7,13]. However, its effectiveness is easily affected by the environment, and the sensors that need to be matched are relatively expensive [14], which is difficult for large-scale promotion.

The model-based estimation method only uses the common low-cost sensors of vehicles to measure or estimate the dynamic response change on the wheel or car body caused by the change in the tyre-road friction coefficient and then calculates the tyre-road friction coefficient [7]. Due to the wide applicability and high precision of this kind of algorithm, scholars have conducted a large body of research, which can be roughly divided into four kinds. The first is the tyre-road friction coefficient estimation method based on the slip-slope relationship. Gustafsson F. [15,16] proposed a tyre-road friction coefficient estimation method based on slip-slope, which has high accuracy only when the slip rate is less than 0.05. Wang J. et al. [17] improved the literature [15,16] and accomplished the coefficient identification of large-scale, slip rate driving conditions, but the estimation results cannot be updated at very low slip rates. The second is the estimation method based on nonlinear formula fitting. Germann S. et al. [18] fitted the nonlinear function based on the linear function, and Castro R.D. et al. [19] fitted the nonlinear equation based on the feedforward neural network (FFNN), both combined with the recursive least squares estimator to accomplish the online estimation of the tyre-road friction coefficient. This method is simple in principle, but it is difficult to guarantee the real-time performance. The third is the estimation method based on road state characteristic factors. Wang B. et al. [20] accomplished road recognition by constructing an eigenvalue that can represent typical road characteristic parameters. The method covers almost all the roads where cars normally travel and requires less sensor signals, but it is limited to straight braking conditions.

The fourth is the estimation method of the tyre-road friction coefficient based on tyre model, which is divided into two categories. The first category is based on the relationship between the tyre mechanical properties and the tyre-road friction coefficient, the slip rate in tyre model. By observing the tyre mechanical state (such as longitudinal force and lateral force), the parameters characterizing the tyre-road friction coefficient in tyre models (such as Dugoff [21,22], LuGre [23,24], Brush [25,26]) are calculated. In the second category, the tyre force is normalized based on the tyre model (such as Dugoff [27,28], Hsri [29], MF [30,31], Uni-tyre [32]), and the tyre-road friction coefficient is separated from the normalized force, which is suitable for establishing the system equation. The tyre-road friction coefficient is estimated by combining the filtering algorithm or iterative algorithm. This method based on the tyre model is called the normalized method according to the following reason. Since there are a large number of mature tyre model studies that can be referred to, different tyre models can be used to study different tyre dynamics fields, which offers great potential for the normalization methods based on tyre model. However, the estimation method based on a certain tyre model is limited in terms of accuracy, adaptability, and real-time performance.

The problems in the above model-based research are: most are based on longitudinal or lateral studies, with little regard to longitudinal- and lateral-coupling processes; most only consider a certain segment of the slip rate interval [0, 1]; they do not consider the corresponding relationship between the slip rate and the estimated tyre-road friction coefficient; the robustness and accuracy of the estimation algorithm cannot be guaranteed; as some novel algorithms only rely on Simulink, CarSim, and other simulation software to verify and do not use real vehicle verification, the actual feasibility is not verified; although some estimation algorithms have high estimation accuracy, they have large amounts of calculations and cannot guarantee real-time performance.

It can be seen from the above content that the characteristics of the estimation algorithm should be simple and practical and should have strong robustness, fast convergence, and strong incentive sensitivity. In the model-based estimation method, the principle of the estimation method based on the tyre model is easy to understand, the estimation accuracy is controllable, and the plasticity is strong. Among them, the normalization method based on a tyre model is simple and has a standard estimation process, which has the potential to apply to use most of the tyre models in the field of vehicle dynamics and makes this method most likely to have the above four excellent characteristics at the same time.

However, the performance of the normalization method based on a tyre model depends on the type of tyre model, and the algorithm can achieve the best performance by matching the high-precision tyre model in the research field. However, not all tyre models can be directly used for normalization. Usually, the more accurate the tyre model is, the more complex it is. Few simple tyre models can accurately reproduce the friction performance of the tyre-road interface while maintaining a simple form [33]. Therefore, a high precision tyre model is difficult to be used in this method.

In view of the difficulties faced by the normalized method based on tyre model, a new normalized strategy is proposed in this paper. This strategy establishes a direct connection between the tyre force output from the tyre model and the tyre-road peak friction coefficient according to the equal ratio relationship between tyre-road peak friction coefficient and the utilization of the friction coefficient on the adjacent typical roads. Normalization is achieved by introducing parameters from outside to avoid complex internal functions. The normalized tyre model is combined with the filtering algorithm and vehicle dynamics model to estimate the tyre-road peak friction coefficient. The normalized strategy can be applied to most tyre models in the field of vehicle dynamics, which means that almost all tyre models can be used to estimate the tyre-road peak friction coefficient. Different tyre models have high-fitting accuracy in different fields, which greatly expands the application scope of the tyre-road friction coefficient estimation algorithm based on tyre model.

In addition, most estimation algorithms can obtain accurate results only when the slip rate is within optimal range [0.15, 0.2], but the actual value of slip rate rarely reaches the optimal level in the vehicle-driving process. Additionally, at the optimal slip ratio stage, the road excitation on the tyre is too intense, which will negatively affect the vehicle-handling stability and comfort [5]. The system error in the non-optimal slip rate stage can be avoided using the equal ratio relationship. Therefore, in full slip rate range conditions, this algorithm can obtain accurate estimation results, and the robustness of the algorithm and high sensitivity to road excitation are ensured.

Finally, the classic MF tyre model is selected as the representative of complex tyre models. Combining the vehicle dynamics model and the extended Kalman filter, the tyre-road peak friction coefficient is estimated. The above algorithm is verified in simulation and in a real vehicle test.

The other parts are set as follows: the first part establishes the vehicle dynamics model; the second part mainly introduces the normalized strategy; the third part introduces the extended Kalman filter; the fourth and fifth parts are simulation and experimental verification; the sixth part is the conclusion.

2. Establish Vehicle Dynamics Model

The 3 DOF vehicle dynamics model is established, as shown in Figure 1.

The following motion differential equations are established.

Longitudinal equation:

$$m_z(\dot{v}_x - v_y\gamma) = F_{xfl}\cos\delta_f - F_{yfl}\sin\delta_f + F_{xfr}\cos\delta_f - F_{yfr}\sin\delta_f + F_{xrl} + F_{xrr} \tag{1}$$

Lateral equation:

$$m_z(\dot{v}_y + v_x\gamma) = F_{xfl}\sin\delta_f + F_{yfl}\cos\delta_f + F_{xfr}\sin\delta_f + F_{yfr}\cos\delta_f + F_{yrl} + F_{yrr} \tag{2}$$

Yaw equation:

$$\dot{\gamma} I_z = a\left(F_{xfl}\sin\delta_f + F_{yfl}\cos\delta_f + F_{xfr}\sin\delta_f + F_{yfr}\cos\delta_f\right) + \frac{T}{2}\left(F_{xrr} + F_{xfr}\cos\delta_f - F_{yfr}\sin\delta_f\right) - \frac{T}{2}\left(F_{xfl}\cos\delta_f - F_{yfl}\sin\delta_f + F_{xrl}\right) - \left(F_{yrl} + F_{yrr}\right)b \quad (3)$$

The load on each wheel can be expressed as:

$$F_{zfl} = \frac{b}{2L}m_z g - \frac{bh_g}{LT}m_b a_y - \frac{m_b a_x h_g}{2L} + k_\psi \psi_b \quad (4)$$

$$F_{zfr} = \frac{b}{2L}m_z g + \frac{bh_g}{LT}m_b a_y - \frac{m_b a_x h_g}{2L} + k_\psi \psi_b \quad (5)$$

$$F_{zrl} = \frac{a}{2L}m_z g - \frac{ah_g}{LT}m_b a_y + \frac{m_b a_x h_g}{2L} + k_\psi \psi_b \quad (6)$$

$$F_{zrr} = \frac{a}{2L}m_z g + \frac{ah_g}{LT}m_b a_y + \frac{m_b a_x h_g}{2L} + k_\psi \psi_b \quad (7)$$

where F_{xfl}, F_{xfr}, F_{xrl}, F_{xrr}, F_{yfl}, F_{yfr}, F_{yrl}, and F_{yrr} are the longitudinal and lateral forces of the four wheels, respectively; v_x and v_y are the longitudinal and lateral velocities of the vehicle centroid, respectively; γ, ψ_b, and δ_f are the yaw rate, roll angle, and front wheel angle, respectively; g is 9.8 m/s^2, a_x and a_y are the longitudinal and lateral accelerations of the vehicle centroid, respectively. The main parameters of the vehicle dynamics model are shown in Table 1.

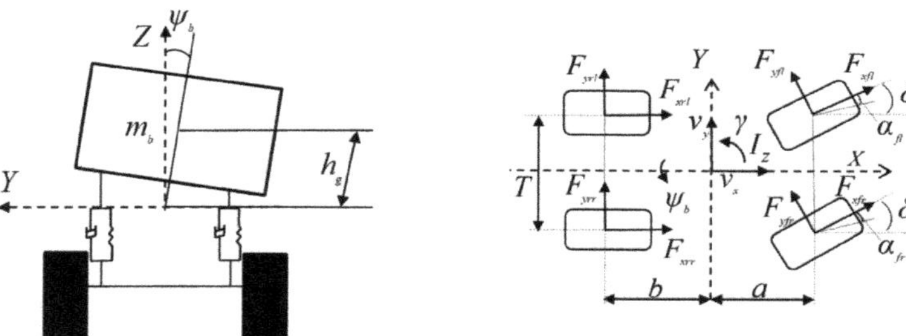

Figure 1. 3 DOF vehicle dynamics model.

Table 1. The main parameters of the test vehicle.

Symbol	Value	Notes
m_z	880 kg	Vehicle mass
m_b	788 kg	Sprung mass
L	2.040 m	Wheel base
a	1.145 m	Distance from centroid to front axle
b	0.895 m	Distance from centroid to rear axle
h_g	0.54 m	Centroid height
T	1.3 m	Wheel track width
I_z	832.3 kg·m^2	Moment of inertia about the z-axis
K_ψ	25,041 N/rad	Tyre slip angle stiffness

3. Normalized Strategy

3.1. Estimation Algorithm Process

The overall estimation algorithm process is shown in Figure 2.

The sensor signals from CarSim or real vehicle tests are processed to obtain the required parameters. Based on the Kiencke tyre model [34], the equal ratio relationship is proposed. Normalization of tyre model can be accomplished by this relationship. The normalized strategy framework is shown in Figure 3. The MF tyre model [35–37] is selected as the representative of high precision and high complexity tyre model. The normalized tyre model is matched with the vehicle dynamics model, and the estimated value of the tyre-road peak friction coefficient is obtained by the extended Kalman filter [38,39].

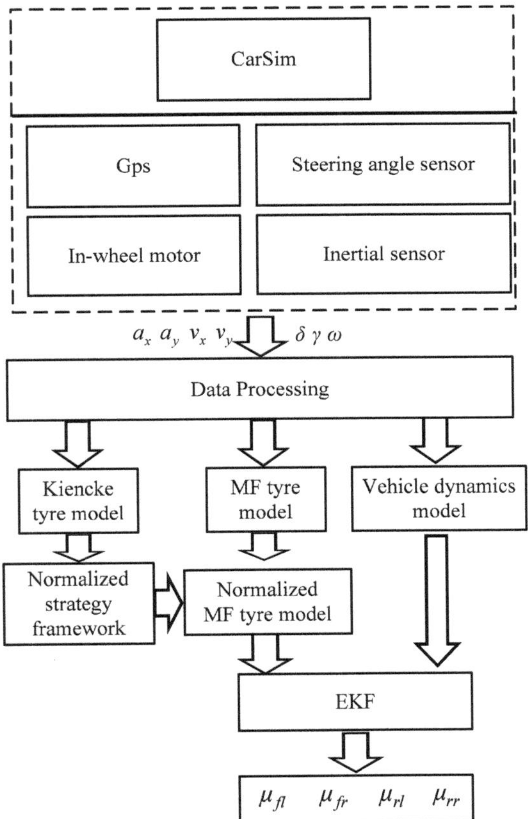

Figure 2. Estimation algorithm flow chart.

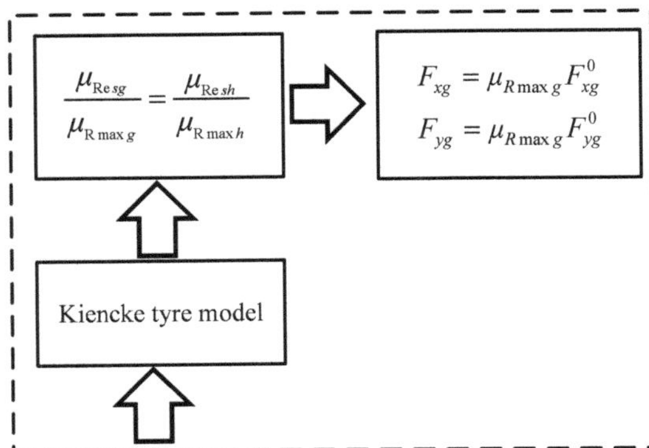

Figure 3. Normalized strategy framework.

3.2. Construction of Normalized Strategy

3.2.1. Kiencke Tyre Model

The Kiencke tyre model optimized the Buckhardt tyre model [40], as shown in Equation (8).

$$\mu_{Res}(s_{Res}) = \left(c_1\left(1 - e^{-c_2 s_{Res}}\right) - c_3 s_{Res}\right) \cdot e^{-c_4 \cdot s_{Res} \cdot v_c} \cdot \left(1 - c_5 F_Z^2\right) \tag{8}$$

where μ_{Res} is the tyre-road utilization friction coefficient, s_{Res} is slip rate, and v_c is the speed of the vehicle center of gravity. F_z is the vertical load on the vehicle. c_1, c_2, and c_3 change with road conditions. The parameter values of six typical roads are given in Table 2 [40]. The value of c_4 is between 0.002 s/m and 0.004 s/m, and the value of c_5 is 0.00015 $(1/\text{kN})^2$ [34].

Table 2. c_2 and c_3 fitting coefficient.

Road Surface Type	c_1	c_2	c_3
Dry asphalt	1.2801	23.99	0.52
Wet asphalt	0.857	33.822	0.347
Cement	1.1973	25.168	0.5373
Wet pebbles	0.4004	33.7080	0.1204
Ice	0.05	306.39	0
Snow	0.1946	94.129	0.0646

3.2.2. Similarity Analysis

According to the Kiencke tyre model, the relationship between the tyre-road utilization friction coefficient and the slip rate on the typical roads can be obtained, as shown in Figure 4.

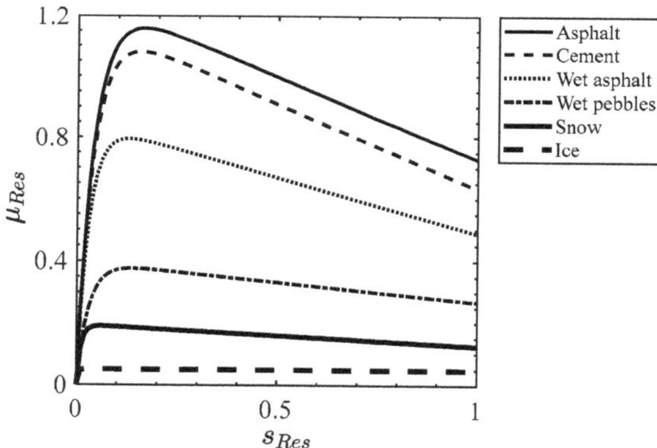

Figure 4. Variation curve of utilization friction coefficient of typical road with slip rate.

It can be seen from Figure 4 that under six typical roads, the change trend of the curve between the tyre-road utilization friction coefficient and slip rate is similar, especially between adjacent typical roads, such as asphalt and cement and wet pebbles and snow. Therefore, the relationship between the tyre-road utilization friction coefficient and the peak friction coefficient can be expressed as [41]:

$$\frac{\mu_{Resg}}{\mu_{Rmaxg}} = \frac{\mu_{Resh}}{\mu_{Rmaxh}} \tag{9}$$

assuming that road g and h are adjacent, and they are the target road and the adjacent road, respectively. μ_{Resg} and μ_{Resh} are the tyre-road utilization friction coefficients of road g and

h, respectively. $\mu_{\text{Rmax}g}$ and $\mu_{\text{Rmax}h}$ are the tyre-road peak friction coefficients of road g and h, respectively.

3.3. Tyre Model

The tyre force driving on the known road can be obtained by the tyre model. There are many tyre models in the field of vehicle dynamics, such as Dugoff, MF, LuGre, and Uni-Tyre. Therefore, in the control process, we can select the tyre model with the highest accuracy according to the tyre dynamic field studied.

The MF tyre model is widely used in vehicle dynamics simulation and analysis due to its high simulation accuracy and wide application range [7]. Because of its complex form and numerous and interrelated parameters, the MF tyre model is difficult to use directly for the normalized estimation algorithm.

To verify the normalized strategy, this paper will take the MF tyre model as an example to study the estimation of the tyre-road peak friction coefficient.

3.3.1. MF Tyre Model

In a single condition, the general expression of the longitudinal tyre force, F_{x0}, and the lateral tyre force, F_{y0}, is

$$F_{x0} = D_x \sin\left\{ C_x \tan^{-1}\left[B_x \kappa_x - E_x (B_x \kappa_x - \tan^{-1}(B_x \kappa_x)) \right] \right\} + S_{Vx} \tag{10}$$

$$F_{y0} = D_y \sin\left\{ C_y \tan^{-1}\left[B_y \alpha_y - E_y (B_y \alpha_y - \tan^{-1}(B_y \alpha_y)) \right] \right\} + S_{Vy} \tag{11}$$

Under combined conditions, the longitudinal tyre force and lateral tyre force can be expressed as

$$\begin{cases} F_x = \frac{|\sigma_x|}{\sigma} F_{x0} \\ F_y = \frac{|\sigma_y|}{\sigma} F_{y0} \end{cases} \tag{12}$$

The factors can be expressed as

$$\begin{cases} \sigma_x = \frac{\kappa}{1+\kappa} \\ \sigma_y = \frac{\tan \alpha}{1+\kappa} \\ \sigma = \sqrt{\sigma_x^2 + \sigma_x^2} \end{cases} \tag{13}$$

Longitudinal slip rate can be expressed as

$$\kappa_x = \frac{\omega r_e - v_{tx}}{v_{tx}} \tag{14}$$

where r_e is the effective rolling radius of the wheel.

The tyre sideslip angle can be expressed as

$$\alpha = \delta - \tan^{-1}\left(\frac{v_{ty}}{|v_{tx}|}\right) \tag{15}$$

where v_{tx} is the longitudinal wheel speed and v_{ty} is lateral wheel speed. For other parameters, see Appendix A.

3.3.2. Normalization of Tyre Model

Under pure longitudinal or pure lateral conditions, the MF tyre model can be expressed as

$$F = D \sin\left\{ C \tan^{-1}\left[B\kappa - E(B\kappa - \tan^{-1}(B\kappa)) \right] \right\} + S_v \tag{16}$$

Tyre force can be expressed as [36]

$$F = \mu_{Rmaxg} F_z \tag{17}$$

Combined with Equation (9), it can be extended to adjacent typical roads with different friction coefficients [40], which is

$$\frac{F_g}{\mu_{Rmaxg}} = \frac{F_h}{\mu_{Rmaxh}} \tag{18}$$

F_g is the tyre force when the vehicle runs on the target road. F_h is the tyre force when the vehicle runs on the adjacent road.

Equation (18) is simply transformed to

$$F_g = \frac{\mu_{Rmaxg} F_h}{\mu_{Rmaxh}} \tag{19}$$

Among them, μ_{Rmaxg} is the tyre-road peak friction coefficient which is to be identified.

In summary, for pure longitudinal conditions and pure lateral conditions, the tyre force can be expressed as:

$$F_{xg} = \frac{\mu_{Rmaxg} F_{xh}}{\mu_{Rmaxh}} \tag{20}$$

$$F_{yg} = \frac{\mu_{Rmaxg} F_{yh}}{\mu_{Rmaxh}} \tag{21}$$

where F_{xg} is the tyre force when the vehicle is in the pure longitudinal condition and runs on the target road. F_{yg} is the tyre force when the vehicle is in the pure lateral condition and runs on the target road.

According to Equations (10)–(12), the tyre force in the combined conditions can be expressed as

$$F_x = \mu_{Rmaxg} F_x^0 \tag{22}$$

$$F_x^0 = \frac{|\sigma_x|}{\mu_{Rmaxh}\sigma} \left\{ D_x \sin[C_x \tan^{-1}[B_x\kappa_x - E_x(B_x\kappa - \tan^{-1}(B_x\kappa_x))]] + S_{Vx} \right\} \tag{23}$$

$$F_y = \mu_{Rmaxg} F_y^0 \tag{24}$$

$$F_y^0 = \frac{|\sigma_y|}{\mu_{Rmaxh}\sigma} \left\{ D_y \sin[C_y \tan^{-1}[B_y\alpha_y - E_y(B_y\alpha_y - \tan^{-1}(B_y\alpha_y))]] + S_{Vy} \right\} \tag{25}$$

where F_x^0 and F_y^0 are the longitudinal and lateral normalized forces, respectively, independent of the tyre-road peak friction coefficient to be identified.

3.3.3. Establish System Equation

The following equations are used to estimate the tyre-road peak friction coefficient and are according to Equations (1)–(3), (22) and (24)

$$\dot{v}_x - v_y\gamma = \mu_{fl}\left(\frac{F_{xfl}^0 \cos\delta_f}{m_z} - \frac{F_{yfl}^0 \sin\delta_f}{m_z}\right) + \mu_{rl}\frac{F_{xrl}^0}{m_z} + \mu_{fr}\left(\frac{F_{xfr}^0 \cos\delta_f}{m_z} - \frac{F_{yfr}^0 \sin\delta_f}{m_z}\right) + \mu_{rr}\frac{F_{xrr}^0}{m_z} \tag{26}$$

$$\dot{v}_y + v_x\gamma = \mu_{rl}\frac{F_{yrl}^0}{m_z} + \mu_{rr}\frac{F_{yrr}^0}{m_z} + \mu_{fl}\left(\frac{F_{xfl}^0 \sin\delta_f}{m_z} + \frac{F_{yfl}^0 \cos\delta_f}{m_z}\right) + \mu_{fr}\left(\frac{F_{xfr}^0 \sin\delta_f}{m_z} + \frac{F_{yfr}^0 \cos\delta_f}{m_z}\right) \tag{27}$$

$$\dot{\gamma} = \mu_{fl}\left(\frac{a}{I_z}F_{xfl}^0 \sin\delta_f - \frac{T}{2I_z}F_{xfl}^0 \cos\delta_f + \frac{a}{I_z}F_{yfl}^0 \cos\delta_f + \frac{T}{2I_z}F_{yfl}^0 \sin\delta_f\right)$$
$$+ \mu_{fr}\left(\frac{a}{I_z}F_{xfr}^0 \sin\delta_f + \frac{T}{2I_z}F_{xfr}^0 \cos\delta_f + \frac{a}{I_z}F_{yfr}^0 \cos\delta_f - \frac{T}{2I_z}F_{yfr}^0 \sin\delta_f\right)$$
$$- \mu_{rl}\left(\frac{T}{2I_z}F_{xrl}^0 + \frac{b}{I_z}F_{yrl}^0\right) + \mu_{rr}\left(\frac{T}{2I_z}F_{xrr}^0 - \frac{b}{I_z}F_{yrr}^0\right) \tag{28}$$

The state equation and measurement equation can be obtained by Equations (26)–(28). The state equation is:

$$\begin{pmatrix} \mu_{fl}(n+1) \\ \mu_{fr}(n+1) \\ \mu_{rl}(n+1) \\ \mu_{rr}(n+1) \end{pmatrix} = \begin{pmatrix} 1 & 0 & 0 & 0 \\ 0 & 1 & 0 & 0 \\ 0 & 0 & 1 & 0 \\ 0 & 0 & 0 & 1 \end{pmatrix} \begin{pmatrix} \mu_{fl}(n) \\ \mu_{fr}(n) \\ \mu_{rl}(n) \\ \mu_{rr}(n) \end{pmatrix} + w(t) \tag{29}$$

The measurement equation can be expressed as:

$$\begin{pmatrix} (\dot{v}_x - v_y\gamma) \\ \dot{v}_y + v_x\gamma \\ \dot{\gamma} \end{pmatrix} = \begin{pmatrix} H(1,1) & H(1,2) & H(1,3) & H(1,4) \\ H(2,1) & H(2,2) & H(2,3) & H(2,4) \\ H(3,1) & H(3,2) & H(3,3) & H(3,4) \end{pmatrix} \begin{pmatrix} \mu_{fl} \\ \mu_{fr} \\ \mu_{rl} \\ \mu_{rr} \end{pmatrix} + v(t) \tag{30}$$

$$\begin{cases} H(1,1) = \dfrac{F^0_{xfl}\cos\delta_f}{m_z} - \dfrac{F^0_{yfl}\sin\delta_f}{m_z} \\[2mm] H(1,2) = \dfrac{F^0_{xfr}\cos\delta_f}{m_z} - \dfrac{F^0_{yfr}\sin\delta_f}{m_z} \\[2mm] H(1,3) = \dfrac{F^0_{xrl}}{m_z},\ H(1,4) = \dfrac{F^0_{xrr}}{m_z} \\[2mm] H(2,1) = \dfrac{F^0_{xfl}\sin\delta_f}{m_z} + \dfrac{F^0_{yfl}\cos\delta_f}{m_z} \\[2mm] H(2,2) = \dfrac{F^0_{xfr}\sin\delta_f}{m_z} + \dfrac{F^0_{yfr}\cos\delta_f}{m_z} \\[2mm] H(2,3) = \dfrac{F^0_{yrl}}{m_z},\ H(2,4) = \dfrac{F^0_{yrr}}{m_z} \\[2mm] H(3,1) = \dfrac{a}{I_z}F^0_{xfl}\sin\delta_f - \dfrac{T}{2I_z}F^0_{xfl}\cos\delta_f + \dfrac{a}{I_z}F^0_{yfl}\cos\delta_f + \dfrac{T}{2I_z}F^0_{yfl}\sin\delta_f \\[2mm] H(3,2) = \dfrac{a}{I_z}F^0_{xfr}\sin\delta_f + \dfrac{T}{2I_z}F^0_{xfr}\cos\delta_f + \dfrac{a}{I_z}F^0_{yfr}\cos\delta_f - \dfrac{T}{2I_z}F^0_{yfr}\sin\delta_f \\[2mm] H(3,3) = -\dfrac{T}{2I_z}F^0_{xrl} - \dfrac{b}{I_z}F^0_{yrl} \\[2mm] H(3,4) = \dfrac{T}{2I_z}F^0_{xrr} - \dfrac{b}{I_z}F^0_{yrr} \end{cases} \tag{31}$$

Among them, $\mu_{xij}(ij = fl, fr, rl, rr)$ represents the peak friction coefficient between the four tyres and the target road, and the random variables, $w(t)$ and $v(t)$, are process noise and measurement noise, respectively.

3.4. Determination of Adjacent Road

According to the existing research and experimental data [2], the tyre-road friction coefficient in Figure 4 is higher than the actual value. However, this does not affect the equal ratio relationship between the tyre-road utilization friction coefficient and the peak friction coefficient.

In the simulation part, the tyre-road peak friction coefficient is set to 0.85 and 0.9, respectively. The real vehicle test road is dry asphalt road; thus, the adjacent road is cement road.

4. EKF Estimation Algorithm

The extended Kalman filter estimation process [38] is shown in Figure 5.

The initial value in the filtering process can be expressed as the measurement noise covariance, $R = 0.03 \times I_{3\times3}$, and the process noise covariance is $P = 0.01 \times I_{4\times4}$, the initial covariance matrix is $P_0 = 0.02 \times I_{4\times4}$, and the initial estimate states matrix is $\mu_{R\max} = [0,0,0,0]^T$.

The systematic equations are illustrated in Section 3.3.3.

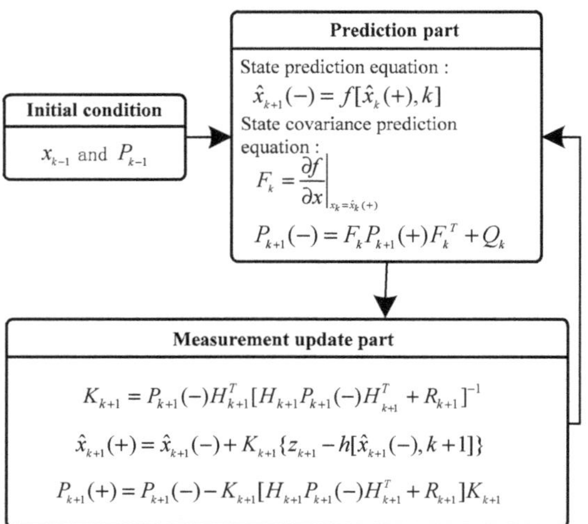

Figure 5. EKF flow chart.

5. Simulation Analysis and Verification

In this paper, Carsim and Matlab/Simulink are used for the simulation of the linear-braking condition and the curve-braking condition.

5.1. Simulation on High Adhesion Road

5.1.1. Linear-Braking Condition

The tyre-road peak friction coefficient is set to 0.85, and the initial velocity is 120 km/h. The simulation [42] results, shown in Figure 6a–c, are based on the four wheels of the car on the road with the same friction coefficient road, while considering the length of the article and taking the right front wheel as an example.

It can be seen from Figure 6a–c that the braking deceleration is close to 5.5 m/s^2. The slip rate of the right front wheel remains around 0.08, which is not enough to reach the range [0.15, 0.20] of slip rate corresponding to sufficient road excitation. However, the tyre-road peak friction coefficient converges to 0.85 before 0.4 s, and the overall situation is stable.

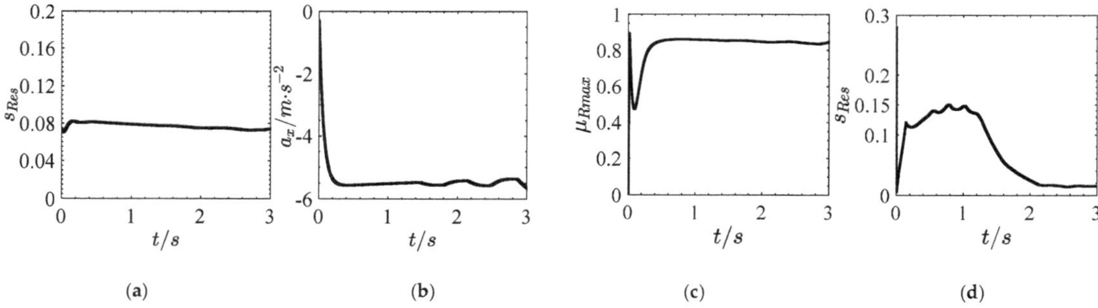

    (a)   (b)   (c)   (d)

Figure 6. *Cont.*

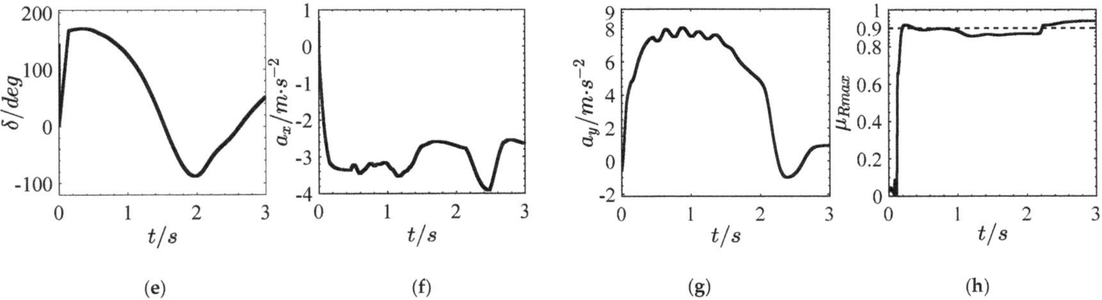

(e)　　　　　　　(f)　　　　　　　(g)　　　　　　　(h)

Figure 6. Simulation results on high adhesion road. (**a**) Slip rate. (**b**) Longitudinal acceleration. (**c**) Estimation results under linear braking condition. (**d**) Slip rate. (**e**) Steering wheel angle. (**f**) Longitudinal acceleration. (**g**) Lateral acceleration. (**h**) Estimation results of curve braking combined condition.

5.1.2. Curve-Braking Combined Condition

The annular road [43] with 33 m radius is set, the tyre-road peak friction coefficient is 0.9, and the initial speed is 60 km/h. Taking the right front wheel as an example, the simulation results are shown in Figure 6d–h.

It can be seen from Figure 6d–h that the maximum steering wheel angle is close to 170 degrees, the maximum braking deceleration is close to 4 m/s^2, and the slip rate is close to [0.15–0.20] at 0.2–1.4 s. At this time, the road excitation is close to sufficient. Under this condition, the estimated value of the tyre-road peak friction coefficient converges to 0.9 at about 0.2 s, remains stable to 1 s, then decreases to 0.87, lasts to 2.3 s, and then rises to 0.93. The overall value is maintained at about 0.9, and the error is maintained within [−0.04, 0.04].

5.2. Simulation on Low Adhesion Road

5.2.1. Linear-Braking Condition

The tyre-road peak friction coefficient is set to 0.2, and the initial velocity is 120 km/h. Taking the right front wheel as an example, the simulation results are shown in Figure 7a–c.

It can be seen from Figure 7a–c that the slip rate remains in [0.15, 0.24], which shows the road excitation is sufficient relatively. The braking deceleration is close to 2 m/s^2. Additionally, the tyre-road peak friction coefficient converges to 0.2 at about 0.2 s and then remains basically stable.

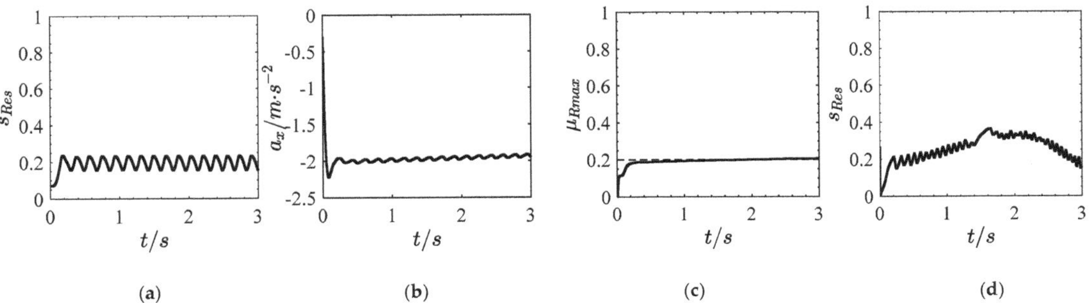

(a)　　　　　　　(b)　　　　　　　(c)　　　　　　　(d)

Figure 7. *Cont.*

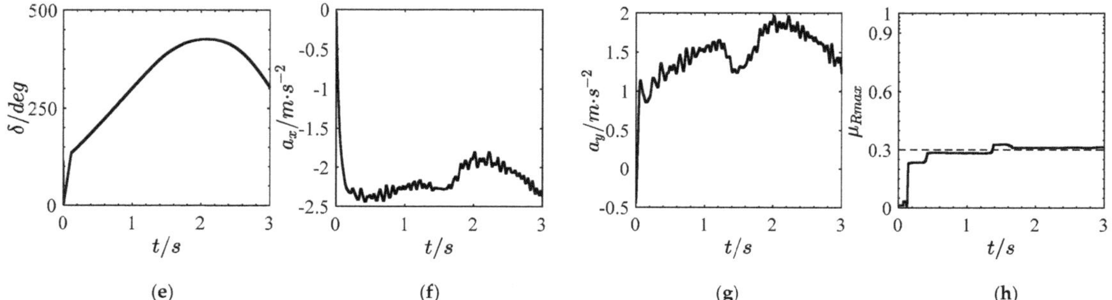

(e) (f) (g) (h)

Figure 7. Simulation results on low adhesion road. (**a**) Slip rate. (**b**) Longitudinal acceleration. (**c**) Estimation results under linear braking condition. (**d**) Slip rate. (**e**) Steering wheel angle. (**f**) Longitudinal acceleration. (**g**) Lateral acceleration. (**h**) Estimation results of curve braking combined condition.

5.2.2. Curve-Braking Combined Condition

The tyre-road peak friction coefficient is set to 0.3. Under turning conditions on low adhesion road, the initial speed is reduced to 35 km/h. Taking the right front wheel as an example, the simulation results are shown in Figure 7d–h.

It can be seen from Figure 7d–h that the slip rate fluctuates between [0.15 and 0.36] and mostly lies outside the optimum interval, which indicates the road excitation level is insufficient. The maximum longitudinal deceleration can reach 2.5 m/s^2, and the maximum lateral acceleration can reach 2 m/s^2. The estimation result converges to 0.3 before 0.45 s, and the estimation error is maintained within [-0.05, 0.05].

6. Test Verification

6.1. Calibration Test of Tyre-Road Peak Friction Coefficient

The BM-III pendulum friction coefficient tester [44] was used to calibrate the tyre-road peak friction coefficient of the test asphalt road surface. The test road is a 100 m straight road; the test results are shown in Figure 8.

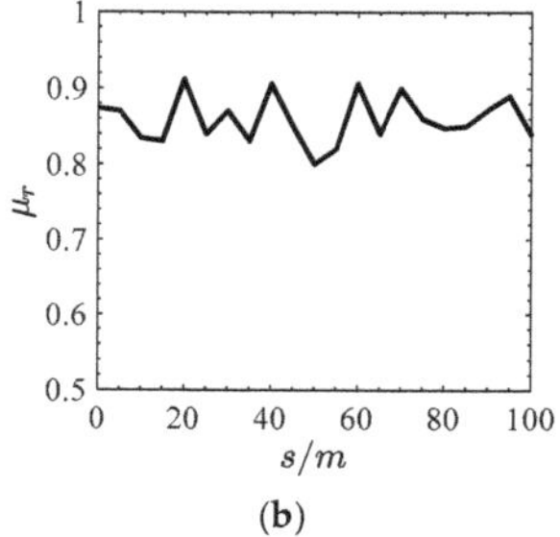

(a) (b)

Figure 8. Calibration test results. (**a**) Dry asphalt test road. (**b**) Test results.

Figure 8 shows that the actual value range of the tyre-road peak friction coefficient on the measured dry asphalt test road is [0.8, 0.92].

6.2. Real Vehicle Test

As shown in Figure 9, the test platform is a wire-controlled, modified UTV (Utility Vehicle), and the drive mode is four-wheel independent drive. The vehicle is equipped with a variety of sensors to check the test results. Sensors include GPS, inertial navigation, steering wheel angle sensors, etc.

Figure 9. Real vehicle test platform.

6.2.1. Straight Line Test

The road of the straight-line test [42] is dry asphalt road, as shown in Figure 8a. The speed is variable, and the average speed is 35 km/h. Taking the right front wheel as an example, the test results are shown in Figure 10a–c.

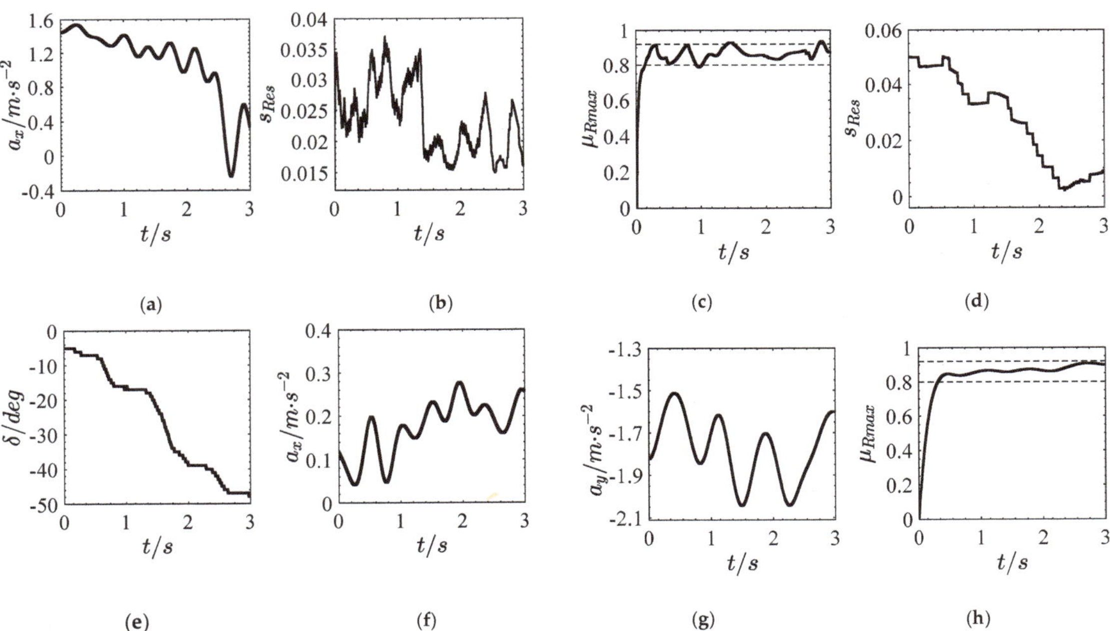

Figure 10. Real vehicle test results. (**a**) Longitudinal acceleration. (**b**) Slip rate. (**c**) Estimation results of straight-line test. (**d**) Slip rate. (**e**) Steering wheel angle. (**f**) Longitudinal acceleration. (**g**) Lateral acceleration. (**h**) Estimation results of steady-state-turning test.

It can be seen from Figure 10a–c that the slip rate fluctuates between 0.015 and 0.038 during the whole straight-driving stage. Under the insufficient road excitation, the peak friction coefficient converges to 0.8 at 0.2 s, then fluctuates within the range of [0.8, 0.92], and produces weak fluctuation errors at 1 s, 1.2 s, and 2.8 s, respectively.

6.2.2. Steady-State-Turning Test

The steady-state-turning test road [43] is a dry asphalt ring road with a radius of 33 m. The speed is variable, and the average speed is 40 km/h. The real vehicle test results are shown in Figure 10d–h.

Figure 10d–h shows that the maximum slip rate can reach 0.05 in the process of turning. Under insufficient road excitation, the estimated value of tyre-road peak friction coefficient converges to 0.8 before 0.4 s, and then stabilizes in [0.8, 0.92].

7. Conclusions

Based on the equal ratio relationship between the peak friction coefficient and the utilization friction coefficient on the adjacent typical roads, the novel normalized strategy is proposed. According to the strategy, the normalization process which is applicable to most tyre models in the field of vehicle dynamics is accomplished. In this paper, the normalized MF tyre model is combined with the vehicle dynamics model and EKF to estimate the tyre-road peak friction coefficient.

According to the simulation and real vehicle test results, when the vehicle is running on the dry asphalt road or the low adhesion road, the general braking or acceleration conditions cannot ensure that sufficient road excitation is triggered, which makes it hard to obtain an accurate estimation using most of the estimation methods based on vehicle dynamics. After the treatment of the normalized strategy and even in the case of insufficient road excitation, the estimation algorithm can also obtain accurate estimated results in time. The universality and high incentive sensitivity of the normalized strategy are verified.

In summary, the new normalized strategy proposed in this paper has great inclusiveness for tyre model, and the normalized estimation algorithm has strong sensitivity to road excitation. It greatly expands the application scope of the normalized estimation algorithm based on the tyre model and improves the robustness of the algorithm. In addition, the algorithm is simple and quick. It plays a great role in promoting the formation of a perfect tyre-road friction coefficient estimation algorithm and plays a positive role in promoting the development of the vehicle active safety system.

Highlights

1. The proposed strategy can improve the estimation algorithm's compatibility for the tyre model and expand the application scope.
2. The proposed strategy can improve the sensitivity to road excitation and improve adaptability to vehicle-driving conditions.
3. Satisfactory estimation results are obtained in both simulation and real vehicle tests.

Author Contributions: Conceptualization, Y.L.; formal analysis, Y.L.; funding acquisition, Y.L.; investigation, Y.H., N.C. and H.W.; methodology, Y.H. and Y.L.; resources, Y.L.; software, Y.H.; supervision, Y.L.; validation, Y.H., Y.L. and H.W.; writing—original draft, Y.H.; writing—review and editing, Y.H., Y.L. and N.C. All authors have read and agreed to the published version of the manuscript.

Funding: This research was funded by National Natural Science Foundation of China, grant number 12072204. National Natural Science Foundation of China, grant number 11572207. Natural Science Foundation of Hebei Province, grant number A2020210039. Independent Subject of State Key Laboratory of Mechanical Behavior and System Safety of Traffic Engineering Structures, grant number ZZ2020–32. And The APC was funded by National Natural Science Foundation of China, grant number 12072204.

Acknowledgments: This work is supported by: National Natural Science Foundation of China (Grant Nos. 12072204, 11572207), Natural Science Foundation of Hebei Province (Grant No.A2020210039), Independent Subject of State Key Laboratory of Mechanical Behavior and System Safety of Traffic Engineering Structures (ZZ2020–32).

Conflicts of Interest: The authors declare no conflict of interest.

Appendix A

The coefficients in the MF tyre model are expressed as follows:

Table A1. Longitudinal force under pure straight line working conditions.

$F_{x0} = D_x \sin[C_x \tan^{-1}\{B_x\kappa_x - E_x(B_x\kappa_x - \tan^{-1}(B_x\kappa_x))\}] + S_{VX}$	(A1)
$\kappa_x = \kappa + S_{Hx}$	(A2)
$\gamma_x = \gamma \cdot \lambda_{\gamma x}$	(A3)
$C_x = p_{Cx1} \cdot \lambda_{Cx}$	(A4)
$D_x = \mu_x \cdot F_z \cdot \zeta_1$	(A5)
$\mu_x = (p_{Dx1} + p_{Dx2}df_2) \cdot (1 - p_{Dx3}\gamma_x^2) \cdot \lambda_{\mu x}$	(A6)
$df_z = \frac{F_z - F_{z0}}{F_{z0}}$	(A7)
$E_x = (p_{Ex1} + p_{Ex2}df_z + p_{Ex3}df_z^2) \cdot \{1 - p_{Ex4}\mathrm{sgn}(\kappa_x)\} \cdot \lambda_{Ex}$	(A8)
$K_x = F_z \cdot (p_{Kx1} + p_{Kx2}df_z) \cdot \exp(p_{Kx3}df_z) \cdot \lambda_{Kx}$	(A9)
$B_x = K_x/(C_xD_x)$	(A10)
$S_{Hx} = (p_{Hx1} + p_{Hx2}df_z) \cdot \lambda_{Hx}$	(A11)
$S_{Vx} = F_z \cdot (p_{Vx1} + p_{Vx2}df_z) \cdot \lambda_{Vx} \cdot \lambda_{\mu x} \cdot \zeta_1$	(A12)

Table A2. Lateral force under steady-state pure-turning condition.

$F_{y0} = D_y \sin[C_y \tan^{-1}\{B_y\alpha_y - E_y(B_y\alpha_y - \tan^{-1}(B_y\alpha_y))\}] + S_{Vy}$	(A13)
$\alpha_y = \alpha + S_{Hy}$	(A14)
$\gamma_y = \gamma \cdot \lambda_{\gamma y}$	(A15)
$C_y = p_{Cy1} \cdot \lambda_{Cy}$	(A16)
$D_y = \mu_y \cdot F_z \cdot \zeta_2$	(A17)
$\mu_y = (p_{Dy1} + p_{Dy2}df_2) \cdot (1 - p_{Dy3}\gamma_y^2) \cdot \lambda_{\mu y}$	(A18)
$E_y = (p_{Ey1} + p_{Ey2}df_z) \cdot \{1 - (p_{Ey3} + p_{Ex4}\gamma_y)\mathrm{sgn}(\alpha_y)\} \cdot \lambda_{Ey}$	(A19)
$K_{y0} = p_{Ky1} \cdot F_{z0} \cdot \sin[2\arctan(\frac{F_z}{p_{ky2}F_0\lambda_{Fz0}})] \cdot \lambda_{Fz0} \cdot \lambda_{Ky}$	(A20)
$K_y = K_{y0} \cdot (1 - p_{Ky3}|\gamma_y|) \cdot \zeta_3$	(A21)
$B_y = K_y/(C_yD_y)$	(A22)
$S_{Hy} = (p_{Hy1} + p_{Hy2}df_z) \cdot \lambda_{Hy} + p_{Hy3}\gamma_y\zeta_0 + \zeta_4 - 1$	(A23)
$S_{Vy} = F_z \cdot [(p_{Vy1} + p_{Vy2}df_z) \cdot \lambda_{Vy} + (p_{Vy3} + p_{Vy4} \cdot df_z) \cdot \gamma_y] \cdot \lambda_{\mu y} \cdot \zeta_4$	(A24)

References

1. Guo, C.; Wang, X.; Su, L.; Wang, Y. Safety distance model for longitudinal collision avoidance of logistics vehicles considering slope and road adhesion coefficient. *Proc. Inst. Mech. Eng. Part D J. Automob. Eng.* **2020**, *235*, 498–512. [CrossRef]
2. Rajamani, R. *Vehicle Dynamics and Control*, 2nd ed.; Springer Science: London, UK, 2012.
3. Hu, J.; Rakheja, S.; Zhang, Y. Tire–road friction coefficient estimation based on designed braking pressure pulse. *Proc. Inst. Mech. Eng. Part D J. Automob. Eng.* **2021**, *235*, 1876–1891. [CrossRef]
4. Ma, B.; Lv, C.; Liu, Y.; Zheng, M.; Yang, Y.; Ji, X. Estimation of Road Adhesion Coefficient Based on Tire Aligning Torque Distribution. *J. Dyn. Syst. Meas. Control* **2017**, *140*, 051010. [CrossRef]
5. Manuel, A.; Stratis, K.; Mike, B. Road friction virtual sensing: A review of estimation techniques with emphasis on low excitation approaches. *Appl. Sci.* **2017**, *7*, 1230.
6. Peng, Y.; Chen, J.; Ma, Y. Observer-based estimation of velocity and tire-road friction coefficient for vehicle control systems. *Nonlinear Dyn.* **2019**, *96*, 363–387. [CrossRef]
7. Khaleghian, S.; Emami, A.; Taheri, S. A technical survey on tyre-road friction estimation. *Friction* **2017**, *5*, 123–146. [CrossRef]
8. Meyer, W.E.; Walter, J.D. *Frictional Interaction of Tire and Pavement*; ASTM International: Philadelphia, PA, USA, 1983.
9. Ping, X.; Cheng, S.; Yue, W.; Du, Y.; Wang, X.; Li, L. Adaptive estimations of tyre–road friction coefficient and body's sideslip angle based on strong tracking and interactive multiple model theories. *Proc. Inst. Mech. Eng. Part D J. Automob. Eng.* **2020**, *234*, 3224–3238. [CrossRef]
10. Leng, B.; Jin, D.; Xiong, L.; Yang, X.; Yu, Z. Estimation of tire-road peak adhesion coefficient for intelligent electric vehicles based on camera and tire dynamics information fusion. *Mech. Syst. Signal Process.* **2020**, *150*, 107275. [CrossRef]

11. Hong, S.; Erdogan, G.; Hedrick, K.; Borrelli, F. Tyre-road friction coefficient estimation based on tyre sensors and lateral tyre deflection: Modelling, simulations and experiments. *Veh. Syst. Dyn.* **2013**, *51*, 627–647. [CrossRef]
12. Alonso, J.; López, J.; Pavón, I.; Recuero, M.; Asensio, C.; Arcas, G.; Bravo, A. On-board wet road surface identification using tyre/road noise and Support Vector Machines. *Appl. Acoust.* **2014**, *76*, 407–415. [CrossRef]
13. Xiong, Y.; Yang, X. A review on in-tire sensor systems for tire-road interaction studies. *Sens. Rev.* **2018**, *38*, 231–238. [CrossRef]
14. Zhang, Z.; Zheng, L.; Wu, H.; Zhang, Z.; Li, Y.; Liang, Y. An estimation scheme of road friction coefficient based on novel tyre model and improved SCKF. *Veh. Syst. Dyn.* **2021**. [CrossRef]
15. Gustafsson, F. Slip-based tire-road friction estimation. *Automatica* **1997**, *33*, 1087–1099. [CrossRef]
16. Gustafsson, F. Monitoring tire-road friction using the wheel slip. *IEEE Control Syst.* **1998**, *18*, 42–49.
17. Wang, J.; Alexander, L.; Rajamani, R. Friction Estimation on Highway Vehicles Using Longitudinal Measurements. *J. Dyn. Syst. Meas. Control* **2004**, *126*, 265–275. [CrossRef]
18. Germann, S.; Wurtenberger, M.; Daiss, A. Monitoring of the friction coefficient between tyre and road surface. In Proceedings of the Third IEEE Conference on Control Applications, Scotland, UK, 24–26 August 1994; Volume 1, pp. 613–618.
19. de Castro, R.; Araujo, R.E.; Cardoso, J.S.; Freitas, D. A new linear parametrization for peak friction coefficient estimation in real time. In Proceedings of the 2010 IEEE Vehicle Power and Propulsion Conference, Lille, France, 1–3 September 2010. [CrossRef]
20. Wang, B.; Guan, H.; Lu, P.; Zhang, A. Road surface condition identification approach based on road characteristic value. *J. Terramech.* **2014**, *56*, 103–117. [CrossRef]
21. Ghandour, R.; Victorino, A.; Doumiati, M.; Charara, A. Tire/road friction coefficient estimation applied to road safety. In Proceedings of the 18th Mediterranean Conference on Control and Automation, MED'10, Marrakech, Morocco, 23–25 June 2010; pp. 1485–1490.
22. Chen, L.; Bian, M.; Luo, Y.; Li, K. Maximum tire road friction estimation based on modified Dugoff tire model. In Proceedings of the International Conference on Mechanical and Automation Engineering IEEE, Jiujang, China, 21–23 July 2013; pp. 56–61.
23. Xin, W.; Liang, G.; Mingming, D.; Xiaolei, L. State estimation of tire-road friction and suspension system coupling dynamic in braking process and change detection of road adhesive ability. *Proc. Inst. Mech. Eng. Part D J. Automob. Eng.* **2021**. [CrossRef]
24. Sharifzadeh, M.; Senatore, A.; Farnam, A.; Akbari, A.; Timpone, F. A real-time approach to robust identification of tyre–road friction characteristics on mixed-μ roads. *Veh. Syst. Dyn.* **2018**, *57*, 1338–1362. [CrossRef]
25. Zhao, J.; Zhang, J.; Zhu, B. Coordinative traction control of vehicles based on identification of the tyre–road friction coefficient. *Proc. Inst. Mech. Eng. Part D J. Automob. Eng.* **2016**, *230*, 1585–1604. [CrossRef]
26. Gao, L.; Xiong, L.; Lin, X.; Xia, X.; Liu, W.; Lu, Y.; Yu, Z. Multi-sensor Fusion Road Friction Coefficient Estimation during Steering with Lyapunov Method. *Sensors* **2019**, *19*, 3816. [CrossRef]
27. Wu, Z.C. *Research on the Algorithm of the Road Friction Coefficient Estimation Based on the Extended Kalman Filter*; Jilin University: Changchun, China, 2008.
28. Li, G.; Fan, D.S.; Wang, Y.; Xie, R.C. Study on vehicle driving state and parameters estimation based on triple cubature Kalman filter. *Int. J. Heavy Veh. Syst.* **2020**, *27*, 126–144. [CrossRef]
29. Zong, C.; Hu, D.; Zheng, H. Dual extended Kalman filter for combined estimation of vehicle state and road friction. *Chin. J. Mech. Eng.* **2013**, *26*, 313–324. [CrossRef]
30. Ren, Y.; Yin, G.; Li, G.; Xia, T.; Liang, J.; Meyer, V. Tire-Road Friction Coefficient Estimators for 4WID Electric Vehicles on Diverse Road Conditions. *J. Mech. Eng.* **2019**, *55*, 80–92. (In Chinese)
31. Yi, K.; Hedrick, K.; Lee, S.C. Estimation of Tire-Road Friction Using Observer Based Identifiers. *Veh. Syst. Dyn. Int. J. Veh. Mech. Mobil.* **2010**, *31*, 233–261. [CrossRef]
32. Zhang, L. *The Research of Vehicle Stability Control Test and Control Algorithm in Ashered Limit State*; China Agricultural University: Beijing, China, 2016. (In Chinese)
33. Chen, L.; Bian, M.; Luo, Y.; Li, K. Real-time identification of the tyre–road friction coefficient using an unscented Kalman filter and mean-square-error-weighted fusion. *Proc. Inst. Mech. Eng. Part D J. Automob. Eng.* **2015**, *230*, 788–802. [CrossRef]
34. Kiencke, U.; Nielsen, L. *Automotive Control Systems: For Engine, Driveline, and Vehicle*; Springer: Berlin/Heidelberg, Germany, 2005.
35. Kuiper, E.; Van Oosten, J.J.M. The PAC2002 advanced handling tyre model. *Veh. Syst. Dyn.* **2007**, *45*, 153–167. [CrossRef]
36. Bakker, E.; Nyborg, L.; Pacejka, H.B. *Tyre Modeling for Use in Vehicle Dynamics Studies*; SAE International Congress and Exposition: Detroit, MI, USA, 1987. [CrossRef]
37. Song, Y.; Shu, H.; Chen, X.; Luo, S. Direct-yaw-moment control of four-wheel-drive electrical vehicle based on lateral tyre–road forces and tyre slip angle observer. *IET Intell. Transp. Syst.* **2019**, *13*, 303–312. [CrossRef]
38. Yoon, J.H.; Eben Li, S.; Ahn, C. Estimation of vehicle sideslip angle and tire-road friction coefficient based on magnetometer with GPS. *Int. J. Automot. Technol.* **2016**, *17*, 427–435. [CrossRef]
39. Hu, J.; Rakheja, S.; Zhang, Y. Real-time estimation of tire–road friction coefficient based on lateral vehicle dynamics. *Proc. Inst. Mech. Eng. Part D J. Automob. Eng.* **2020**, *234*, 2444–2457. [CrossRef]
40. Burckhardt, M. *Wheel Slip Control Systems*; Vogel Verlag: Würzburg, Germany, 1993.
41. Yuan, C.C.; Zhang, L.F.; Chen, L.; He, Y.; Shen, J.; Bei, S. A Research on the Algorithm for Identifying the Peak Adhesion Coefficient of Road Surface. *Automot. Eng.* **2017**, *39*, 1268–1273. (In Chinese)
42. *ISO 21994:2007*; Passenger Cars—Stopping Distance at Straight-Line Braking with ABS—Openloop Test Method. International Organization for Standardization: Geneva, Switzerland, 2007.

43. *ISO 4138:2004*; Passengers Cars—Steady-State Circular Driving Behaviour—Open-Loop Test Methods. International Organization for Standardization: Geneva, Switzerland, 2012.
44. Jafari, K.; Toufigh, V. Interface between Tire and Pavement. *J. Mater. Civ. Eng.* **2017**, *29*, 04017123. [CrossRef]

actuators

Article

A New Torque Distribution Control for Four-Wheel Independent-Drive Electric Vehicles

Dejun Yin [1], Junjie Wang [1,*], Jinjian Du [1], Gang Chen [1] and Jia-Sheng Hu [2]

1   School of Mechanical Engineering, Nanjing University of Science and Technology, Nanjing 210094, China; yin@njust.edu.cn (D.Y.); jinjiandu@njust.edu.cn (J.D.); chengang@njust.edu.cn (G.C.)
2   Department of Greenergy, National University of Tainan, Tainan 700, Taiwan; jogson@mail.nutn.edu.tw
*   Correspondence: wjj2719@njust.edu.cn

Abstract: Torque distribution control is a key technique for four-wheel independent-drive electric vehicles because it significantly affects vehicle stability and handling performance, especially under extreme driving conditions. This paper, which focuses on the global yaw moment generated by both the longitudinal and the lateral tire forces, proposes a new distribution control to allocate driving torques to four-wheel motors. The proposed objective function not only minimizes the longitudinal tire usage, but also make increased use of each tire to generate yaw moment and achieve a quicker yaw response. By analysis and a comparison with prior torque distribution control, the proposed control approach is shown to have better control performance in hardware-in-the-loop simulations.

Keywords: electric vehicles; independent drive; direct yaw control; torque distribution

Citation: Yin, D.; Wang, J.; Du, J.; Chen, G.; Hu, J.-S. A New Torque Distribution Control for Four-Wheel Independent-Drive Electric Vehicles. *Actuators* **2021**, *10*, 122. https://doi.org/10.3390/act10060122

Academic Editors: Olivier Sename, Van Tan Vu and Thanh-Phong Pham

Received: 23 April 2021
Accepted: 2 June 2021
Published: 6 June 2021

Publisher's Note: MDPI stays neutral with regard to jurisdictional claims in published maps and institutional affiliations.

1. Introduction

Electric vehicles (EVs) are enjoying a wide distribution in road transportation not only thanks to their benefits for the environment [1], but also owing to their better dynamic performance [2].

Of the current EVs, four-wheel independent-drive electric vehicles (4WIDEVs), with motors installed in each wheel, have great advantages in generating both traction and braking torque quickly, accurately, and independently. These merits make 4WIDEVs an ideal platform for active chassis control, especially for direct yaw moment control (DYC). The DYC system, in contrast to four-wheel steering (4WS) and active front-wheel steering (AFS), utilizes the yaw moment directly generated by a reasonable distribution of longitudinal forces to adjust vehicle motion [3,4]. Therefore, as the basis of a DYC system, torque distribution control plays a key role in maintaining vehicle stability [5–7].

The early torque distribution control method for 4WIDEVs adopted a rule-based distribution method. Considering the tire characteristics, Shan formulated new rules to arrange the execution of actuators in a certain order [8]. Park took both the characteristic of independent wheel motor and tire friction circle into account and proposed a novel torque distribution algorithm based on daisy-chaining allocation [9]. Although it is easy to achieve this implementation, this kind of method, based on specific rules, had weak adaptability to the environment and low allocation accuracy. It faced difficulties in satisfying performance requirements under actual various driving conditions.

For this purpose, current research works have been adopting the optimal control theory to conduct torque distribution control to improve the control performance of DYC. For optimal control, it is very important to find the suitable objective function and constraints.

Joa and Feng proposed integration methods to minimize the allocation error, unintended braking, and tire slip [10,11]. However, in the critical situation in which DYC operates, it is more important to keep vehicles stable while passing through a curve quickly than to minimize tire dissipation or unintended deceleration.

Hori and Peng considered the sum of squares of longitudinal and lateral tire forces as an index to optimize torque distribution [12]. This method is very close to the concept of tire usage. Mokhiamar and Abe proposed the concept of tire workload usage first and built up an objective function. In the subsequent research works, a weight coefficient and more constraints were involved to improve the control performance [13,14]. The method-based tire usage rate relies on the idea that, the smaller the tire usage, the larger the margin left for lateral force and the more stable the vehicle. Ono introduced the tire grip margin coefficient, which minimized and equaled the usage of each tire. Additionally, this research proved the convergence of the proposed objective function [15]. Ignoring the uncontrollable lateral tire force, Yu took the constraints of the motor peak torque and road contact surface into account and defined a new objective function including the longitudinal tire force and weight coefficient [16]. Based on this research, Yang gave consideration to the relation between the lateral and longitudinal tire forces [17]. Wang added the constraint condition of longitudinal tire forces [18], and Guo also considered wheel slip ratio control for emergency conditions [19]. Li proposed a multifunctional optimization approach to simultaneously minimize the errors of force and moment at the center of gravity, actuator control efforts, and tire usage [20]. In addition to the driving safety object function, Huang also took drive system efficiency into account in their controller design [21]. Hu decoupled four-wheel torque vectoring and innovated a two-level distribution formula to reduce energy consumption while ensuring handling stability [22].

Nevertheless, these control designs based on tire usage devote the most effort to a single tire rather than on the rigid characteristics of 4WIDEVs. For example, in the curve scenario, even with the same tire usage, the left wheels have obvious differences from the right wheels in terms of their potential and contribution to global yaw moment. In the same way, the front and rear wheels also have different efficiencies in the generation of yaw moment. The simple consideration of tire usage cannot make full use of each tire to generate yaw moment. Thus, there is some space left for improving the DYC performance in 4WIDEVs.

Therefore, this paper, considering tire usage as well as the efficiency of global yaw moment generation, focuses on the development of a new torque distribution control system for 4WIDEVs. This system includes the models of yaw moment generation constructed for each tire and involves a new objective function to improve the DYC performance.

The rest of the paper is organized as follows. Section 2 describes the vehicle's dynamic model. Section 3 proposes the optimal torque distribution control approach. Section 4 validates the effectiveness and real-time performance of the proposed approach in a hierarchical DYC system. Section 5 analyzes its implementation, compares it with a typical method on the basis of the optimal tire usage, and analyzes the reason for its higher performance in depth.

2. Vehicle Dynamic Model

A seven-degree-of-freedom (7-DOF) vehicle dynamic model—including the longitudinal, lateral, and yaw motion of the chassis as well as the rotation of the four wheels—was constructed for controller design. The chassis plane motion model is presented in Figure 1. Table 1 displays the definition of the notation used in the model.

The corresponding equations of vehicle planar motion are as follows:

$$m\left(\dot{u} - vr\right) = \left(F_{x1} + F_{x2}\right)\cos\delta + \left(F_{x3} + F_{x4}\right) - \left(F_{y1} + F_{y2}\right)\sin\delta \tag{1}$$

$$m\left(\dot{v} - ur\right) = \left(F_{x1} + F_{x2}\right)\sin\delta + \left(F_{y3} + F_{y4}\right) - \left(F_{y1} + F_{y2}\right)\cos\delta \tag{2}$$

$$I_z\dot{r} = l_f\left(F_{y1} + F_{y2}\right)\cos\delta - l_r\left(F_{y3} + F_{y4}\right) + \frac{d_f\left(F_{y1} - F_{y2}\right)\sin\delta}{2} + \frac{d_r}{2}\left(F_{x4} - F_{x3}\right) \\ + \frac{d_f}{2}\left(F_{x2} - F_{x1}\right)\cos\delta + l_f\left(F_{x1} + F_{x2}\right)\sin\delta \tag{3}$$

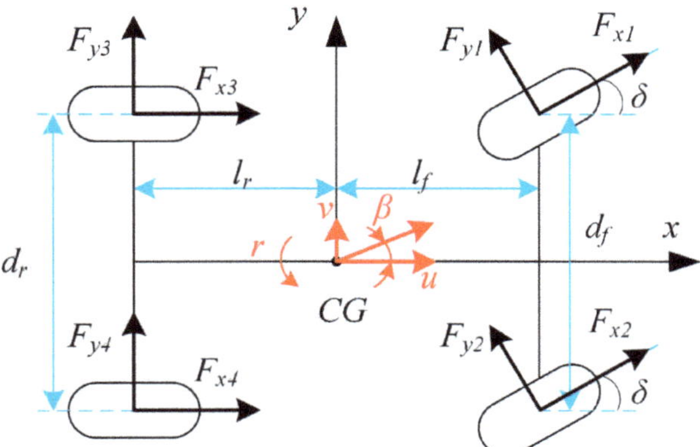

Figure 1. Vehicle plane motion model.

Table 1. Definitions of symbols used in modeling.

Symbols	Definitions	Unit
CG	Center of gravity	
C_f	Cornering stiffness of front wheels	N/rad
C_r	Cornering stiffness of rear wheels	N/rad
d_f	Front track width	m
d_r	Rear track width	m
F_{xi}	Longitudinal force of the ith tire	N
F_{yi}	Lateral force of the ith tire	N
I_w	Rotational inertia of the wheel	kg·m^2
I_z	Yaw moment of inertia of the vehicle	kg·m^2
l_f	Distance from CG to front axle	m
l_r	Distance from CG to rear axle	m
l	Distance from front axle to rear axle	m
m	Vehicle mass	kg
R_{eff}	Wheel effective radius	m
T_{wi}	Motor torque on the ith wheel	N·m
u	Vehicle longitudinal velocity	m/s
v	Vehicle lateral velocity	m/s
r	Yaw rate	rad/s
β	Sideslip angle	rad
δ	Steering wheel angle	rad
ω_i	Wheel rotational speed	rad/s

The tire rotation dynamic equations can be described as

$$I_w \dot{\omega}_i = T_{wi} - F_{xi} R_{eff} \tag{4}$$

3. Controller Design

As illustrated in Figure 2, this study employed a hierarchical DYC system comprising three layers: a parameter estimator, yaw moment controller, and torque distribution controller. The parameter estimator uses measurable sensor signals to estimate the sideslip angle and tire forces [23,24]. The measured parameters include the longitudinal acceleration a_x, lateral acceleration a_y, yaw rate r, wheel angular velocity ω_i, and steering wheel angle δ. The measured and estimated parameters are input to the upper yaw motion controller.

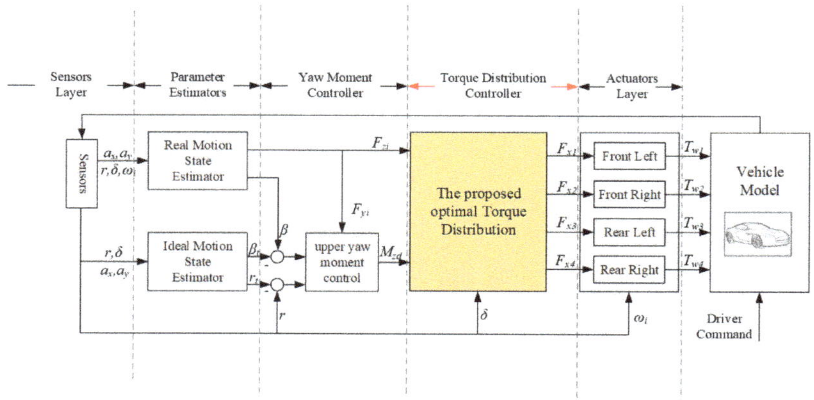

Figure 2. Hierarchical direct yaw moment control (DYC) system.

The upper yaw moment controller calculates the global yaw moment requirement on the CG to follow the desired sideslip angle and yaw rate and sends it as the equality constraint of the torque distribution controller.

Finally, the torque distribution controller allocates the optimal driving torque command to the four in-wheel motors to comply with the global yaw moment requirement.

3.1. Yaw Moment Controller

Thanks to its high robustness to sensor noise and variation in the vehicle state parameters, the sliding mode control method is easy to implement and widely used in vehicle stability controllers [25,26]. This study takes advantage of the sliding mode control method to design the yaw moment controller. The sliding surface is designed as

$$S = k_3(r - r_t) + k_4(\beta - \beta_t) \tag{5}$$

where r_t and β_t are the target yaw rate and sideslip angle, respectively, which can be obtained from a 2-DOF vehicle model [27]. k_3 and k_4 are the weight coefficients, and r_t and β_t are calculated as

$$r_t = \frac{1}{1 - \frac{mu^2}{2l^2}\frac{l_f C_f - l_r C_r}{C_f C_r}}\frac{u}{l}\delta \tag{6}$$

$$\beta_t = \frac{1 - \frac{mu^2}{2l^2}\frac{l_f}{l_r C_r}}{1 - \frac{mu^2}{2l^2}\frac{l_f C_f - l_r C_r}{C_f C_r}}\frac{l_r}{l}\delta \tag{7}$$

where the parameters in (6) and (7) are listed in Table 1. For the convenience of calculation, the tire stiffness is replaced by an approximate fixed value. Owing to the limitation of road adhesion, the target yaw rate and sideslip angle have an upper limitation, which can be expressed as

$$r_{max} = 0.85\left|\frac{\mu g}{u}\right| \tag{8}$$

$$\beta_{max} = \tan^{-1}(0.02\mu g) \tag{9}$$

where μ is the road friction coefficient and is assumed to be a constant.

The switching control law is designed as follows:

$$\dot{S} = -k_1 sgn(S) - k_2 S \tag{10}$$

The control law presented in (10) eliminates the system chattering caused by the sign switching function $sgn(S)$ at high frequencies.

The sliding surface (5) is derived as

$$\dot{S} = k_3 (\dot{r} - \dot{r}_t) + k_4 \left(\dot{\beta} - \dot{\beta}_t \right) \tag{11}$$

The output of the upper yaw moment controller is set to M_{zd}. The yaw moment generated by longitudinal tire forces is easy to control directly, so it is suitable as the output of the upper yaw moment controller. Combined with Formulas (3), (10), and (11), the output is as follows:

$$
\begin{aligned}
M_{zd} = {} & I_z \dot{r} - \left[l_f (F_{y1} + F_{y2}) \cos \delta + \tfrac{d_f}{2} (F_{y1} - F_{y2}) \sin \delta - l_r (F_{y3} + F_{y4}) \right] \\
= {} & I_z \left\{ \dot{r}_t + \tfrac{1}{k_3} \left[-k_1 sgn(S) - k_2 S - k_4 \left(\dot{\beta} - \dot{\beta}_t \right) \right] \right\} \\
& - \left[l_f (F_{y1} + F_{y2}) \cos \delta + \tfrac{d_f}{2} (F_{y1} - F_{y2}) \sin \delta - l_r (F_{y3} + F_{y4}) \right]
\end{aligned}
\tag{12}
$$

According to the Lyapunov stability theory, in order to make the system stable, k_1 and k_2 are positive constants. The smaller k_1 is, the smaller the chattering is. For a good balance between response and stability, the values of the four control parameters (k_1, k_2, k_3, and k_4) were tuned as 0.01, 50, 1.0, and −0.5 in the simulation, respectively.

Finally, the stability of the system using Formula (10) as the control law is analyzed. The stability is proven as follows:

Consider the Lyapunov function as follows:

$$V = \frac{1}{2} S^2 \tag{13}$$

By substituting the control law of Formula (10), the following can be obtained:

$$\dot{V} = S\dot{S} = S(-k_1 sgn(S) - k_2 S) = -k_1 |S| - k_2 S^2 < 0 \tag{14}$$

3.2. Torque Distribution Controller

In order to make full use of the lateral and longitudinal tire forces to generate the yaw moment, this paper proposes a new nonlinear optimal torque distribution control approach, with the objective function shown in (15). The ratio of the yaw moment generated by the longitudinal tire force to the global yaw moment, as well as the tire usage to be minimized, indicates that, in addition to the advantages of tire usage method, use is made of the lateral tire force to contribute as large a yaw moment as possible, and the rigid characteristics of 4WIDEVs have also been fully considered.

$$min\ J = \sum_{i=1}^{4} \left(\frac{F_{xi}}{F_{zi}} \cdot \frac{M_{xi}}{M_{zi}} \right)^2 \tag{15}$$

where F_{xi} is the longitudinal force of the ith wheel (i = 1, 2, 3, and 4), F_{zi} is the vertical load of the ith wheel (i = 1, 2, 3, and 4), M_{xi} is the yaw moment generated by the ith in-wheel motor driving force, M_{yi} is the yaw moment generated from the ith lateral tire force, and M_{zi} is the sum of M_{xi} and M_{yi}.

According to the 7-DOF vehicle dynamic model, M_{xi} and M_{yi} (i = 1, 2, 3, and 4) in (15) can be described as

$$
\begin{aligned}
M_{x1} &= F_{x1} \left(l_f \sin \delta - \tfrac{d_f}{2} \cos \delta \right) \\
M_{x2} &= F_{x2} \left(l_f \sin \delta + \tfrac{d_f}{2} \cos \delta \right) \\
M_{x3} &= -F_{x3} \tfrac{d_r}{2} \\
M_{x4} &= F_{x4} \tfrac{d_r}{2}
\end{aligned}
\tag{16}
$$

$$
\begin{aligned}
M_{y1} &= F_{y1}\left(l_f \cos\delta + \tfrac{d_f}{2}\sin\delta\right) \\
M_{y2} &= F_{y2}\left(l_f \cos\delta - \tfrac{d_f}{2}\sin\delta\right) \\
M_{y3} &= -F_{y3}l_r \\
M_{y4} &= -F_{y4}l_r
\end{aligned}
\tag{17}
$$

Although the relation between the longitudinal and lateral tire forces can be approximately expressed as a friction ellipse, a simplified circle model with a safety factor s is proposed to reduce the computation for actual implementation. s is set to 0.8 in the formula to indicate that, even when the tire slip angle is large, the lateral force is not over estimated. The tire circle model is expressed as

$$
F_{yi} = sgn(\delta)\cdot\sqrt{\left(s\mu F_{zi}\right)^2 - F_{xi}{}^2}
\tag{18}
$$

where the ith tire lateral force, F_{yi}, and traction/braking force, F_{xi}, are restricted by the friction coefficient μ multiplied by the vertical load F_{zi}.

In the equality constraints (19), the sum of yaw moment generated by the longitudinal tire forces is designed to meet the requirement of global yaw moment from the upper yaw moment controller.

$$
\sum_{i=1}^{4} M_{xi} = M_{zd}
\tag{19}
$$

The inequality constraints, including the motor peak torque and road adhesion constraints, can be expressed as follows:

$$
|F_{xi}| \le \frac{T_{imax}}{R_{eff}}
\tag{20}
$$

$$
|F_{xi}| \le \mu F_{zi}
\tag{21}
$$

where T_{imax} is the peak torque of the ith in-wheel motor.

4. Simulation and Results

4.1. HIL Simulation System

This paper used an HIL simulation to verify the effectiveness of the proposed optimal torque distribution approach. As illustrated in Figure 3, the HIL system comprises three subsystems: an NI PXI Express engine, an electronic control unit (ECU), and a host personal computer (PC).

- NI PXI Express Engine and models: The NI PXI Express engine contains different modular slots to simulate a vehicle model and sensor model. Detailed parameters of the PXI Express engine are provided in Table 2.
- ECU: The ECU is based on STM32F407ZGT6. The C code files of the yaw moment and torque distribution controllers are embedded in the ECU and calculate the target yaw moment and optimal motor torque exerted on each wheel. The step time is set to 5 ms.
- Host PC: The host PC is connected to the PXI Express by an Ethernet cable. The user interface on the PC is used to send a test command and display the vehicle state information.

Table 2. Parameters of the NI PXI engine.

Product	Module	Specification
PXIe-1071	PXI Chassis	Four-Slot, up to 3 GB/s
PXIe-8821	Controller	2.6 GHz dual-core processer
PXI-8512	CAN Interface	Flexible data rate, high-speed
PXIe-6738	Analog Output	16 bit, 32 channel, 1 MS/s

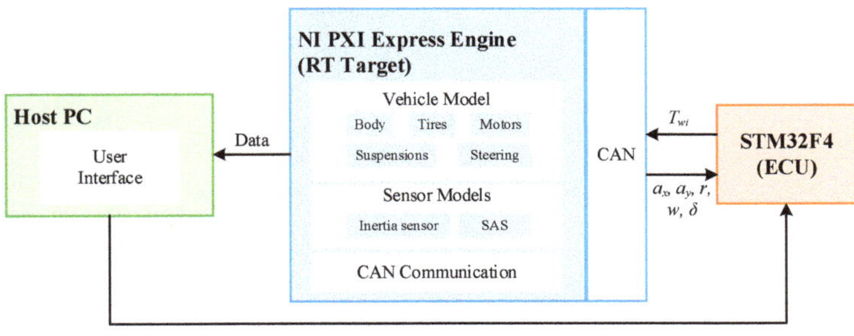

Figure 3. Structure of the HIL simulation system. ECU, electronic control unit; PC, personal computer.

This study develops a common C-class hatchback vehicle model in CarSim, which includes a chassis motion model, a steering system model, suspension rack models, tire models, and motor models. The vehicle parameters are listed in Table 3. The sensor models comprise a 6-DOF inertia sensor and a steering angle sensor model. Band-limited white noise signals are injected into the sensor models to simulate the noise in real sensors shown in Table 4.

Table 3. Vehicle parameters.

Parameters	Values
Vehicle mass	1412 kg
Sprung mass	1270 kg
Height of center of gravity (CG)	0.540 m
Wheel base	2.910 m
Distance from CG to front axle	1.015 m
Distance from CG to rear axle	1.895 m
Track width	1.675 m
Vehicle yaw inertia	1536.7 kg·m^2
Wheel inertia	0.9 kg·m^2
Wheel effective radius	0.325 m

Table 4. Noise signals.

Signal	Amplitude	Reference
δ	6.3°	HiTech SAS (Steering Angle Sensor), HiRain
a_x, a_y	0.049 m/s^2	Technologies Co., Ltd., Beijing, China
r	1 deg/s	TAMAGAWA AU7428N200, TAMAGAWA
ω	10 rpm	SEIKI Co., Ltd., Nagano Prefecture, Japan

4.2. HIL Simulation Results

4.2.1. Sine with Dwell

The Sine with Dwell (SWD) maneuver in the 126 requirements of the American Federal Motor Vehicle Safety Standard (FMVSS) was used to verify the effectiveness of the proposed optimal torque distribution method. The initial speed was set to 80 km/h and the friction coefficient was 0.8.

Figure 4 shows the yaw rate and sideslip angle responses of the vehicles with and without the proposed optimal torque distribution approach. As illustrated in Figure 4a, without control, the yaw rate was larger than 35% of its peak value. In contrast, with the proposed control, the yaw rate followed the variation in the steering wheel angle well, with its value reaching 20% of the peak value at 0.6 s after steering was completed. Table 5 shows the comparative evaluation of the SWD test. According to the FMVSS 126 requirements, it can be concluded that the vehicle with the proposed approach passed the test.

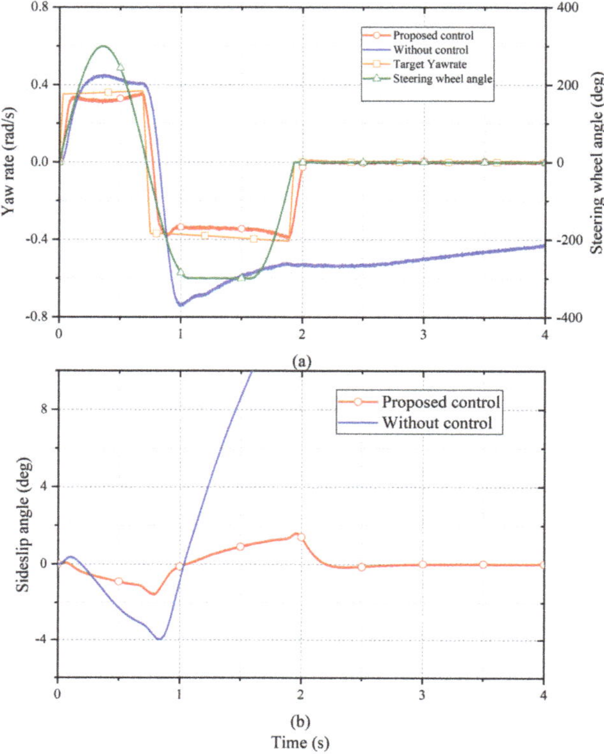

Figure 4. Comparison of the vehicles with and without the proposed optimal torque distribution control: (**a**) yaw rate responses; (**b**) sideslip angle responses.

Table 5. Evaluation of the Sine with Dwell (SWD) test.

Yaw Rate	Vehicle with the Proposed Control Method	Vehicle without Any Control
Peak value	−0.431 rad/s	−0.736 rad/s
35% of the peak value	−0.151 rad/s	−0.258 rad/s
1 s after completing steering	−0.003 rad/s	−0.505 rad/s
20% of the peak value	−0.086 rad/s	−0.147 rad/s
1.75 s after completing steering	+0.003 rad/s	−0.450 rad/s

4.2.2. Double Lane Change

Closed-loop simulations were conducted at a constant speed of 60 km/h under road conditions with $\mu = 0.8$.

Figure 5a displays the vehicle's trajectory. Figure 5b,c display the yaw rate and sideslip angle responses of the vehicles with and without the proposed optimal torque distribution approach. As illustrated in Figure 5a, the vehicle without control was not able to follow the expected trajectory. As illustrated in Figure 5b, without control, the yaw rate was constantly changing owing to the failure to follow the expected trajectory, meaning that the vehicle lost stability. By contrast, under the proposed control approach, the change of yaw rate was able to track the target yaw rate quickly and accurately, thus achieving vehicle stability control.

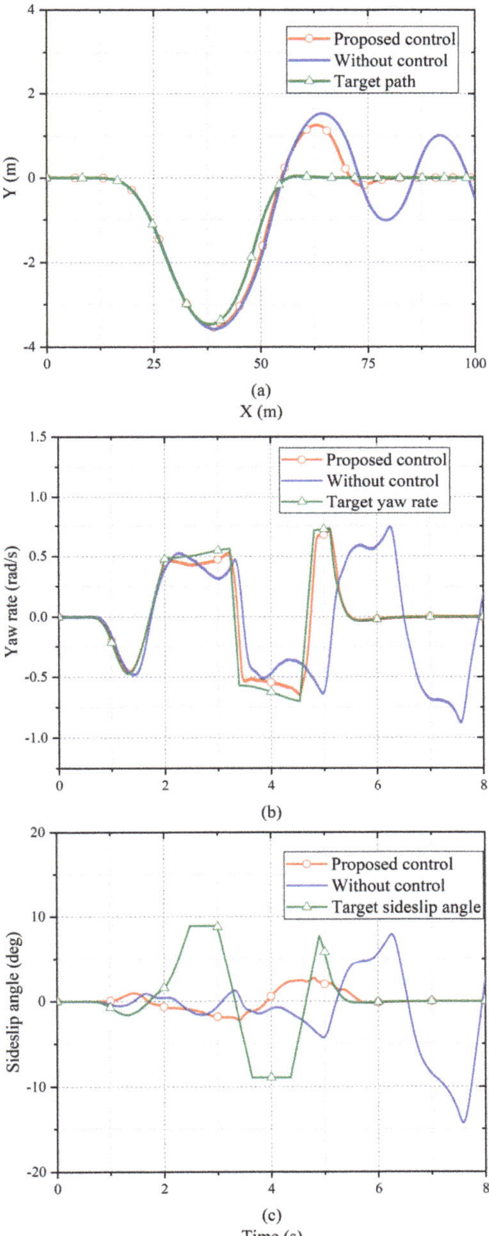

Figure 5. Comparison of the vehicles with and without the proposed optimal torque distribution control method: (**a**) path tracking; (**b**) yaw rate responses; (**c**) sideslip angle responses.

The detailed comparison in Figure 5c reveals that, when the control was not used, the sideslip angle increased rapidly after 5 s and ultimately exceeded the vehicle stability boundary. However, with the proposed control approach, the amplitude of the sideslip angle was always below $3°$ and changed smoothly, which means the vehicle could be easily handled by the driver.

5. Analysis and Discussion

5.1. Global Optimal Proof

A global optimal solution can be obtained for convex optimization problems. In order to prove that the proposed algorithm is a convex optimization problem, it is necessary to prove that the proposed objective function is a convex function. It is assumed that the arm of yaw moments generated by longitudinal tire force and lateral tire force are A_i and B_i. The specific values of A_i and B_i are derived from (16) and (17). The objective function can be clearly expressed as (22), which can also be rewritten as (23). In the process of optimization, A_i, B_i, and F_{zi} are constants, where $F_{xi} = [F_{x1}\ F_{x2}\ F_{x3}\ F_{x4}]^T$.

$$min\ J = \sum_{i=1}^{4} \left(\frac{F_{xi}}{F_{zi}}\right)^2 \left(\frac{A_i F_{xi}}{A_i F_{xi} + B_i F_{yi}}\right)^2 \tag{22}$$

$$min\ J = \sum_{i=1}^{4} \left(\frac{A_i}{F_{zi}}\right)^2 \left(\frac{F_{xi}^2}{A_i F_{xi} + B_i \sqrt{(s\mu F_{zi})^2 - F_{xi}^2}}\right)^2 \tag{23}$$

Taking F_{x1} as an example, in order to prove that (24) is a convex function in a simple and clear manner, the image of $g(F_{x1})$ is described in Figure 6, which indicates that the proposed objective function is clearly a convex function in a feasible region. Similarly, it can be proved that the inequalities F_{x2}, F_{x3}, and F_{x4} are also true. In conclusion, the proposed objective optimization problem is a convex optimization problem and represents a suitable result for any driving condition.

$$g(F_{x1}) = \left(\frac{A_1}{F_{z1}}\right)^2 \left(\frac{F_{x1}^2}{A_1 F_{x1} + B_1 \sqrt{(s\mu F_{zi})^2 - F_{x1}^2}}\right)^2 \tag{24}$$

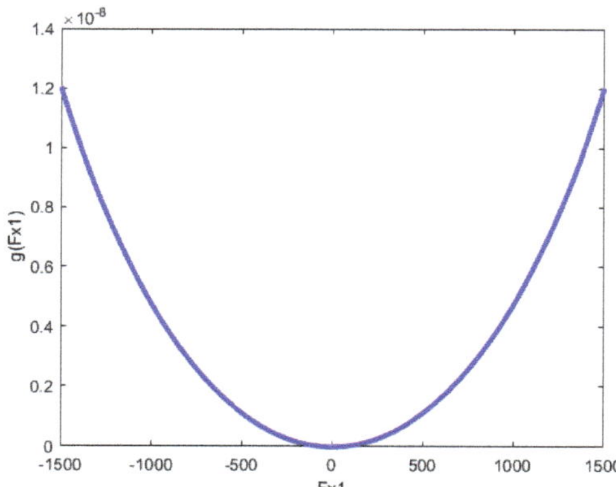

Figure 6. Image of $g(F_{x1})$.

5.2. Control Performance

This paper compares the proposed approach with the method based on the optimal tire usage rate [16] to discuss the reason behind the higher performance of the proposed approach.

$$minJ = \sum_{i=1}^{4} C_i \frac{F_{xi}^2}{(\mu F_{zi})^2} \tag{25}$$

The simulation condition was a double-lane-change maneuver performed at 50 km/h, where the road adhesion coefficient was set to 0.6 for more comparable results. Figure 7 presents the analysis of the yaw moment generated by the motor driving forces and lateral tire forces. The results show that the proposed approach utilizes the lateral forces to generate yaw moment more fully and quickly, as well as to reduce the torque output of the four in-wheel motors. Figure 8 compares the CG lateral force, sideslip angle and driving trajectory to prove that the vehicle with the proposed torque distribution approach can follow the target trajectory better than with tire usage rate control.

Figure 7c,d display the yaw moments (M_{z-x} and M_{z-y}) generated by the motor driving forces and lateral tire forces, respectively. According to the results, the difference in the front-left and rear-right motor driving forces obtained for the two controllers is mainly caused by the yaw moment generated by the lateral forces. As shown in the Figure 7d, during 1.6–2.2 s, the proposed approach can use a larger M_{z-y} to compensate for the yaw moment generated from the motor driving forces when the sign of M_{z-y} is the same as that of M_{z-x}. Therefore, the M_{z-x} used to track the target yaw moment from the upper controller is smaller, and the torque output of the four in-wheel motors is also reduced. In addition, by comparing the curves of M_{z-x} and M_{z-y} during 1.2–1.5 s and 2.6–3.8 s, the proposed approach outputs a lower M_{z-x} when M_{z-x} and M_{z-y} have opposite signs, owing to the lower M_{z-y}.

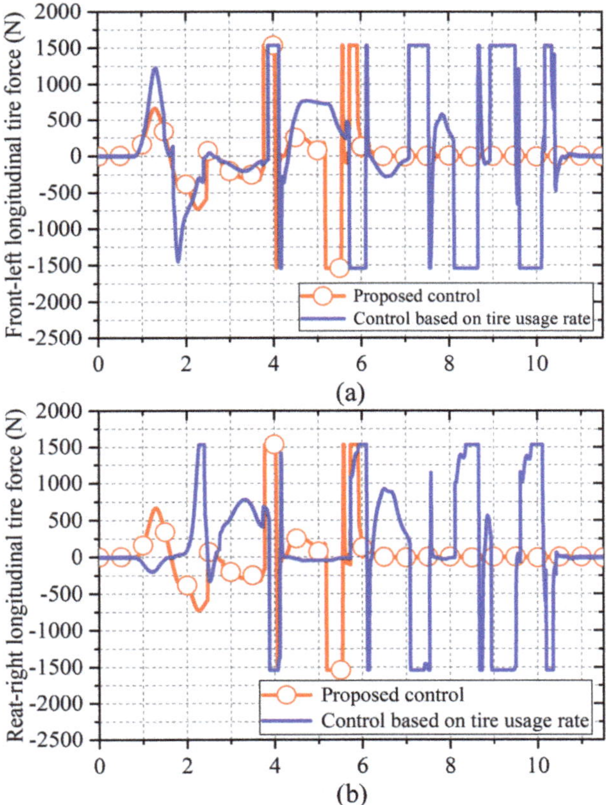

Figure 7. *Cont.*

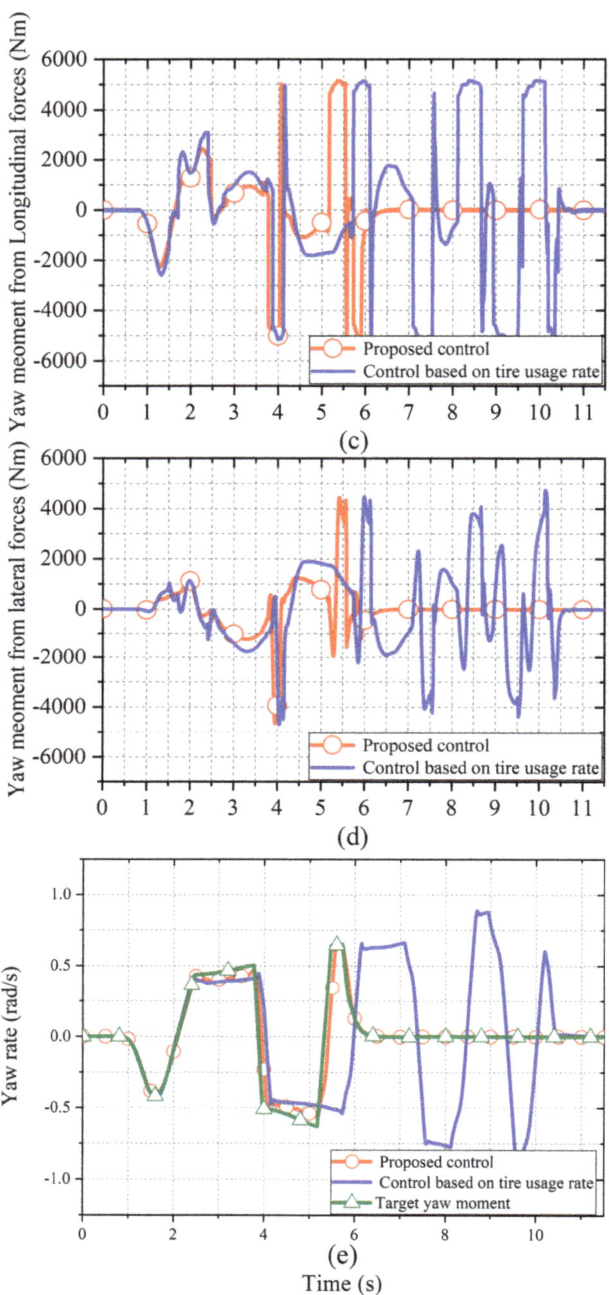

Figure 7. Comparison of the simulation results between the vehicle with the use of the proposed control and the optimal tire usage rate control: (**a**,**b**) front−left and rear−right motor driving forces; (**c**,**d**) center of gravity (CG) yaw moment from the motor driving forces and the lateral tire forces; (**e**) yaw rate responses.

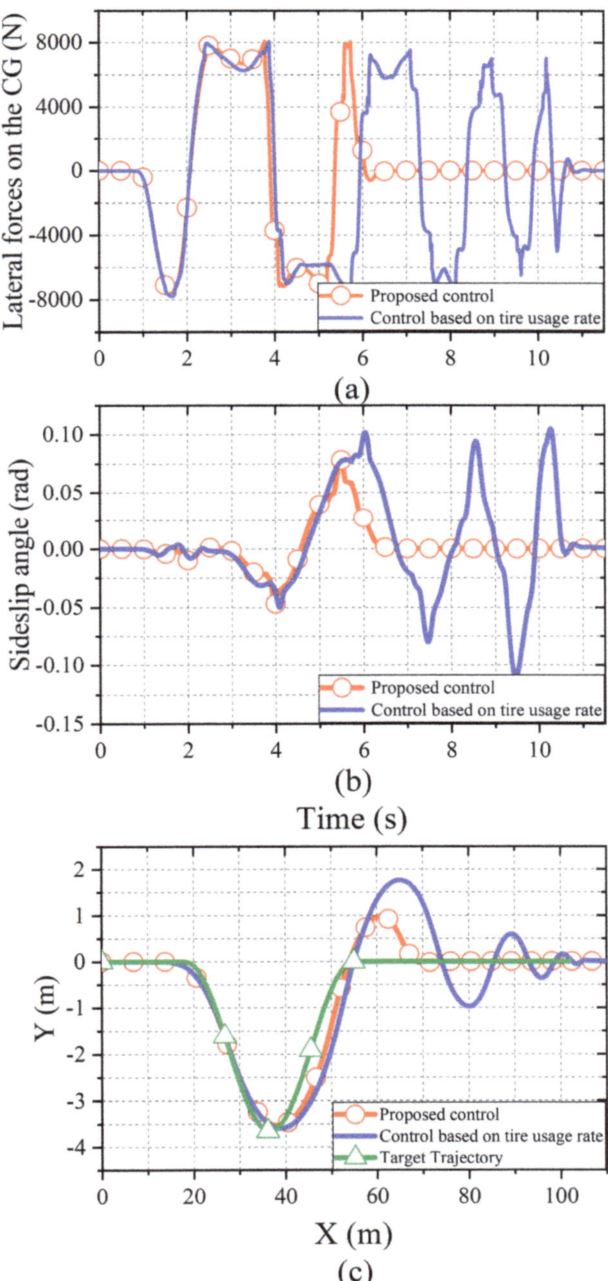

Figure 8. Comparison of the vehicles using the proposed control and the optimal tire usage rate control: (**a**) CG lateral force; (**b**) sideslip angle responses; (**c**) vehicle trajectory.

Figure 7e presents a comparison of the yaw rate responses. Under the proposed method, the yaw rate can track the target value very well. The curve obtained under the tire usage rate control clearly has a delay and a larger fluctuation around 5.2 s. Moreover, it takes longer to converge to the stable state than in the proposed approach. In conclusion, the proposed approach makes better use of the lateral tire forces to generate the yaw moment and improves the vehicle yaw response.

Figure 8 illustrates the effect of CG lateral force on vehicle stability. As shown in Figure 8c, with the proposed control, the vehicle path is maintained within a smaller range and converges to the target trajectory more rapidly. Figure 8b, showing the comparative sideslip angle, shows the same conclusion. Figure 8a reveals that, with the proposed control method, the lateral force responds more quickly at 4.5 s, which causes the vehicle trajectory and sideslip angle to converge rapidly to a stable state.

In summary, the comparisons presented in Figures 7 and 8 reveal that the proposed optimal torque distribution control approach makes increased use of the motor driving forces and the lateral tire forces to improve the vehicle's yaw responses and trajectory-following ability.

6. Conclusions

For the purpose of improving the efficiency of yaw moment generation, this paper took the rigid characteristics of 4WIDEVs into full consideration and designed a control for torque distribution. This controller employed a new objective function, which considered tire usage and the efficiency of yaw moment generation.

The SWD results based on the HIL simulation demonstrated the effectiveness of our approach. The further analysis proved that this method gives suitable results for any driving condition, and the comparative simulation results in DLC experiments showed that the proposed method made quicker and fuller use of lateral force to generate yaw moment and gained better vehicle stability.

This approach has considerable value for distributed-drive EVs and can improve handling stability when negotiating curves. Moreover, the in-wheel motor with restrained torque output can also be used to achieve functions that enable great handling stability.

Author Contributions: D.Y. proposed the method and refined the manuscript; J.W. performed the experiments and prepared the original draft; J.D. and G.C. reviewed and supervised the manuscript; J.-S.H. provided software and experimental equipment support. All authors have read and agreed to the published version of the manuscript.

Funding: This work is supported by "Natural Science Foundation of Jiangsu Province", BK20201307.

Institutional Review Board Statement: Not applicable.

Informed Consent Statement: Not applicable.

Data Availability Statement: Not applicable.

Conflicts of Interest: The authors declare no conflict of interest.

References

1. Lin, C.-L.; Hung, H.-C.; Li, J.-C. Active Control of Regenerative Brake for Electric Vehicles. *Actuators* **2018**, *7*, 84. [CrossRef]
2. Savitski, D.; Ivanov, V.; Augsburg, K.; Emmei, T.; Fuse, H.; Fujimoto, H.; Fridman, L.-M. Wheel Slip Control for the Electric Vehicle with In-Wheel Motors: Variable Structure and Sliding Mode Methods. *IEEE Trans. Ind. Electron.* **2020**, *67*, 8535–8544. [CrossRef]
3. Zhang, L.; Ding, H.; Huang, Y.; Chen, H.; Guo, K.; Li, Q. An Analytical Approach to Improve Vehicle Maneuverability via Torque Vectoring Control: Theoretical Study and Experimental Validation. *IEEE Trans. Veh. Technol.* **2019**, *68*, 4514–4526. [CrossRef]
4. Hang, P.; Xia, X.; Chen, X. Handling Stability Advancement With 4WS and DYC Coordinated Control: A Gain-Scheduled Robust Control Approach. *IEEE Trans. Veh. Technol.* **2021**, *70*, 3164–3174. [CrossRef]
5. Kasinathan, D.; Kasaiezadeh, A.; Wong, A.; Khajepour, A.; Chen, S.-K.; Litkouhi, B. An Optimal Torque Vectoring Control for Vehicle Applications via Real-Time Constraints. *IEEE Trans. Veh. Technol.* **2016**, *65*, 4368–4370. [CrossRef]
6. Shi, K.; Yuan, X.; Huang, G.; Liu, Z. Compensation-Based Robust Decoupling Control System for the Lateral and Longitudinal Stability of Distributed Drive Electric Vehicle. *IEEE/ASME Trans. Mechatron.* **2019**, *24*, 2768–2778. [CrossRef]
7. Tan, D.; Lu, C. The Influence of the Magnetic Force Generated by the In-Wheel Motor on the Vertical and Lateral Coupling Dynamics of Electric Vehicles. *IEEE Trans. Veh. Technol.* **2016**, *65*, 4655–4668. [CrossRef]

8. Shan, D.; Yin, D. A New Hierarchical Yaw Stability Control Approach for Four-Wheel Independent Driving Electric Vehicles. In Proceedings of the 35th Chinese Control Conference (CCC), Chengdu, China, 27–29 July 2016; pp. 4653–4657.
9. Park, G.; Han, K.; Nam, K.; Kim, H.; Choi, S.-B. Torque Vectoring Algorithm of Electronic-Four-Wheel Drive Vehicles for Enhancement of Cornering Performance. *IEEE Trans. Veh. Technol.* **2020**, *69*, 3668–3679. [CrossRef]
10. Joa, E.; Park, K.; Koh, Y.; Yi, K.; Kim, K. A Tyre Slip-Based Integrated Chassis Control of Front/Rear Traction Distribution and Four-Wheel Independent Brake from Moderate Driving to Limit Handling. *Veh. Syst. Dyn.* **2018**, *56*, 579–603. [CrossRef]
11. Feng, J.; Chen, S.; Qi, Z. Coordinated Chassis Control of 4WD Vehicles Utilizing Differential Braking, Traction Distribution and Active Front Steering. *IEEE Access* **2020**, *8*, 81055–81068. [CrossRef]
12. Peng, H.; Hori, Y. Optimum Traction Force Distribution for Stability Improvement of 4WD EV in Critical Driving Condition. In Proceedings of the 9th IEEE International Workshop on Advanced Motion Control, Istanbul, Turkey, 27–29 March 2006; pp. 596–600.
13. Abe, M.; Mokhiamar, O. An Integration of Vehicle Motion Controls for Full Drive-by-Wire Vehicle. *Proc. Ins. Mech. Eng. Part K J. Multi-Body Dyn.* **2007**, *221*, 117–127. [CrossRef]
14. Mokhiamar, O.; Abe, M. Combined Lateral Force and Yaw Moment Control to Maximize Stability as well as Vehicle Responsiveness During Evasive Maneuvering for Active Vehicle Handling Safety. *Veh. Syst. Dyn.* **2016**, *37*, 246–256. [CrossRef]
15. Ono, E.; Hattori, Y.; Muragishi, Y.; Koibuchi, K. Vehicle Dynamics Integrated Control for Four-Wheel-Distributed Steering and Four-Wheel-Distributed Traction/Braking Systems. *Veh. Syst. Dyn.* **2007**, *44*, 139–151. [CrossRef]
16. Xiong, L.; Yu, Z.; Wang, Y.; Yang, C.; Meng, Y. Vehicle Dynamics Control of Four In-Wheel Motor Drive Electric Vehicle Using Gain Scheduling Based on Tyre Cornering Stiffness Estimation. *Veh. Syst. Dyn.* **2012**, *50*, 831–846. [CrossRef]
17. Yang, P. A Study on Multi-Actuators Control Allocation of Distributed Drive Electric Vehicle for Handling and Stability. Ph.D. Thesis, Tongji University, Shanghai, China, 2015.
18. Wang, X. Optimal Torque Distribution of Four-Wheel Independent Drive Electric Vehicle Under Steering Condition. Master's Thesis, Jilin University, Changchun, China, 2017.
19. Guo, L.; Ge, P.; Sun, D. Torque Distribution Algorithm for Stability Control of Electric Vehicle Driven by Four In-Wheel Motors Under Emergency Conditions. *IEEE Access* **2019**, *7*, 104737–104748. [CrossRef]
20. Li, B.; Goodarzi, A.; Khajepour, A.; Chen, S.-K.; Litkouhi, B. An Optimal Torque Distribution Control Strategy for Four-Independent Wheel Drive Electric Vehicles. *Veh. Syst. Dyn.* **2015**, *53*, 1172–1189. [CrossRef]
21. Huang, J.; Liu, Y.; Liu, M.; Cao, M.; Yan, Q. Multi-Objective Optimization Control of Distributed Electric Drive Vehicles Based on Optimal Torque Distribution. *IEEE Access* **2019**, *7*, 16377–16394. [CrossRef]
22. Hu, X.; Chen, H.; Li, Z.; Wang, P. An Energy-Saving Torque Vectoring Control Strategy for Electric Vehicles Considering Handling Stability Under Extreme Conditions. *IEEE Trans. Veh. Technol.* **2020**, *69*, 10787–10796. [CrossRef]
23. Li, X.; Chan, C.; Wang, Y. A Reliable Fusion Methodology for Simultaneous Estimation of Vehicle Sideslip and Yaw Angles. *IEEE Trans. Veh. Technol.* **2016**, *65*, 4440–4458. [CrossRef]
24. Rezaeian, A.; Zarringhalam, R.; Fallah, S.; Melek, W.; Khajepour, A.; Chen, S.-K.; Moshchuck, N.; Litkouhi, B. Novel Tire Force Estimation Strategy for Real-Time Implementation on Vehicle Applications. *IEEE Trans. Veh. Technol.* **2015**, *64*, 2231–2241. [CrossRef]
25. Ni, J.; Hu, J.; Xiang, C. Envelope Control for Four-Wheel Independently Actuated Autonomous Ground Vehicle Through AFS/DYC Integrated Control. *IEEE Trans. Veh. Technol.* **2017**, *66*, 9712–9726. [CrossRef]
26. Hu, J.; Tao, J.; Xiao, F.; Niu, X.; Fu, C. An Optimal Torque Distribution Control Strategy for Four-Wheel Independent Drive Electric Vehicles Considering Energy Economy. *IEEE Access* **2019**, *7*, 141826–141837. [CrossRef]
27. Wang, Q.; Zhao, Y.; Deng, Y.; Xu, H.; Deng, H.; Lin, F. Optimal Coordinated Control of ARS and DYC for Four-Wheel Steer and In-Wheel Motor Driven Electric Vehicle with Unknown Tire Model. *IEEE Trans. Veh. Technol.* **2020**, *69*, 10809–10819. [CrossRef]

Article

Extension Coordinated Multi-Objective Adaptive Cruise Control Integrated with Direct Yaw Moment Control

Hongbo Wang [1,2], **Youding Sun** [1], **Zhengang Gao** [3,*] **and Li Chen** [1]

[1] Department of Vehicle Engineering, School of Automobile and Transportation Engineering, Hefei University of Technology, Hefei 230009, China; wanghongbo@hfut.edu.cn (H.W.); 2018110879@mail.hfut.edu.cn (Y.S.); chenli@hbwe.edu.cn (L.C.)

[2] Anhui Intelligent Vehicle Engineering Laboratory, Hefei 230009, China

[3] Department of Mechanical and Transportation Engineering, Ordos Institute of Technology, Ordos 017000, China

* Correspondence: gzg198510@oit.edu.cn

Abstract: An adaptive cruise control (ACC) system can reduce driver workload and improve safety by taking over the longitudinal control of vehicles. Nowadays, with the development of range sensors and V2X technology, the ACC system has been applied to curved conditions. Therefore, in the curving car-following process, it is necessary to simultaneously consider the car-following performance, longitudinal ride comfort, fuel economy and lateral stability of ACC vehicle. The direct yaw moment control (DYC) system can effectively improve the vehicle lateral stability by applying different longitudinal forces to different wheels. However, the various control objectives above will conflict with each other in some cases. To improve the overall performance of ACC vehicle and realize the coordination between these control objectives, the extension control is introduced to design the real-time weight matrix under a multi-objective model predictive control (MPC) framework. The driver-in-the-loop (DIL) tests on a driving simulator are conducted and the results show that the proposed method can effectively improve the overall performance of vehicle control system and realize the coordination of various control objectives.

Keywords: advanced driver assistant systems; adaptive cruise control; direct yaw moment control; extension control; model predictive control

Citation: Wang, H.; Sun, Y.; Gao, Z.; Chen, L. Extension Coordinated Multi-Objective Adaptive Cruise Control Integrated with Direct Yaw Moment Control. *Actuators* **2021**, *10*, 295. https://doi.org/10.3390/act10110295

Academic Editors: Peng Hang, Xin Xia and Xinbo Chen

Received: 20 September 2021
Accepted: 3 November 2021
Published: 6 November 2021

Publisher's Note: MDPI stays neutral with regard to jurisdictional claims in published maps and institutional affiliations.

1. Introduction

1.1. Background

An adaptive cruise control system is a key basic function of the advanced driver assistant systems (ADAS) developed to enhance driving comfort, reduce driving errors, improve safety, increase traffic capacity and reduce fuel consumption [1]. The ACC system is developed from the conventional cruise control (CC) system. It measures the distance and relative longitudinal speed between the host vehicle and preceding vehicle by range sensors (such as radar, lidar or video camera), then the throttle and brake will be controlled by ACC algorithm to realize the longitudinal motion control of the vehicle. As the ACC system takes over the longitudinal motion control of vehicle, the driver workload is largely reduced.

1.2. Literature Review and Analysis

A lot of research has been done on improving the longitudinal car-following performance of ACC vehicles. Moon et al. proposed a multiple-target tracking adaptive cruise control system to improve the system performance [2]. Martinez and Canudas-de-Wit proposed a novel reference model-based control approach for automotive longitudinal control [3]. Ganji et al. proposed an adaptive cruise control for a hybrid electric vehicle based on a sliding mode controller which can deal with the problem of variable set-point of ACC [4]. Lin proposed an adaptive neuro-fuzzy predictor-based control approach to enhance the fuel efficiency [5]. Althoff proposed an exchangeable nominal controller to ensure comfort [6].

In recent years, in order to save energy, reduce emission and improve the passenger comfort, in addition to improving the longitudinal car-following performance of ACC system, some scholars have also considered the fuel economy, longitudinal ride comfort, safety into the design of ACC system. Moser proposed a stochastic model predictive control (MPC) to optimize the fuel consumption in a vehicle following context [7]. Luo et al. proposed an adaptive cruise control algorithm with multiple objectives based on a model predictive control framework [8]. Li et al. proposed a novel vehicular adaptive cruise control system to comprehensively address the issues of tracking ability, fuel economy and driver desired response [9]. Luo et al. proposed a novel ACC system for intelligent HEVs to improve the energy efficiency and control system integration [10]. Ren et al. proposed a hierarchical adaptive cruise control system to get a balance among the driver's expectation, collision risk and ride comfort [11]. Asadi and Vahidi proposed a method which used the upcoming traffic signal information within the vehicle's adaptive cruise control system to reduce idle time at stop lights and fuel consumption [12].

Most of the above studies usually assumed that the vehicle was running along the straight lane. With the development of radar detection range and V2 X technology, it enables ACC vehicle to detect the preceding vehicle on the curved road. Thus, in order to expand the application of ACC system, some studies have been done under the condition that the ACC vehicle runs on a curved road. D. Zhang et al. presented a curving adaptive cruise control system to coordinate the direct yaw moment control system and considered both longitudinal car-following capability and lateral stability on curved roads [13]. Cheng et al. proposed a multiple-objective ACC integrated with direct yaw moment control to ensure vehicle dynamics stability and improve driving comfort on the premise of car following performance [14]. Idriz et al. proposed an integrated control strategy for adaptive cruise control with auto-steering for highway driving [15]. The references above have considered the car-following performance, longitudinal ride comfort, fuel economy and lateral stability of ACC vehicle. However, when an ACC vehicle drives on a curved road, these control objectives usually conflict with each other. For example, in order to obtain better car-following performance, ACC vehicles usually tend to adopt larger acceleration and acceleration rate to adapt to the preceding vehicle, which will lead to poor longitudinal ride comfort. Moreover, in order to ensure vehicle lateral stability, the differential braking forces generated by the DYC system are usually applied to track the desired vehicle sideslip angle and yaw rate, whereas the additional braking forces will make the car-following performance worse, especially when the ACC vehicle is in an accelerating process. Meanwhile, to ensure the car-following performance when the additional braking force acts on the wheel, the ACC vehicles will increase the throttle opening to track the desired longitudinal acceleration, which usually means the increase of fuel consumption. The traditional constant weight matrix MPC has been unable to adapt to various complex conditions. In this paper, the extension control is introduced to design the real-time weight matrix under the MPC framework to coordinate the control objectives including longitudinal car-following capability, lateral stability, fuel economy and longitudinal ride comfort and improve the overall performance of vehicle control system.

Extension control is developed from the extension theory founded by Wen Cai. It is a new type of intelligent control that combines extenics and control. It can imitate people's ability to summarize, study and solve the incompatible issue [16]. Its basic idea is to deal with the control problems from the perspective of information conversion. In other words, the qualified degree (dependent degree) of control input information is used as the basis to determine the correction value of control output, then the controlled information will be converted to the qualified range [17]. Extension control is a cross-discipline method which has been applied into various engineering control domain. Currently, the extension control has also been applied into vehicle stability control [18–20], and these studies showed that the extension control could improve the performance of control system effectively. In this paper, the extension control is used to supervise the control effect of longitudinal car-following distance error and the risk of losing vehicle lateral stability and then adjust the weight matrix in the MPC framework.

1.3. Contribution and Organization

The main contribution of this paper is as follows.

The extension sets are designed to supervise the control effect of longitudinal car-following distance error and the risk of losing vehicle lateral stability. Both the control effect and the risk can be reflected by the corresponding extension distance. Then, the control system is designed by the following purpose. That is, on the premise of ensuring longitudinal car-following performance and lateral stability, the fuel economy and longitudinal ride comfort should be improved as much as possible.

Based on the system integrating ACC with DYC, this paper introduces the extension control to design the real-time weight matrix under a multi-objective MPC framework to solve the contradiction among the control objectives above. It can coordinate various control objectives and improve the comprehensive performance of vehicle control system under different conditions. Then, the DIL tests are carried out to validate the effectiveness of the proposed control strategy.

The rest of this paper is organized as follows: the vehicle models are established in Section 2. The design of control system is presented in Section 3. The DIL tests and results are shown in Section 4, and the conclusions are drawn in Section 5.

2. Vehicle Models

2.1. Longitudinal Dynamics Model

Newton's second law is applied to establish the vehicle longitudinal dynamics model. As shown in Figure 1, the longitudinal forces acting on the vehicle are expressed as the acceleration, rolling, gravitational, and drag [4]. The longitudinal dynamics equation is shown in Equation (1).

$$F_d = ma_x + mgf \cos \theta + mg \sin \theta + F_w, \tag{1}$$

where F_d represents the net traction force, m is the vehicle mass, a_x is vehicle longitudinal acceleration, g is the gravitational acceleration, f denotes the rolling coefficient, θ is the grade of road, and F_w is the aerodynamic drag as shown in Equation (2).

$$F_w = \frac{1}{2}\rho C_D A v_x^2, \tag{2}$$

where ρ is the air density, C_D is the drag coefficient, A is the windward area of the vehicle, and v_x represents vehicle longitudinal speed.

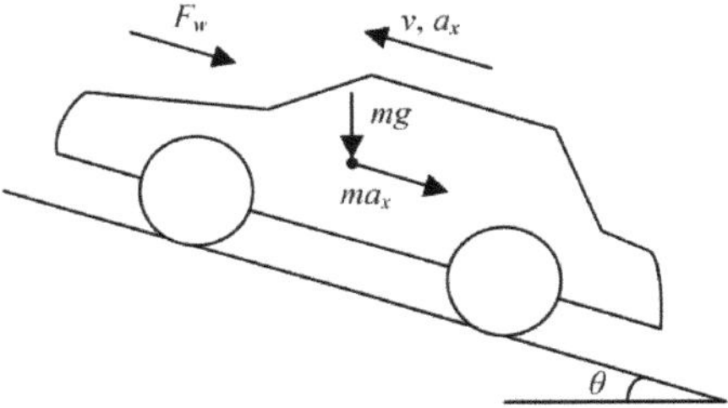

Figure 1. Vehicle longitudinal dynamics model.

2.2. Four-Wheel Vehicle Dynamics Model

In this paper, the longitudinal, lateral and yaw motion of vehicle are considered, and the pitch, roll and vertical motion of the vehicle are neglected. The simplified four-wheel vehicle dynamics model [14] is established as shown in Figure 2, where F_{xi} and F_{yi} are the longitudinal and lateral forces of the four wheels respectively, and the subscript i is 1, 2, 3, and 4, representing the front-left, front-right, rear-left and rear-right wheel respectively; δ_f is the front wheel steering angle, l_f and l_r are the distance from vehicle gravity center to the front axle and rear axle, respectively; l is the wheelbase, and T is the track width. The longitudinal, lateral and yaw motion are presented as follows:

$$m\left(\dot{v}_x - v_y\omega\right) = F_{x3} + F_{x4} - \left(F_{y1} + F_{y2}\right)\sin\delta_f + \left(F_{x1} + F_{x2}\right)\cos\delta_f \tag{3}$$

$$mv_x\left(\dot{\beta} + \omega\right) = F_{y3} + F_{y4} + \left(F_{y1} + F_{y2}\right)\cos\delta_f + \left(F_{x1} + F_{x2}\right)\sin\delta_f, \tag{4}$$

$$I_z\dot{\omega} = \left(F_{y1} + F_{y2}\right)l_f\cos\delta_f + \left(F_{y1} - F_{y2}\right)\tfrac{T}{2}\sin\delta_f - \left(F_{y3} + F_{y4}\right)l_r + \left(F_{x1} + F_{x2}\right)l_f\sin\delta_f - \left(F_{x1} - F_{x2}\right)\tfrac{T}{2}\cos\delta_f - \left(F_{x3} - F_{x4}\right)\tfrac{T}{2}, \tag{5}$$

where v_y represents vehicle lateral velocity, β and ω represent vehicle sideslip angle and yaw rate respectively, and I_z represents the inertia moment.

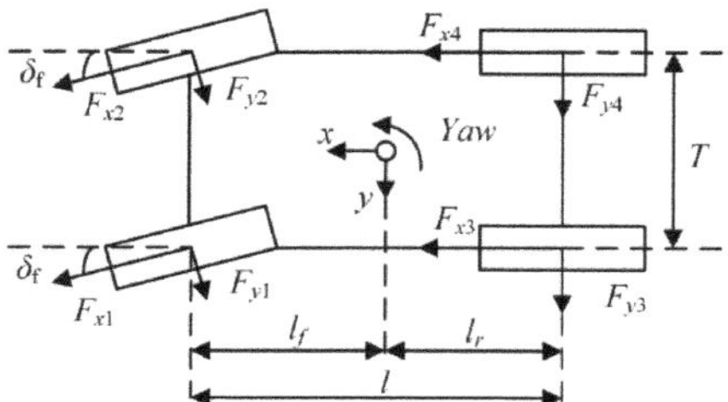

Figure 2. Vehicle dynamics model.

2.3. Tire Model

In this paper, Pacejka's magic formula [21] is used to describe the dynamics of tire. The longitudinal and lateral tire force can be calculated by Pacejka's magic formula. It can be depicted as follows:

$$\left\{ \begin{array}{c} y = D\sin[C\arctan\{Bx - E(Bx - \arctan Bx)\}] \\ Y(X) = y(x) + S_V \\ x = X + S_H \end{array} \right., \tag{6}$$

where Y represent longitudinal force F_x, lateral force F_y or aligning torques M_z, X is wheel slip ratio or wheel sideslip angle, B is stiffness coefficient, C is shape coefficient, D is peak value, E is curvature coefficient, S_H is horizontal offset, and S_V is vertical offset.

3. Control System Design

In the car-following process, the host vehicle sometimes needs to consider the lateral stability. For example, when the preceding vehicle drives away from the curve and accelerates into the straight lane, the host vehicle may still run on the curve, and it will also be accelerated to ensure the car-following performance. At this moment, the acceleration, steering and high longitudinal speed of host vehicle will increase the risk of losing lateral stability. Thus, it is necessary to consider the car-following performance and lateral stability simultaneously. Moreover, to improve driver satisfaction and reduce fuel consumption, the longitudinal ride comfort and fuel economy should also be considered into the control system design.

The extension control is introduced to design the weight matrix under the multi-objective MPC framework to coordinate the above control objectives and improve the overall performance of vehicle control system. The framework of control system is shown in Figure 3.

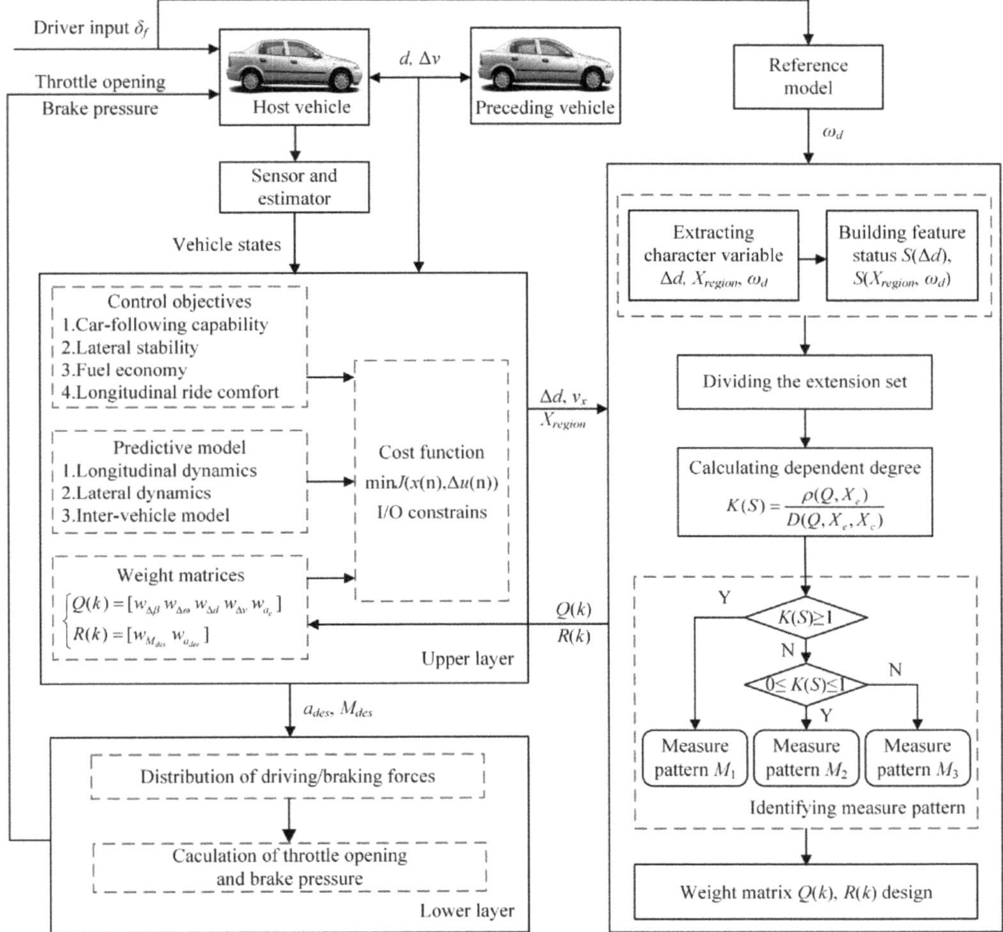

Figure 3. Framework of the proposed control.

This paper mainly focuses on the design of coordinated control system and there are many studies have been done on estimation of the key variables [22–24]. Therefore, it is assumed that vehicle states such as sideslip angle, sideslip angle rate and road friction coefficient can be estimated accurately.

The purpose of the vehicle control system in this paper is as follows:

1. On the premise of ensuring vehicle lateral stability, the additional yaw moment should be as small as possible to reduce the impact on longitudinal car-following performance and improve the fuel economy.

2. On the premise of ensuring the longitudinal car-following performance, the longitudinal acceleration and its change rate should be as small as possible to improve the longitudinal ride comfort.

3.1. Predictive Model

3.1.1. Longitudinal Car-Following Model

The function of ACC is to take over the longitudinal motion control of host vehicle to make it run at the driver's preset longitudinal speed or car-following distance. The longitudinal kinematic diagram of host vehicle and preceding vehicle is shown in Figure 4.

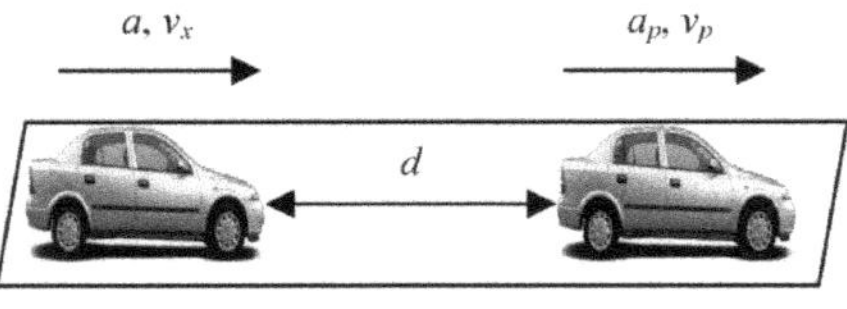

Figure 4. Vehicle following model.

The desired car-following distance between the host vehicle and the preceding vehicle is calculated by using the constant time headway, as shown in Equation (7).

$$d_{des} = T_h v_x + d_0, \tag{7}$$

where d_{des} is the desired car-following distance, T_h is the time headway, v_x is the longitudinal speed of host vehicle, and d_0 is the static inter-vehicle distance. Here, $T_h = 2$, $d_0 = 10$.

Usually, the longitudinal car-following performance can be represented by relative speed Δv and the car-following distance error Δd between the actual car-following distance d and the desired car-following distance ddes, as shown in Equation (8).

$$\begin{aligned} \Delta d &= d - d_{des} \\ \Delta v &= v_p - v_x \end{aligned}, \tag{8}$$

where v_p is the longitudinal speed of the preceding vehicle. The derivative of Equation (8) can be derived as follows:

$$\begin{aligned} \dot{\Delta d} &= \Delta v - T_h a_x \\ \dot{\Delta v} &= a_p - a_x \end{aligned}, \tag{9}$$

where a_p is the longitudinal acceleration of preceding vehicle.

The relationship between the desired acceleration and the actual longitudinal acceleration can be described by the first-order inertial system [13], as shown in the following Equation (10).

$$a_x = \frac{1}{T_{ax}s + 1} a_{des}, \tag{10}$$

where a_x and a_{des} are the actual longitudinal acceleration and desired longitudinal acceleration of host vehicle respectively, T_{ax} is time-constant and $T_{ax} = 0.45$.

3.1.2. Lateral Dynamic Model

A 2-DOF vehicle model is usually used to design the lateral stability controller [25], as shown in Equation (11).

$$\begin{aligned} \dot{\beta} &= -\frac{C_f + C_r}{m v_x}\beta + \left(\frac{l_r C_r - l_f C_f}{m v_x^2} - 1\right)\omega + \frac{C_f}{m v_x}\delta_f \\ \dot{\omega} &= \frac{l_r C_r - l_f C_f}{I_z}\beta - \frac{l_f^2 C_f + l_r^2 C_r}{I_z v_x}\omega + \frac{l_f C_f}{I_z}\delta_f + \frac{M_{des}}{I_z} \end{aligned}, \tag{11}$$

where C_f and C_r are the cornering stiffness of the front wheel and rear wheel, respectively, and M_{des} is the desired additional yaw moment.

The desired values of yaw rate ω_d and side slip angle β_d are defined according to vehicle parameters, longitudinal speed, and front steering angle δ_f directly manipulated by driver's steering action [26], as shown in Equation (12).

$$
\begin{aligned}
\omega_d &= \frac{v_x l}{1+m\left(l_f/C_r-l_r/C_f\right)v_x^2}\delta_f \\[2ex]
\beta_d &= \frac{l_r-\dfrac{l_f m v_x^2}{2C_r\left(l_f+l_r\right)}}{l_f+l_r+\dfrac{m v_x^2\left(l_r C_r-l_f C_f\right)}{2C_f C_r\left(l_f+l_r\right)}}\delta_f
\end{aligned}
\tag{12}
$$

Considering the road friction limitation, the desired yaw rate and side slip angle are modified as follows:

$$
\begin{aligned}
\omega_d &= \min\left(\frac{v_x l}{1+m\left(l_f/C_r-l_r/C_f\right)v_x^2}\delta_f,\ \frac{\mu g}{v_x}\right) \\[2ex]
\beta_d &= \min\left(\frac{l_r-\dfrac{l_f m v_x^2}{2C_r\left(l_f+l_r\right)}}{l_f+l_r+\dfrac{m v_x^2\left(l_r C_r-l_f C_f\right)}{2C_f C_r\left(l_f+l_r\right)}}\delta_f,\ \tan^{-1}(0.02\,\mu g)\right)
\end{aligned}
\tag{13}
$$

The yaw rate error and the vehicle sideslip angle error can be calculated by Equation (14).

$$
\begin{aligned}
\Delta\beta &= \beta - \beta_d \\
\Delta\omega &= \omega - \omega_d
\end{aligned}
\tag{14}
$$

The error between the desired value and the actual value reflects the stability of the vehicle. When the error of yaw rate or vehicle sideslip angle is small, the vehicle is in a steady status. When the error is large, it means that the vehicle is out of control or loses its stability.

3.1.3. Model Discretization

By combining (7)–(10), the state-space equation can be obtained as shown in Equation (15).

$$
\dot{x} = Ax + Bu + D\omega_1,
\tag{15}
$$

where $x = \begin{bmatrix} \beta & \omega & \Delta d & \Delta v & a_x \end{bmatrix}^T$, $u = \begin{bmatrix} M_{des} & a_{des} \end{bmatrix}^T$, $\omega = \begin{bmatrix} \delta_f & a_p \end{bmatrix}^T$, and the matrix expression is shown in Equations (16)–(18).

$$
A = \begin{bmatrix}
-\dfrac{C_f+C_r}{mv_x} & \dfrac{l_r C_r-l_f C_f}{mv_x^2}-1 & 0 & 0 & 0 \\[2ex]
\dfrac{l_r C_r-l_f C_f}{I_z} & -\dfrac{l_f^2 C_f+l_r^2 C_r}{I_z v_x} & 0 & 0 & 0 \\[2ex]
0 & 0 & 0 & 1 & -\tau_h \\[1ex]
0 & 0 & 0 & 0 & -1 \\[1ex]
0 & 0 & 0 & 0 & -\dfrac{1}{\tau_{ax}}
\end{bmatrix},
\tag{16}
$$

$$
B = \begin{bmatrix}
0 & 0 \\[1ex]
\dfrac{1}{I_z} & 0 \\[1ex]
0 & 0 \\[1ex]
0 & 0 \\[1ex]
0 & \dfrac{1}{\tau_{ax}}
\end{bmatrix},
\tag{17}
$$

$$D = \begin{bmatrix} \frac{C_f}{mv_x} & 0 \\ \frac{l_f C_f}{I_z} & 0 \\ 0 & 0 \\ 0 & 1 \\ 0 & 0 \end{bmatrix}, \tag{18}$$

In order to get the numerical solution of rolling optimization, the Taylor expansion method is applied to discretize Equation (15) to obtain the discrete state-space equation as shown in Equation (19).

$$x(k+1) = A_d x(k) + B_d u(k) + D_d \omega(k), \tag{19}$$

where A_d, B_d, D_d can be calculated by Taylor expansion method, as shown in Equation (20).

$$\begin{cases} A_d = I + T_s \cdot \partial f(x, u, \omega) / \partial x \\ B_d = T_s \cdot \partial f(x, u, \omega) / \partial u \\ D_d = T_s \cdot \partial f(x, u, \omega) / \partial \omega \end{cases}, \tag{20}$$

where T_s is the sampling time and I is the unit matrix.

3.2. Performance Index

3.2.1. Longitudinal Car-Following Performance

The longitudinal car-following performance of ACC system is usually evaluated by the distance error and relative speed between host vehicle and preceding vehicle. To ensure the longitudinal car-following performance, the distance error and relative speed are used to build the cost function for longitudinal car-following capability, as shown in Equation (21).

$$J_{ACC} = w_{\Delta d}\left(\Delta d - \Delta d_{ref}\right)^2 + w_{\Delta v}\left(\Delta v - \Delta v_{ref}\right)^2 + w_{ae}\left(a_x - a_{x,ref}\right)^2 + w_{a_{des}} a_{des}^2, \tag{21}$$

where the reference value of Δd_{ref}, Δv_{ref}, $a_{x,ref}$ are set as zero.

3.2.2. Lateral Dynamics Stability

Vehicle yaw rate error and sideslip angle error are usually used to describe vehicle lateral stability. When the error is small, it means that the vehicle status is in a stability area; when the error is large, it means that the vehicle loses control or loses the stability. The DYC system is usually applied to ensure the lateral stability of vehicle. However, the additional yaw moment required by DYC system is usually generated by the braking pressure of different wheels, the additional yaw moment will affect the longitudinal car-following performance and fuel economy of ACC vehicles. Therefore, on the premise of ensuring the vehicle lateral stability, the additional yaw moment is expected to be as small as possible. The quadratic form of $\Delta \omega$, $\Delta \beta$ and the additional yaw moment M_{des} is used to form the cost function for lateral stability, as shown in Equation (22).

$$J_{VLS} = w_{\Delta \omega} \Delta \omega^2 + w_{\Delta \beta} \Delta \beta^2 + w_{M_{des}} M_{des}^2, \tag{22}$$

3.2.3. Longitudinal Ride Comfort

In order to improve driver satisfaction and ensure the longitudinal ride comfort, the absolute value of longitudinal acceleration and jerk caused by the change of longitudinal acceleration are used to describe the longitudinal ride comfort performance index of ACC vehicle. Therefore, the absolute value of longitudinal acceleration and jerk are set as the constrains to ensure the longitudinal ride comfort, as shown in Equation (23).

$$\begin{aligned} |a_x| &\leq a_{max} \\ |a_x(k) - a_x(k-1)| &\leq j_{max} \end{aligned}, \tag{23}$$

3.2.4. Cost Function Design

By combining Equations (21) and (22), the cost function for the multi-objective control is formed as shown in Equation (24).

$$J = w_{\Delta d}\left(\Delta d - \Delta d_{ref}\right)^2 + w_{\Delta v}\left(\Delta v - \Delta v_{ref}\right)^2 + w_{ae}\left(a_x - a_{x,ref}\right)^2 + w_{a_{des}}a_{des}^2 + \\ w_{\Delta \omega}\Delta \omega^2 + w_{\Delta \beta}\Delta \beta^2 + w_{M_{des}}M_{des}^2, \tag{24}$$

Then, the predictive expression of the cost function can be obtained, as shown in Equation (25).

$$J = \sum_{n=0}^{N_p-1} ||x(k+n|n) - x_{ref}(k+n|n)||_{Q(k)}^2 + \sum_{n=0}^{N_c-1} ||u(k+n|n)||_{R(k)}^2 \tag{25}$$

where N_p and N_c denote the predictive horizon and control horizon, respectively. $Q(k)$ and $R(k)$ are non-negative weight matrices, as shown in Equation (26). x_{ref} is the reference value of MPC, and $x_{ref} = [\beta_d \ \omega_d \ 0 \ 0 \ 0]^T$.

$$\begin{cases} Q(k) = \begin{bmatrix} w_{\Delta \beta} \ w_{\Delta \omega} \ w_{\Delta d} \ w_{\Delta v} \ w_{a_e} \end{bmatrix} \\ R(k) = \begin{bmatrix} w_{M_{des}} \ w_{a_{des}} \end{bmatrix} \end{cases}, \tag{26}$$

Then the desired longitudinal acceleration and additional yaw moment can be obtained by minimizing the cost function as shown in Equation (25) subject to the car-following model, vehicle dynamics model, and the constraints as shown in Equation (23).

3.3. Extension Control Design

In order to improve the performance of MPC and make it adapt to various conditions, that is, the deceleration process, constant speed process and acceleration process of host vehicle in the curve. The extension control is introduced to design the real-time weight matrix under the framework of MPC. The design process is as follows:

3.3.1. Extracting Character Variable

In terms of the longitudinal car-following performance, due to the drivers are more sensitive to the distance error than the relative speed during the car-following process [13], this paper selects the car-following distance error to adjust the weight $w_{\Delta d}$ of the distance error, and sets the weight $w_{\Delta v}$ of the relative speed as a constant, then the distance error is selected to form the longitudinal car-following feature status $S(\Delta d)$.

In terms of vehicle lateral stability, the phase plane method composed of the sideslip angle and the sideslip angle rate is usually used to judge the lateral stability of vehicle [25] because its good identification of vehicle stability condition. The phase plane method can be expressed as Equation (27).

$$X_{region} = \left| B_1 \dot{\beta} + B_2 \beta \right| \leq 1, \tag{27}$$

where B_1 and B_2 are the parameters related to the road friction coefficient μ, here $B_1 = 0.064$ and $B_2 = 0.214$ [27].

The vehicle phase plane can be divided into 'stability region' and 'instability region' by Equation (27), as shown in Figure 5. The area 'stability region' means vehicle status in this area is safe and stable, while the remaining area is 'instability region', which means vehicle status in this area is risky in losing stability and unsafe [25].

It is also a challenge to ensure vehicle lateral stability when the driver desired yaw rate is in a large range. Thus, the value of X_{region} and the desired yaw rate ω_d are selected as the character variables of lateral stability to form the feature status $S(X_{region}, \omega_d)$.

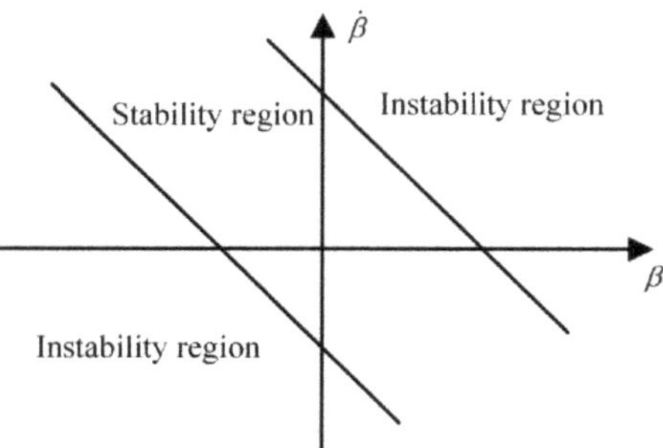

Figure 5. Sideslip angle phase plane division region.

3.3.2. Dividing the Extension Set

The one-dimensional (1-D) extension set of the longitudinal car-following distance error is shown in the Figure 6, where Δd_1 and Δd_2 are the boundaries of the classic domain and the extension domain, respectively. The distance error should be in driver's permissible longitudinal car-following range to reduce the driver intervention. The boundary of extension domain reflects the boundary of permissible region and impermissible region. Therefore, Δd_2 is set to the driver's maximum permissible value. The driver's permissible longitudinal car-following range [13] is shown in Equation (28).

$$- \Delta d_{max} \cdot SDE^{-1} \leq \Delta d \leq \Delta d_{max} \cdot SDE^{-1},\tag{28}$$

where SDE is the driver's sensitivity to distance error. The boundary of extension domain is calculated as $\Delta d_2 = \Delta d_{max} \cdot SDE^{-1}$. The SDE^{-1} is calculated as follows:

$$SDE^{-1} = k_{SDE} v_x + d_{SDE},\tag{29}$$

Figure 6. 1-D extension set of car-following distance error.

The parameters in Equations (28) and (29) are identified by driver experiment data in highway and city road traffic conditions [13]. Here, $\Delta d_{max} = 7.2$ m, $k_{SDE} = 0.06$, and $d_{SDE} = 0.12$. The boundary of classic domain Δd_1 is set to a relatively small value and $\Delta d_1 = 0.1 \times \Delta d_2$.

The lateral stability is represented by a two-dimensional (2-D) extension set, including classic domain, extension domain and non-domain. In the classic domain, it indicates the vehicle is stable; in the extension domain, it indicates the vehicle is transiting from stability to instability, and the vehicle state can be converted into the stable state by control; while in the non-domain, the vehicle is instable. The x-axis is desired yaw rate, and the y-axis is X_{region}, as shown in the Figure 7, where ω_1 and ω_2 are the boundaries of the classic domain and the extension domain in the x-axis direction, $X_{region1}$ and $X_{region2}$ are the boundaries of the classic domain and the extension domain in the y-axis direction, respectively. Here, $X_{region1}$ and $X_{region2}$ are set to 0.1 and 1 respectively. The extension boundary ω_2 in the x-axis direction reflects the boundary under large steering condition. Based on the experience and previous works [25], 0.2 μrad/s is set as the threshold of large steering condition. Therefore, the boundary ω_2 is set as 0.2 μrad/s. The classic boundary ω_1 is set as $0.1 \times \omega_2$.

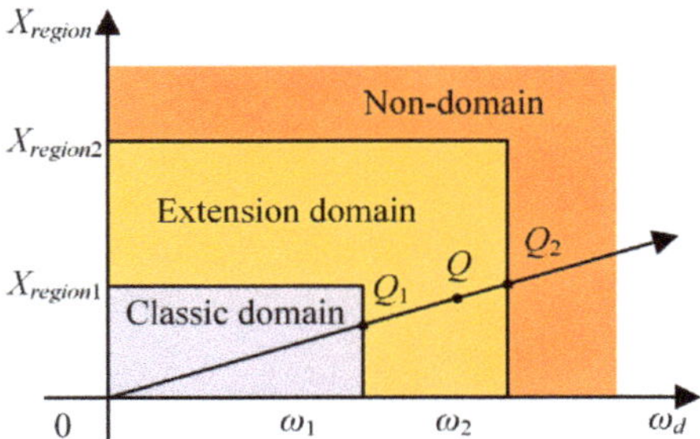

Figure 7. 2-D extension set of lateral stability.

Here, the "classic domain" and "extension domain" correspond to the stability region, and the "non-domain" corresponds to the instability region of vehicle. The "extension domain" can be understand as a transition domain.

3.3.3. Calculating Dependent Degree

Compared with the result of whether the vehicle status is in the stable region or not, it will help to improve the performance of control system if more detailed information, i.e., the degree of vehicle lateral status is known, and then we can design the control method according to that degree. In extension control, the "degree" above was defined as "dependent degree". The ideal point in the extension set is the original point O which represents the longitudinal car-following distance error, X_{region} and ω_d are zero. The point Q is supposed as a point in the extension domain. Connecting the point O with the point Q, the intersection points of the line OQ and the domains' boundaries are the points Q_1 and Q_2, respectively. Obviously, in 1-D extension set of car-following distance error, the points Q_1 and Q_2 correspond to Δd_1 and Δd_2 respectively. As shown in the Figure 7, the line segment OQ is the shortest distance for the point Q to approach the ideal point O. In the extension sets, the extension distance is defined as the distance from a point to a set, which is defined in a 1-D coordinate system. Therefore, it is required to convert the extension distance of 2-D extension set of lateral stability to a 1-D extension form, as shown in Figure 8.

Figure 8. 1-D extension set.

Set the classic domain $<O, Q_1> = X_c$, the extension domain $<Q_1, Q_2> = X_e$. The extension distance from the point Q to classic domain is represented as $\rho(Q, X_c)$, and the extension distance from point Q to extension domain is represented as $\rho(Q, X_e)$. The extension distance can be calculated as follows:

$$\rho(Q, X_c) = \left\{ \begin{array}{l} -|OQ_1|, Q \in \langle O, Q_1 \rangle \\ |OQ_1|, Q \in \langle Q_1, +\infty \rangle \end{array} \right. , \tag{30}$$

$$\rho(Q, X_e) = \left\{ \begin{array}{l} -|OQ_2|, Q \in \langle O, Q_2 \rangle \\ |OQ_2|, Q \in \langle Q_2, +\infty \rangle \end{array} \right. , \tag{31}$$

Thus, the dependent degree $K(S)$, also known as correlation function, can be calculated as follows:

$$\left\{ \begin{array}{l} K(S) = \frac{\rho(Q,X_e)}{D(Q,X_e,X_c)} \\ D(Q, X_e, X_c) = \rho(Q, X_e) - \rho(Q, X_c) \end{array} \right. , \tag{32}$$

3.3.4. Identifying Measure Pattern

The dependent degree of any point Q in the extension set can be described quantitatively by the dependent degree $K(S)$. The measure pattern can be divided as follows:

$$\left\{ \begin{array}{l} M_1 = \{S|K(S) > 1\} \\ M_2 = \{S|0 \leq K(S) \leq 1\} \\ M_3 = \{S|K(S) < 0\} \end{array} \right. , \tag{33}$$

The classic domain, extension domain and non-domain correspond to the measure pattern M_1, M_2 and M_3, respectively.

3.3.5. Weight Matrix Design

After the dependent degree $K(S)$ is calculated, it is used to design the real-time weight matrix because it can reflect the degree of longitudinal car-following distance error and the risk of losing lateral stability. The weights for $w_{\Delta\beta}$, $w_{\Delta\omega}$ and $w_{\Delta d}$ are set as the real-time weights which are adjusted by the corresponding values of the dependent degree $K(S)$, and the other weights $w_{\Delta v}$, w_{ae}, w_{Mdes}, w_{ades} are set as constants.

When the car-following distance error belongs to the measure pattern M_1, it means that the distance error is in a small range, and it is not necessary to increase the corresponding weight. When the car-following distance error belongs to the measure pattern M_2, the distance error is in a relatively large range, and it is possible to exceed the driver's sensitivity limit of the distance error if the corresponding weight is not adjusted timely. When the car-following distance error belongs to the measure pattern M_3, the distance error exceeds the driver's sensitivity limit, and the corresponding weight should be maximized to reduce the distance error by control. The real-time weight for longitudinal car-following distance is designed as follows:

$$w_{\Delta d} = \left\{ \begin{array}{ll} 0.3, & K_{ACC}(S) > 1 \\ 0.3 + 0.4 \cdot k_{ACC}, & 0 \leq K_{ACC}(S) \leq 1 \\ 0.7, & K_{ACC}(S) < 0 \end{array} \right. , \tag{34}$$

where $k_{ACC} = 1 - K_{ACC}(S)$, k_{ACC} and $K_{ACC}(S)$ are defined as the adjustment factor and dependent degree for vehicle longitudinal control.

Similarly, when the lateral stability status belongs to the measure pattern M_1, it indicates that the vehicle lateral stability status is in a stability region, and it is not necessary to adjust the corresponding weight. When the lateral stability status belongs to the measure pattern M_2, the lateral stability status is in the area between stability region and instability region and the vehicle may lose stability if the corresponding weight is not adjusted timely. When the lateral stability status belongs to the measure pattern M_3, the lateral stability status is in the instability region, the corresponding weight should be maximized to maintain vehicle lateral stability by control. The real-time weights for lateral stability are designed as follows:

$$w_{\Delta\beta}, w_{\Delta\omega} = \left\{ \begin{array}{ll} 0, & K_{VLS}(S) > 1 \\ 0.5 \cdot k_{VLS}, & 0 \leq K_{VLS}(S) \leq 1 \\ 0.5, & K_{VLS}(S) < 0 \end{array} \right. , \tag{35}$$

where $k_{VLS} = 1 - K_{VLS}(S)$, k_{VLS} and $K_{VLS}(S)$ are defined as the adjustment factor and dependent degree for vehicle stability control.

The real-time weight matrices of the proposed control are designed as follows:

$$\left\{ \begin{array}{l} Q(k) = \begin{bmatrix} w_{\Delta\beta} & w_{\Delta\omega} & w_{\Delta d} & 1 & 1 \end{bmatrix} \\ R(k) = \begin{bmatrix} 0.001 & 2 \end{bmatrix} \end{array} \right. , \tag{36}$$

To show the effectiveness of the proposed control, a constant weight ACC and a constant weight ACC&DYC are used for comparison. The constant weight matrices of conventional ACC are shown in Equation (37).

$$\begin{cases} Q(k) = [0 \quad 0 \quad 0.5 \quad 1 \quad 1] \\ \quad R(k) = [0.001 \quad 2] \end{cases} \tag{37}$$

The constant weight matrices of conventional ACC&DYC are shown in (38).

$$\begin{cases} Q(k) = [0.5 \quad 0.5 \quad 0.5 \quad 1 \quad 1] \\ \quad R(k) = [0.001 \quad 2] \end{cases}, \tag{38}$$

3.4. Lower Layer Design

After the desired longitudinal acceleration and additional yaw moment are obtained from MPC in the upper layer, the lower layer of the control system is to realize these objectives by the throttle opening and brake pressure. The desired longitudinal force on rear wheels can be obtained by Equation (39).

$$\begin{cases} F_3 + F_4 = F_d \\ (F_4 - F_3) \cdot T/2 = M_{des} \end{cases}, \tag{39}$$

Simplifying the power-train, consider the constant efficiencies at final drive, transmission, torque converter and neglecting the slip in wheels [4].

The fuel consumption is shown in Figure 9a. The desired longitudinal acceleration is realized by the engine map which is shown in Figure 9b. The throttle opening is determined through the look-up table by utilizing engine speed and desired engine torque. The brake pressure of rear wheel is calculated by inverse brake system. As ACC and DYC system usually does not need too large braking deceleration, it can be considered that there is a linear relationship between the brake pressure P_b and braking torque T_b at the wheels [28], as shown in Equation (40).

$$T_b = 150 \cdot P_b \tag{40}$$

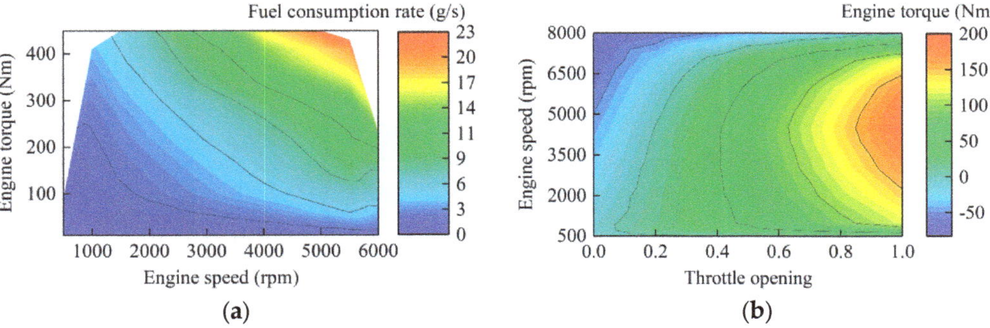

Figure 9. Fuel consumption rate and map of engine: (a) fuel consumption rate of engine; (b) engine map.

The relationship between F_d and engine output torque T_e is as follows [14].

$$R \cdot F_d = \eta f(\omega_t / \omega_e) i_g i_o T_e \tag{41}$$

where η represents mechanical efficiency, $f(\omega_t / \omega_e)$ is the torque characteristic function of torque converter, and i_g and i_o denote the transmission ratio of the gearbox and main reducer, respectively.

4. DIL Test Results and Analysis

As ACC and DYC systems always work with the driver, a closed DIL evaluation would be more effective than the open-loop simulations because of the real action of drivers' steering behavior [13]. Therefore, a driving simulator is used in the DIL tests for coordinated multi-objective ACC, as shown in Figure 10. In the simulator, the vehicle model is built in the vehicle simulation software, CarMaker. The coordinated multi-objective ACC controller is implemented with MATLAB/Simulink.

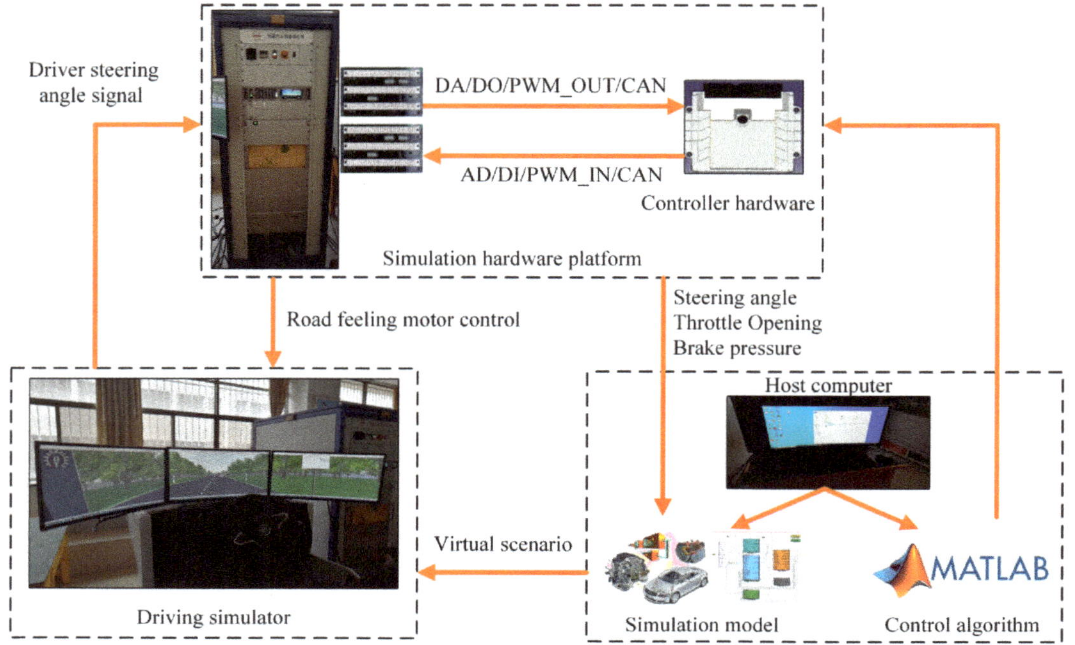

Figure 10. DIL test hardware platform.

The driving simulator contains the steering wheel, monitor, brake pedal, accelerator pedal and a road feeling motor. The driver's steering angle signal is obtained by steering wheel, the virtual scenario in CarMaker is displayed on the monitor. Due to the ACC system takes over the longitudinal control, the brake pedal and accelerator pedal are not used here. The road feeling motor can make the driver perceive the road feeling information of the vehicle through the steering wheel. The simulation hardware platform contains the controller hardware, board card, CAN card, NI PXI real-time processor and the platform is used to simulate all the input signals required by the normal operation of the controller to be tested, and collect the control commands from the controller.

The traditional constant weight ACC and the constant weight ACC and DYC are denoted as "ACC" and "ACC&DYC" in the simulation results, respectively. The parameters in the simulation model are shown in Table 1. The type of tire model used in CarMaker is magic formula tire model "MF_205_60R15". The values "205", "60" in name "MF_205_60R15" represent the tread width and flat ratio of tire. The letter "R15" indicates that the tire is a radial tire and "15" is the outer diameter of rim.

Table 1. Parameters in the simulation model.

Parameter	Symbol	Value
Vehicle mass	m	1301 kg
Gravitational acceleration	g	9.8 m/s^2
Inertial of z axis	I_z	1600 kg·m^2
Track width	T	1.544 m
Distance from vehicle gravity center to front axle	l_f	0.97 m
Distance from vehicle gravity center to rear axle	l_r	1.567 m
Road adhesion coefficient	μ	0.6
Maximum acceleration	a_{max}	2.5 m/s^2
Maximum jerk	j_{max}	0.5 m/s^3

A common scenario is conducted to show the effectiveness of the proposed controller. The preceding vehicle and host vehicle go through a curved path which is shown in Figure 11. Before entering the curve, the preceding vehicle drives at a constant speed 110 km/h. Then the preceding vehicle slows into the curve with a deceleration of –1 m/s^2 and drives at a low constant speed 54 km/h in the curve, as shown in Figure 12a. Finally, the preceding vehicle speeds up with an acceleration of 1 m/s^2 to drive away from the curve. During the driver in the loop test, in order to reduce the influence of driver's subjective factors on the results, the driver is not told what kind the controller is, and the steering wheel angle from driver is shown in Figure 12b. It can be seen that the driver's steering angle under the three controllers is almost the same as a whole, and the driver's steering wheel angle is little different with the different three controllers.

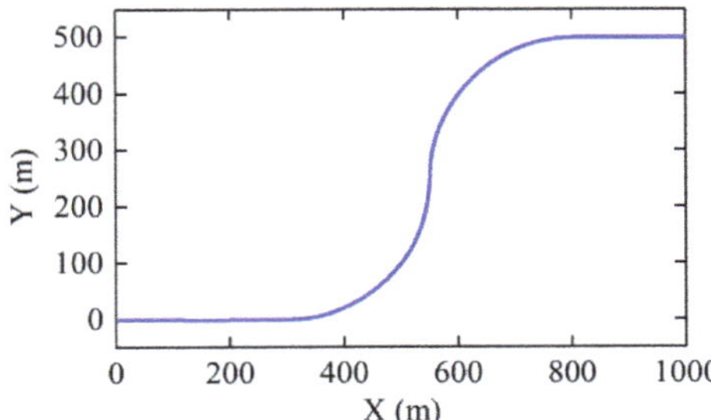

Figure 11. Curve path in the simulation model.

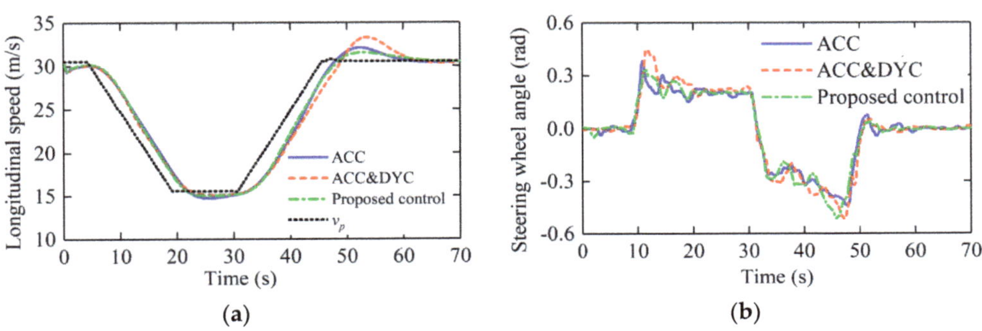

Figure 12. Longitudinal speed and steering wheel angle: (**a**) longitudinal speed; (**b**) steering wheel angle.

The longitudinal car-following errors, lateral stability error and phase plane of errors are shown in Figures 13–15, respectively.

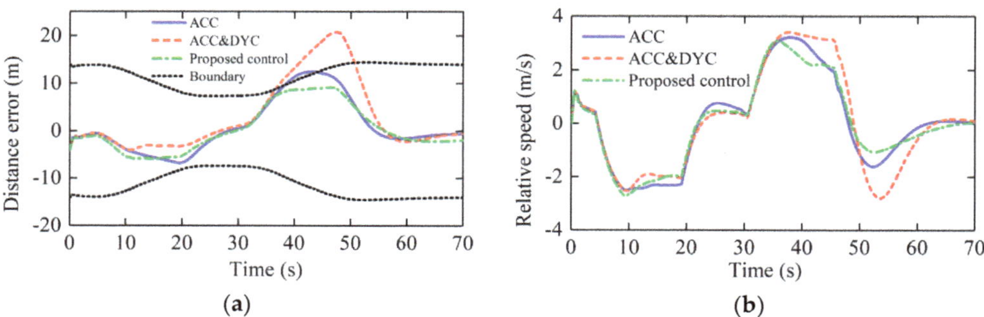

Figure 13. Longitudinal car-following errors, (**a**) Longitudinal car-following distance error, (**b**) Relative speed.

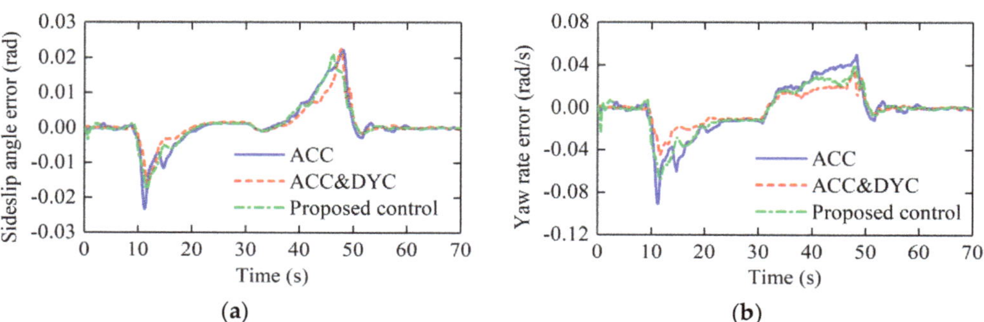

Figure 14. Lateral stability errors: (**a**) vehicle sideslip angle error; (**b**) yaw rate error.

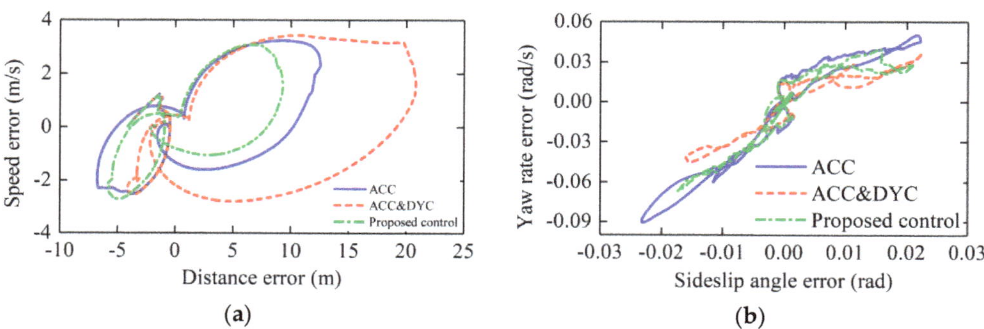

Figure 15. Phase plane of errors: (**a**) phase plane of longitudinal car-following errors; (**b**) phase plane of lateral errors.

It can be seen from Figure 13 that for the constant weight ACC&DYC, when the host vehicle decelerates, the additional braking force will make the longitudinal car-following errors smaller than the constant weight ACC and proposed control, but when the host vehicle is in an accelerating process, the additional braking force will make the longitudinal car-following capability worse. When the distance error is close to the limit value, the proposed control will increase its weight according to the degree of approaching the limit value to keep the distance error within the limit value as far as possible. As can be seen in Figure 13a, the distance error with the proposed control is kept within the driver sensitivity limit, while the errors with the other two controllers exceed the limit value.

In terms of lateral stability control, as can be seen in Figure 14, the maximum yaw rate errors with constant weight ACC, constant ACC&DYC and the proposed control are about

0.091 rad/s, −0.045 rad/s and 0.067 rad/s, and the maximum sideslip angle errors with constant weight ACC, constant ACC&DYC and the proposed control are about 0.023 rad, −0.023 rad and 0.021 rad. With the proposed control, the maximum yaw rate error and sideslip angle error are both in a small range. As shown in Figure 15, in the aspect of lateral stability errors, ACC&DYC has the best control effect. However, in the aspect of longitudinal tracking errors, the maximum distance error with ACC&DYC has exceeded 20 m which is too large for driver's sensitivity limit. Although the lateral stability errors are smaller with ACC&DYC, it sacrifices too much longitudinal car-following performance. By supervising the risk of losing lateral stability and then apply the corresponding control strength of DYC system, the proposed control realized coordination of car-following performance and lateral stability, so as to ensure that the car-following errors and lateral stability errors are both in a relatively acceptable range.

The proposed control determines the weight of the distance error by the dependent degree $K_{ACC}(S)$ which can reflect the control effect of the longitudinal control. The proposed control determines the weights of sideslip angle error and yaw rate error by the dependent degree $K_{VLS}(S)$ which can reflect the risk of losing vehicle lateral stability. As can be seen from Figure 16b, when the longitudinal distance error increases, the KACC will be increased to adjust the weights and ensure the longitudinal car-following capability; when the value of driver steering wheel angle and X_{region} increase, the K_{VLS} will be increased to adjust the weights and ensure the lateral stability. The maximum errors and X_{region} with three controllers are shown in Table 2. Obviously, the overall performance of the control system is improved. The proposed control can intelligently determine the weight matrices by the control effect of the longitudinal distance error and the risk of losing lateral stability. Thus, on the premise of ensuring the car-following performance and lateral stability, the fuel economy and longitudinal ride comfort are improved as much as possible.

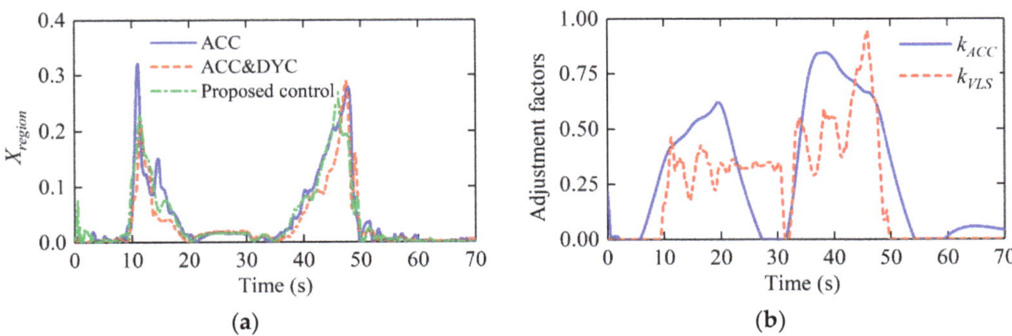

Figure 16. X_{region} and adjustment factors: (**a**) X_{region}; (**b**) adjustment factors.

Table 2. Maximum errors and X_{region} with the three controllers.

	ACC	ACC&DYC	Proposed Control
Δd (m)	12.539	20.836	9.311
Δv (m/s)	3.232	3.425	3.092
$\Delta \beta$ (rad)	0.021	0.021	0.020
$\Delta \omega$ (rad/s)	0.090	0.045	0.067
X_{region}	0.321	0.289	0.273

From the perspective of fuel economy and longitudinal ride comfort, the fuel consumption with ACC is the lowest because of zero additional yaw moment, i.e., additional braking forces are zero. The fuel consumption with ACC&DYC is the highest because of the biggest control strength of DYC system. The fuel consumption with the proposed control is in a medium range so that the vehicle can improve the fuel economy as much as possible on the premise of ensuring the lateral stability.

As can be seen in Figure 17, the longitudinal acceleration of host vehicle with the proposed control increases rapidly at about 35 s. The reason is that the distance error is about to reach the sensitivity limit. Therefore, it is necessary to increase the weights and the control strength in the longitudinal control. When the distance error is in a relatively small range, the proposed control will decrease the weights to improve the longitudinal ride comfort as much as possible. The control outputs and brake pressure are shown in Figures 18 and 19, respectively. It can be seen from the Figures 18 and 19 that the throttle opening of ACC&DYC control method is greater than 0 between 10 s and 20 s, while the throttle opening of the other two methods is 0. This is because the ACC&DYC method needs to provide additional relatively large yaw moment during deceleration, which is the same between 40 s and 50 s. This is also one of the reasons for the high fuel consumption of ACC&DYC, because some fuel energy is converted into heat energy during differential braking process, which is also reflected both in larger throttle opening and brake pressure. This is the conflict between ACC (car-following performance) and DYC (lateral stability). The proposed control gets a balance between car-following performance and lateral stability. Meanwhile, the fuel economy has been improved by reducing such conflict.

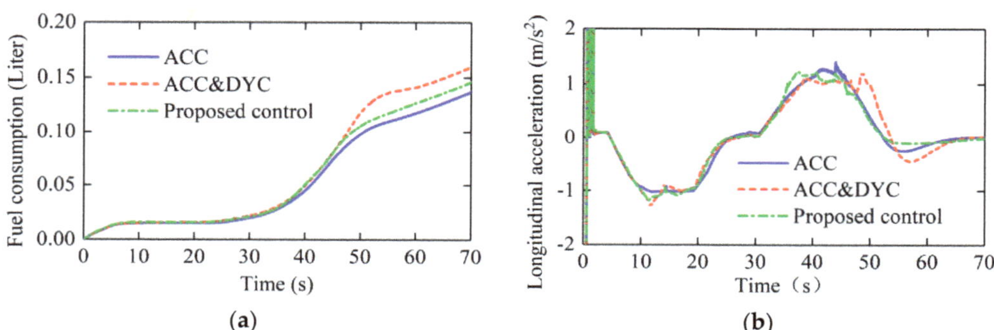

Figure 17. Fuel consumption and longitudinal acceleration: (**a**) fuel consumption; (**b**) longitudinal acceleration.

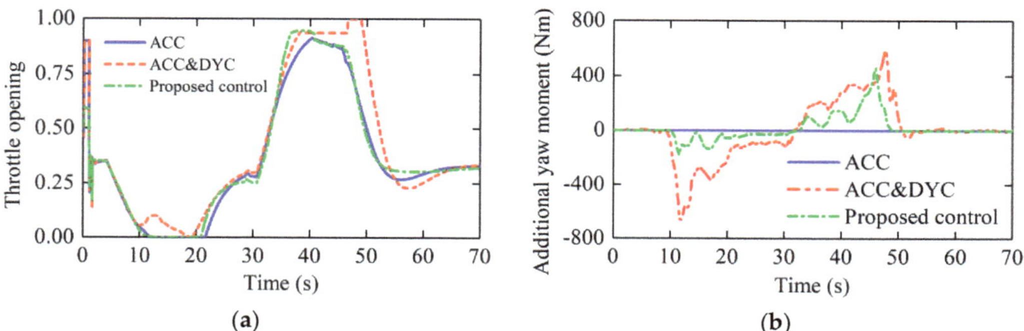

Figure 18. Control outputs: (**a**) throttle opening; (**b**) additional yaw moment.

To summarize, compared with the traditional constant weight ACC and ACC&DYC, the proposed control can both ensure the car-following performance and lateral stability by intelligently designing the real-time weight matrices. It solves the problem of excessive sacrifice of other performances when improving one performance during car-following process.

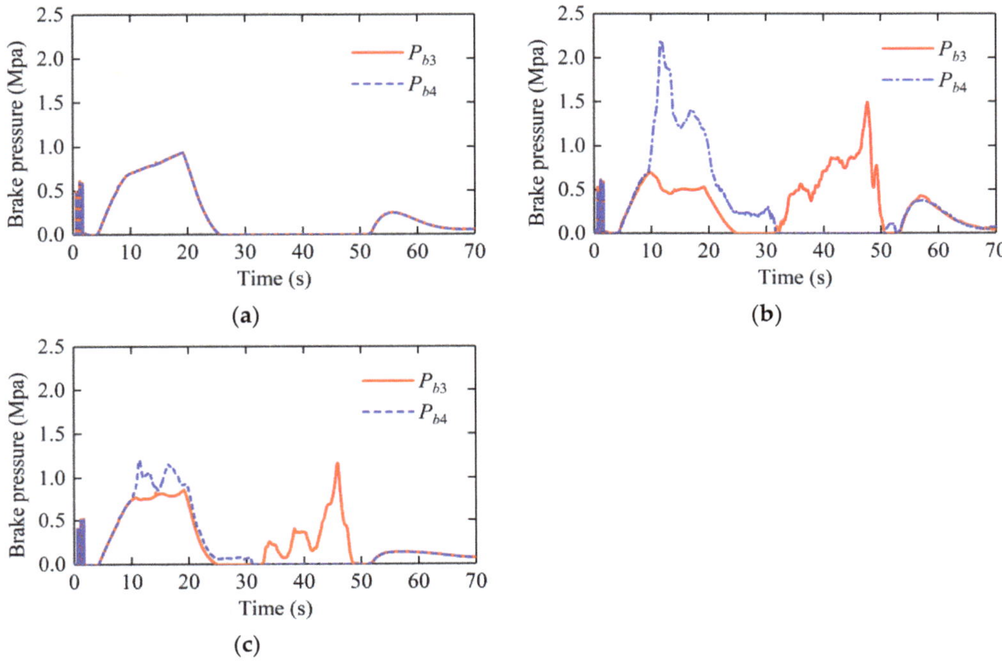

Figure 19. Brake pressure on wheels: (**a**) brake pressure with ACC; (**b**) brake pressure with ACC&DYC; (**c**) brake pressure with proposed control.

5. Conclusions

A coordinated multi-objective ACC integrated with DYC under the MPC framework was proposed in this paper. The extension control is introduced into the real-time weight matrix design to realize the coordination of various control objectives. The extension control can intelligently adjust the weight matrix by evaluating the control effect of ACC and the risk of losing lateral stability.

The longitudinal car-following performance, lateral stability, fuel economy and longitudinal ride comfort are considered in the control design. On the premise of ensuring longitudinal car-following performance and lateral stability, the fuel economy and longitudinal ride comfort are improved as much as possible.

With the proposed control, the longitudinal car-following distance error was kept within the driver sensitivity limit. The lateral stability was ensured by applying DYC system. Compared with the other two constant weight-matrix MPCs, the proposed control can improve the overall performance of vehicle control system and realize the coordination of longitudinal car-following capability, lateral stability, fuel economy and longitudinal ride comfort. The application of extension coordinated control enables ACC vehicles to deal with the problem of multi-objective coordinated control on curved roads. From a practical point of view, it is conducive to reduce traffic accidents, reduce energy consumption and improve driver comfort and vehicle safety.

Author Contributions: Conceptualization, H.W. and Y.S.; methodology, H.W. and Y.S.; software, Y.S.; validation, Y.S.; formal analysis, H.W., Y.S. and Z.G.; investigation, Y.S. and Z.G.; resources, H.W.; data curation, Y.S. and L.C.; writing—original draft preparation, H.W., Y.S. and Z.G.; writing—review and editing, H.W. and Y.S.; visualization, Y.S. and L.C.; supervision, H.W. All authors have read and agreed to the published version of the manuscript.

Funding: This research was supported by National Natural Science Foundation of China (U1564201), Science Fund of Anhui Intelligent Vehicle Engineering Laboratory (PA2018AFGS0026), Natural Science Foundation of Inner Mongolia (2021MS05051), Research Program of science and technology at Universities of Inner Mongolia Autonomous Region (NJZZ21058) and Key Technology Research Project of Inner Mongolia Science and Technology Department (2020GG0200).

Institutional Review Board Statement: Not applicable.

Informed Consent Statement: Not applicable.

Data Availability Statement: The data presented in this study are available on request from the corresponding author. The data are not publicly available due to privacy.

Conflicts of Interest: The authors declare no conflict of interest.

References

1. Xiao, L.; Gao, F. A comprehensive review of the development of adaptive cruise control systems. *Veh. Syst. Dyn.* **2010**, *48*, 1167–1192. [CrossRef]
2. Moon, I.-K.; Yi, K.; Caveney, D.; Hedrick, J.K. A multi-target tracking algorithm for application to adaptive cruise control. *J. Mech. Sci. Technol.* **2005**, *19*, 1742–1752. [CrossRef]
3. Martinez, J.J.; Canudas-de-Wit, C. A safe longitudinal control for adaptive cruise control and stop-and-go scenarios. *IEEE Trans. Control Syst. Technol.* **2007**, *15*, 246–258. [CrossRef]
4. Ganji, B.; Kouzani, A.Z.; Khoo, S.Y.; Shams-Zahraei, M. Adaptive cruise control of a HEV using sliding mode control. *Expert Syst. Appl.* **2014**, *41*, 607–615. [CrossRef]
5. Lin, Y.; Nguyen, H.L.T. Adaptive Neuro-Fuzzy Predictor-Based Control for Cooperative Adaptive Cruise Control System. *IEEE Trans. Intell. Transp. Syst.* **2020**, *21*, 1054–1063. [CrossRef]
6. Althoff, M.; Maierhofer, S.; Pek, C. Provably-Correct and Comfortable Adaptive Cruise Control. *IEEE Trans. Intell. Veh.* **2021**, *6*, 159–174. [CrossRef]
7. Moser, D.; Schmied, R.; Waschl, H.; del Re, L. Flexible Spacing Adaptive Cruise Control Using Stochastic Model Predictive Control. *IEEE Trans. Control Syst. Technol.* **2018**, *26*, 114–127. [CrossRef]
8. Luo, L.; Liu, H.; Li, P.; Wang, H. Model predictive control for adaptive cruise control with multi-objectives comfort, fuel-economy, safety and car-following. *J. Zhejiang Univ. Sci. A* **2010**, *11*, 191–201. [CrossRef]
9. Li, S.; Li, K.; Rajamani, R.; Wang, J. Model Predictive Multi-Objective Vehicular Adaptive Cruise Control. *IEEE Trans. Control Syst. Technol.* **2011**, *19*, 556–566. [CrossRef]
10. Luo, Y.; Chen, T.; Zhang, S.; Li, K. Intelligent Hybrid Electric Vehicle ACC With Coordinated Control of Tracking Ability, Fuel Economy, and Ride Comfort. *IEEE Trans. Intell. Transp. Syst.* **2015**, *16*, 2303–2308. [CrossRef]
11. Ren, Y.; Zheng, L.; Yang, W.; Li, Y. Potential field-based hierarchical adaptive cruise control for semi-autonomous electrc vehicle. *Proc. Inst. Mech. Eng. Part D J. Automob. Eng.* **2019**, *233*, 2479–2491. [CrossRef]
12. Asadi, B.; Vahidi, A. Predictive Cruise Control: Utilizing Upcoming Traffic Signal Information for Improving Fuel Economy and Reducing Trip Time. *IEEE Trans. Control Syst. Technol.* **2011**, *19*, 707–714. [CrossRef]
13. Zhang, D.; Li, K.; Wang, J. A curving ACC system with coordination control of longitudinal car following and lateral stability. *Veh. Syst. Dyn.* **2012**, *50*, 1085–1102. [CrossRef]
14. Cheng, S.; Li, L.; Mei, M.; Nie, Y.-L.; Zhao, L. Multiple-Objective Adaptive Cruise Control System Integrated With DYC. *IEEE Trans. Veh. Technol.* **2019**, *68*, 4550–4559. [CrossRef]
15. Idriz, A.F.; Rachman, A.S.; Baldi, S. Integration of auto-steering with adaptive cruise control for improved cornering behaviour. *IET Intell. Transp.Syst.* **2017**, *11*, 667–675. [CrossRef]
16. Cai, W. The extension set and non-compatible problems. In *Advances in Applied Mathematics and Mechanics in China*; International Academic Publishers: Lausanne, Switzerland, 1990; Volume 2, pp. 1–21.
17. Li, J.; Wang, S. Primary research on extension control. In Proceedings of the AMSE International Conference on Information and Systems, New York, NY, USA, 16–18 December 1991; Volume 1, pp. 392–395.
18. Chen, W.; Wang, H. Function allocation based vehicle suspension/steering system extension control and stability analysis. *J. Mech. Eng.* **2013**, *49*, 67–75. [CrossRef]
19. Zhao, W.; Fan, M.; Wang, C.; Jin, Z.; Li, Y. H∞/extension stability control of automotive active front steering system. *Mech. Syst. Signal Process.* **2019**, *115*, 621–636. [CrossRef]
20. Chen, W.; Liang, X.; Wang, Q.; Zhao, L.; Wang, X. Extension coordinated control of four wheel independent drive electric vehicles by AFS and DYC. *Control Eng. Pract.* **2020**, *101*, 1–12. [CrossRef]
21. Pacejka, H. *Tire and Vehicle Dynamics*; Butterworth Heinemann: Oxford, UK, 2012.
22. Hu, J.; Rakheja, S.; Zhang, Y. Real-time estimation of tire—Road friction coefficient based on lateral vehicle dynamics. *Proc. IMechE Part D J. Automob. Eng.* **2020**, *234*, 2444–2457. [CrossRef]
23. Kim, D.; Min, K.; Kim, H.; Huh, K. Vehicle sideslip angle estimation using deep ensemble-based adaptive Kalman filter. *Mech. Syst. Signal Process.* **2020**, *144*, 1–17. [CrossRef]

24. Bascetta, L.; Baur, M.; Ferretti, G. A simple and reliable technique to design kinematic-based sideslip estimators. *Control Eng. Pract.* **2020**, *96*, 1–17. [CrossRef]
25. Liang, Y.; Li, Y.; Yu, Y.; Zheng, L. Integrated lateral control for 4WID/4WIS vehicle in high-speed condition considering the magnitude of steering. *Veh. Syst. Dyn.* **2020**, *58*, 1711–1735. [CrossRef]
26. Rajamani, R. *Vehicle Dynamics and Control*; Springer Science: Boston, MA, USA, 2006.
27. Inagaki, S.; Kushiro, I.; Yamamoto, M. Analysis on vehicle stability in critical cornering using phase-plane method. *JSAE Rev.* **1995**, *16*, 287–292.
28. Zhan, J. Research on model of brake system based on ACC system. *China Mech. Eng.* **2005**, *16*, 450–452.

 *actuators*

Article

Automatic Lane-Changing Decision Based on Single-Step Dynamic Game with Incomplete Information and Collision-Free Path Planning

Hongbo Wang [1,2,3,*] , Shihan Xu [1,2,*] and Longze Deng [1,2]

1. School of Automotive and Transportation Engineering, Hefei University of Technology, Hefei 230009, China; 2018180261@mail.hfut.edu.cn
2. Anhui Intelligent Vehicle Engineering Laboratory, Hefei 230009, China
3. School of Mechanical and Mechatronic Engineering, University of Technology, Sydney, NSW 2007, Australia
* Correspondence: wanghongbo@hfut.edu.cn (H.W.); hansx@mail.hfut.edu.cn (S.X.)

Citation: Wang, H.; Xu, S.; Deng, L. Automatic Lane-Changing Decision Based on Single-Step Dynamic Game with Incomplete Information and Collision-Free Path Planning. *Actuators* **2021**, *10*, 173. https://doi.org/10.3390/act10080173

Academic Editors: Peng Hang, Xin Xia and Xinbo Chen

Received: 18 June 2021
Accepted: 22 July 2021
Published: 24 July 2021

Abstract: Traffic accidents are often caused by improper lane changes. Although the safety of lane-changing has attracted extensive attention in the vehicle and traffic fields, there are few studies considering the lateral comfort of vehicle users in lane-changing decision-making. Lane-changing decision-making by single-step dynamic game with incomplete information and path planning based on Bézier curve are proposed in this paper to coordinate vehicle lane-changing performance from safety payoff, velocity payoff, and comfort payoff. First, the lane-changing safety distance which is improved by collecting lane-changing data through simulated driving, and lane-changing time obtained by Bézier curve path planning are introduced into the game payoff, so that the selection of the lane-changing start time considers the vehicle safety, power performance and passenger comfort of the lane-changing process. Second, the lane-changing path without collision to the forward vehicle is obtained through the constrained Bézier curve, and the Bézier curve is further constrained to obtain a smoother lane-changing path. The path tracking sliding mode controller of front wheel angle compensation by radical basis function neural network is designed. Finally, the model in the loop simulation and the hardware in the loop experiment are carried out to verify the advantages of the proposed method. The results of three lane-changing conditions designed in the hardware in the loop experiment show that the vehicle safety, power performance, and passenger comfort of the vehicle controlled by the proposed method are better than that of human drivers in discretionary lane change and mandatory lane change scenarios.

Keywords: autonomous vehicles; lane-changing; decision-making; path planning

1. Introduction

1.1. Background

With the increase in the number of vehicles, the fatality in traffic accidents keeps rising. According to a survey by the World Health Organization (WHO), approximately 1.24 million people were killed in road traffic accidents in 2010 [1]; this number has soared to 1.35 million in 2016 [2] and has remained stubbornly high in recent years. Furthermore, more than 90% of traffic accidents are caused by human error [3]. The drivers' inaccurate estimation of traffic status or illegal operation under lane-changing conditions are the main factors of various traffic accidents [4,5]. Therefore, the safety of lane-changing has attracted extensive attention in the vehicle and traffic fields.

Vehicle lane-changing is a complex condition [6]. Successful lane-changing requires the driver to find an appropriate insertion position in the target lane, control the distance between the vehicle and the front vehicle, and maintain a safe driving position. The function of a lane-changing assistance system is to select an appropriate lane-changing time, plan a reasonable lane-changing path, and further coordinate the vehicle dynamic performance to realize lane-changing operation. The core of the control system includes three parts: early warning and decision-making, path planning, and path tracking. The analysis of the research status will be carried out from the above three aspects.

1.2. Literature Review and Analysis

A lane-changing early warning and decision-making system mainly determines when the vehicle changes its lane and directly affects the vehicle lane-changing safety. Dang et al. [7] realized the lane-changing warning function by vehicle-to-vehicle (V2V) communication. However, the maturity and popularization of V2V communication still need more time. Song et al. [8] used global positioning system and real-time kinematic (GPS-RTK) positioning technology to achieve high-precision vehicle positioning, which could be used to calculate the vehicles' interval. Zhu et al. [9] classified and recognized drivers' driving characteristics based on machine learning, and introduced the parameters considering drivers' characteristics into the following vehicle safety distance to adjust lane-changing decision time. Nevertheless, a long time of data accumulation is needed to realize the recognition of driving characteristics. Butakov and Ioannou [10] collected a large amount of lane-changing related data to understand the reaction characteristics of drivers and vehicles in different driving environments before and during lane-changing, which provided a data basis for the relevant research. By comparing the relative distance between vehicles when drivers change lanes with the traditional headway safety distance, it can be found that the distance between the driver and front vehicle when drivers change the lanes may be less than the safety time distance. Because the distance between vehicles is often less than the safety distance in the real driving environment, the lane-changing decision based on vehicle interval is obviously not in line with the real scene. When there is a car behind the target lane, the lane-changing can be described as the game between the host car and the car behind the target lane. Yu et al. [11] introduced driver aggressiveness into game theory to design lane-changing decision-making to simulate human drivers. Meng et al. [12] combined receding horizon control into game theory and proved the effectiveness of the method through traffic case simulation. Cao et al. [13] established lane-level link performance function to evaluate the driving efficiency of the lane-changing behavior, to improve the macroscopic traffic flow efficiency.

In terms of path planning, there are two common methods, i.e., stochastic and kinematic methods [10]. Although the stochastic method can be used for dynamic planning according to the traffic environment, its planning results are difficult to accurately solve the physical parameters such as the expected yaw rate, which is not conducive to the design of the path tracking controller. The main methods based on kinematic are Polynomial curve [14,15], Clothoid curve [16], Bézier curve [17], and B-spline curve [18]. The kinematic method can describe the lane-changing path in the form of the equation which makes up for the defect of stochastic method, but it is difficult to constrain the path through the vehicle position relationship in the traffic environment. Bae et al. [19] designed a lane-changing path based on the quintic Bézier curve and compared it with the cubic Bézier curve. However, the constraints on the control points only consider the vehicle driving parameters of the starting point and the end point, and therefore there is a lack of basis for the selection of other constraint points. In addition, Mukai and Kawabe [20] used the multiparameter programming method to solve the problem of lane-changing decision and optimal path generation of model predictive control, but the scene construction was relatively simple and could not represent the real driving situation. Hu et al. [21] designed several cost functions to realize real-time path planning and optimal path selection, and the proposed method achieves real-time path planning and speed planning.

In the path tracking, the representative control methods are model predictive control, intelligent control, and sliding mode control. Falcone et al. [22] used the model predictive control to design an active steering path tracking controller. The LTV-MPC method achieved similar performance with nonlinear model predictive control at a lower hardware cost. Naranjo et al. [23] designed an overtaking system with path tracking and lane-changing functions using fuzzy controller. Ren et al. [24] designed the lane-changing path for the curve road and used the nonsingular terminal sliding mode controller to track the lane-changing path. Wu et al. [25] combined sliding mode control with active disturbance rejection control for path tracking, and compared with model predictive control, the effectiveness of the method was verified.

In the vehicle lane-changing system, the upper-level system's decision on lane-changing instruction will have an impact on vehicle power performance and driving comfort when changing lane. Although Yu et al. [11], Meng et al. [12], and Cao et al. [13] verified that the designed decision-making method could improve the efficiency of macroscopic traffic flow, it lacked consideration of automobile power performance and driving comfort in the decision-making method. Liniger et al. [26] formulated three different racing games to study the game between automatic racing cars, which are of great value to promote the research of lane-changing games, and it was necessary to modify the focus of the payoff when applied to the passenger car system. In the aspect of path planning, the frequently used quintic polynomial path planning method [27] can easily obtain the yaw rate and lane-changing path. However, it can only adjust the lane-changing path by changing the whole process time, and therefore it is difficult to restrain the driving path through the distance relationship among the surrounding vehicles. The Bézier curve can be adjusted by constraint points, which is more flexible than the polynomial. Bulut et al. [28] compared the cubic Bézier curve with the quintic Bézier curve. The results showed that compared with the cubic Bézier curve, the variation of the quintic Bézier curve on velocity, lateral acceleration, longitudinal, and lateral jerk was more reasonable.

1.3. Paper Contribution and Organization

In order to improve the above shortcomings, the automatic lane-changing decision-making based on game theory with Bézier curve path planning is proposed. This paper has made the following contributions to the automatic vehicle lane-changing system. (I) In this paper, the existing safety distance model is improved, and the key parameters of the model are optimized through the lane-changing data of the human driver. (II) Through a single-step dynamic game with incomplete information, the safety payoff, comfort payoff, and velocity payoff are taken into account for the lane-changing system and achieve the balance in the optimization. (III) The lane-changing time calculated by the path planning layer is taken as an important parameter to ensure the safety and comfort of lane-changing, which realizes the strong coupling of decision layer and planning layer. The proposed algorithm can better adapt to both discretionary lane change (DLC) and mandatory lane change (MLC). This paper mainly studies the typical expressway driving environments as shown in Figure 1, in which Car.0 is the host car, Car.1 is the front car in the current lane, Car.2 is the front car in the target lane, and Car.3 is the rear car in the target lane.

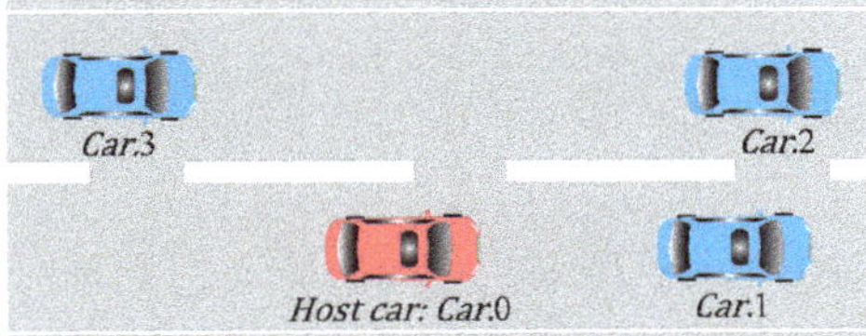

Figure 1. Driving environment and vehicle code name.

The rest of this paper is organized as follows. In Section 2, the lane-changing safety distance is improved by collecting lane-changing data. In Section 3, the payoff function of the lane-changing decision-making method based on game theory is analyzed. In Section 4, the control points of the quintic Bézier curve are constrained to obtain the lane-changing path without forward collision. Based on the vehicle linear two-degree-of-freedom model, the path tracking sliding mode controller of front wheel angle compensation by radical basis function (RBF) neural network is designed. In Section 5, the model-in-the-loop (MIL) simulation, driving simulation test, and hardware-in-the-loop (HIL) verification are carried out. Section 6 is the discussion, and the conclusions are drawn in Section 7.

2. Lane-Changing Safety Distance

There are some driving parameters in the widely used Gipps car following model [29], which is difficult to obtain in the actual environment, so the headway safety distance model [30] is selected. The headway safety distance model is based on the time difference between the adjacent front and rear vehicles passing through the specified point in turn. When the relative speed of the two cars is small, there is an approximately linear relationship between the headway time and the distance of the two cars. Based on this, the following models are established:

$$D_t = V_m t_d + l \tag{1}$$

where D_t is the safety distance based on headway, t_d is the headway time, generally 1.2–2.0 s, V_m is the rear car speed, and l is the safety margin, generally taken as 2–5 m [31].

Considering that the speed of the rear car often changes with the speed of the front car under the car-following condition, a safety distance margin is needed. When the braking distance model of the driver for emergency braking is applied to the car-following scenario, we get

$$D_a = \left(V_m - V_f\right)\tau + \frac{t_i}{2} + \frac{\left(V_m - V_f\right)^2}{2a_{max}} + l \tag{2}$$

where V_f is the front car speed; D_a is the braking distance; τ is the sum of driver reaction time and brake system coordination time, generally taken as 0.8–1.0 s; t_i is the growth time of braking deceleration, generally taken as 0.1–0.2 s; and a_{max} is the maximum braking deceleration that can be achieved during braking, generally taken as 6–8 m/s^2.

It can be seen that the main influencing factor of the safety distance model based on the headway is the host car speed without considering the relative speed of adjacent vehicles. The braking distance model focuses on the relationship with the relative speed and is not sensitive to the speed of the host car. In order to make up for the shortcomings of the two models and achieve complementary advantages, the fusion safety distance (FSD) can be obtained as follows.:

$$D_c = \begin{cases} q_1 D_t + q_2 D_a & V_m > V_f \\ q_1 D_t + q_2 l & V_m \leq V_f \end{cases} \tag{3}$$

where D_c is the fusion of safety distance, q_1 is the weight coefficient of safety distance based on headway, and q_2 is the weight coefficient of braking distance model. When $V_m \leq V_f$, the safety margin is used to replace the braking distance model.

Left lane-changing and right lane-changing often occur in the real driving process, and the steering characteristics of left lane-changing and right lane-changing are similar [32]. On the basis of the specific left lane-changing scene, the specific analyses of the safety distance between *Car*.0 and other cars are carried out as follows.

Based on (3), analyzing *Car*.0 and *Car*.1 may have a rear-end collision during the lane-changing process. It is necessary to consider the safety distance under emergency braking and lane changing, and the safety distance to avoid the rear-end collision is

$$S_{01} = \begin{cases} MAX\left(\begin{array}{c} V_0 - V_1\frac{t_{lc}}{2} + L + \frac{W}{2}sin\theta, \\ q_1 V_0 t_d + q_2\left((V_0 - V_1)\tau + \frac{t_i}{2} + \frac{V_0 - V_1^2}{2a_{max}}\right) + l \end{array} \right), & V_0 > V_1 \\ q_1 V_0 t_d + l, & V_0 \leq V_1 \end{cases} \tag{4}$$

where V_0 is the speed of *Car*.0, V_1 is the speed of *Car*.1, S_{01} is the safety distance between *Car*.0 and *Car*.1, t_{lc} is the time for *Car*.0 to change its lanes, L is the length of both car, W is the width of both car, and θ is the *Car*.0's heading angle when it collides with *Car*.1. As *Car*.0 is generally near the center line of the lane at this time, $\theta = \theta_{max}$.

The collision between *Car*.0 and *Car*.2 occurs after the beginning of the lane-changing. Therefore, while ensuring the safety distance for the lane-changing, a safe car-following distance should be reserved for *Car*.0 after entering the target lane to avoid subsequent rear-end collisions. According to the analysis of possible collisions based on the FSD, it can be concluded that the safety distance of *Car*.0 to avoid rear-end collision or side scraping with *Car*.2 as shown in (5), where S_{02} is the FSD between *Car*.0 and *Car*.2, and V_2 is the speed of *Car*.2.

$$S_{02} = \begin{cases} V_0 - V_2\frac{t_{lc}}{2} + L - \frac{W}{2}\sin\theta + q_1V_0t_d + q_2\left((V_0 - V_2)\cdot\tau + \frac{t_i}{2} + \frac{V_0-V_2^2}{2a_{max}}\right), V_0 > V_2 \\ q_1V_0t_h + l, \; V_0 \leq V_2 \end{cases} \tag{5}$$

The situation where *Car*.3 collides with *Car*.0 occurs at the end of lane-changing. At this time, *Car*.0 has entered into the target lane, the steering wheel angle is about to return to zero, and the heading angle θ of *Car*.0 is very small. Considering the above conditions, based on the FSD, the possible collision with *Car*.3 is analyzed, and the safety distance to avoid the rear-end collision of *Car*.3 can be expressed as (6), where S_{03} the safety distance between *Car*.0 and *Car*.3, and V_3 is the speed of *Car*.3.

$$S_{03} = \begin{cases} V_3 - V_0\frac{3t_{lc}}{4} + L + q_1V_3t_d + q_2\left((V_3 - V_0)\cdot\tau + \frac{t_i}{2} + \frac{V_3-V_0^2}{2a_{max}}\right), V_3 > V_0 \\ q_1V_3t_h + l, \; V_3 \leq V_0 \end{cases} \tag{6}$$

Lane-changing habits of human driver is affected by age, gender, experience, etc. [33]. In order to ensure the authenticity of lane-changing related parameters in the model, 83 drivers with different driving experience and ages are specially invited for lane-changing operation on the driving simulator. Due to the difference between the subjective feeling of driving simulator and real vehicle, each driver is given a period of operation training.

After completing the training, 83 drivers changed the lanes left and right for a total of 672 times. The final results are shown in Table 1, where $\bar{t}_{lc}$ is the average lane-changing time, SD_t is the standard deviation corresponding to t_{lc}, $\bar{\theta}_{max}$ is the average maximum heading angle, and SD_θ is the standard deviation corresponding to the θ_{max}. The last column in Table 1 is the total number of left and right lane-changing, and $\bar{t}_{lc}$, SD_t, $\bar{\theta}_{max}$, SD_θ, corresponding to the total lane-changing number. From Table 1, it can be concluded that the average value of the maximum heading angle is 3.20°, which indicates that the assumption that the maximum heading angle appears during lane-changing process is reasonable, and the average lane-changing time of 5.17 s is close to the 5.48 s [34]. The standard deviation corresponding to the lane-changing time and the maximum heading angle are small, which implies that the data are concentrated. According to the statistics of this lane-changing simulation data, 86.41% of the lane-changing time are in the interval (3.7, 6.3), so the change range of t_{lc} is taken as an integer of (3, 7), and the speed change range is set to 16~33 m/s. The changing rule of the lane-changing safety distance with t_{lc} is shown in Figure 2.

Table 1. Results of driving simulation lane-changing experiment.

	Left Lane-Changing	Right Lane-Changing	General
Number of times	347	325	672
$\bar{t}_{lc}(s)$	5.11	5.23	5.17
SD_t	0.8002	0.9403	0.8703
$\bar{\theta}_{max}$ (deg)	3.33	3.10	3.20
SD_θ	0.6410	0.6122	0.6266

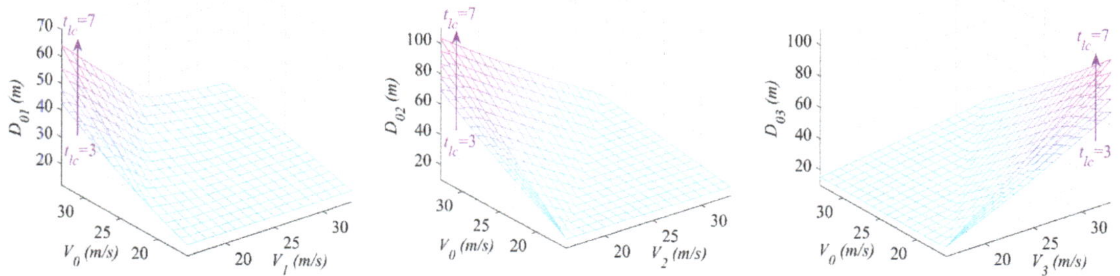

Figure 2. Variation of lane-changing safety distance with t_{lc}.

Through Figure 2, it can be found out that the greater the expected lane-changing time t_{lc}, the greater the lane-changing safety distance will be. Combining the data in Table 1, $t_{lc} = 5.17$ s is used to determine the lane-changing safety distance. The relevant parameters in the lane-changing safety distance model have been determined shown in Table 2.

Table 2. Parameters of lane-changing safety distance model.

Parameter	Value	Unit
Vehicle length, L	4.2	m
Vehicle width, W	1.8	m
Safety margin, l	3	m
Maximum heading angle, θ_{max}	3.20	deg
Expected time for lane $-$ changing, t_{lc}	5.17	
Headway time, t_d	1.2	s
Braking deceleration increase time, t_i	0.15	s
Driver's reaction time and brake delay time, τ	0.9	s
Weight of the safety distance based on the headway, q_1	0.65	-
Weight of braking distance model, q_2	0.35	-
Maximum braking deceleration, a_{max}	7	m/s^2

3. Lane-Changing Decision-Making Based on Single-Step Dynamic Game with Incomplete Information

If *Car*.3 refuses the lane-changing behavior of *Car*.0, it may cause a rear-end collision or side scraping. If *Car*.3 accepts the lane-changing behavior of *Car*.0, *Car*.3 will slow down and avoid *Car*.0. Therefore, there is a strong interaction between *Car*.0 and *Car*.3 in lane-changing scenarios. Game theory is a powerful tool to study the interaction between decision-makers [11]. The relationship between two cars can be regarded as players playing lane-changing games. Game behavior can be defined as a definite mathematical object which mainly includes three essential elements: player, strategy, and payoff [35]. First of all, in the process of driving, the strategies of both players will be adjusted according to the change of traffic environment, and the result of the game is determined once, so the game type between players belongs to single-step dynamic game. Second, assuming the relative distance and speed of the surrounding vehicles can be obtained by radar, but for vehicles without V2V communication function, only the controlled vehicle (*Car*.0) can obtain the payoff function of both players in the game, that is, the information obtained by both players in the game is not complete. Finally, a single-step dynamic game with incomplete information is selected to model the game relationship between *Car*.0 and *Car*.3 in the lane-changing scene.

As shown in Figure 3, C_i is the player, and d_i^j is the corresponding strategy. When two players play a single-step dynamic game, the dotted line connects the two possible behaviors of C_1 which means that C_2 does not know what decision C_1 will make. This game is equivalent to C_1 and C_2 making a lane-changing decision at the same time, so that the extended game problem under incomplete information can be transformed into a static game problem for solution [36]. The pure strategies produced by this game can be written as shown in Table 3.

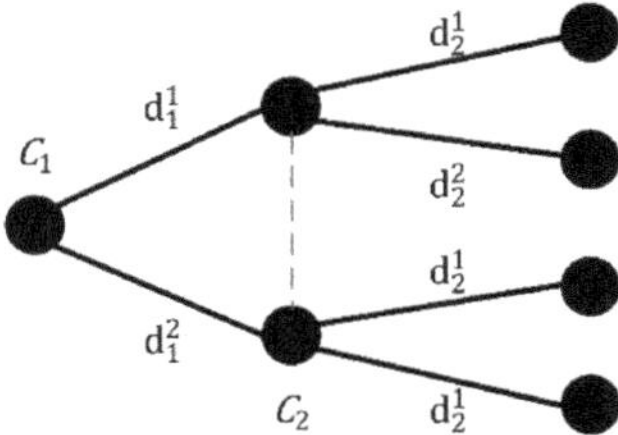

Figure 3. The extended schematic diagram of single-step game.

Table 3. Pure strategies of the lane-changing game.

Player	Car.3	
	Lane-Changing/Accept	**Lane Keeping/Refuse**
Car.0	(U_0^1, U_3^1)	(U_0^2, U_3^2)

In Table 3, U_0^1 is lane-changing payoff of *Car*.0, U_0^2 is lane keeping payoff of *Car*.0, U_3^1 is the payoff of *Car*.3 when *Car*.3 accepts *Car*.0's lane-changing, and U_3^2 is the payoff of *Car*.3 when *Car*.3 refuses *Car*.0 to change its lane.

3.1. Safety Payoff

Taking U_{safety} as the safety payoff, the safety payoff function can be described as

$$U_{safety}^{i,p} = \begin{cases} 1 & D_k \geq S_k \\ \dfrac{\ln\left(\frac{D_k}{S_k}+1\right)}{\ln 2} & l < D_k < S_k \\ -\infty & D_k \leq l \end{cases} \tag{7}$$

where $i \in \{0,3\}$, 0 is corresponding to *Car*.0, and 3 is corresponding to *Car*.3; $p \in \{1,2\}$, 1 means lane-changing or accepting lane-changing, 2 means lane keeping or refusing lane-changing; $k \in \{02, 01, 03, 23\}$, 02 represents the relationship between *Car*.0 and *Car*.2, 01 represents the relationship between *Car*.0 and *Car*.1, 03 represents the relationship between *Car*.0 and *Car*.3, and 23 represents the relationship between *Car*.2 and *Car*.3; The numbers in i, p, and k play the role of codes, D_k is the distance between vehicles, and S_k is the lane-changing safety distance.

It can be seen that when $D_k < l$, the safety payoff will directly reach $-\infty$, prompting *Car*.0 to change its lane immediately. Similarly, when *Car*.3 is close to *Car*.0, the payoff of *Car*.3 will directly reach $-\infty$ when it agrees *Car*.0 changes the lane. When *Car*.0 is close to *Car*.3, the lane-changing behavior of *Car*.0 will not be carried out. If *Car*.0 is close to *Car*.1, *Car*.0 will change its lane immediately. Assuming that the speed of all vehicles is constant, when the above two situations occur at the same time, the collision loss caused by lane-changing or not is judged by combining other payoff functions.

3.2. Velocity Payoff

For both players of the game, the vehicle's current speed of the player is set as the threshold that the player can continue to obtain, and the speed difference between the front vehicle and the player's own vehicle is the payoff variable. According to this setting, the velocity payoff U_v can be described as follows.

$$U_v^{i,p} = \begin{cases} 1 & v_k \geq 2v_i \\ \frac{v_k - v_i}{v_i} & 0 < v_k < 2v_i \\ -1 & v_k = 0 \end{cases} \tag{8}$$

When there is a stationary car ahead or vehicle is detected currently on the ramp and needs to change its lane, the velocity payoff reaches the minimum value of -1. When the target speed is equal to the current speed, the velocity payoff is 0. Because the speed of each car on the road is basically within the speed limit range when driving at high speed, the velocity gain reaches 1 when the target speed is twice the current speed.

The significance of the velocity payoff setting is that when the speed of *Car*.1 is greater than the current speed or the expected speed of *Car*.0, *Car*.0 does not need to change its lanes. When the speed of *Car*.1 is less than the current speed or the expected speed of *Car*.0, the speed of *Car*.1 will be compared with the speed of *Car*.2. If the speed of *Car*.2 can better meet the velocity payoff of *Car*.0, the lane-changing demand will be generated. That is to say, the earlier the lane-changing is completed, the better the vehicle power performance will be.

3.3. Comfort Payoff

When the speed of the preceding car is lower than the speed of the host car, the host car choosing to brake or change its lane to avoid the collision is needed. However, when the relative speed difference is large, the small lane-changing time is needed, the short lane-changing time will cause large lateral acceleration. Which is bad to the comfort of passengers. Therefore, connecting the lane changing time obtained from (14) and the comfort payoff. The comfort payoff $U_{comfort}$ is

$$U_{comfort}^{i,p} = \begin{cases} \frac{2}{1+e^{-t_{ca}}} - 2 & v_i > v_k \\ 0 & v_i \leq v_k \end{cases} \tag{9}$$

where t_{ca} is half of the total lane-changing time shown in (14). The big speed difference between the preceding car and the host car will cause a small t_{ca}, and the time reserved for the driver to change the lanes is also very short. That is, the big lateral acceleration will be generated during lane-changing, leading to the worse comfort of passengers. The changing trend of $U_{comfort} - t_{ca}$ is shown in Figure 4. It is indicated that when t_{ca} is more than 2.5 s, the change of $U_{comfort}$ is relatively gentle, while when t_{ca} is less than 2.5 s, the comfort payoff decreases sharply, which is consistent with the collected average lane-changing time of human drivers. The change trend of $U_{comfort} - t_{ca}$ meets the influence rule of lane-changing time on lateral comfort.

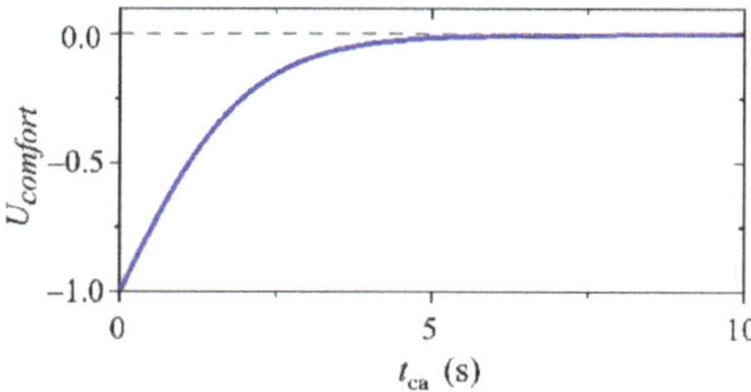

Figure 4. The change trend of $U_{comfort} - t_{ca}$.

3.4. Total Payoff and Game Solution

The total payoff is a linear combination of safety payoff, velocity payoff, and comfort payoff. α, β, and γ in (10) are the weights of corresponding payoff.

$$U_{payoff} = \alpha U_{safety} + \beta U_v + \gamma U_{comfort} \tag{10}$$

Excessive consideration of safety will reduce the lane-changing possibility, leading to system conservativeness increased. On the contrary, weakening the consideration of safety will increase the risk of collision [37]. This paper takes $\alpha = 0.5$, $\beta = 0.3$, and $\gamma = 0.2$.

The game model has at least one equilibrium point, which can be calculated by the change in the payoff function while driving. The problem of solving the equilibrium point is transformed into a problem of extreme points for solving, as in (11).

$$\begin{cases} C_2'(d_1) \triangleq d_2' \\ d_2' = \left\{ U_3^{d_1,d_2'} \geq U_3^{d_1,d_2}, d_1 \in C_1, d_2 \in C_2 \right\} \\ \hat{d}_1 = argmax\left(U_0^{d_1}, U_0^{d_2'} \right) \\ \hat{d}_2 = argmax\left(U_3^{\hat{d}_1}, U_3^{d_2} \right) \end{cases} \tag{11}$$

where d_1 and d_2 are the decisions of $Car.0$ and $Car.3$, respectively; C_1 and C_2 are the strategy sets of the two cars; $C_2'(d_1)$ is the decision set of $Car.3$ after the decision of $Car.0$; d_2' is the decision under the decision set; and $\hat{d}_1$, $\hat{d}_2$ are the final decisions of the two vehicles.

In order to verify the effectiveness of the proposed game method, the MLC scenario shown in Figure 5 designed in [11] is used for the simulation verification. Suppose the decision-making method used in that paper is M_1, and the decision-making method used in this paper is M_2. The comparison results are shown in Table 4. It can be seen from the Table 4 that the method of M_1 and M_2 adopt the same lane-changing cut-in position in the two scenarios of Test 1 and Test 2. In Test 3, the distance between $Car.3$ and $Car.4$ is only 10 m. At this time, $Car.0$ still choosing to insert gap 2 will have a greater impact on the velocity payoff of $Car.4$, and it may even cause rear-end collision. Therefore, the proposed method chooses to change the lane immediately when $Car.4$ overtakes $Car.0$. The price of delaying the lane-changing operation is to produce greater lateral acceleration, that is, to ensure driving safety by sacrificing part of the comfort. It is indicated that the proposed game lane-changing decision is effective.

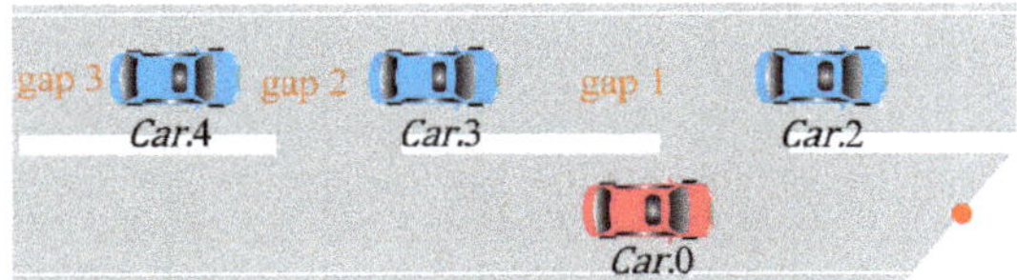

Figure 5. MLC scenario.

Table 4. Lane-changing cut-in positions comparison.

		Test 1	Test 2	Test 3
Parameter	v_0 (m/s^2)	10	10	10
	D_{01} (m)	50	50	50
	D_{02} (m)	10	10	10
	v_2 (m/s^2)	15	15	15
	D_{03} (m)	−20	−8	−8
	v_3 (m/s^2)	15	15	15
	D_{04} (m)	−30	−25	−18
	v_4 (m/s^2)	15	15	15
Decision	M_1	gap 1	gap 2	gap 2
	M_2	gap 1	gap 2	gap 3

4. Lane-Changing Path Planning and Tracking Control

The lane-changing path needs the lateral speed and lateral acceleration of the starting point and the ending point to be continuous, and the path can be constrained through the control points according to the traffic environment. The first part of this section will complete the lane-changing path planning by constraining the control points of the quintic Bézier curve. In the case of MLC, the high accuracy of the path tracking controller is required to control the vehicle interval accurately. There are some assumptions and simplifications in vehicle modeling, which will inevitably lead to the decline of tracking control accuracy [38]. Therefore, in the second part of this section, RBF neural network is used to compensate the vehicle front wheel angle modeling error under sliding mode control (SMC).

4.1. Lane-Changing Path Planning Based on Bézier Curve

The Bézier curve was invented by Pierre Bézier and has been widely used in computer graphics and animation [19]. Taking the lane direction as the coordinate X and the vertical lane direction as the coordinate Y, the lane change path can be given in the form of the parametric equation as

$$
\begin{cases}
f_x(j) = \sum_{i=0}^{5} \begin{pmatrix} 5 \\ i \end{pmatrix} (1-j)^{5-i} j^i P_{xi} \, , \ (0 \le j \le 1) \\
f_y(j) = \sum_{i=0}^{5} \begin{pmatrix} 5 \\ i \end{pmatrix} (1-j)^{5-i} j^i P_{yi} \, , \ (0 \le j \le 1)
\end{cases}
\tag{12}
$$

where P_{xi} and P_{yi} are the horizontal and vertical coordinates of the control point $P_i(P_{xi}, P_{yi})$, respectively.

In order to meet the requirements of lane-changing lateral velocity and lateral acceleration at the starting and ending to be continuous, the lane width $h = 3.75$ m, then $P_{y0} = P_{y1} = P_{y2} = 0; P_{y3} = P_{y4} = P_{y5} = h$.

Setting the lane-changing starting point $P_{x0} = 0$, the vehicle-mounted radar is installed on the top of the vehicle. Considering that *Car.*0 may collide to *Car.*1 during the lane-changing process (shown in Figure 6), the horizontal coordinates P_{x2} and P_{x3} of the midpoint of the path are restricted shown in (13).

$$
\begin{cases}
P_{x2} = P_{x3} = v_0 t_{c1} - D_i \\
t_{c1} = \frac{D_{01}}{v_0 - v_1} \\
D_i \approx L_i \cos\left(\arctan\left(\frac{W}{2L_f}\right) - \theta\right)
\end{cases}
\tag{13}
$$

where t_{c1} is the time for *Car.*0 to rear-end *Car.*1, D_i is the longitudinal distance between the vehicle-mounted radar of *Car.*0 and the closest point of *Car.*1, and L_i is the linear distance between the vehicle-mounted radar and the closest point of *Car.*1. As θ is small, $L_i \approx l + L_a$ and $L_a \approx L_f$, where L_f is the distance from the vehicle-mounted radar to the front bumper of *Car.*0.

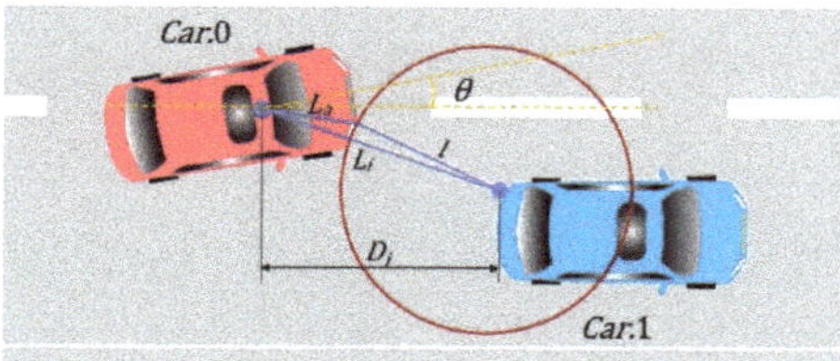

Figure 6. The relationship between *Car.*0 and *Car.*1 avoiding collision.

Then, t_{ca} for the vehicle to reach the lane-changing midpoint can be obtained as

$$t_{ca} = \frac{P_{x2}}{v_0 - v_1} \tag{14}$$

Because $P_{x5} = 2P_{x2}$, the control point P_{x5} has also been constrained. Setting that P_{x1} and P_{x4} are symmetrical about $\left(P_{x2}, \frac{h}{2}\right)$, so we get

$$\begin{cases} P_{x1} = \frac{P_{x2}-P_{x0}}{i} \ (i \geq 1) \\ P_{x4} = P_{x5} - \frac{P_{x5}-P_{x3}}{i} \ (i \geq 1) \end{cases} \tag{15}$$

Take i as an integer between [1,10] to draw Bézier curve shown in Figure 7a. It can be found out that when $P_{x5} = 2P_{x2}$, $h_1 = h_2$, and no matter what the value of i is, the curve will always pass $\left(P_{x2}, \frac{h}{2}\right)$, which also verifies the constraints setting of $P_0\left(P_{x0}, P_{y0}\right)$, $P_2\left(P_{x2}, P_{y2}\right)$, $P_3\left(P_{x3}, P_{y3}\right)$, and $P_5\left(P_{x5}, P_{y5}\right)$ are correct and reasonable. In addition, the curve gradually tends to be flat with i increasing. Therefore, it can be judged that with the increase of i, the maximum yaw rate generated by the vehicle in the tracking process is decreased. In order to verify this conjecture, according to (16), the change trend of yaw rate corresponding to different values of i is obtained shown in Figure 7b.

$$\begin{cases} k(j) = \dfrac{\dot{f}_x(j)\ddot{f}_y(j)-\dot{f}_y(j)\ddot{f}_x(j)}{\left(\dot{f}_x^{\,2}(j)+\dot{f}_y^{\,2}(j)\right)^{\frac{3}{2}}} \\ \omega(j) = \dot{f}_x(j)k(j) \end{cases} \tag{16}$$

where $k(j)$ is the curvature of the path and $\omega(j)$ is the yaw rate; $j \in [0, 1]$.

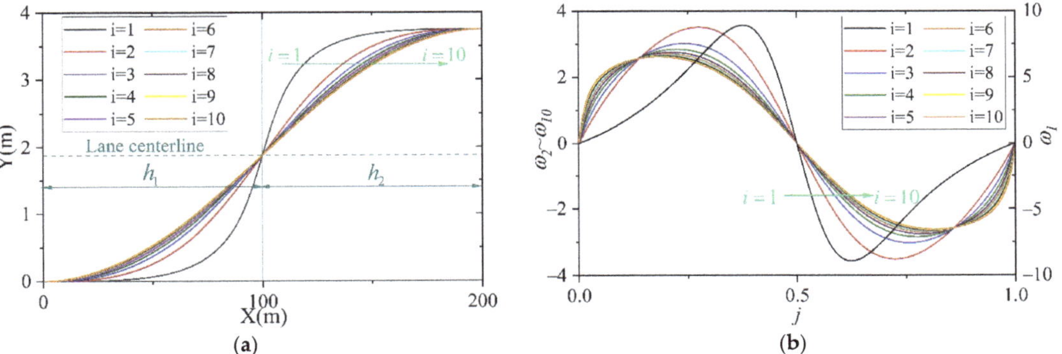

Figure 7. Influence of different i values on lane-changing path. (**a**) Comparison of driving path. (**b**) Comparison of yaw rate.

In Figure 7b, the maximum yaw rate that is generated by the planned path decreases with the increase of i, which is conducive to providing better dynamism and comfort during the lane-changing process. However, with increases in i, a greater rate of change in yaw will be generated near the starting and ending points of the lane-changing curve, which may cause passengers to become anxious when subjected to an instantaneously increasing lateral force. When $i > 5$, the decrease in yaw rate caused by the increase of i is small, so $i = 5$ is chosen to constrain P_{x1} and P_{x4} in this paper.

4.2. Path Tracking Controller

In the design of the path tracking controller, the following reasonable assumptions are put forward: (1) Ignore the roll, pitch, and vertical movement of the vehicle. (2) The vehicle lateral acceleration during the lane-changing is small, and the tires can be assumed working in the linear region. (3) The controller directly controls the front wheel angle. (4) The road surface is flat. Based on the above assumptions, the vehicle linear two-degree-of-freedom model built shown in (17).

$$
\begin{bmatrix} \dot{v}_y \\ \dot{\omega}_c \end{bmatrix} = \begin{bmatrix} -\dfrac{C_f + C_r}{m v_x} & \dfrac{a C_f - b C_r}{m v_x} - v_x \\ \dfrac{-a C_f + b C_r}{I_z v_x} & -\dfrac{a^2 C_f + b^2 C_r}{I_z v_x} \end{bmatrix} \begin{bmatrix} v_y \\ \omega_c \end{bmatrix} + \begin{bmatrix} \dfrac{C_f}{m} \\ \dfrac{a C_f}{I_z} \end{bmatrix} \delta_f
\tag{17}
$$

where v_x is the longitudinal velocity; v_y is the lateral velocity; ω_c is the yaw rate; δ_f is the turning angle of the front wheels; m is the mass of the vehicle; C_f and C_r are the cornering stiffnesses of the front and rear axles, respectively; and a and b are the distances from the center of mass to the front and rear axles, respectively.

There are some unavoidable disturbances during vehicle driving. In order to improve the robustness of the path tracking controller, the design of the sliding mode surface is

$$
s = ce + \dot{e} , \ c > 0
\tag{18}
$$

where $e = \varphi - \varphi_r$, φ is the actual yaw angle of the vehicle, and φ_r is the desired yaw angle.

$$
\dot{s} = -\eta sgn(s) - ks , \ \eta > 0, k > 0
\tag{19}
$$

Differentiating (18) and combining with (17) and (19), we get

$$
\delta_{eq} = \left(\dot{\omega}_r + c\omega_r + (f_1 - c)\dot{\omega}_c - f_2 v_y - \eta sgn(s) - ks \right) / f_3
\tag{20}
$$

where $f_1 = \left(a^2 C_f + b^2 C_r \right) / I_z v_x$; $f_2 = \left(-a C_f + b C_r \right) / I_z v_x$; $f_3 = a C_f / I_z$; ω_r is the desired yaw angle velocity; $\dot{\omega}_c$ is the actual yaw angle acceleration.

In order to reduce the chattering of the path tracking system, the saturation function shown in (21) is used instead of the symbolic function.

$$
sat(s) = \begin{cases} s \ |s| < 1 \\ sgn(s) \ |s| \geq 1 \end{cases}
\tag{21}
$$

Past research has shown that any nonlinear function over a compact set with arbitrary accuracy can be approximated by an RBF neural network [39], and the solution is hard to fall into the local optimal. Considering the inevitable error of the built model, the RBF neural network is used to compensate the front wheel angle by the sliding mode control. The input value of the RBF neural network is $X = \begin{bmatrix} s & \dot{s} \end{bmatrix}$, and the performance index function of the RBF neural network is $E = s\dot{s}$. The number of neurons in the hidden layer of the neural network is m, and the output layer has one neuron. The Gaussian radial basis function of the hidden layer is

$$
h_n = \exp \left(-\frac{||X - c_n||^2}{2 b_n{}^2} \right) \ n = 1, 2, \ldots, m.
\tag{22}
$$

The compensation value of RBF neural network to the front wheel angle is obtained as follows:

$$
\delta_{sw} = W^T H
\tag{23}
$$

where W is the neural network weight vector, $W = [w_1 \ w_2 \ldots \ w_m]^T$; $H = [h_1 \ h_2 \ldots \ h_m]^T$.

Then, the front wheel angle control law with the path tracking SMC compensated by RBF (SMC-RBF) is as follows:

$$
\delta_f = \delta_{eq} + \delta_{sw}
\tag{24}
$$

In order to verify the effectiveness of the proposed tracking control method, the lane-changing path tracking results of SMC and SMC-RBF are compared in MIL simulation environment. The vehicle speed is 25 m/s, the total lane changing time is 5.1 s, and the simulation time is 10 s. The results are compared in Figure 8.

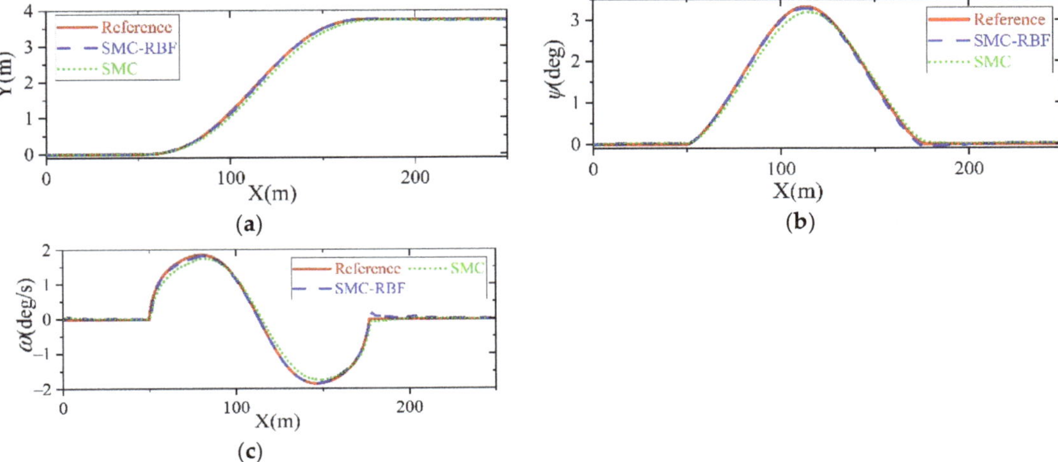

Figure 8. Comparison of path following controller with and without angle compensation control. (**a**) Comparison of driving path. (**b**) Comparison of heading angle. (**c**) Comparison of yaw rate.

Figure 8a shows that although the sliding mode controller can complete path tracking without compensation by RBF, vehicle lateral error reaches 0.1524 m, which will have a greater impact in emergency lane changing scenes. In Figure 8b,c, the errors between the heading angle, the yaw rate and the reference values are reduced under RBF angle compensation. After the RBF is applied to compensate the angle, the accuracy of path tracking is greatly improved, and the maximum lateral error is only 0.0317 m. It shows that the designed path tracking sliding mode controller of front wheel angle compensation by RBF greatly improves the accuracy of path tracking.

5. Simulation and Result Analysis

In the Section 3, the effectiveness of the game lane-changing decision-making method is verified by simulation. In Section 4, the lane-changing curve is obtained by controlling the constraint points of Bézier curve, and SMC-RBF is used for path tracking. In this section, the advantages of the proposed method will be analyzed and discussed considerably by compared with the traditional method (decision-making by time to collision and quintic polynomial curve path planning) through MIL, and human driver through HIL.

5.1. MIL Simulation

In MIL simulation, a game decision-making method, a time-based decision-making method, Bézier curve path planning, and quintic polynomial path planning will be combined and applied to further prove the effectiveness of game lane-changing decision-making and Bézier curve path planning method (GT-B), and the influence of different decision-making methods and lane-changing paths on passenger comfort will be discussed. MIL simulation is carried out in Simulink/Carsim environment. Vehicle parameters are set in Carsim shown in Table 5.

Table 5. Parameters of simulation vehicle.

Parameter	Value	Units
Distance from the center of mass to the front axis, a	1.232	m
Distance from the center of mass to the rear axis, b	1.468	m
Vehicle mass, m	1520	kg
Total lateral stiffness of front axle, C_f	66,900	N/rad
Total lateral stiffness of rear axle, C_r	62,700	N/rad
The moment of inertia, I_z	3965	kg·m^2

Build a traffic scene in Carsim (Figure 9), and perform a simulation with 15 s.

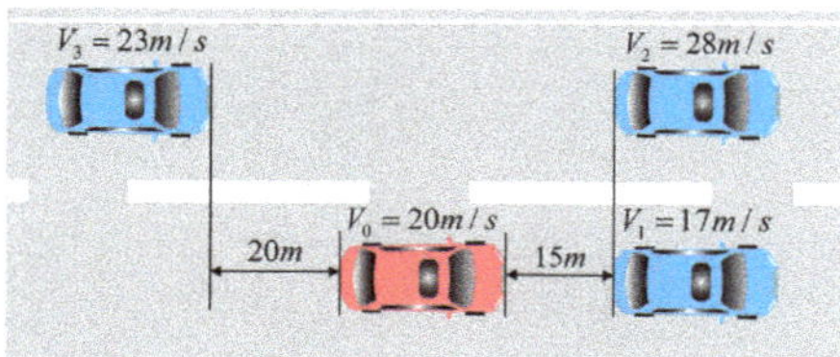

Figure 9. Traffic environment in MIL.

In Figure 10, GT-Bay is the lateral acceleration generated by tracking the Bézier curve under game decision-making; GT-Pay is the lateral acceleration generated by tracking the quintic polynomial curve under game decision-making; T-Bay is the lateral acceleration generated by tracking Bézier curve based on time to collision decision-making; T-Pay is the lateral acceleration generated by tracking the quintic polynomial curve based on the time to collision decision-making. The lateral acceleration division is shown in Table 6 where $a_{max} = 0.4g$ [40].

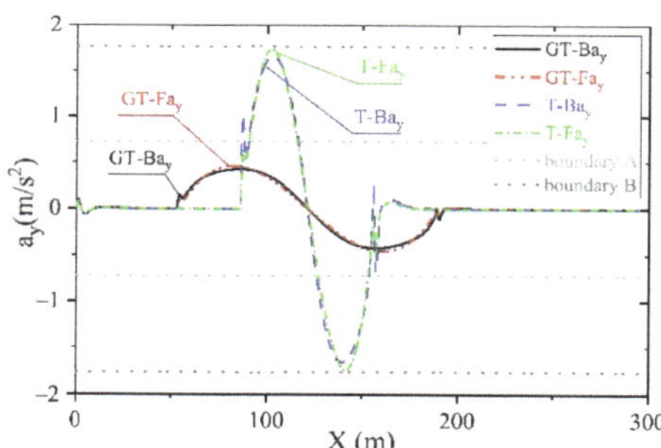

Figure 10. Comparison of lateral acceleration under different combinations of decision-making and path planning methods.

Table 6. Classification of lateral acceleration intensity.

Level	Lateral Acceleration Range (m/s^2)	Code Name
Normal-level	$0 \leq a_y < (0.1 - 0.0013v_x)g$	boundary A
Strong-level	$(0.1 - 0.0013v_x)g \leq a_y < (0.22 - 0.002v_x)g$	boundary B
Restricted-level	$(0.22 - 0.002v_x)g \leq a_y < 0.67a_{max}$	boundary C
Maximum-level	$0.67a_{max} \leq a_y < 0.85a_{max}$	boundary D

Compared with the lane-changing start time based on game theory and time to collision decision-making in Figure 10, it can be found out that by the decision-making method based on time to collision delays the lane-changing starting time is at 1.7 s. Owing to the close distance to *Car*.1 at this time, the first half of the lane-changing has to be compressed to 1.8 s, making maximum lateral acceleration of *Car*.0 reach 1.6595 m/s^2. However, the vehicle by game decision-making starts to change its lane when the speed of *Car*.1 is lower than *Car*.0, and the distance between *Car*.3 and *Car*.0 is large. As the lane-changing start time is earlier, the first half of lane change can be controlled to 3.4 s when the vehicle maximum lateral acceleration is only 0.4249 m/s^2, reduced by 74.40%. Earlier lane-changing can also enable *Car*.0 to reach the desired speed as soon as possible, so that improving the vehicle velocity payoff. Comparing GT-Pay with GT-Bay, it can be concluded that with the Bézier curve lane-changing path planning the maximum lateral acceleration is reduced by 8.03% than the quintic polynomial lane-changing path planning, and T-Bay also reduces 4.46% than T-Pay. The lane-changing path planned by the Bézier curve can provide higher passenger comfort. Compared with T-Pay, which is close to boundary B, the maximum lateral acceleration with GT-Bay is reduced by 75.54%, which greatly improves the passenger comfort.

5.2. HIL Experiment

In order to analyze the difference of the decision-making and behavior differences between the proposed lane-changing method and the human driver, and to collect some lane-changing data of human drivers, it is necessary to invite drivers to conduct a real-car driving test in the same scene. However, due to the low safety, poor repeatability, and difficult scene modeling of real-car driving, it is decided to use driving simulator for HIL driving experiment. The principle of driving simulator is shown in Figure 11a. The core component of the system is the PXIe-8840RT real-time processor. NI VeriStand realizes the communication between the computer and PXI real-time processor through the network cable. The maximum simulation frequency in HIL experiment can reach 1000 Hz. In an HIL experiment, the steering wheel angle is controlled by the control program through the feedback information from Carmaker. In the driving experiment, the road and traffic scenes are displayed to the driver on the monitors, and the driver directly controls the steering wheel angle to realize the vehicle lateral movement.

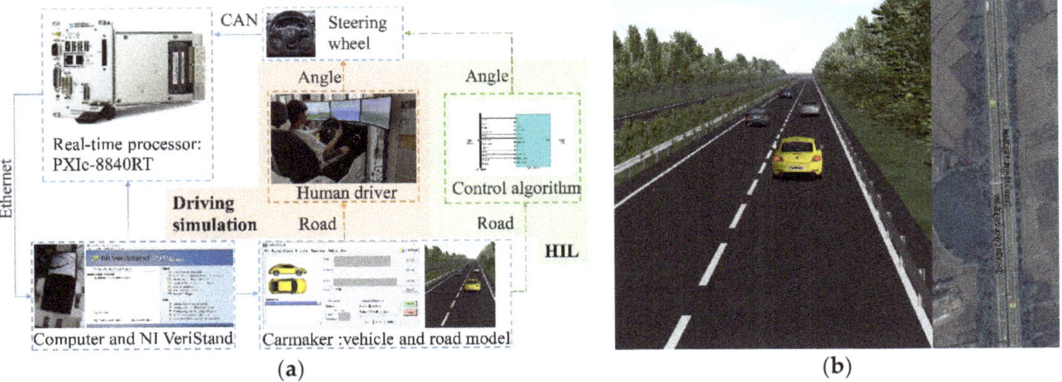

Figure 11. Experimental equipment framework and road modeling. (**a**) Driving simulation and HIL implementation. (**b**) Hefei-Xuzhou Section of G3 Jingtai Expressway in China in Carmaker.

In the HIL experiment, in order to be close to the real road scene, the Hefei-Xuzhou Section of G3 Jingtai Expressway in China is simulated in Carmaker to build a one-way two-lane road with a total length of 700 m shown in Figure 11b. Three different traffic environments are set shown in Figure 12. From the setting of the relationship between the speed and distance of each vehicle, the software indicates that Case I represents a kind of condition without driving danger. If the driver wants to reach a higher speed, the lane needs to be changed timely; Case II represents an emergency driving situation, that is, the speed difference between the preceding car and the host car suddenly increases, and the driver can choose to lane-changing or brake for car-following. Case III is a common lane merging situation where the driver must perform lane-changing operations.

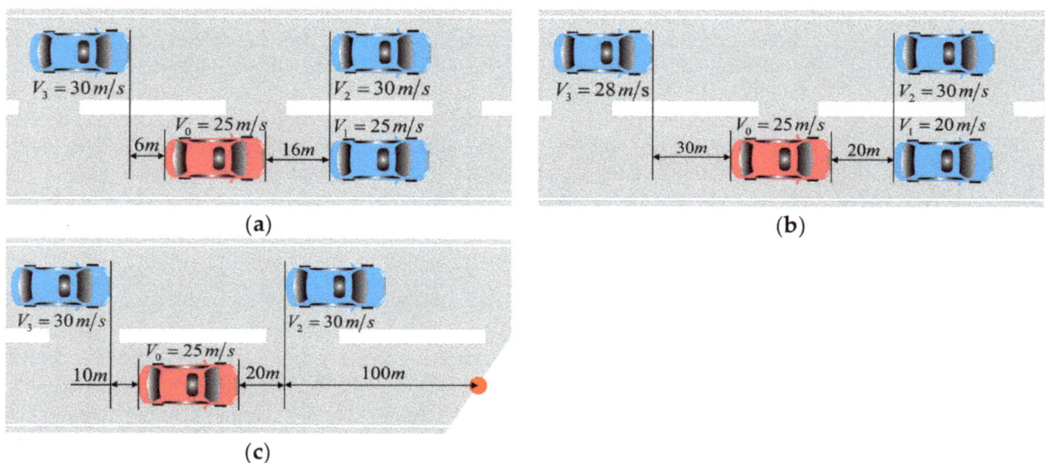

Figure 12. HIL traffic environment settings. (**a**) Traffic environment Case I. (**b**) Traffic environment Case II. (**c**) Traffic environment Case III.

Five people are randomly selected from 83 people who perform the simulated lane-changing operations to conduct the driving tests in three scenarios. Before the simulated driving, the driver is only informed of the host car (*Car*.0) speed and expected speed, and is not given any imply to the driver's operation. To facilitate the description of the simulation results, the method proposed in this paper is referred to as GT-B in short. The code names of the driving experiments in the three cases are A, B, C, D, and E, which do not represent a specific driver. In order to simulate the speed fluctuation of the real high-speed vehicle during stable driving and the detection error of the radar equipment, the speed of the vehicle in the experiment is fluctuates in a sine curve with a fluctuation range of ± 1 km/h, and *Car*.0 cannot detect this small range of the speed fluctuations. That is to say, the speed obtained in the lane-changing game decision-making is not accurate.

5.2.1. Case I

In Case I, the target speed of *Car*.0 is set as 33 m/s; at 2 s after the start of the simulation, *Car*.3 surpasses *Car*.0 to become the new *Car*.2, and *Car*.2 surpasses *Car*.1 at 6 s. The relationship between decision time and vehicle distances of different drivers under Case I is shown in Table 7, where D_{01min} is the shortest distance between *Car*.0 and *Car*.1 when *Car*.0 crosses the lane, and D_{02min} is the shortest distance between *Car*.0 and *Car*.2 when *Car*.0 crosses the lane. In order to express the process of *Car*.2 surpassing *Car*.1 more visually, take the distance between vehicles calculated in Figure 13a as the relative distance among vehicle-mounted radars, the relationship with the shortest distance among vehicles is $D_{radar} = D_{min} + L$. It can be seen from Figure 13a and Table 7 that the speed difference between *Car*.1 and *Car*.2 is large, *Car*.0 chooses to follow up when *Car*.2 has not completely surpassed *Car*.1 under GT-B control. Further, a longer total lane-changing time is calculated by GT-B under the premise of ensuring a safety distance from the front vehicle, and the passenger comfort of *Car*.0 is improved and the expected speed can be achieved earlier. Three drivers choose to follow up when *Car*.2 does not completely surpass *Car*.1. This decision-making result is the same as GT-B. Driver C quickly follows *Car*.3 and changes the lane when the speed of *Car*.3 is judged to be high. Although the expected speed can be reached early, the driving safety of *Car*.0 will decrease if the distance to the vehicle in front is closer. Before driver E starts to change lanes, D_{02} is already greater than S_{02}, indicating that lane change starting time of driver E is too conservative. The analysis shows that decision-making of GT-B is more in line with the human driver's perception of the driving environment in Case I.

Table 7. Driving parameters of different drivers under Case I.

Driver	Begin Time (s)	Finish Time (s)	D_{02min} (m)	D_{01min} (m)
A	5.4	10.3	11.5	13.5
B	5.9	11.7	13.9	14.4
C	4.3	13.2	6.3	14.8
D	6.7	13.4	17.9	14.4
E	11.5	19.9	45.0	17.5
GT-B	5.3	13.3	11.0	13.5

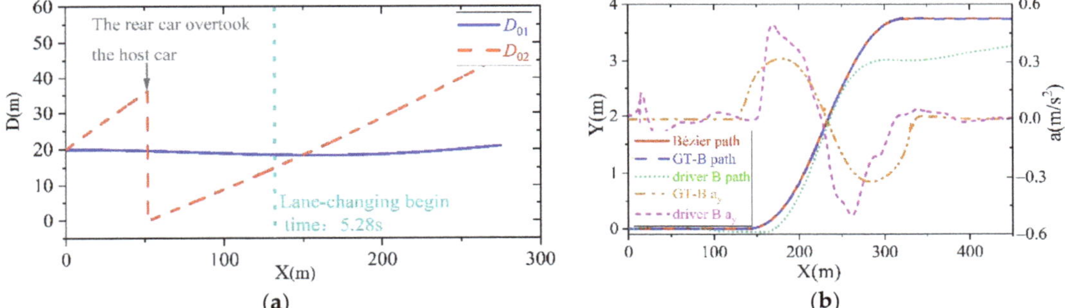

Figure 13. Vehicles' interval, vehicle path and lateral acceleration in Case I. (**a**) Decision-making time and distance. (**b**) Vehicle path and lateral acceleration.

The vehicle path of driver B is compared with GT-B and shown in Figure 13b. The traffic setting in Case I is no risk of collision when lane is changed in the current scene, so GT-B chooses a longer lane-changing time, thus reducing the maximum lateral acceleration. In Figure 13b, the maximum lateral acceleration of the vehicle under the control of GT-B is only 0.32 m/s², which is decreased by 36% than the maximum lateral acceleration of driver B of 0.5 m/s² in the simulation, so that the comfort of passengers is improved.

5.2.2. Case II

Unlike Case I, the difference between v_1 and v_0 in Case II is relatively large. In order to achieve higher speed or maintain the current speed, the lane-changing is needed for *Car*.0. The expected speed of *Car*.0 in Case II is set as 25 m/s. In Case II, two drivers choose to brake and follow *Car*.1, so there is no relevant comparison in Table 8. The remaining three drivers quickly judge the traffic situation after simulation beginning, and then all decisively change their lanes. Driver A makes full use of the distance with *Car*.1, and compared with driver B whose lane changing beginning time is close to driver A, the lateral acceleration of driver A during lane-changing is greatly reduced. However, D_{01min} generated by driver A and B is less than the safety distance margin, and a certain safety hazard has existed.

Table 8. Driving parameters of different drivers under Case II.

Driver	Begin Time (s)	Midpoint Time (s)	Finish Time (s)	D_{01min} (m)	a_{ymax} (m/s^2)
A	1.3	3.4	6.2	2.3	1.4571
B	1.5	3.4	5.4	2.3	2.1989
C	0.6	2.5	5.9	7.2	2.1240
GT-B	0.2	3.1	6.1	4.0	0.6031

Figure 14a indicates that *Car.3* starts to slow down gradually after *Car.0* starts to change its lanes to guarantee a safety distance between different cars. Under the same deceleration with braking, the different lane-changing starting time of each driver leads to final different car-following distances of *Car.3*. In the same way, if the car-following distance is kept constant and the lane-changing starts later, *Car.3* requires greater deceleration with braking. Therefore, *Car.0* starting lane-changing operation early has smaller impact on other vehicles in traffic environment built in Case II.

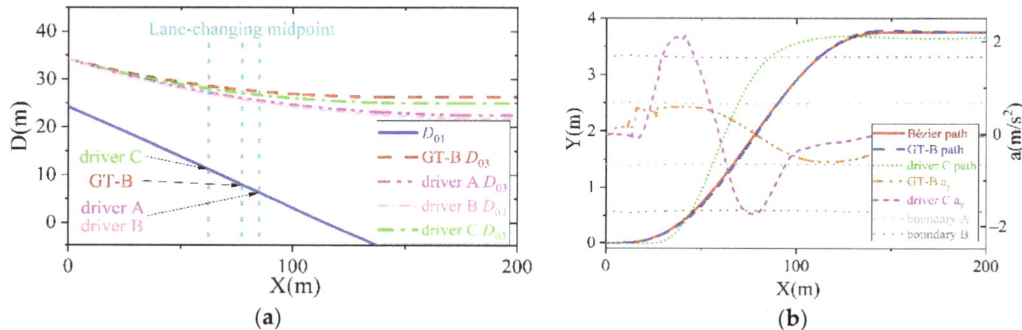

Figure 14. Vehicles' interval, path and lateral acceleration in Case II. (**a**) Vehicle interval during lane-changing by different drivers. (**b**) Vehicle path and lateral acceleration.

The lane-changing time of driver C is close to the lane-changing time under GT-B control, so driver C and GT-B are selected for comparison shown in Figure 14. It can be seen that at the lane-changing midpoint, the distance between *Car.0* under the control of GT-B and *Car.1* is relatively close. Although there is no collision, it may have negative impact on the psychology of the passengers in *Car.0*. Table 8 shows that the time from beginning to midpoint is less than the time from midpoint to finish by human drivers, and human drivers may be more accustomed to cutting into the target lane and then overtaking, rather than overtaking during lane-changing. However, under dangerous situations, overtaking during lane-changing is obviously a benefit choice for both vehicle safety and passenger comfort. Figure 14b shows that the lateral acceleration of the vehicle driven by driver C has exceeded boundary B and reached 2.1240 m/s^2. Under the control of GT-B, the lateral acceleration of *Car.0* during lane-changing is still within boundary A, and its maximum value is only 0.6055 m/s^2, which is decreased by 71.49% than that by driver C.

5.2.3. Case III

Vehicle lane-changing conditions can be divided into (1) MLC due to environmental constraints and (2) DLC to improve driving environment [41]. In Case III, the vehicle is driving to a road condition that needs to be merged, such as driving into an underpass or driving into a main road from a ramp, so the scene of Case III can be regarded as MLC.

Because $v_3 > v_0$ in Case III, and the distance between *Car.3* and *Car.0* is small, *Car.0* can not quickly change its lanes. Therefore, *Car.0* can only choose lane-changing after *Car.3* overtakes *Car.0* and become the new *Car.2*. In the first test of driving experiment, all five participants collide with *Car.3* or the end of the road due to the emergency of Case

III. As a result of the emergency of MLC, it will increase the mental load of drivers [42]. Therefore, under the suggestions and requirements of the participants, each participant is given three opportunities to adapt to Case III, and the final data is obtained from the third test results. Even so, there is still a driver who collides with *Car.3* during the lane-changing. The lane-changing time and vehicle distance of the remaining four drivers and GT-B control are shown in Table 9.

Table 9. Driving parameters of different drivers under Case III.

Driver	Begin Time (s)	Midpoint Time (s)	Finish Time (s)	D_{01min} (m)	D_{02min} (m)
A	1.7	3.7	5.8	6.5	3.4
B	1.5	3.8	5.3	4	3.8
C	2.0	3.9	5.5	1.5	4.3
D	2.5	3.9	7.0	1.5	4.3
GT-B	2.0	3.7	5.4	6.5	3.4

Table 9 shows that no matter when the driver starts to change the lanes, their crossing time is basically the same as that of vehicles under GT-B control. However, D_{01min} generated by driver C and driver D is only 1.5 m. Although the road space is fully utilized, this lane-changing has certain risk. Only D_{01min} generated by driver A and GT-B is more reasonable. Judging from the lane-changing end time, the end time by GT-B is close to driver B and driver C, so they can reach the expected speed ahead of time compared with other experimental participants.

It can be determined from Table 9 and Figure 15a that as the front lane is about to end, vehicle safety, power performance, and human comfort can be coordinated by GT-B, so that *Car.0* drives to the adjacent lane with 6.5 m away from the end of the lane. Driver C with the same lane-changing start time is selected for comparison with GT-B shown in Figure 15b. Although the vehicle lateral acceleration under the control of GT-B is at the edge of boundary B, the lateral acceleration of vehicle driven by driver C already exceeds boundary C. In addition, the shorter lane-changing time will lead to the larger heading angle while crossing the lane. During turning the steering wheel to right after reaching the target lane, the maximum lateral acceleration of 2.9676 m/s^2 is generated by driver C, and the maximum vehicle lateral acceleration under the control of GT-B is 1.8156 m/s^2, which is 38.82% lower than that of driver C. The lateral deviation distance of the vehicle relative to the target lane center line by driver C reaches 0.375 m, and the close distance with the left guardrail will aggravate the deviation of the vehicle owing to the wind pressure in the case of high-speed driving, resulting in difficulty of the vehicle direction control. The maximum deviation distance of the vehicle relative to the lane centerline by SMC-RBF during the lane-changing is 0.125 m, and the vehicle can quickly return to the lane center.

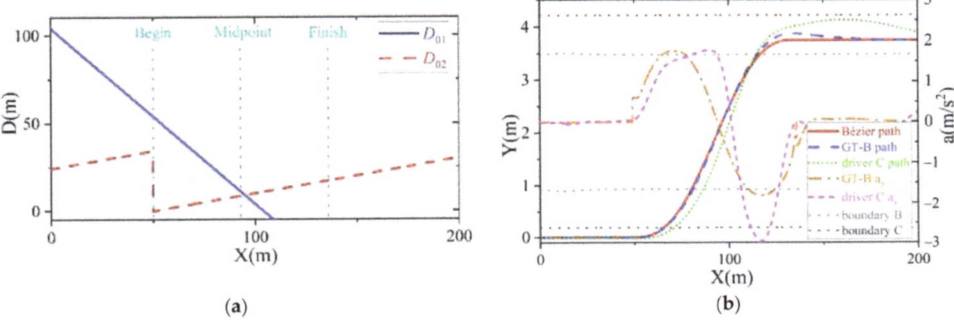

Figure 15. Vehicles' interval, path and lateral acceleration in Case III. (**a**) Relative distance between other vehicles and *Car. 0*; (**b**) Vehicle path and lateral acceleration.

6. Discussion

Comparing the analysis of lane-changing process under the control of GT-B with that of human drivers, it is obvious that vehicle safety, power performance, and human comfort can be coordinated well by the GT-B method. In the common traffic situation (Case I), part of vehicle power performance is sacrificed to guarantee driving safety, and the lane-changing time is appropriately extended to obtain good human comfort. In emergency situation (Case II), the vehicle's performance is improved while guaranteeing its safety. In a very emergency situation (Case III), part of the power performance and human comfort are both sacrificed to make driving safety guaranteed. Moreover, vehicle performance by GT-B method is superior to that by human drivers in lane-changing path planning and vehicle lateral control. From the experiment results, it can be seen that GT-B method can balance the payoffs generated by lane-changing, and obtain more rapid, smoother, and safer lane-changing.

In order to simplify the lane-changing model, this paper makes some reasonable assumptions, such as the speed changing in a small range and only considering lane-changing in straight road. However, there are two conditions that may be needed to be noticed: (I) When the surrounding vehicles decelerate or accelerate suddenly, the lane-changing decision and the longitudinal motion of the host car will be affected. (II) Although lane-changing in a curved road is not recommended, it is still necessary to design an automatic lane-changing system that is safe for curved roads. When the vehicle is driving in the curved road to avoid the obstacle, the appropriate decision results are very important to ensure the vehicle and passengers' safety. In the follow-up related research, the decision-making and path planning method will be studied and optimized based on the lane-changing in the curved road and consider longitudinal acceleration of each car in the traffic flow. Herrmann et al. [43] optimized the velocity on the available paths for the racing cars, which inspired velocity planning of host car in future study. The difference is that the racing cars need to fully utilize the maximum possible tire forces, whereas the passenger cars need to consider the impact of speed planning on ride comfort.

At present, the system is in the principle verification stage, so the vehicle distance signal obtained in the simulation is accurate value, however any distance measurement method has the error and noise. Vehicle state variables also need to be acquired through sensors, and sensor signals are bound to have delays and noises. If the system designed in this paper is to be applied to the actual vehicle in the future, it is necessary to study the sensor signal fusion technology and the vehicle state parameter estimation system. Although it is difficult to apply this method in the actual driving scene at present, the experimental data presented in this paper would promote the development of autonomous lane-changing systems and the further research based on this paper will help to reduce the number of traffic accidents caused by lane-changing.

7. Conclusions

A game of lane-changing decision-making with Bézier curve path planning is proposed in this paper which considers driving safety, power performance, and passenger comfort comprehensively. Lane-changing safety distance is obtained by using 83 driver lane-changing data. The lane-changing safety distance and lane-changing time calculated by the path planning layer are considered in game payoff to enhance safety considerations, which realizes the strong coupling between path planning layer and decision layer. The results of the planning layer are returned to the decision layer as the input, which can improve the security of the decision results. In addition, a detailed constrained optimization method is proposed for Bézier curves, which improves the safety, traceability, and comfort of the planned path. In the MIL simulation, it is proved that the method proposed in this paper greatly improves the vehicle safety and passenger comfort. The HIL experimental results indicate that the method proposed in this paper is superior to human drivers in the selection of lane change time, the control of vehicle interval, and can achieve the balance among vehicle safety, power performance and passenger comfort. In the HIL verification, there are not many driving scenarios, but the three lane-changing scenarios are representative that can be used for comparative tests.

It is obvious that the method proposed in this paper can meet the requirements well in the decision-making of lane-changing starting time, the total lane-changing time, the lane-changing planning path, and the tracking control of planned path in both scenarios of DLC and MLC. The results of this paper will accumulate experience for the further research.

Author Contributions: Conceptualization, H.W. and S.X.; methodology, H.W. and S.X.; software, S.X.; validation, S.X.; formal analysis, H.W., S.X. and L.D.; investigation, S.X.; resources, H.W.; data curation, S.X.; writing—original draft preparation, H.W. and S.X.; writing—review and editing, H.W. and S.X.; visualization, S.X. and L.D.; supervision, H.W. All authors have read and agreed to the published version of the manuscript.

Funding: This research was funded by National Nature Science Foundation of China, grant number U1564201, and Science Fund of Anhui Intelligent Vehicle Engineering Laboratory, grant number PA2018AFGS0026.

Data Availability Statement: The data presented in this study are available on request from the corresponding author. The data are not publicly available due to privacy.

Conflicts of Interest: The authors declare no conflict of interest.

References

1. World Health Organization. *Global Status Report on Road Safety 2013*; Injury Prevention: London, UK, March 2013; pp. 4–10. Available online: https://www.who.int/violence_injury_prevention/road_safety_status/2013/en/ (accessed on 14 March 2021).
2. World Health Organization. *Global Status Report on Road Safety 2018*; Injury Prevention: London, UK, June 2018; pp. 4–7. Available online: https://www.who.int/publications/i/item/9789241565684 (accessed on 14 March 2021).
3. Singh, S. *Critical Reasons for Crashes Investigated in The National Motor Vehicle Crash Causation Survey*; DOT HS 812 115; Nat. Highway Traffic Safety Admin.: Washington, DC, USA, 2015.
4. Kim, J.; Kim, J.; Jang, G.J.; Lee, M. Fast learning method for convolutional neural networks using extreme learning ma-chine and its application to lane detection. *Neural Netw.* **2017**, *87*, 109–121. [CrossRef]
5. Zhao, D.; Lam, H.; Peng, H.; Bao, S.; LeBlanc, D.J.; Nobukawa, K.; Pan, C.S. Accelerated Evaluation of Automated Vehi-cles Safety in Lane-Change Scenarios Based on Importance Sampling Techniques. *IEEE Trans. Intell. Trans.-Portation Syst.* **2016**, *18*, 595–607. [CrossRef]
6. Ali, Y.; Haque, M.M.; Zheng, Z.; Washington, S.; Yildirimoglu, M. A hazard-based duration model to quantify the impact of connected driving environment on safety during mandatory lane-changing. *Transp. Res. Part C Emerg. Technol.* **2019**, *106*, 113–131. [CrossRef]
7. Dang, R.; Ding, J.; Su, B.; Yao, Q.; Tian, Y.; Li, K. A lane change warning system based on V2V communication. In Proceedings of the 17th International IEEE Conference on Intelligent Transportation Systems (ITSC), Qingdao, China, 8–11 October 2014; pp. 1923–1928.
8. Song, Z.; Ji, J.; Zhang, R.; Cao, L. Development of a test equipment for rating front to rear-end collisions based on C-NCAP-2018. *Int. J. Crashworthiness* **2020**, 1–11. [CrossRef]
9. Zhu, B.; Yan, S.; Zhao, J.; Deng, W. Personalized Lane-Change Assistance System with Driver Behavior Identification. *IEEE Trans. Veh. Technol.* **2018**, *67*, 10293–10306. [CrossRef]
10. Butakov, V.A.; Ioannou, P. Personalized Driver/Vehicle Lane Change Models for ADAS. *IEEE Trans. Veh. Technol.* **2015**, *64*, 4422–4431. [CrossRef]
11. Yu, H.; Tseng, H.E.; Langari, R. A human-like game theory-based controller for automatic lane changing. *Transp. Res. Part C Emerg. Technol.* **2018**, *88*, 140–158. [CrossRef]
12. Meng, F.; Su, J.; Liu, C.; Chen, W.-H. Dynamic decision making in lane change: Game theory with receding horizon. In Proceedings of the 2016 UKACC 11th International Conference on Control (CONTROL), Belfast, UK, 31 August–2 September 2016; pp. 1–6.
13. Cao, P.; Xu, Z.; Fan, Q.; Liu, X. Analysing driving efficiency of mandatory lane change decision for autonomous vehicles. *IET Intell. Transp. Syst.* **2019**, *13*, 506–514. [CrossRef]
14. Berglund, T.; Brodnik, A.; Jonsson, H.; Staffansson, M.; Soderkvist, I. Planning Smooth and Obstacle-Avoiding B-Spline Paths for Autonomous Mining Vehicles. *IEEE Trans. Autom. Sci. Eng.* **2009**, *7*, 167–172. [CrossRef]
15. Stahl, T.; Wischnewski, A.; Betz, J.; Lienkamp, M. Multilayer Graph-Based Trajectory Planning for Race Vehicles in Dynamic Scenarios. In Proceedings of the 2019 IEEE Intelligent Transportation Systems Conference (ITSC), Paris, France, 9–12 June 2019.
16. Broggi, A.; Medici, P.; Zani, P.; Coati, A.; Panciroli, M. Autonomous vehicles Control in the VisLab Intercontinental Au-tonomous Challenge. *Annu. Rev. Control* **2012**, *36*, 161–171. [CrossRef]

17. Choi, J.-W.; Curry, R.; Elkaim, G. Path Planning Based on Bézier Curve for Autonomous Ground Vehicles. In Proceedings of the Advances in Electrical and Electronics Engineering—IAENG Special Edition of the World Congress on Engineering and Computer Science 2008, San Francisco, CA, USA, 22–24 October 2008; pp. 158–166. [CrossRef]
18. Glaser, S.; Vanholme, B.; Mammar, S.; Gruyer, D.; Nouveliere, L. Maneuver-Based Trajectory Planning for Highly Autonomous Vehicles on Real Road With Traffic and Driver Interaction. *IEEE Trans. Intell. Transp. Syst.* **2010**, *11*, 589–606. [CrossRef]
19. Bae, I.; Moon, J.; Park, H.; Kim, J.H.; Kim, S. Path generation and tracking based on a Bézier curve for a steering rate controller of autonomous vehicles. In Proceedings of the 16th International IEEE Conference on Intelligent Transportation Systems (ITSC 2013), Kurhaus, The Hague, 6–9 October 2013; pp. 436–441.
20. Mukai, M.; Kawabe, T. Model predictive control for lane change decision assist system using hybrid system representation. In Proceedings of the 2006 SICE-ICASE International Joint Conference, Busan, Korea, 18–21 October 2006; pp. 5120–5125. [CrossRef]
21. Hu, X.; Chen, L.; Tang, B.; Cao, D.; He, H. Dynamic path planning for autonomous driving on various roads with avoidance of static and moving obstacles. *Mech. Syst. Signal Process.* **2018**, *100*, 482–500. [CrossRef]
22. Falcone, P.; Borrelli, F.; Asgari, J.; Tseng, H.E.; Hrovat, D. Predictive Active Steering Control for Autonomous Vehicle Systems. *IEEE Trans. Control. Syst. Technol.* **2007**, *15*, 566–580. [CrossRef]
23. Naranjo, J.E.; Gonzalez, C.; Garcia, R.; De Pedro, T. Lane-Change Fuzzy Control in Autonomous Vehicles for the Overtaking Maneuver. *IEEE Trans. Intell. Transp. Syst.* **2008**, *9*, 438–450. [CrossRef]
24. Ren, D.; Zhang, J.; Zhang, J.; Cui, S. Trajectory planning and yaw rate tracking control for lane changing of intelligent vehicle on curved road. *Sci. China Ser. E Technol. Sci.* **2011**, *54*, 630–642. [CrossRef]
25. Wu, Y.; Wang, L.; Zhang, J.; Li, F. Path Following Control of Autonomous Ground Vehicle Based on Nonsingular Terminal Sliding Mode and Active Disturbance Rejection Control. *IEEE Trans. Veh. Technol.* **2019**, *68*, 6379–6390. [CrossRef]
26. Liniger, A.; Lygeros, J. A Noncooperative Game Approach to Autonomous Racing. *IEEE Trans. Control. Syst. Technol.* **2019**, *28*, 884–897. [CrossRef]
27. Yao, W.; Zhao, H.; Davoine, F.; Zha, H. Learning lane change trajectories from on-road driving data. In Proceedings of the 2012 IEEE Intelligent Vehicles Symposium, Madrid, Spain, 3–7 June 2012; pp. 885–890.
28. Bulut, V. Path planning for autonomous ground vehicles based on quintic trigonometric Bézier curve. *J. Braz. Soc. Mech. Sci. Eng.* **2021**, *43*, 1–14. [CrossRef]
29. Gipps, G. A behavioural car-following model for computer simulation. *Transp. Res. Board* **1981**, *15*, 105–111. [CrossRef]
30. Aycin, M.F.; Benekohal, R.E. Linear Acceleration Car-Following Model Development and Validation. *Transp. Res. Rec. J. Transp. Res. Board* **1998**, *1644*, 10–19. [CrossRef]
31. Seiler, P.; Song, B.; Hedrick, J.K. Development of a Collision Avoidance System. *SAE Tech. Paper Ser.* **1998**, *98PC417*. [CrossRef]
32. Salvucci, D.D.; Liu, A. The time course of a lane change: Driver control and eye-movement behavior. *Transp. Res. Part F Traffic Psychol. Behav.* **2002**, *5*, 123–132. [CrossRef]
33. Sharma, A.; Ali, Y.; Saifuzzaman, M.; Zheng, Z.; Haque, M. *Human Factors in Modelling Mixed Traffic of Traditional, Connected, and Au-tomated Vehicles. Advances in Intelligent Systems and Computing*; Springer: Berlin, Germany, 2017; pp. 262–273.
34. Long, X.; Zhang, L.; Liu, S.; Wang, J. Research on Decision-Making Behavior of Discretionary Lane-Changing Based on Cumulative Prospect Theory. *J. Adv. Transp.* **2020**, *2020*, 1291342. [CrossRef]
35. Nash, J. Non-Cooperative Games. *Ann. Math.* **1951**, *54*, 286–295. [CrossRef]
36. Başar, T.; Olsder, G.J. *Dynamic Noncooperative Game Theory*, 2nd ed.; SIAM Publications: Philadelphia, PA, USA, 1998. [CrossRef]
37. Lindgren, A.; Chen, F. State of the art analysis: An overview of advanced driver assistance systems (ADAS) and possible human factors issues. In Proceedings of the Human Factors and Economic Aspects on Safety, Linköping, Sweden, 5–7 April 2007; Volume 38, pp. 38–50.
38. Zhang, K.; Cui, S.M.; Wang, J.F. Intelligent vehicle's path tracking control based on self-adaptive RBF network compensation. *Control Decision.* **2014**, *29*, 627–631.
39. Sonfack, L.L.; Kenné, G.; Fombu, A.M. An improved adaptive RBF neuro-sliding mode control strategy: Application to a static synchronous series compensator controlled system. *Int. Trans. Electr. Energy Syst.* **2019**, *29*, e2835. [CrossRef]
40. Jr, N.H.S. An Investigation of Vehicle Critical Speed and its Influence on Lane-Change Trajectories. Ph.D. Thesis, University of Texas at Austin, Austin, TX, USA, 1997.
41. Toledo, T.; Koutsopoulos, H.N.; Ben-Akiva, M.E. Modeling Integrated Lane-Changing Behavior. *Transp. Res. Rec. J. Transp. Res. Board* **2003**, *1857*, 30–38. [CrossRef]
42. Plankermann, K. Human Factors as Causes for Road Traffic Accidents in the Sultanate of Oman Under Consideration of Road Construction Designs. Ph.D.'s Thesis, University of Regensburg, Regensburg, Germany, 2014. [CrossRef]
43. Herrmann, T.; Wischnewski, A.; Hermansdorfer, L.; Betz, J.; Lienkamp, M. Real-Time Adaptive Velocity Optimization for Autonomous Electric Cars at the Limits of Handling. *IEEE Trans. Intell. Veh.* **2020**. [CrossRef]

Article

Pressure Estimation of the Electro-Hydraulic Brake System Based on Signal Fusion

Biaofei Shi [1,2], Lu Xiong [1,2,*] and Zhuoping Yu [1,2]

[1] School of Automotive Studies, Tongji University, Shanghai 201804, China; shi_biaofei@tongji.edu.cn (B.S.); yuzhuoping@tongji.edu.cn (Z.Y.)

[2] Institute of Intelligent Vehicle, Clean Energy Automotive Engineering Center, Tongji University, Shanghai 201804, China

* Correspondence: xiong_lu@tongji.edu.cn

Abstract: At present, the master cylinder pressure estimation algorithm (MCPE) of electro-hydraulic brake systems (EHB) based on vehicle dynamics has the disadvantages of poor condition adaptability, and there are delays and noise in the estimated pressure; however, the MCPE based on the characteristics of an EHB (i.e., the pressure–position relationship) is not robust enough to prevent brake pad wear. For the above reasons, neither method be applied to engineering. In this regard, this article proposes a MCPE that is based on signal fusion. First, a five-degree-of-freedom (5-DOF) vehicle model that includes longitudinal motion, lateral motion, yaw motion, and front and rear wheel rotation is established. Based on this, an algebraic expression for MCPE is derived, which extends the MCPE from a straight condition to a steering condition. Real vehicle tests show that the MCPE based on the 5-DOF vehicle model can effectively estimate the brake pressure in both straight and steering conditions. Second, the relationship between the hydraulic pressure and the rack position in the EHB is tested under different brake pad wear levels, and the results show that the pressure–position relationship will change as the brake pad is worn down, so the pressure estimated by the pressure–position model based on fixed parameters is not robust. Third, a MCPE based on the fusion the above two MCPEs through the recursive least squares algorithm (RLS) is proposed, in which the pressure-position model can be updated online by vehicle dynamics and the final estimated pressure is calculated based on the updated pressure–position model. Finally, several simulations based on vehicle test data demonstrate that the fusion-based MCPE can estimate the brake pressure accurately and smoothly with little delay and is robust enough to prevent brake pad wear. In addition, by setting the enabling conditions of RLS, the fusion-based MCPE can switch between driving and parking smoothly; thus, the fusion-based MCPE can be applied to all working conditions.

Keywords: electro-hydraulic brake system; master cylinder pressure estimation; five-degree-of-freedom vehicle model; pressure–position model; recursive least square

Citation: Shi, B.; Xiong, L.; Yu, Z. Pressure Estimation of the Electro-Hydraulic Brake System Based on Signal Fusion. *Actuators* **2021**, *10*, 240. https://doi.org/10.3390/act10090240

Academic Editor: Andrea Vacca

Received: 12 August 2021
Accepted: 14 September 2021
Published: 16 September 2021

Publisher's Note: MDPI stays neutral with regard to jurisdictional claims in published maps and institutional affiliations.

1. Introduction

Under the global trend of electrification and intelligence, automobile braking systems have undergone new changes. Traditional braking systems are increasingly unable to meet the new demands, and brake-by-wire systems (BBW) have come into existence. BBW are mainly divided into electro-mechanical brake systems (EMB) and electro-hydraulic brake systems (EHB). EMB, in which the motor drives the reduction gears to directly push the pad to clamp the disc, cancels the hydraulic components and can control the clamping force accurately and quickly. It is considered to be the supreme form of BBW; however, the braking capacity of EMB depends on a 42 V power supply system, which is not equipped on most vehicles. More importantly, the EMB does not meet the requirements of current regulations for brake system failure backup. Therefore, although some companies and universities worldwide have developed EMB prototypes [1–3], such as Bosch, Akipollo, Hanyang University, etc., EMB have not yet entered the market. In contrast, EHB retains the hydraulic brake circuit and adopts the motor and reduction gears to push the master cylinder piston to build pressure. It has a lower cost, and it is easier to realize failure backup. Moreover, EHB can also achieve satisfactory brake control through a suitable pressure control algorithm. In addition, since the area of the wheel cylinder piston is larger than that of the master cylinder piston, the hydraulic circuit can amplify the thrust at the master cylinder so that the 12 V on-board power supply system can meet the power requirements of the EHB. Therefore, EHB is considered to be the first approach to BBW [4]. Currently, EHB have been mass-produced, such as Bosch's i-Booster [5] and Hitachi's e-ACT [6].

From the perspective of vehicle dynamics, the essence of brake control is the control of the braking force. Due to the fact that it is difficult to measure the braking force acting on the wheels, EHB usually implements closed-loop pressure control by installing a pressure sensor in the master cylinder, thus indirectly controlling the braking force. As the core technology of EHB, the master cylinder pressure control algorithm (MCPC), which ensures that the EHB can realize high-performance regenerative braking control and active braking control, has been extensively studied, including aspects such as friction compensation technology [7–10], multi-closed-loop control architecture [11–14], robust control algorithms [15–17], etc. However, as one of the critical safety components of automobiles, once the pressure sensor fails, the function of MCPC, which is based on the pressure sensor, will be seriously affected. Some products have adopted two pressure sensors in the master cylinder for mutual inspection as a solution for failure detection and backup, which has led to a further increase in cost [18]. For this reason, master cylinder pressure estimation (MCPE) is a promising solution to the above-mentioned problems.

At present, according to different models, MCPE of EHB can be classified into three categories: (1) MCPE based on the characteristics of EHB (i.e., pressure–position relationship and EHB dynamics); (2) MCPE based on vehicle dynamics; and (3) MCPE based on intelligent algorithms.

1.1. MCPE Based on EHB's Own Characteristics

Under braking, the master cylinder piston squeezes the brake fluid in the brake circuit to generate hydraulic pressure. During this process, the master cylinder piston position, which can be obtained from the motor rotational angle and the transmission ratio of the reduction gears, and the hydraulic pressure will form a nonlinear relationship, which is known as the so-called pressure–position relationship. In addition, the moving parts of EBH satisfy the force balance equation, namely the dynamics of the EHB, which mainly include the motor force term, the friction force term, and the hydraulic pressure term. The existing literature mainly focuses on the above two aspects to estimate the brake pressure. Refs. [9,11,14] obtained pressure–position models by polynomial fitting and by a look-up table. However, due to the hysteresis and time-varying characteristics of the pressure–position relationship, the above methods were not accurate and robust. In [19], the pressure was estimated based on EHB dynamics, and simulation results showed that drastic fluctuation occurred when the piston moved forward and back due to the nonlinearity of friction. To this end, Ref. [13] proposed an interconnected pressure estimation method in which the key characteristic parameter of the pressure–position curve, namely the nonlinearly parameterized perturbations, could be estimated via EHB dynamics based on the LuGre friction model. For this method, though the pressure–position model could be updated online, the friction model, which depended on the piston position, was not robust when the pressure–position curve changes.

The MCPE of EHB is a novel topic, and there is not much research related to this topic at present; however, some of the previous research conducted on EMB can be instructive. In fact, EMB and EHB have certain similarities in the friction of the reduction gears and load characteristics (i.e., pressure–position relationship for EHB and the clamping force–motor angle relationship for EMB). In addition, limited by the cost and installation space of the clamping force sensor, EMB also needs to estimate the clamping force. Ref. [20] developed a clamping force estimation algorithm based on EMB dynamics. To avoid the need for a friction model, a high-frequency low-amplitude sinusoid was superimposed on the gross angular motion from the motor. This served to force the motor to pass the same location in a short period of time between a clamping and a releasing action. Using this method, the friction term could be cancelled out due to a sign change from clamping to releasing and vice versa, and the clamping force could be calculated. Ref. [21] obtained a first-order clamping force–motor angle model through system identification. Taking into account the time-varying characteristics of the clamping force–motor angle relationship caused by brake pad wear, when the vehicle is in a parking position, the method of [20] can be used to adapt the clamping force–motor angle model based on least squares (LS). The major issue with the method of [20,21] when applied in EHB is that the friction in EHB is not symmetrical because the friction in the pressurization process is larger than that in the depressurization process at the same location, so the friction cannot be thoroughly cancelled out [17].

It can be seen from the above references that the brake pressure can be simply and directly estimated by an EHB pressure–position model, but this method is not robust enough to prevent brake pad wear. For this reason, the pressure–position model can be updated based on EHB dynamics. However, due to the lack of a robust friction model, the MCPE based on the fusion of EHB characteristics cannot guarantee robustness either.

1.2. MCPE Based on Vehicle Dynamics

During the braking process, the hydraulic pressure pushes the pad to clamp the disc to force the vehicle to decelerate. Therefore, the longitudinal deceleration of the vehicle can reflect the pressure value to a certain extent. Ref. [22] proposed a MCPE based on vehicle longitudinal dynamics and wheel rotational dynamics for the first time. However, the brake linings' coefficient of friction (BLCF) was regarded as constant. In fact, the BLCF is greatly affected by vehicle speed, brake pressure, and the temperature of the brake lining [23]. Ref. [24] introduced the evolution of BLCF at different initial temperatures, different initial vehicle speeds, and different brake pressures through real vehicle tests. The results show that under normal driving conditions, the evolution of BLCF is mainly related to vehicle speed; thus, a revised BLCF model is proposed. As expected, the accuracy of the estimated pressure was further improved after the adoption of the revised BLCF model. In addition, by introducing the inertial measurement unit (IMU), the MCPE based on vehicle dynamics could be extended from level roads to slope roads in [24]. Although the MCPE based on vehicle dynamics avoids the nonlinear and time-varying characteristics of EHB, it is limited by many restrictions. First, according to the principle of the algorithm, when the vehicle is stopped on a flat road, the longitudinal acceleration measured by the IMU is zero, and the MCPE based on vehicle dynamics is invalid. Second, the existing literature only studies the MCPE under straight conditions, and the research regarding braking with steering conditions, which is very common in daily driving, has not yet been conducted. Third, the estimated pressure is directly calculated based on the sensor signals and the vehicle model, and there is a lot of noise (especially when encountering bad roads and even speed bumps). Finally, the IMU is installed on the vehicle body, and it measures the motion state of the vehicle body. However, in the braking process, the hydraulic pressure first decelerates the wheel speed, and the deceleration is then transmitted to the vehicle body. That is, the signal of the IMU lags behind the hydraulic pressure, which results in the estimated pressure lagging behind the actual pressure.

1.3. MCPE Based on Intelligent Algorithms

In recent years, machine learning has increasingly been applied to the state estimation of vehicles due to the availability of large amounts of training data. The ability of machine learning to learn from data and to self-optimize behavior makes it well suited to estimate vehicle state in complex and dynamic environments [25,26]. Ref. [27] proposed a brake pressure estimation method based on a multilayer artificial neural network (ANN) with a Levenberg–Marquardt backpropagation (LMBP) training algorithm. Real vehicle tests were conducted on a chassis dynamometer under the new European driving cycle (NEDC). Experimental data for the vehicle and powertrain systems were collected to train the developed multilayer ANN. The results show that the proposed method can accurately estimate the brake pressure. However, the training method for conventional back propagation suffers from the problems of overfitting, a vanishing gradient as well as higher computational complexity in training. To this end, in [28], a deep neural network (DNN) was structured and was trained using deep-learning training techniques, such as dropout and rectified units, and a more accurate estimation was finally obtained. In [29], a time-series model based on multivariate deep recurrent neural networks (RNN) with long short-term memory (LSTM) units was developed for brake pressure estimation. This model also included a vehicle speed estimation module, which contributed to a more precise pressure estimation. Test data show that the proposed method was able to estimate the brake pressure for the next 2s in the future with a root mean square error (RMSE) of 5bar. In all of the above research, the training and model verification were conducted offline. For the possibility of being applied in real vehicles, the robustness of the algorithm needs to be further verified. In addition, only the training data include vehicle signals and powertrain signals without EHB signals and IMU signals. Therefore, the estimation model based on intelligent algorithms lacks theoretical data and persuasiveness.

As summarized by the above literature, the existing MCPEs are not able to simultaneously solve the problems of poor robustness, poor working condition adaptability, signal noise, and delay. In this regard, this paper proposes a MCPE that integrates vehicle dynamics and the pressure–position relationship. Two main contributions make this work distinctive from the previous studies: (1) a MCPE based on the five-degree-of-freedom (5-DOF) vehicle model is proposed so that the pressure can be estimated under steering conditions, and (2) a pressure estimation method realized by fusing the vehicle dynamics-based MCPE and the pressure–position-based MCPE through the recursive least squares (RLS) is proposed, in which the robustness of the pressure–position-based MCPE has been improved, and the adaptability of the working conditions of a vehicle dynamics-based MCPE has been strengthened, and the noise and delay have been reduced. The rest of this article is organized as follows: The test vehicle is introduced in Section 2. The MCPE based on a 5-DOF vehicle model is proposed and verified via a vehicle test in Section 3. The pressure–position relationship under different brake pad wear levels is tested, and a novel dynamic pressure–position model is introduced in Section 4. The MCPE based on signal fusion is proposed in Section 5 and includes the principle of the RLS with a forgetting factor, initial state setting, and update condition setting. Simulations based on experimental data are conducted to verify the proposed fusion-based MCPE in Section 6. Section 7 concludes the article.

2. Test Vehicle

The test vehicle and the in-vehicle network system have been elaborated in the author's previous research [24], in which signals of the anti-lock brake system (ABS) (i.e., wheel speeds), electric power steering system (EPS) (i.e., wheel steering angle), IMU (i.e., absolute longitudinal acceleration, absolute lateral acceleration, and yaw rate), and EHB (i.e., rack position and master cylinder pressure) can be obtained by the EHB controller. In order to save space, this article only presents the picture and parameters of the of the test vehicle as shown in Figure 1 and Table 1, respectively.

Figure 1. Picture of the test vehicle.

Table 1. Configuration and parameters of the test vehicle.

Item	Value
Vehicle type	SUV-class electric vehicle
Powertrain type	Front-wheel drive
Braking system type	Electro-hydraulic brake
Steering system type	Electric power steering
Vehicle mass	1580 kg
Rolling radius of all wheels	0.3183 m
Wheelbase	2.56·m
Distance between the front axle and the center of gravity	1.33·m
Distance between the rear axle and the center of gravity	1.23·m
Yaw inertia moment	2370 kg·m^2
Braking force distribution coefficient	0.78

3. MCPE Based on 5-DOF Vehicle Model

In order to ensure safety and comfort, braking is often applied when steering in daily driving. In the literature, MCPE on flat roads [22] and sloped roads [24] without steering based on the longitudinal vehicle dynamics have been studied. In order to estimate the brake pressure under steering conditions, this article proposes an MCPE based on a 5-DOF vehicle model.

A vehicle dynamic model is usually used to describe the dynamics of vehicles. It is mainly derived through Newton's law [30]. By considering the accuracy and complexity of the model, this article selects a 5-DOF vehicle model that includes longitudinal motion, lateral motion, yaw motion, and front and rear wheel rotation for the purposes of this research, as shown in Figure 2.

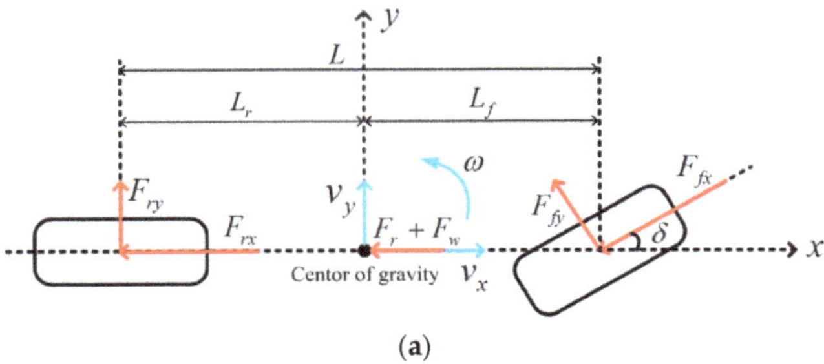

(**a**)

Figure 2. *Cont.*

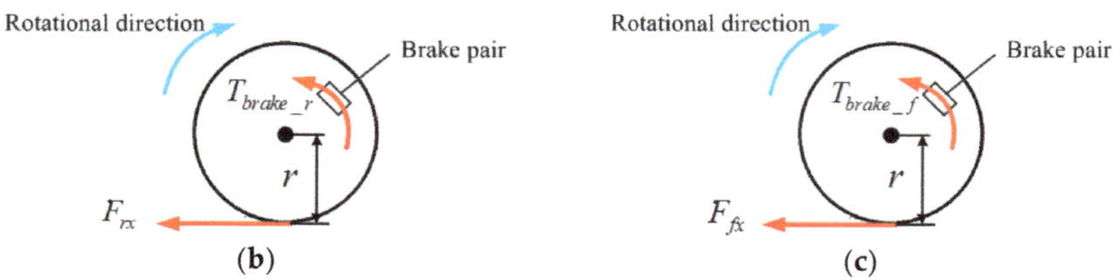

Figure 2. Scheme of the 5-DOF vehicle model: (**a**) denotes the whole vehicle. (**b,c**) denote the rear and front wheels, respectively.

There are some assumptions of the 5-DOF vehicle model that are considered in this article.

1. Ignoring the Ackerman steering principle, the left front wheel and the right front wheel share the same steering angle, that is, the vehicle is symmetrical.
2. The rolling resistance and the wind resistance are very small compared to the braking force; therefore, the rolling resistance and the wind resistance are all in the longitudinal dynamics, and there is no projection in the lateral dynamics under steering conditions.
3. The moment of inertia of the wheels is ignored so that the longitudinal tire force is the same as the friction braking force of each wheel.
4. All of the wheels share the same rolling radius, which is a reasonable assumption for most vehicles [22,31].
5. This work investigates the MCPE in ordinary braking scenarios. The ABS must not work, and all of the wheels share the same pressure.
6. The master cylinder pressure is the same as that of the wheel cylinders. In other words, the throttling effect of ABS is ignored.

Based on the above assumptions, the longitudinal, lateral, and yaw dynamics of the vehicle can be derived from Equations (1)–(3), according to Newton's law:

$$M(\dot{v}_x - \omega v_y) = -F_{fx}\cos\delta - F_{fy}\sin\delta - F_{rx} - F_r - F_w, \tag{1}$$

$$M(\dot{v}_y + \omega v_x) = F_{fy}\cos\delta - F_{fx}\sin\delta - F_{fy}\sin\delta + F_{ry}, \tag{2}$$

$$I\dot{\omega} = \left(F_{fy}\cos\delta - F_{fx}\sin\delta\right)L_f - F_{ry}L_r, \tag{3}$$

where M denotes the mass of the vehicle, kg; v_x and v_y denote the longitudinal speed and the lateral speed of the vehicle, respectively, m/s; ω denotes the yaw rate of the vehicle, rad/s; F_{fx} and F_{fy} denote the longitudinal tire force and the lateral tire force of the front wheel, N; F_{rx} and F_{ry} denote the longitudinal tire force and the lateral tire force of the rear wheel, N; δ denotes the steering angle of the front wheel, rad; F_r and F_w denote the rolling resistance and the wind resistance, respectively, N; I denotes the yaw inertia moment of the vehicle, $kg \cdot m^2$; L_f denotes the distance between the front axle and the center of gravity of the vehicle, m; L_r denotes the distance between the rear axle and the center of gravity of the vehicle, m.

The sum of the rolling resistance and the wind resistance (i.e., $F_r + F_w$) in Equation (1) is the so-called driving resistance, which can be obtained through the coasting test. For specific principles, the test procedures, and the test results of the coasting test, please refer to [24].

The rotational dynamics of the front wheel and the rear wheel are expressed by Equation (4) and Equation (5), respectively.

$$F_{fx} = \frac{p\left(k_{fl} + k_{fr}\right)}{r}, \tag{4}$$

$$F_{rx} = \frac{p(k_{rl} + k_{rr})}{r}, \tag{5}$$

where p denotes the pressure in the hydraulic circuit, *bar*; r denotes the rolling radius of all wheels, *m*; k_{fl}, k_{fr}, k_{rl} and k_{rr}, which are related to the time-varying BLCF and other time-invariant parameters of the brakes, denote the pressure–torque conversion factor of the front left wheel, front right wheel, rear left wheel, and rear right wheel, respectively, Nm/bar. Ref. [24] noted that under normal driving conditions, the BLCF is mainly related to the relative speed of the pad and disc, and the sum of the pressure–torque conversion factors of all of the wheels is given as Equation (6):

$$k_{fl} + k_{fr} + k_{rl} + k_{rr} = \begin{cases} 70 - \frac{70-53}{25}u & ,u \leq 25 \\ 53 & ,u > 25 \end{cases}, \tag{6}$$

where u denotes the average wheel speed under both straight and steering conditions, km/h.

For most passenger cars, the ratio of the braking force between the front and rear brakes is a fixed value [32], as shown in Equation (7):

$$\frac{F_{fx}}{F_{fr}} = \frac{\frac{p(k_{fl}+k_{fr})}{r}}{\frac{p(k_{rl}+k_{rr})}{r}} = \frac{k_{fl} + k_{fr}}{k_{rl} + k_{rr}} = \frac{\beta}{1 - \beta}, \tag{7}$$

where $\beta = \frac{F_{fx}}{F_{fx}+F_{fr}}$ denotes the braking force distribution coefficient. Thus, $\left(k_{fl} + k_{fr}\right)$ and $(k_{rl} + k_{rr})$ can be determined based on Equations (6) and (7).

The IMU is mounted on the vehicle body and can measure the absolute longitudinal acceleration, absolute lateral acceleration, and yaw rate of the vehicle, as shown in Equations (8)–(10):

$$\dot{v}_x - \omega v_y = a_{x\_IMU}, \tag{8}$$

$$\dot{v}_y + \omega v_x = a_{y\_IMU}, \tag{9}$$

$$\omega = \omega_{IMU}, \tag{10}$$

where a_{x_IMU}, a_{y_IMU}, and ω_{IMU} denote the absolute longitudinal acceleration, absolute lateral acceleration, and yaw rate of the vehicle measured by the IMU, respectively.

Substituting Equations (4), (5) and (8)–(10) into Equations (1)–(3), we can derive Equations (11)–(13).

$$Ma_{x\_IMU} = -\frac{p\left(k_{fl} + k_{fr}\right)}{r}\cos\delta - F_{fy}\sin\delta - \frac{p(k_{rl} + k_{rr})}{r} - F_r - F_w, \tag{11}$$

$$Ma_{y\_IMU} = F_{fy}\cos\delta - \frac{p\left(k_{fl} + k_{fr}\right)}{r}\sin\delta - F_{fy}\sin\delta + F_{ry}, \tag{12}$$

$$I\dot{\omega}_{IMU} = \left(F_{fy}\cos\delta - \frac{p\left(k_{fl} + k_{fr}\right)}{r}\sin\delta\right)L_f - F_{ry}L_r, \tag{13}$$

where $\dot{\omega}_{IMU}$ can be obtained by the difference of ω_{IMU}. Note that Equations (11)–(13) are linear and unrelated to the three unknown variables (i.e., p, F_{fy}, and F_{ry}), so there is a unique solution to the equation set consisting of Equations (11)–(13). The algebraic expression of the pressure estimation algorithm based on the 5-DOF vehicle model can be derived by solving the above-mentioned equation set.

$$p = \frac{(-Ma_{x\_IMU} - F_r - F_w)L\cos\delta - Ma_{y\_IMU}L_r\sin\delta - I\dot{\omega}_{IMU}\sin\delta}{\left(k_{fl} + k_{fr}\right)L + (k_{rl} + k_{rr})L\cos\delta}r, \tag{14}$$

where $L = L_f + L_r$ denotes the wheelbase of the vehicle. If $\delta = 0$, Equation (14) degenerates to Equation (15), as in Ref [24].

$$p = \frac{-Ma_{x\_IMU} - F_r - F_w}{k_{fl} + k_{fr} + k_{rl} + k_{rr}} r, \tag{15}$$

From Equations (14) and (15), we can see that with input signals of sensors (i.e., IMU and wheel steering angle), vehicle parameters (i.e., M, L and, etc.), driving resistance and pressure-torque conversion factors, the brake pressure can be estimated online. Besides, Equation (14) is applicable to both straight and steering condition while Equation (15) is only applicable to straight condition.

Vehicle tests under steering conditions were conducted. In order to highlight the superiority of the MCPE based on a 5-DOF vehicle model (MCPE 2), it was compared to the MCPE based on longitudinal vehicle dynamics (MCPE 1). The test results are shown in Figure 3, where the vehicle speed is calculated by the average wheel speed, the steering angle is obtained from the EPS, the actual pressure is obtained from the master cylinder pressure sensor, and where MCPE 1 and MCPE 2 correspond to Equation (15) and Equation (14), respectively.

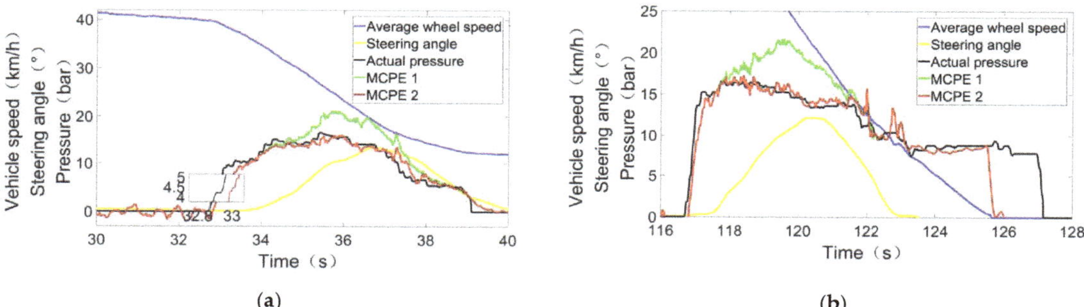

Figure 3. Test results of MCPEs based on difference vehicle models: (**a**,**b**) represent different brake conditions.

In Figure 3a, before braking, the vehicle is in a coasting state. At this time, the longitudinal acceleration measured by the IMU is the exact driving resistance. Therefore, according to Equation (15), theoretically, the estimated pressure at this time is zero. However, affected by the noise of the IMU signals, the estimated pressure jitters around zero, with a peak-to-peak value of about 3bar. After the start of braking, since the signal of the IMU lags behind the brake pressure, the estimated pressure lags behind the actual pressure, and the lag time is about 100ms. When the vehicle starts steering, as the steering angle increases, MCPE 1 deviates from the actual pressure, while MCPE 2 tracks the actual pressure well, thus proving that the MCPE based on the 5-DOF vehicle model can effectively estimate the brake pressure under both straight and steering conditions.

In Figure 3b, at about 122 s, the noise of the IMU increases due to the uneven road surface, and there is a large jitter in the estimated pressure, with a peak-to-peak value of about 5bar. When the vehicle speed is reduced to zero at about 126 s, the output signal of the IMU keeps zero, and both MCPE 1 and MCPE 2 are invalid.

From the above analysis, it can be seen that although the MCPE based on vehicle dynamics can be extended to steering conditions by adopting the 5-DOF vehicle model, it is still limited by signal noise, road conditions, and algorithm principles. There are still jitter, delay, and condition limitations in the estimated pressure.

4. Pressure-Position Model

The scheme of the EHB is shown in Figure 4 [24]. Under normal braking, the permanent magnet synchronous motor (PMSM) is adopted as the power source, which pushes the master cylinder piston to build pressure through the worm–worm gear and pinion–rack reductions. The electronic control unit (ECU) analyzes the target pressure according to the pedal strokes and performs closed-loop control of the master cylinder pressure based on the master cylinder pressure sensor [33].

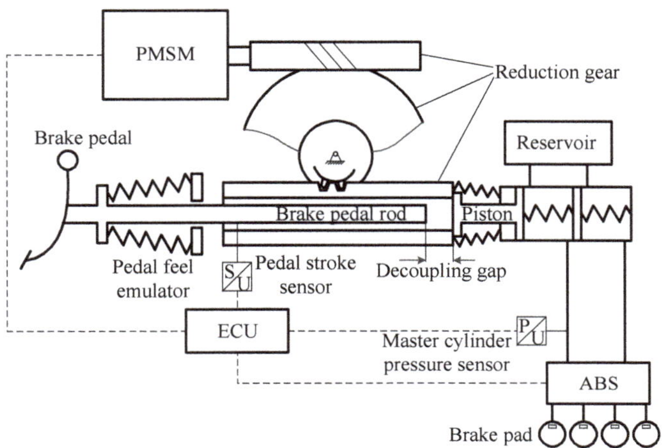

Figure 4. Scheme of the EHB [24].

Thanks to the angular position sensor of the rotor built in the PMSM, we can estimate the pressure of the master cylinder based on the derived rack position and the pressure–position relationship of the hydraulic circuit.

During the braking process, the pipelines expend [34], the caliper deforms [35], and the free gas in the brake fluid is compressed and dissolved [36], which contributes to the pressure–position relationship. The pressure–position relationship is affected by many factors, which are difficult to accurately modeled. Existing studies have shown that the pressure–position relationship has strong nonlinearity (i.e., hysteresis) and time-varying characteristics (brake pad wear, rack speed, etc.). In this article, the pressure–position relationship of a light commercial vehicle (not the test vehicle in Figure 1) under different brake pad wear levels is tested, as shown in Figure 5.

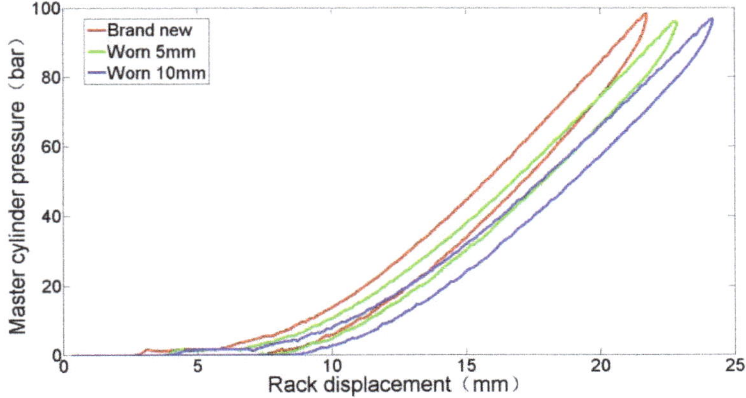

Figure 5. Pressure–position relationship under different brake pad wear levels.

As it can be seen, there is a dead zone of about 6mm in the pressure–position relationship due to gaps in the hydraulic circuit. As the brake pad wears, the pressure–position relationship "softens", and it takes greater rack displacement to build the same pressure. Moreover, the pressure–position relationship shows hysteresis characteristics. Under the same rack position, the pressure in the pressurization process is greater than that in the depressurization process.

In Ref [37], tests under different rates of motor torque were conducted, and a novel dynamic pressure–position model was proposed as Equation (16).

$$p = a + bx + cx^2 + d\dot{x} \tag{16}$$

where a, b, c, and d denote the coefficient; x denotes the rack position; $\dot{x}$ denotes the rack speed; $a + bx + cx^2$ represents the "average value" or the "static part" of the pressure–position curve; and $d\dot{x}$ represents the "hysteresis" or the "dynamic part" caused by different rack speed.

Experimental results show that compared to the traditional pressure–position model shown in Equation (17), which is adopted by almost all previous studies in the literature, the dynamic model can render hysteresis and speed influence effect more accurately. In addition, when the rack position and rack speed are used as input, the output pressure of the dynamic model has a faster response speed than the static model [37].

$$p = a + bx + cx^2 \tag{17}$$

Although the state-of-the-art pressure–position model can characterize hysteresis and the speed influence effect, when the brake pad is worn down, the "average value" of the pressure–position relationship changes, and the dynamic pressure–position model with a fixed coefficient will not be robust.

5. MCPE Based on Signal Fusion

The vehicle dynamics-based MCPE (VD-based MCPE) has many limitations (sensor noise, delay, road conditions, vehicle speed being zero, etc.), but the "average value" of the estimated pressure tracks the actual value very well. Although the pressure–position model based MCPE (PP-based MCPE) is simple and straightforward, it is not robust enough to prevent brake pad wear. Therefore, this article proposes a MCPE based on signal fusion (fusion-based MCPE), in which the coefficients of the pressure–position model are updated by the pressure estimated by the VD-based MCPE based on RLS, and the updated pressure–position model is finally adopted to estimate the brake pressure.

5.1. Principle of the RLS

Suppose $x = \begin{bmatrix} 1 & x & x^2 & \dot{x} \end{bmatrix}$, $\phi = \begin{bmatrix} a \\ b \\ c \\ d \end{bmatrix}$, the pressure–position model can be expressed by Equation (18):

$$p = x\phi \tag{18}$$

Suppose $\hat{p}_{VD}$ denotes the estimated pressure based on VD-based MCPE, as shown in Equation (19):

$$\hat{p}_{VD} = \frac{(-Ma_{x\_IMU} - F_r - F_w)L\cos\delta - Ma_{y\_IMU}L_r\sin\delta - I\dot{\omega}_{IMU}\sin\delta}{\left(k_{fl} + k_{fr}\right)L + (k_{rl} + k_{rr})L\cos\delta}r \tag{19}$$

Since the "average value" of $\hat{p}_{VD}$ is accurate, we hope that the fitted pressure–position model $x\hat{\phi}$ is as close to $\hat{p}_{VD}$ as possible. This article adopts the LS method [38] to solve this problem. In a linear system, this is equivalent to finding $\hat{\phi}(k)$ which causes the target function $V\big(\hat{\phi}(k), k\big)$ to obtain the smallest value, as shown in Equation (20):

$$V\big(\hat{\phi}(k), k\big) = \frac{1}{2}\sum_{i=1}^{k}\big(\hat{p}_{VD}(i) - x(i)\hat{\phi}(k)\big)^2 \tag{20}$$

where k denotes the current sampling time. When Equation (20) obtains its smallest value, $\hat{\phi}(k)$ can be solved as Equation (21):

$$\hat{\phi}(k) = \big(X(k)^T X(k)\big)^{-1} X(k)^T Y(k), \tag{21}$$

where $X(k)^T = \begin{bmatrix} x(1) \\ \vdots \\ x(k) \end{bmatrix}$, $Y(k) = \begin{bmatrix} \hat{p}_{VD}(1) \\ \vdots \\ \hat{p}_{VD}(k) \end{bmatrix}$. There are two things that need to be pointed out. First, $\hat{\phi}(k)$ is the optimal solution to all of the historic data for x and $\hat{p}_{VD}$; therefore, when the actual pressure–position relationship changes, a certain amount of new data (equivalent to a certain amount of time) is needed for $\hat{\phi}(k)$ to converge to the real value. That is, the convergence speed of the LS is slow. Second, with the increase of k, the calculation burden of $\hat{\phi}(k)$ will be heavier, so the storage capacity and computing capacity of the controller are very demanding, and it is not feasible to be applied in engineering.

The first problem can be solved by adopting a LS with a forgetting factor. By adding the forgetting factor to the LS, the data that are farther away from the current moment will occupy a smaller proportion; thus, the convergence speed of the LS is improved. The target function is represented as Equation (22).

$$V\big(\hat{\phi}(k), k\big) = \frac{1}{2}\sum_{i=1}^{k}\lambda^{k-i}\big(\hat{p}_{m\_pre}(i) - x(i)\hat{\phi}(k)\big)^2 , \tag{22}$$

For the second problem, the calculation of $\hat{\phi}(k)$ can be transformed into a recursive form. The recursive least square with a forgetting factor can be expressed as Equation (23):

$$\hat{\phi}(k) = \hat{\phi}(k-1) + K(k)\big[\hat{p}_{VD}(k) - x(k)\hat{\phi}(k-1)\big]$$
$$K(k) = \frac{P(k-1)x^T(k)}{\lambda + x(k)P(k-1)x^T(k)} \tag{23}$$
$$P(k) = \frac{1}{\lambda}[I - K(k)x(k)]P(k-1)$$

where $K \in \mathbb{R}^{4\times1}$ denotes the gain; $P \in \mathbb{R}^{4\times4}$ denotes the covariance matrix; and $\lambda \in (0,1)$ denotes the forgetting factor. Finally, the updated pressure–position model is adopted to estimate the brake pressure, as seen in Equation (24):

$$\hat{p}_{final}(k) = x(k)\hat{\phi}(k) \tag{24}$$

5.2. Initial Value of the RLS

It can be seen from Equation (23) that the operation of the RLS requires a given initial value of $\hat{\phi}(k)$ and $P(k)$, that is, $\hat{\phi}(0)$ and $P(0)$. An appropriate $\hat{\phi}(0)$ and $P(0)$ can speed up the convergence speed of the algorithm. To this end, this article uses the collected data from the real vehicle test in [24] to fit the pressure–position model by means of LS to obtain $\hat{\phi}(0)$ and $P(0)$. The collected data, including vehicle speed, brake pressure, rack position, and rack speed with a sampling time of 10ms, are shown in Figure 6.

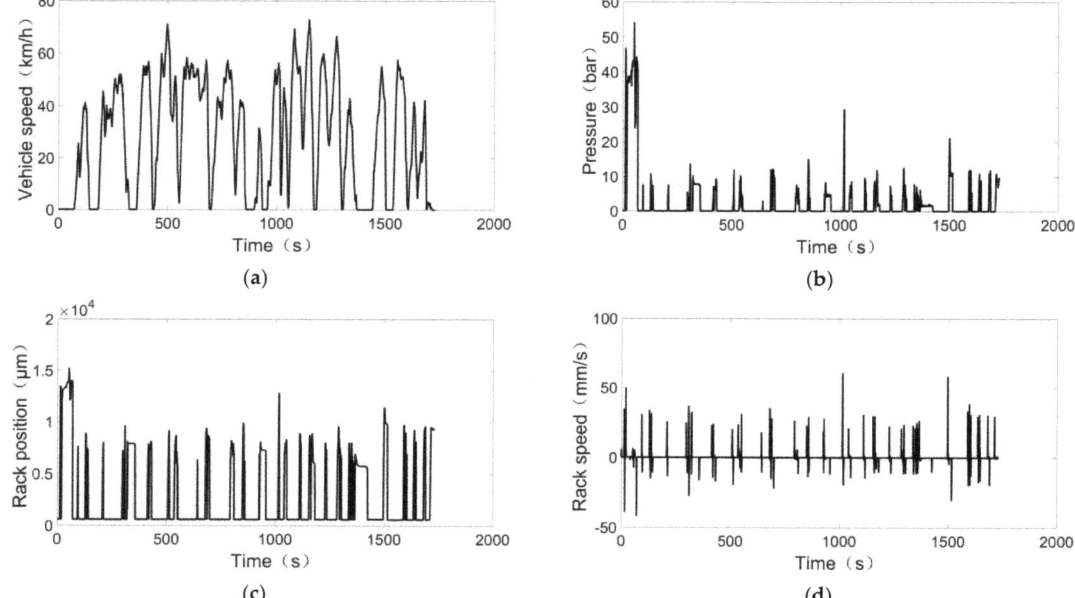

Figure 6. The collected data from the real vehicle test: (**a–d**) denote the vehicle speed, brake pressure, rack position, and rack speed, respectively.

The data for the brake pressure, rack position, and rack speed should be selected when the brake pressure is greater than zero, and Equations (25) and (26) should be used to calculate $\hat{\phi}(0)$ and $P(0)$.

$$\hat{\phi}(0) = \left(X(n)^T X(n)\right)^{-1} X(n)^T Y(n), \tag{25}$$

$$P(0) = \left(X(n)^T X(n)\right)^{-1}, \tag{26}$$

where $X(n)^T = \begin{bmatrix} x(1) \\ \vdots \\ x(n) \end{bmatrix}$, $Y(n) = \begin{bmatrix} p(1) \\ \vdots \\ p(n) \end{bmatrix}$, n denotes the number of the selected data. The results are shown in Equation (27) and Equation (28), respectively.

$$\hat{\phi}(0) = \begin{bmatrix} -1.261 \\ -9.396 \times 10^{-4} \\ 2.469 \times 10^{-7} \\ 0.5436 \end{bmatrix}, \tag{27}$$

$$P(0) = \begin{bmatrix} 5.418 \times 10^{-21} & 5.418 \times 10^{-21} & 5.418 \times 10^{-21} & 5.418 \times 10^{-21} \\ 5.418 \times 10^{-21} & 5.418 \times 10^{-21} & 5.418 \times 10^{-21} & 5.418 \times 10^{-21} \\ 5.357 \times 10^{-21} & 5.357 \times 10^{-21} & 5.357 \times 10^{-21} & 5.357 \times 10^{-21} \\ 5.022 \times 10^{-21} & 5.022 \times 10^{-21} & 5.022 \times 10^{-21} & 5.022 \times 10^{-21} \end{bmatrix}, \tag{28}$$

The fitted pressure–position model (i.e., $p(k) = x(k)\phi(0)$) and the real data are shown in Figure 7. As it can be seen, the fitted model can essentially represent the average value of the real date. Note that the fitted model in Figure 7 is $p = -1.261 - 9.396 \times 10^{-4}x + 2.469 \times 10^{-7}x^2$, for there is no dimension to add $0.5436\dot{x}$ in Figure 7.

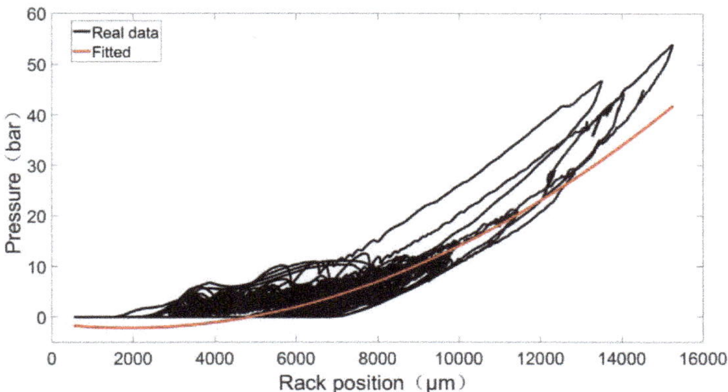

Figure 7. The fitted pressure–position model and the real data.

5.3. Condition Setting for Updating

When $\hat{\phi}(0)$, $P(0)$, and real-time $\hat{p}_{VD}(k)$ and $x(k)$ are given, Equation (23) can continue to run and update $\hat{\phi}(k)$ and $P(k)$. However, the purpose of this article is to fit the coefficients of the pressure model; that is, the data point $(\hat{p}_{VD}(k), x(k))$ can only be used when they are near the actual pressure–position curve. Therefore, just as when calculating $\hat{\phi}(0)$ and $P(0)$ in 5.2, only data with a pressure greater than zero are selected. It is necessary to filter the data point $(\hat{p}_{VD}(k), x(k))$ before updating $\hat{\phi}(k)$. Unfortunately, for EHB without a pressure sensor, the selection criteria of pressure greater than zero are no longer applicable. For this reason, this article proposes a new screening method. For the sake of analysis, suppose the vehicle is on a straight and flat road.

When the vehicle stops, $Ma_{x\_IMU} = 0$, $\hat{p}_{VD}(k) = \frac{-F_r - F_w}{k_{fl}+k_{fr}+k_{rl}+k_{rr}} r < 0$. Obviously, the data point $(\hat{p}_{VD}(k), x(k))$ cannot be used to update $\hat{\phi}(k)$ and $P(k)$.

When coasting, $Ma_{x\_IMU} = -F_r - F_w$, $\hat{p}_{VD}(k) = \frac{-Ma_{x\_IMU} - F_r - F_w}{k_{fl}+k_{fr}+k_{rl}+k_{rr}} r = 0$. According to the control logic of EHB, when the brake pedal is not depressed, the rack will be pushed to the zero position. Therefore, the data point $(\hat{p}_{VD}(k), x(k))$ is not on the effective section of the pressure–position curve at that moment; therefore, the pressure–position model cannot be updated while coasting.

When accelerating, $Ma_{x\_IMU} > -F_r - F_w$, $\hat{p}_{VD}(k) = \frac{-Ma_{x\_IMU} - F_r - F_w}{k_{fl}+k_{fr}+k_{rl}+k_{rr}} r < 0$. This is the same as when the vehicle stops.

When the brake pedal is stepped on and when the rack crosses the dead zone and builds pressure, $Ma_{x\_IMU} < -F_r - F_w$, $\hat{p}_{VD}(k) = \frac{-Ma_{x\_IMU} - F_r - F_w}{k_{fl}+k_{fr}+k_{rl}+k_{rr}} r > 0$. In addition, when the rack has crossed the dead zone and is in the effective zone, the data point $(\hat{p}_{VD}(k), x(k))$ is suitable to update $\hat{\phi}(k)$ and $P(k)$ at this time.

In summary, the pressure–position model is only updated when the vehicle speed is greater than a certain threshold and when the rack position is greater than a certain threshold, as in Equation (29).

$$
\hat{\phi}(k) = \begin{cases} \hat{\phi}(k-1), & u < u_{threshold} \ or \ x < x_{threshold} \\ \hat{\phi}(k-1) + K(k)\left[\hat{p}_{VD}(k) - x(k)\hat{\phi}(k-1)\right], & otherwith \end{cases}
$$

$$
K(k) = \frac{P(k-1)x^T(k)}{\lambda + x(k)P(k-1)x^T(k)} \qquad\qquad (29)
$$

$$
P(k) = \begin{cases} P(k-1), & u < u_{threshold} \ or \ x < x_{threshold} \\ \frac{1}{\lambda}\left[I - K(k)x(k)\right]P(k-1), & otherwith \end{cases}
$$

where $u_{threshold}$ and $x_{threshold}$ denote the vehicle speed threshold and the rack position threshold, respectively. Note that by updating the pressure–position model with the filtered data pairs, a more accurate pressure–position model is expected to be obtained so that when the rack is in the dead zone, the estimated pressure must be negative, as shown in Figure 7. To this end, the estimated pressure is limited by Equation (30).

$$\hat{p}_{final}(k) = \begin{cases} x(k)\hat{\phi}(k), & x(k)\hat{\phi}(k) > 0 \\ 0, & otherwith \end{cases} \tag{30}$$

Finally, the fusion-based MCPE can be represented by Equations (19) and (27)–(30).

6. Validation of the Proposed Fusion-Based MCPE

Based on the experimental data in [24], the proposed fusion-based MCPE is verified by the MATLAB/Simulink platform, and the simulation step is 5 ms. The verification consists of two parts. The first part verifies that the fusion-based MCPE outperforms the VD-based MCPE in terms of smoothness, delay time, robustness to road conditions, and adaptability to parking conditions. The second part verifies that the fusion-based MCPE outperforms the PP-based MCPE in terms of robustness to brake pad wear. The parameter settings of the fusion-based MCPE are as they are seen in Table 2.

Table 2. Parameter settings of the fusion-based MCPE.

Item	Value
Vehicle speed threshold $u_{threshold}$	3.6 km/h
Rack position threshold $x_{threshold}$	6000 μm
Forgetting factor λ	0.999

6.1. Normal Driving Conditions

Simulations based on experimental data under normal driving conditions were conducted, the results are shown in Figure 8.

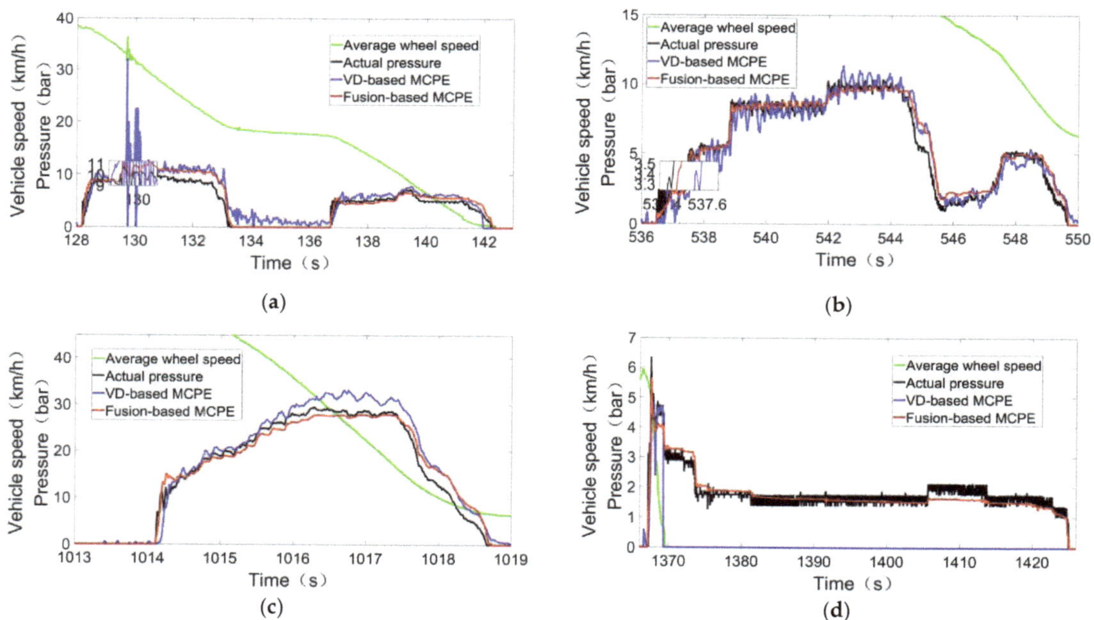

Figure 8. *Cont.*

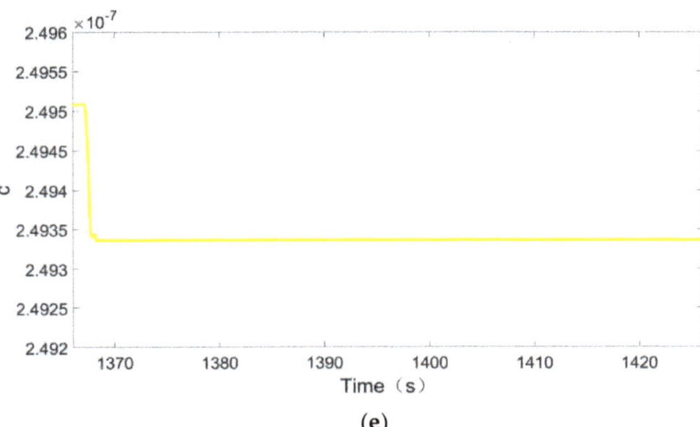

(e)

Figure 8. Simulation results of the fusion-based MCPE and the VD-based MCPE under normal driving conditions: (**a**) represents road with speed bumps. (**b**) represents road without speed bumps. (**c**) represents a brake event of 30bar. (**d**) and (**e**) represent the vehicle stops at a traffic intersection.

In Figure 8a, when the vehicle encounters a speed bump at about 130s, there is severe jitter in the VD-based MCPE with a peak-to-peak value of 60bar. In contrast, the fusion-based MCPE can still work smoothly, with a peak-to-peak value of only 1.2bar. In addition, when the vehicle is coasting, the estimated pressure of the fusion-based MCPE is zero, while the estimated pressure of VD-based MCPE jitters around zero. In Figure 8b, under normal driving conditions, the fusion-based MCPE is much more stable than the VD-based MCPE, and the RMSE of them are 0.3597bar and 0.9182bar, respectively. In addition, in terms of delay time, the fusion based MCPE is much smaller than the VD-based MCPE due to the fast response of the novel dynamic pressure–position model proposed in [37]; the former is only 25ms, and the latter exceeds 100ms. The brake pressure under normal driving conditions is generally not more than 30bar. Figure 8c shows that under 30bar, the proposed fusion-based MCPE can still estimate the brake pressure precisely. Figure 8d represents the condition where the vehicle stops at a traffic intersection. After the vehicle speed is reduced to zero, the driver still brakes with a small amount of brake pressure. It can be seen in Figure 8d that after the vehicle stops, the VD-based MCPE fails, while the fusion-based MCPE can estimate the brake pressure consistently and accurately. The evolution of the coefficient c of the pressure–position model is shown in Figure 8e. Note that c is the coefficient of the square term of the rack position and that its value has a great influence on the pressure–position model. In Figure 8e, when the vehicle speed is about to decrease to zero, the RLS stops updating $\hat{\phi}(k)$, and c remains unchanged, thus ensuring that after the vehicle speed is reduced to zero, the fusion-based MCPE can continue to estimate the brake pressure.

6.2. Brake Pad Wear

In order to verify the robustness of the fusion-based MCPE to brake pad wear, the fusion-based MCPE was compared with the PP-based MCPE with the fixed coefficient as Equation (31).

$$p(k) = x(k)\phi(0) \tag{31}$$

When the brake pad is worn, the pressure–position curve becomes "soft"; that is, the same pressure corresponds to a larger rack position. Therefore, in order to simulate the brake pad wear, the experimental data of the rack position are set as 1.2 times of the original at 540s in the simulation, the result of which is shown in Figure 9.

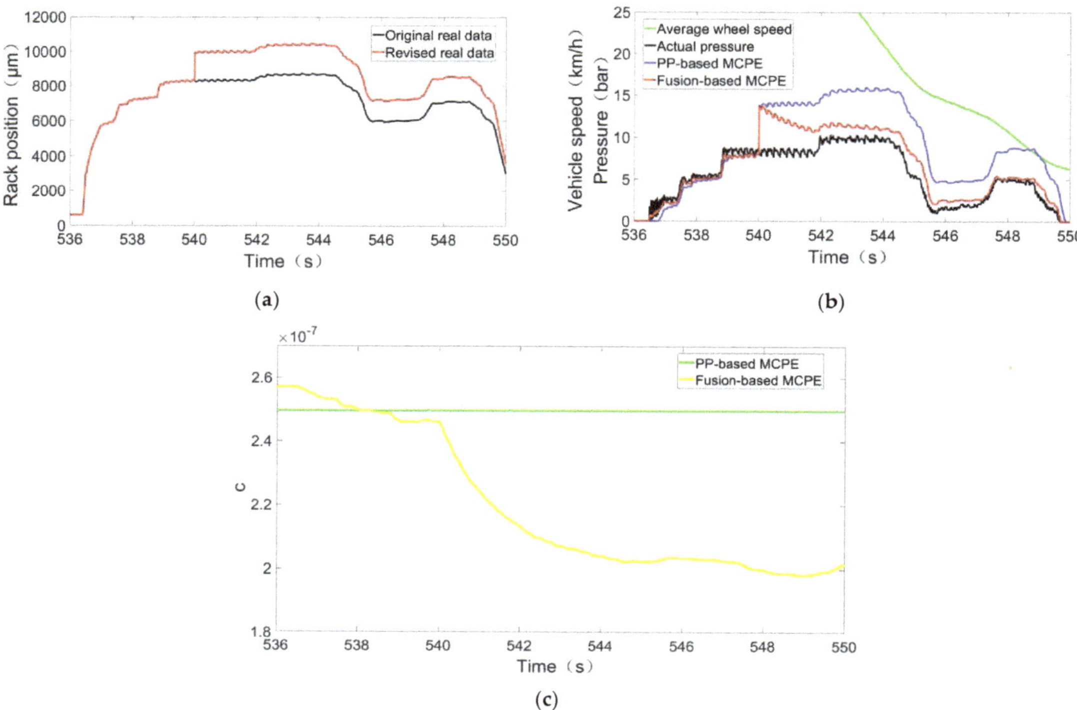

Figure 9. Simulation results of the fusion-based MCPE and the PP-based MCPE when the brake pad wears: (**a**) represents the rack position. (**b**) represents the actual and the estimated brake pressure. (**c**) represents the coefficient c in the pressure-position model.

Figure 9a shows that the original rack position is changed to 1.2 times of the original value at 540 s. In Figure 9b, after 540 s, as expected, the estimated pressure of the PP-based MCPE is always greater than the actual value, while the fusion-based MCPE can gradually converge to the actual value. Figure 9c shows that after 540 s, since the RLS continuously updates $\hat{\phi}(k)$ according to the new data point of $(\hat{p}_{VD}(k), x(k))$, c gradually decreases and converges to a stable value at about 548 s. This proves the robustness of the fusion-based MCPE to brake pad wear.

7. Conclusions

For the problem that a MCPE based on longitudinal vehicle dynamics cannot be used in steering conditions, a MCPE based on a 5-DOF vehicle model was proposed. Real vehicle test showed that the proposed method can effectively estimate the brake pressure in both straight and steering conditions.

Aiming to solve the problem of noise and delay in the VD-based MCPE and the poor robustness of the PP-based MCPE, a fusion-based MCPE was proposed. A RLS with a forgetting factor was adopted to update the coefficients of the pressure–position model, and the brake pressure was then estimated by the updated pressure–position model. Simulations were conducted based on the vehicle test data. The results show that the fusion-based MCPE can estimate the brake pressure accurately, smoothly, and quickly under various working conditions. Specifically, compared to a VD-based MCPE, the RMSE is reduced from 0.9182 bar to 0.3597 bar, and the delay time is reduced from 100ms to 25 ms. In addition, due to the reasonable setting of the enabling conditions of the RLS, the updated pressure–position model is more accurate. Therefore, when the brake is not applied, the rack position is zero. At this time, the estimated pressure is negative which avoids the problem of the VD-based MCPE oscillating near zero. Moreover, when the vehicle speed drops to zero, the RLS stops updating, so the brake pressure can be estimated smoothly and continuously when the vehicle is in stationary mode. Finally, since the fusion algorithm will constantly update the pressure–position model based on new data, when the pressure–position model is changed due to brake pad wear, the RLS will automatically update the pressure–position model to the worn state; therefore, the robustness of the fusion-based MCPE is ensured.

Author Contributions: Conceptualization, B.S.; data curation, B.S.; formal analysis, B.S.; funding acquisition, L.X. and Z.Y.; investigation, B.S.; methodology, B.S.; project administration, L.X. and Z.Y.; resources, B.S.; software, B.S.; supervision, L.X. and Z.Y.; validation, B.S.; visualization, B.S.; writing—original draft, B.S.; writing—review and editing, B.S. All authors have read and agreed to the published version of the manuscript.

Funding: This research was funded by the Research on Development of Electronic Control Chassis System and Active Control Technology (Grant No. 20511104601) and the Program of Shanghai Automotive Industry Science and the Technology Development (Grant No. 1734).

Institutional Review Board Statement: Not applicable.

Informed Consent Statement: Not applicable.

Data Availability Statement: Detailed data are contained within the article. More data that support the findings of this study are available from the author B.S. upon reasonable request.

Acknowledgments: The authors are thankful for the support of the School of Automotive Studies and the IIV (Institute of Intelligent Vehicle) of Tongji University and Tongyu Automobile Technology Co., Ltd.

Conflicts of Interest: The authors declare no conflict of interest.

Abbreviations

BBW	Brake by wire system
EMB	Electro-mechanical brake system
EHB	Electro-hydraulic brake system
MCPC	Master cylinder pressure control algorithm
MCPE	Master cylinder pressure estimation algorithm
LS	Least squares
BLCF	Brake lining coefficient of friction
IMU	Inertial measurement unit
ANN	Artificial neural network
LMBP	Levenberg–Marquardt backpropagation
NEDC	New European driving cycle
DNN	Deep neural network
RNN	Recurrent neural networks
LSTM	Long short-term memory
RMSE	Root mean square error
5-DOF	Five-degree-of-freedom
RLS	Recursive least squares
ABS	Anti-lock brake system
EPS	Electric power steering
PMSM	Permanent magnet synchronous motor
ECU	Electronic control unit

References

1. Harris, A.L.; Fuller, R.G.; Uzzell, R.G.; Mortimer, I. Hydraulic Braking Systems of the Brake-by-Wire Type for Vehicles. U.S. Patent No. 6,189,982, 20 February 2001.
2. Ko, J.; Ko, S.; Son, H.; Yoo, B.; Cheon, J.; Kim, H. Development of Brake System and Regenerative Braking Cooperative Control Algorithm for Automatic-Transmission-Based Hybrid Electric Vehicles. *IEEE Trans. Veh. Technol.* **2015**, *64*, 431–440. [CrossRef]

3. Petersen, I. Wheel Slip Control in ABS Brakes using Gain Scheduled Optimal Control with Constraints. Ph.D. Thesis, Norwegian University of Science and Technology, Tondheim, Norway, 2003.
4. Jonner, W.-D.; Winner, H.; Dreilich, L.; Schunck, E. Electrohydraulic Brake System—The First Approach to Brake-By-Wire Technology. *SAE Tech. Pap. Ser.* **1996**, *960991*. [CrossRef]
5. Kreutz, S. Ideal regeneration with electromechanical brake booster (eBKV) in Volkswagen e-up! and Porsche 918 Spyder. In Proceedings of the 5th International Munich Chassis Symposium, Munich, Germany, 24–25 June 2014.
6. Ohtani, Y.; Innami, T.; Obata, T.; Yamaguchi, T.; Kimura, T.; Oshima, T. Development of an Electrically Driven Intelligent Brake Unit. In Proceedings of the SAE Technical Paper Series, Detroit, MI, USA, 5 January 2011.
7. Han, W.; Xiong, L.; Yu, Z. Braking pressure control in electro-hydraulic brake system based on pressure estimation with nonlinearities and uncertainties. *Mech. Syst. Signal Process.* **2019**, *131*, 703–727. [CrossRef]
8. De Castro, R.; Todeschini, F.; Araujo, R.E.; Savaresi, M.S.; Corno, M.; Feritas, D. Adaptive robust friction compensation in a hybrid brake-by-wire actuator. *Proc. I Mech. E Part I J. Systs. Control Eng.* **2014**, *228*, 769–786. [CrossRef]
9. Yong, J.; Gao, F.; Ding, N.; He, Y. Design and validation of an electro-hydraulic brake system using hardware-in-the-loop real-time simulation. *Int. J. Automot. Technol.* **2017**, *18*, 603–612. [CrossRef]
10. Li, H.; Yu, Z.; Xiong, L.; Han, W. Hydraulic Control of Integrated Electronic Hydraulic Brake System Based on LuGre Friction Model. *SAE Tech. Pap. Ser.* **2017**, *1*. [CrossRef]
11. Todeschini, F.; Corno, M.; Panzani, G.; Savaresi, S.M. Adaptive position–pressure control of a brake by wire actuator for sport motorcycles. *Eur. J. Control* **2014**, *2*, 79–86. [CrossRef]
12. Todeschini, F.; Corno, M.; Panzani, G.; Fiorenti, S.; Savaresi, S.M. Adaptive Cascade Control of a Brake-By-Wire Actuator for Sport Motorcycles. *IEEE ASME Trans. Mechatron.* **2014**, *20*, 1310–1319. [CrossRef]
13. Han, W.; Xiong, L.; Yu, Z. Interconnected Pressure Estimation and Double Closed-Loop Cascade Control for an Integrated Electrohydraulic Brake System. *IEEE ASME Trans. Mechatron.* **2020**, *25*, 2460–2471. [CrossRef]
14. Zhao, J.; Hu, Z.; Zhu, B. Pressure Control for Hydraulic Brake System Equipped with an Electro-Mechanical Brake Booster. *SAE Tech. Pap. Ser.* **2018**. [CrossRef]
15. Wang, Z.; Yu, L.; Wang, Y.; You, C.; Ma, L.; Song, J. Prototype of Distributed Electro-Hydraulic Braking System and its Fail-Safe Control Strategy. In *SAE Technical Paper Series*; SAE: Warrendale, PA, USA, 2013. [CrossRef]
16. Yang, I.-J.; Choi, K.; Huh, K. Development of an electric booster system using sliding mode control for improved braking performance. *Int. J. Automot. Tech.* **2012**, *13*, 1005–1011. [CrossRef]
17. Han, W.; Xiong, L.; Yu, Z. A novel pressure control strategy of an electro-hydraulic brake system via fusion of control signals. *Proc I Mech. E Part D J. Automt. Eng.* **2019**, *233*, 3342–3357. [CrossRef]
18. Ohkubo, N.; Matsushita, S.; Ueno, M.; Akamine, K.; Hatano, K. Application of Electric Servo Brake System to Plug-In Hybrid Vehicle. *SAE Int. J. Passeng. Cars Electron. Electr. Syst.* **2013**, *6*, 255–260. [CrossRef]
19. Han, W.; Xiong, L.; Yu, Z. Pressure Estimation Algorithms in Decoupled Electro-Hydraulic Brake System Considering the Friction and Pressure-Position Relationship. In *SAE Technical Paper Series*; SAE: Warrendale, PA, USA, 2019.
20. Schwarz, R.; Isermann, R.; Böhm, J.; Nell, J.; Rieth, P. Clamping Force Estimation for a Brake-by-Wire Actuator. In Proceedings of the International Congress & Exposition, Detroit, MI, USA, 1–4 March 1999.
21. Saric, S.; Bab-Hadiashar, A.; Hoseinnezhad, R. Clamp-Force Estimation for a Brake-by-Wire System: A Sensor-Fusion Approach. *IEEE Tran. Veh. Technol.* **2008**, *57*, 778–786. [CrossRef]
22. Han, W.; Xiong, L.; Yu, Z.; Xu, S. Vehicle Validation for Pressure Estimation Algorithms of Decoupled EHB Based on Actuator Characteristics and Vehicle Dynamics. In *SAE Technical Paper Series*; SAE: Warrendale, PA, USA, 2020. [CrossRef]
23. Ricciardi, V.; Travagliati, A.; Schreiber, V.; Klomp, M.; Ivanov, V.; Augsburg, K.; Faria, C. A novel semi-empirical dynamic brake model for automotive applications. *Tribol. Int.* **2020**, *146*, 106223. [CrossRef]
24. Shi, B.; Xiong, L.; Yu, Z. Pressure Estimation Based on Vehicle Dynamics Considering the Evolution of the Brake Linings' Coefficient of Friction. *Actuators* **2021**, *10*, 76. [CrossRef]
25. Meng, B.; Yang, F.; Liu, J.; Wang, Y. A Survey of Brake-by-Wire System for Intelligent Connected Electric Vehicles. *IEEE Access* **2020**, *8*, 225424–225436. [CrossRef]
26. Guo, H.; Cao, D.; Hong, C.; Lv, C.; Wang, H.; Yang, S. Vehicle dynamic state estimation: State of the art schemes and perspectives. *J. Acta Autom. Sin.* **2018**, *5*, 418–431. [CrossRef]
27. Lv, C.; Xing, Y.; Zhang, J.; Na, X.; Li, Y.; Liu, T.; Cao, D.; Wang, F. Levenberg–Marquardt Backpropagation Training of Multilayer Neural Networks for State Estimation of A Safety Critical Cyber-Physical System. *IEEE Trans. Ind. Inf.* **2017**, *14*, 3436–3446. [CrossRef]
28. Al-Sharman, M.; Murdoch, D.; Cao, D.; Lv, C.; Zweiri, Y.; Rayside, D.; Melek, W. A Sensorless State Estimation for A Safety-Oriented Cyber-Physical System in Urban Driving:Deep Learning Approach. *J. Acta Autom. Sin.* **2019**, *8*, 169–178.
29. Xing, Y.; Lv, C. Dynamic State Estimation for the Advanced Brake System of Electric Vehicles by using Deep Recurrent Neural Networks. *IEEE Trans. Ind. Electron.* **2019**, *67*, 9536–9547. [CrossRef]
30. Bai, G.; Liu, L.; Meng, Y.; Lou, W.; Gu, Q.; Ma, B. Path Tracking of Mining Vehicles Based on Nonlinear Model Predictive Control. *Appl. Sci.* **2019**, *9*, 1372. [CrossRef]

31. Ricciardi, V.; Acosta, M.; Augsburg, K.; Kanarachos, S.; Ivanov, V. Robust brake linings friction coefficient estimation for enhancement of ehb control. In Proceedings of the 2017 XXVI International Conference on Information, Communication and Automation Technologies (ICAT), Sarajevo, Bosnia and Herzegovina, 26–28 October 2017.
32. Yu, Z. *Automobile Theory*, 2nd ed.; China Machinery Industry Press: Beijing, China, 1990.
33. Xiong, L.; Han, W.; Yu, Z.; Lin, J.; Xu, S. Master cylinder pressure reduction logic for cooperative work between electro-hydraulic brake system and anti-lock braking system based on speed servo system. *Proc. Inst. Mech. Eng. Part D J. Autom. Eng.* **2020**, *234*, 3042–3055. [CrossRef]
34. Antanaitis, D.; Riefe, M.; Sanford, J. Automotive Brake Hose Fluid Consumption Characteristics and Its Effects on Brake System Pedal Feel. *SAE Int. J. Passenger Cars Mech. Syst.* **2010**, *3*, 113–130. [CrossRef]
35. Mo, J.-O. Numerical and Experimental Investigation of Compressible-Brake-Fluid Flow Characteristics and Brake-Bleeding Performance in EPB Caliper. *Int. J. Automot. Technol.* **2021**, *22*, 315–325. [CrossRef]
36. Gholizadeh, H.; Bitner, D.; Burton, R.; Schoenau, G. Modelling and Experimental Validation of the Effective Bulk Modulus of a Mixture of Hydraulic Oil and Air. *J. Dyn. Syst. Meas. Contr.* **2014**, *136*, 051013. [CrossRef]
37. Shi, B.; Xiong, L.; Yu, Z. Master Cylinder Pressure Estimation of the Electro-hydraulic Brake System Based on Fusion of the Friction Character and the Pressure-position Character. *IEEE Trans. Veh. Technol.* (Submitted).
38. Gibbs, B.P. *Advanced Kalman Filtering, Least-Squares and Modeling: A Practical Handbook*; Wiley: Hoboken, NJ, USA, 2011.

Article

A Multi-Semantic Driver Behavior Recognition Model of Autonomous Vehicles Using Confidence Fusion Mechanism

Hongze Ren, Yage Guo, Zhonghao Bai * and Xiangyu Cheng

The State Key Laboratory of Advanced Design and Manufacturing for Vehicle Body, Hunan University, Changsha 410082, China; hnurhz@hnu.edu.cn (H.R.); guogeya@hnu.edu.cn (Y.G.); hnucxy@hnu.edu.cn (X.C.)
* Correspondence: baizhonghao@163.com

Abstract: With the rise of autonomous vehicles, drivers are gradually being liberated from the traditional roles behind steering wheels. Driver behavior cognition is significant for improving safety, comfort, and human–vehicle interaction. Existing research mostly analyzes driver behaviors relying on the movements of upper-body parts, which may lead to false positives and missed detections due to the subtle changes among similar behaviors. In this paper, an end-to-end model is proposed to tackle the problem of the accurate classification of similar driver actions in real-time, known as MSRNet. The proposed architecture is made up of two major branches: the action detection network and the object detection network, which can extract spatiotemporal and key-object features, respectively. Then, the confidence fusion mechanism is introduced to aggregate the predictions from both branches based on the semantic relationships between actions and key objects. Experiments implemented on the modified version of the public dataset Drive&Act demonstrate that the MSRNet can recognize 11 different behaviors with 64.18% accuracy and a 20 fps inference time on an 8-frame input clip. Compared to the state-of-the-art action recognition model, our approach obtains higher accuracy, especially for behaviors with similar movements.

Keywords: intelligent electric vehicles; driver behavior recognition; multi-semantic description; confidence fusion

Citation: Ren, H.; Guo, Y.; Bai, Z.; Cheng, X. A Multi-Semantic Driver Behavior Recognition Model of Autonomous Vehicles Using Confidence Fusion Mechanism. *Actuators* **2021**, *10*, 218. https://doi.org/10.3390/act10090218

Academic Editors: Peng Hang, Xin Xia and Xinbo Chen

Received: 27 July 2021
Accepted: 29 August 2021
Published: 31 August 2021

1. Introduction

Driver-related factors (e.g., distraction, fatigue, and misoperation) are the leading causes of unsafe driving, and it is estimated that 36% of vehicle accidents can be avoided if no driver engages in distracting activities [1,2]. Secondary activities such as talking with cellphones, consuming food, and interacting with in-vehicle devices lead to the significant degradation of driving skills, and increases in reaction times in emergency events [3]. With the rise of autonomous vehicles, drivers are gradually being liberated from the traditional roles behind steering wheels, thereby more freedom may contribute to complex behaviors [4]. As full automation could be decades away, driver behavior recognition is essential for autonomous vehicles with partial or conditional automation, where drivers have to be ready for requests for intervention [5].

With the growing demand for analyses of driver behaviors, driver behavior recognition has rapidly gained attention. Previous studies mainly adopted machine learning algorithms, such as random forest [6], Adaboost [7], and support vector machine [8], to detect distracted drivers. Deep learning technology hastens the parturition of outstanding driver behavior recognition models due to its powerful studying and generalizing ability. A typical pipeline of driver behavior recognition models based on deep learning is presented in Figure 1. First, driver movements are captured by cameras and fed into the data processing part in sequences of frames. The next step is to extract deep features and assign corresponding labels to these features. During this process, classification accuracy is critical to the model's performance. In [9], the multi-scale Faster-RCNN [10] is employed in driver's cellphone usage detection with the fusion approach based on features and geometric information. Streiffer et al. [11] propose a deep learning solution for distracted driving detection by means of aggregating the classification results of frame-sequence and IMU-sequence. Baheti et al. [12] adapted the VGG-16 [13] with various regularization techniques (e.g., dropout, L2 regularization, and batch normalization) to perform distracted driver detection.

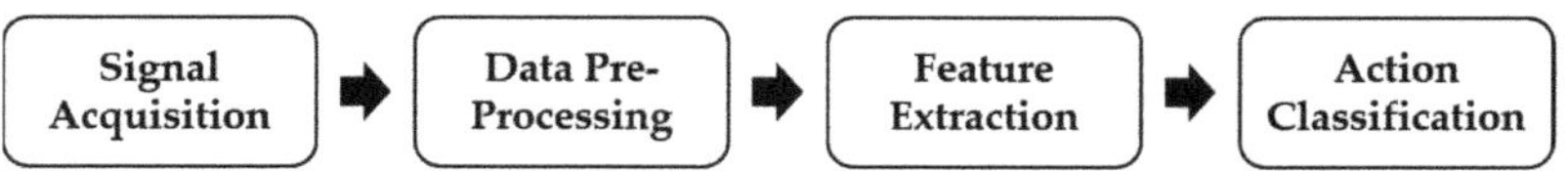

Figure 1. A typical pipeline of driver behavior recognition models based on deep learning.

The 3D-CNN is widely utilized for driver behavior recognition in order to aggregate the deep features from both spatial and temporal dimensions. Martin et al. [14] introduced the large-scale video dataset Drive&Act and provided benchmarks by adopting prominent methods for driver behavior recognition. Reiß et al. [15] adopted the fusion mechanism based on semantic attributes and word vectors to tackle the issue of zero-shot activity recognition. In [16], an interwoven CNN is used to identify driver behaviors by merging the features coming from multi-stream inputs.

In summary, it is ambitious to achieve high accuracy while maintaining runtime efficiency for driver behavior recognition. Existing research mostly analyzes driver behaviors by relying on the movements of upper-body parts, which may lead to false positives and missed detections due to the subtle changes among similar behaviors [17,18]. To tackle this problem, an end-to-end model is proposed, inspired by the human visual cognitive system. When humans understand complex and similar behaviors, our eyes capture not only the action cues, but also the key-object cues, in order to obtain more complete descriptions of behaviors. The example in Figure 2 illustrates our inspiration. Therefore, two parallel branches are presented to perform action classification and object classification, respectively. The action detection network, called ActNet, is used to extract spatiotemporal features from an input clip, and the object detection network called ObjectNet is used to extract key-object features from the key frame. Then, the confidence fusion mechanism (CFM) is introduced to aggregate the predictions from both branches based on the semantic relationships between actions and key-objects. Figure 3 illustrates the overall architecture of the proposed model. Our contributions can be summarized as follows:

- An end-to-end multi-semantic model is proposed to tackle the problem of accurate classification of similar driver behaviors in real-time, which can both characterize driver actions and focus on the key-objects linked with corresponding behaviors;
- The category of Drive&Act in the level of fine-grained activity is adapted to establish the clear relationships between behaviors and key-objects based on hierarchical annotations;
- Experiments implemented on the modified version of the public dataset Drive&Act demonstrate that the MSRNet can recognize 11 different behaviors with 64.18% accuracy and a 20 fps inference time on an 8-frame input clip. Compared to the state-of-the-art action recognition model, our approach obtains higher accuracy, especially for behaviors with similar movements.

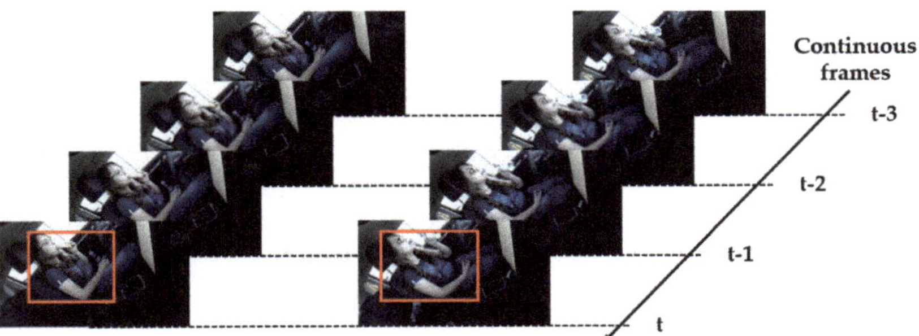

Figure 2. Drinking water or consuming food? Although the region of interest can be effectively obtained, it may not be possible to identify the driver action positively using only action cues. The key-object cues, such as food and bottles, should be integrated to classify which behavior the driver is taking on correctly.

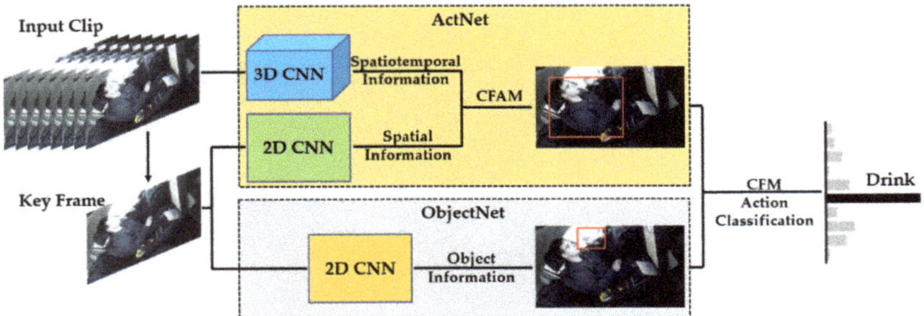

Figure 3. The overall architecture of the proposed model. ActNet is used to extract spatiotemporal features from an input clip and the ObjectNet is used to extract key-object features from the key frame. The predictions from both branches are fed into the CFM to perform confidence fusion and action classification based on the semantic relationships between actions and key objects.

2. Materials and Methods

In this section, the distribution of the fine-grained activity groups in the modified Drive&Act is introduced firstly, in order to facilitate the design, training, and evaluation of the proposed model. Subsequently, an end-to-end model with two parallel branches, called MSRNet, is employed to perform driver behavior recognition. Inspired by the intuition of human vision, the proposed model focuses on both the actions and the objects involved in the actions to derive holistic descriptions of driver behaviors. ActNet is used to extract spatiotemporal features from input clips, which can capture the action cues of driver behaviors. ObjectNet is utilized to extract key-object features from key frames, which mainly concentrates on object cues. The predictions from both branches are merged via the confidence fusion mechanism, based on the semantic relationships between actions and key objects. Overall this ensemble demonstrably improves model accuracy and robustness for driver behavior recognition. Finally, the implementation of MSRNet is described briefly.

2.1. Dataset

In this paper, experiments are conducted on the modified version of the public dataset Drive&Act [14], which collects data on the secondary activities of 15 subjects for 12 h (over 9.6 million frames). Drive&Act provides the hierarchical annotations of 12 classes of coarse tasks, 34 categories of fine-grained activities, and 372 groups of atomic action units. In contrast to the first (coarse task) and the third (atomic action unit) levels, the second level (fine-grained activity) can provide sufficient visual details while maintaining clear semantic descriptions. Therefore, the categories of Drive&Act at the level of fine-grained activity are adapted to establish clear relationships between behaviors and key objects based on hierarchical annotations. First, the classes involved in driving preparation activities (e.g., entering/exiting cars, fastening belts) are excluded due to the fact that the solution only focuses on the secondary activities in the running process of autonomous vehicles. In addition, the integrity of behaviors in the temporal dimension is preserved to simplify the correspondence between actions and key objects. For example, the actions of opening bottles, drinking water, and closing bottles are considered as the different stages of the same action. Finally, the 34 categories of Drive&Act are restructured into 11 classes, including nine semantic relationships between behaviors and key objects. Figure 4 illustrates the distribution of the fine-grained activity groups in the modified dataset.

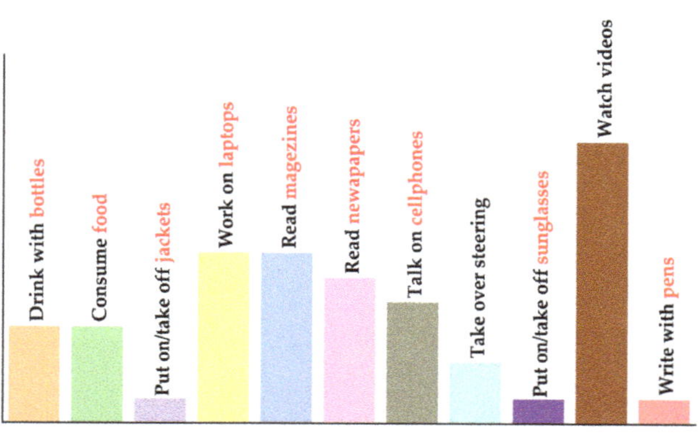

Figure 4. The distribution of the fine-grained activity groups in the modified dataset. The groups are: (1) drink from bottle; (2) consume food; (3) put on or take off jacket; (4) work on laptop; (5) read magazine; (6) read newspaper; (7) talk on cellphone; (8) take over the steering wheel; (9) put on or take off sunglasses; (10) watch videos; (11) write with a pen. The key objects corresponding to actions are colored in red.

2.2. ActNet

Since contextual information is crucial for understanding driver behaviors, the proposed model uses 3D-CNN to extract spatiotemporal features, which is able to capture motion information encoded in multiple consecutive frames. The 3D-CNNs form a cube by stacking multiple consecutive frames, and then apply 3D convolution not only in the space dimension, but also in the time dimension. The feature maps in the convolutional layer are related to the multiple adjacent frames in the upper layer to obtain motion information. YOWO [19] is the state-of-the-art 3D-CNN architecture for real-time spatiotemporal action localization in video streams. In YOWO, a unified network called ActNet is used to obtain the information on driver actions encoded in multiple contiguous frames. ActNet is made up of three major parts. The first part, the 3D branch, extracts spatiotemporal features from an input clip via 3D-CNN. The ResNext-101 is used as the 3D backbone of the 3D branch due to its good performance on kinetics and UCF-101 [20]. The second part, the 2D branch, extracts spatial features from the key frame (i.e., the last frame of an input clip) via 2D-CNN to address the spatial localization issue. Darknet-19 [21] is applied as the 2D backbone of the 2D branch. The concat layer merges the feature maps from the 2D branch and the 3D branch, and feeds them into the third part, the channel fusion and attention mechanism (CFAM), to aggregate the features smoothly from the two branches above.

The prior mechanism proposed in [21] is utilized to bound box regression localization. The final outputs are resized to $[5 \times (11 + 4 + 1) \times H \times W]$, indicating five prior anchors,

11 categories of activities, four coordinates, a confidence score, and the height and width of the images in the grid, respectively. The smooth $L1$ loss [22],

$$\text{smooth}_{L_1}(x) = \begin{cases} 0.5x^2, & \text{if } |x| < 1 \\ |x| - 0.5, & \text{otherwise,} \end{cases} \tag{1}$$

is adopted to calculate the loss of bounding box regression, where x is the difference in the elements between the bounding box and the groundtruth. The focal loss [23],

$$\text{FL}(p_t) = -(1 - p_t)^\gamma \log(p_t), \tag{2}$$

is applied to determine classification loss, where p_t

$$p_t = \begin{cases} p, & \text{if } y = 1 \\ 1 - p, & \text{otherwise,} \end{cases} \tag{3}$$

is the variation in cross-entropy loss, and $(1 - p_t)^\gamma$ is a modulating factor in cross-entropy loss, with a tunable focusing parameter $\gamma \geq 0$.

2.3. ObjectNet

ActNet is able to capture the action cues of driver behaviors from input clips directly, and provide accurate predictions in most situations. However, driver behaviors may be so subtle or similar that they lead to false positives and missed detections. Therefore, ObjectNet is proposed to capture the key-object cues involved in driver actions, such as bottles for drinking, food for eating, and laptops for working. ObjectNet is expected to further filter the predictions of ActNet in order to classify subtle or similar actions. YOLO-v3 [24] is one of the more popular algorithms used for generic object detection, and is successfully adapted to many recognition problems. YOLO-v3 is employed as the basic framework of ObjectNet due to its excellent trade-off between accuracy and efficiency. In order to enhance the performance to detect small objects, ObjectNet extracts features from multiple scales of the key frame, following the same guideline as the feature pyramid network [25]. In detail, the multi-scale outputs of different detection layers are merged to derive the final predictions using non-maximum suppression.

2.4. Confidence Fusion Mechanism

The outputs of ActNet and ObjectNet are reshaped to the same dimension (i.e., class index, four coordinates, and confidence score). For a specific class, the confidence score for each box is defined as

$$\Pr(\text{Class}_i | \text{Object}) * \Pr(\text{Object}) * \text{IOU}_{\text{pred}}^{\text{truth}} = \Pr(\text{Class}_i) * \text{IOU}_{\text{pred}}^{\text{truth}}, \tag{4}$$

which reflects both the probability of the class appearing in the box and how well the predicted box fits the object [26]. To utilize the complementary effects of different items of semantic information, the Confidence Fusion Mechanism (CFM) is introduced to aggregate predictions from both ActNet and ObjectNet based on the semantic relationships between actions and key-objects. The CFM is a decision fusion approach that combines the decisions of multiple classifiers into a common decision about driver behavior. This grounds independence from the type of data source, making it possible to aggregate the information derived from different semantic aspects.

In order to illustrate the implications of the CFM, we consider a simple scenario: there are two binary classifiers (S1 and S2) used to detect whether drivers are drinking water or not. It performs one detection using S1 and S2, and there will be four possible situations, as shown in Table 1. If the results of S1 and S2 are in agreement, it is reasonable to conclude on whether drivers are drinking water or not. Otherwise, the results of the classifier with greater confidence will be preferably accepted.

Table 1. The possible situations of driver drinking detection by two binary classifiers.

Possible Situations	S1	S2
(0, 0)	not_drink (0)	not_drink (0)
(0, 1)	not_drink (0)	drink (1)
(1, 0)	drink (1)	not_drink (0)
(1, 1)	drink (1)	drink (1)

Expanding the simple scenario to our task, ActNet performs driver behavior recognition on a given clip, and outputs N predictions. In general, we can conclude which actions drivers engage in by reference to the maximum confidence score. Figure 5 illustrates the algorithm flowchart of the CFM. First, the N predictions are sorted in order of confidence scores from largest to smallest. Afterwards, the top m predictions are fed into the decision in turn to examine whether they match with the correspondences between actions and key-objects. In this paper, we set m as 3, because the confidence scores of these predictions are generally lower than the threshold when m is beyond 3. If the prediction (i) is compatible with the key-object detected by ObjectNet, it is assumed that the prediction (i) is accurate, and the circulation is ended. Otherwise, this process will continue until all the top m predictions have been examined. In addition, there is a possible situation wherein none of the top m predictions match with the key-object. In this case, the original results of ActNet will be adopted.

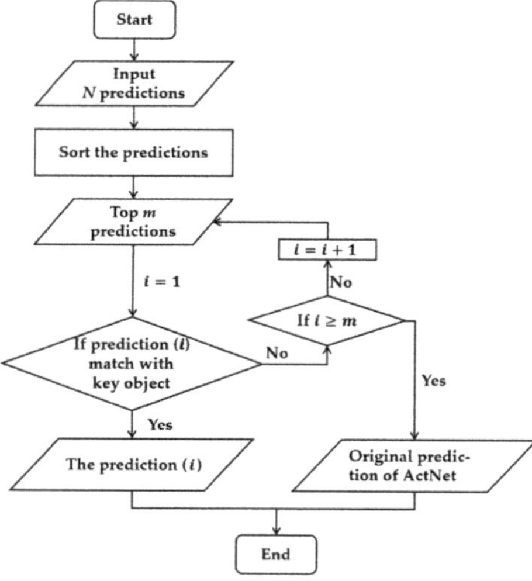

Figure 5. The algorithm flowchart of the confidence fusion mechanism.

2.5. Implementation Details

The publicly released YOLO-v3 [24] model is used for ObjectNet and is fine-tuned on the modified Drive&Act [14] following default configuration. For ActNet, the parameters of the 3D backbone and the 2D backbone are initialized on kinetics [27] and COCO [28], respectively. The training is implemented using stochastic gradient descent with an initial learning rate of 0.0001, which is degraded with a modulating factor of 0.5 after the 30 k, 40 k, 50 k, and 60 k iterations. The weight decay rate is set to 0.0005, and the momentum value is set to 0.9. For the dataset Drive&Act, the training process is converged after five epochs. Both ActNet and ObjectNet are trained and tested using a Tesla V100 GPU with 16 GB RAM. The proposed model is carried out end-to-end in PyTorch.

3. Results and Discussion

In this section, the accuracies of the MSRNet and YOWO are compared to illustrate the improvement in driver behavior recognition by aggregating multi-semantic information. Afterwards, the visualization of the output from different branches is used to determine what is learned by the MSRNet. Finally, some limitations that affect the MSRNet's performance are discussed.

Experiments are implemented on the modified public dataset Drive&Act. As in [14], the datasets for training, validation, and testing are randomly divided based on the identity of subjects; using videos, we assign the data of 10 persons for training, 2 persons for validation, and 3 persons for testing. Each action segment is spilt into 3-s chunks for balancing the various durations of driver behaviors. The standard evaluation metric of accuracy is adopted to measure the performance of the proposed dataset. Table 2 reports the results derived from comparing the accuracy between MSRNet and the state-of-the-art action recognition model YOWO [19]. It is observed that MSRNet performs better in both validation and testing, with significant 4.65% (Val) and 3.16% (Test) improvements in accuracy when recognizing 11 different behaviors on an 8-frame input clip.

Table 2. The results of comparing the accuracy between MSRNet and YOWO.

Model	Val (%)	Test (%)
YOWO	67.71	61.02
Our Model	72.36	64.18

Figure 6 illustrates the activation maps giving a visual explanation of the classification decision made by ActNet and ObjectNet [29]. It can be observed that ActNet mainly focuses on the areas where movements are happening, whereas ObjectNet mainly focuses on the key-objects. Figure 7 gives a precise description of 11 fine-grained activities carried out on the modified Drive&Act by the confusion matrixes. Each row of the confusion matrix represents the instances in an actual label, while each column represents the instances in a predicted label. As can be seen from the confusion matrixes, the proposed model accurately recognizes the majority of classes, with 99% accurate identification of drinking with bottles, 95% accurate identification of working on laptops, and 94% accurate identification of reading magazines. In addition, a significant improvement is made in recognizing similar actions. For example, 16% (drinking with bottles vs. consuming food) and 14% (reading magazines vs. reading newspaper) of the misrecognitions are avoided when using the MSRNet. Our experiments demonstrate the effectiveness of utilizing multi-semantic classification for driver recognition with the confidence fusion mechanism. Although the proposed model shows superiority in solving the problem of interclass similarity, it also suffers from some limitations that degrade its performance. Figure 8 illustrates examples of images for which the MSRNet fails in driver behavior recognition. It is observed that the misrecognition of the proposed model is mainly caused by some challenging situations in Drive&Act, such as occlusion and multi-class visibility.

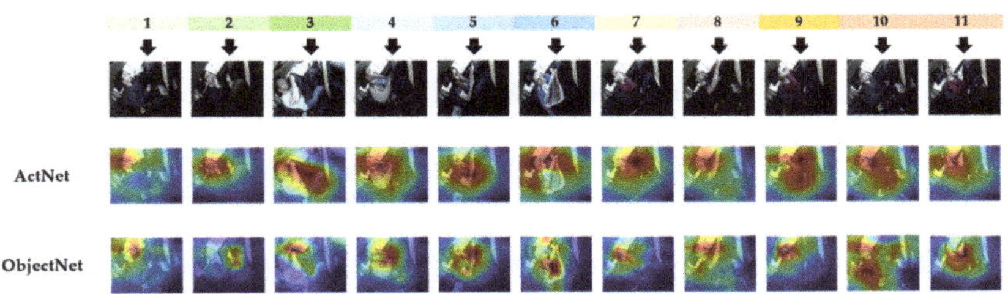

Figure 6. The activation maps giving a visual explanation of the classification decision made by ActNet and ObjectNet.

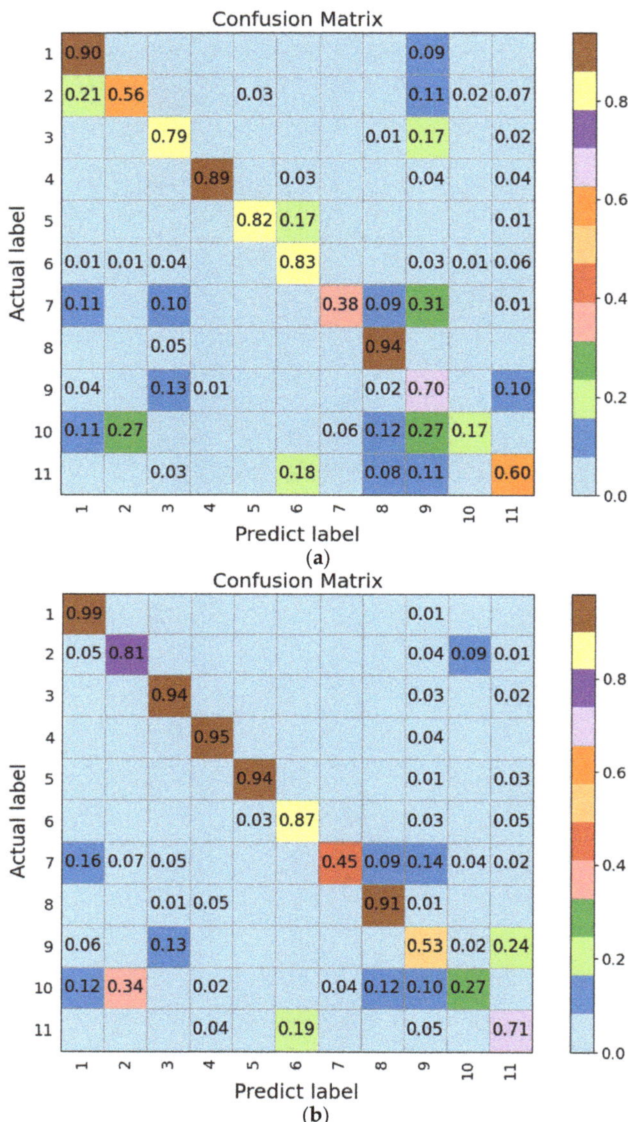

Figure 7. The confusion matrixes for YOWO (**a**) and MSRNet (**b**). The indexes of rows and columns from 1 to 11 represent: (1) drink from a bottle; (2) consuming food; (3) putting on or taking off a jacket; (4) working on a laptop; (5) reading a magazine; (6) reading a newspaper; (7) talking on a cellphone; (8) taking over the steering wheel; (9) putting on or taking off sunglasses; (10) watching videos; (11) writing with a pen.

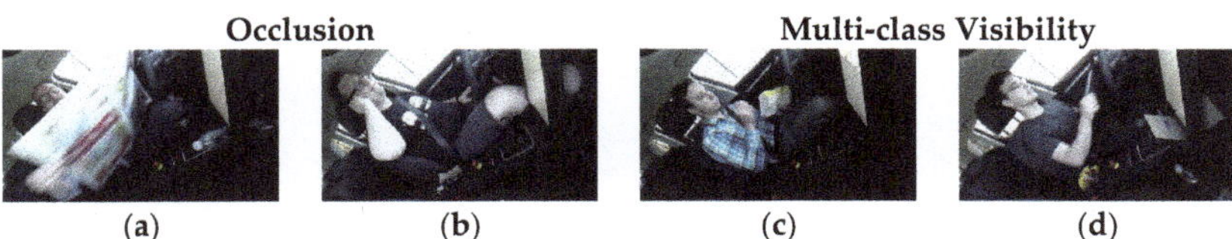

Figure 8. The examples of driver images for which MSRNet fails driver behavior recognition. The challenging situations are: (**a**) the newspaper covers the driver's upper body; (**b**) the cellphone is completely covered by the driver's hand; (**c**) the driver is consuming food while watching a video; (**d**) a bottle, a pen and food are all visible.

4. Conclusions

In this paper, an end-to-end multi-semantic model is proposed for driver behavior recognition, employing a confidence fusion mechanism known as MSRNet. First, the category of Drive&Act at the level of fine-grained activity is adapted to establish the clear relationships between behaviors and key-objects based on hierarchical annotations. This modification facilitates the design, training, and evaluation of the proposed model. Subsequently, MSRNet uses two parallel branches to perform action classification and object classification, respectively. ActNet mainly focuses on areas wherein movements are happening, whereas ObjectNet mainly focuses on key objects. The proposed confidence fusion mechanism aggregates the predictions from both branches based on the semantic relationships between actions and key-objects. The proposed approach can both characterize driver actions and focus on the key-objects linked with behaviors to obtain more complete descriptions of behaviors. Overall, this approach demonstrably improves the model's accuracy and robustness for driver behavior recognition. The experiments have demonstrated that the MSRNet performs better in terms of both validation and testing, with significant 4.65% (Val) and 3.16% (Test) improvements in accuracy when recognizing 11 different behaviors in an 8-frame input clip. The proposed model can perform accurate recognition for the majority of classes, such as 99% accurate identification of drinking from a bottle, 95% accurate identification of working on a laptop, and 94% accurate identification of reading a magazine.

Although the MSRNet shows superiority in solving the problem of interclass similarity, it also suffers from some limitations (e.g., occlusion and multi-class visibility) that degrade its performance. In future work, we would like to try other possible approaches to solving these limitations. As feature extraction from occluded human body parts is rarely possible, it is important to find robust classifiers that can handle the occlusion problem, such as probabilistic approaches. In addition, collecting additional sensor data (e.g., body pose, depth, and infrared) from other sensors mounted on real cars is a potential mitigation strategy. It is considered that this could help in deriving more complete descriptions of driver behavior.

Author Contributions: Conceptualization, H.R. and Z.B.; methodology, H.R. and Y.G.; software, H.R. and Y.G.; validation, H.R. and X.C.; formal analysis, H.R.; resources, H.R. and Z.B.; data curation, H.R.; writing—original draft preparation, H.R.; writing—review and editing, Y.G.; visualization, H.R. and X.C.; supervision, Y.G. and Z.B.; project administration, Z.B.; funding acquisition, Z.B. All authors have read and agreed to the published version of the manuscript.

Funding: This research was funded by Natural Science Foundation of Hunan Province, grant number 2020JJ4184.

Institutional Review Board Statement: Not applicable.

Informed Consent Statement: Not applicable.

Data Availability Statement: Detailed data are contained within the article. More data that support the findings of this study are available from the author R.H. upon reasonable request.

Acknowledgments: The authors appreciate the reviewers and editors for their helpful comments and suggestions in this study.

Conflicts of Interest: The authors declare no conflict of interest.

References

1. Singh, S. *Critical Reasons for Crashes Investigated in the National Motor Vehicle Crash Causation Survey*; National Highway Traffic Safety Administration: Washington, DC, USA, 2015.
2. Dingus, T.A.; Guo, F.; Lee, S.; Antin, J.F.; Perez, M.; Buchanan-King, M.; Hankey, J. Driver crash risk factors and prevalence evaluation using naturalistic driving data. *Proc. Natl. Acad. Sci. USA* **2016**, *113*, 2636–2641. [CrossRef] [PubMed]
3. Dindorf, R.; Wos, P. Analysis of the Possibilities of Using a Driver's Brain Activity to Pneumatically Actuate a Secondary Foot Brake Pedal. *Actuators* **2020**, *9*, 49. [CrossRef]
4. Naujoks, F.; Purucker, C.; Neukum, A. Secondary task engagement and vehicle automation–Comparing the effects of different automation levels in an on-road experiment. *Transp. Res. Part F Traffic Psychol. Behav.* **2016**, *38*, 67–82. [CrossRef]
5. International, S. *Taxonomy and Definitions for Terms Related to Driving Automation Systems for On-Road Motor Vehicles*; SAE; SAE International: Washington, DC, USA, 2018.
6. Ragab, A.; Craye, C.; Kamel, M.S.; Karray, F. A Visual-Based Driver Distraction Recognition and Detection Using Random Forest. In *Proceedings of the Transactions on Petri Nets and Other Models of Concurrency XV*; Springer Science and Business Media LLC: Berlin/Heidelberg, Germany, 2014; pp. 256–265.
7. Seshadri, K.; Juefei-Xu, F.; Pal, D.K.; Savvides, M.; Thor, C.P. Driver cell phone usage detection on Strategic Highway Research Program (SHRP2) face view videos. In Proceedings of the 2015 IEEE Conference on Computer Vision and Pattern Recognition Workshops (CVPRW), Columbus, OH, USA, 23–28 June 2014; IEEE: New York, NY, USA, 2015; pp. 35–43.
8. Liu, T.; Yang, Y.; Huang, G.-B.; Yeo, Y.K.; Lin, Z. Driver distraction detection using semi-supervised machine learning. *Ieee Trans. Intell. Transp. Syst.* **2015**, *17*, 1108–1120. [CrossRef]
9. Le, T.H.N.; Zheng, Y.; Zhu, C.; Luu, K.; Savvides, M. Multiple Scale Faster-RCNN Approach to Driver's Cell-Phone Usage and Hands on Steering Wheel Detection. In Proceedings of the 2016 IEEE Conference on Computer Vision and Pattern Recognition Workshops (CVPRW), Las Vegas, NV, USA, 26 June–1 July 2016; Institute of Electrical and Electronics Engineers (IEEE): New York, NY, USA, 2016; pp. 46–53.
10. Ren, S.; He, K.; Girshick, R.; Sun, J. Faster r-cnn: Towards real-time object detection with region proposal networks. *Adv. Neural Inf. Process. Syst.* **2015**, *28*, 91–99. [CrossRef] [PubMed]
11. Streiffer, C.; Raghavendra, R.; Benson, T.; Srivatsa, M. Darnet: A deep learning solution for distracted driving detection. In Proceedings of the 18th Acm/Ifip/Usenix Middleware Conference: Industrial Track, Las Vegas, NV, USA, 11–15 December 2017; pp. 22–28.
12. Baheti, B.; Gajre, S.; Talbar, S. Detection of Distracted Driver using Convolutional Neural Network. In Proceedings of the 2018 IEEE/Cvf Conference on Computer Vision and Pattern Recognition Workshops, Salt Lake City, UT, USA, 18–22 June 2018; pp. 1145–1151. [CrossRef]
13. Simonyan, K.; Zisserman, A. Very deep convolutional networks for large-scale image recognition. *arXiv* **2014**, arXiv:1409.1556.
14. Martin, M.; Roitberg, A.; Haurilet, M.; Horne, M.; ReiB, S.; Voit, M.; Stiefelhagen, R. Drive&Act: A Multi-Modal Dataset for Fine-Grained Driver Behavior Recognition in Autonomous Vehicles. In Proceedings of the 2019 IEEE/CVF International Conference on Computer Vision (ICCV), Seoul, Korea, 27 October–2 November 2019; Institute of Electrical and Electronics Engineers (IEEE): New York, NY, USA, 2019; pp. 2801–2810.
15. Reis, S.; Roitberg, A.; Haurilet, M.; Stiefelhagen, R. Activity-aware Attributes for Zero-Shot Driver Behavior Recognition. In Proceedings of the 2020 IEEE/CVF Conference on Computer Vision and Pattern Recognition Workshops (CVPRW), Seattle, WA, USA, 14–19 June2020; Institute of Electrical and Electronics Engineers (IEEE): New York, NY, USA, 2020; pp. 3950–3955.
16. Zhang, C.; Li, R.; Kim, W.; Yoon, D.; Patras, P. Driver behavior recognition via interwoven deep convolutional neural nets with multi-stream inputs. *Ieee Access* **2020**, *8*, 191138–191151. [CrossRef]
17. Wang, H.; Zhao, M.; Beurier, G.; Wang, X. Automobile driver posture monitoring systems: A review. *China J. Highw. Transp* **2019**, *2*, 1–18.
18. Wharton, Z.; Behera, A.; Liu, Y.; Bessis, N. Coarse Temporal Attention Network (CTA-Net) for Driver's Activity Recognition. In Proceedings of the 2021 IEEE Winter Conference on Applications of Computer Vision (WACV), New York, NY, USA, 5–9 January2021; IEEE: New York, NY, USA, 2021; pp. 1278–1288.
19. Köpüklü, O.; Wei, X.; Rigoll, G.J.A.P.A. You Only Watch Once: A Unified Cnn Architecture for Real-Time Spatiotemporal Action Localization. *ArXiv* **2019**, arXiv:1911.06644.
20. Hara, K.; Kataoka, H.; Satoh, Y. Can Spatiotemporal 3D CNNs Retrace the History of 2D CNNs and ImageNet? In Proceedings of the 2018 IEEE/CVF Conference on Computer Vision and Pattern Recognition, Salt Lake City, UT, USA, 8–23 June 2018; IEEE: New York, NY, USA, 2018; pp. 6546–6555.
21. Redmon, J.; Farhadi, A. YOLO9000: Better, Faster, Stronger. In Proceedings of the 2017 IEEE Conference on Computer Vision and Pattern Recognition (CVPR), Honolulu, HI, USA, 21–26 July 2017; IEEE: New, York, NY, USA, 2017; pp. 6517–6525.
22. Girshick, R.B. Fast R-CNN. In Proceedings of the 2015 IEEE International Conference on Computer Vision (ICCV), Santiago, Chile, 7–13 December 2015; pp. 1440–1448. [CrossRef]

23. Lin, T.-Y.; Goyal, P.; Girshick, R.B.; He, K.; Dollár, P. Focal Loss for Dense Object Detection. *IEEE Trans. Pattern Anal. Mach. Intell.* **2020**, *42*, 318–327. [CrossRef] [PubMed]
24. Redmon, J.; Farhadi, A. Yolov3: An incremental improvement. *arXiv* **2018**, arXiv:1804.02767.
25. Lin, T.-Y.; Dollar, P.; Girshick, R.; He, K.; Hariharan, B.; Belongie, S. Feature Pyramid Networks for Object Detection. In Proceedings of the 2017, IEEE Conference on Computer Vision and Pattern Recognition (CVPR), Honolulu, HI, USA, 21–26 July 2017; pp. 936–944. [CrossRef]
26. Redmon, J.; Divvala, S.; Girshick, R.; Farhadi, A. You only look once: Unified, real-time object detection. In Proceedings of the 29th IEEE Conference on Computer Vision and Pattern Recognition, Las Vegas, NV, USA, 7–30 June 2016; IEEE: New York, NY, USA, 2016; pp. 779–788.
27. Carreira, J.; Zisserman, A. Quo Vadis, Action Recognition? A New Model and the Kinetics Dataset. In Proceedings of the 2017 IEEE Conference on Computer Vision and Pattern Recognition (CVPR), Honolulu, HI, USA, 21–26 July 2017; pp. 4724–4733.
28. Tsung-Yi, L.; Maire, M.; Belonge, S.; Hays, J.; Perona, P.; Ramanan, D.; Dollár, P.; Zitnick, C.L. Microsoft coco: Common objects in context. In *European Conference on Computer Vision*; Springer: Cham, Switzerland, 2014; pp. 740–755.
29. Zhou, B.; Khosla, A.; Lapedriza, A.; Oliva, A.; Torralba, A. Learning deep features for discriminative localization. In Proceedings of the IEEE Conference on Computer Vision and Pattern Recognition, Las Vegas, NV, USA, 27–30 June 2016; pp. 2921–2929.

Article

Pressure Estimation Based on Vehicle Dynamics Considering the Evolution of the Brake Linings' Coefficient of Friction

Biaofei Shi [1,2], **Lu Xiong** [1,2,*] **and Zhuoping Yu** [1,2]

[1] School of Automotive Studies, Tongji University, Shanghai 201804, China; shi_biaofei@tongji.edu.cn (B.S.); yuzhuoping@tongji.edu.cn (Z.Y.)

[2] Institute of Intelligent Vehicle, Clean Energy Automotive Engineering Center, Tongji University, Shanghai 201804, China

* Correspondence: xiong_lu@tongji.edu.cn

Abstract: To mitigate the issue of low accuracy and poor robustness of the master cylinder pressure estimation (MCPE) of the electro-hydraulic brake system (EHB) by adopting EHB's own information, a MCPE algorithm based on vehicle information considering the evolution of the brake linings' coefficient of friction (BLCF) is proposed. First, the MCPE algorithm was derived combining the vehicle longitudinal dynamics and the wheel dynamics, in which the inertial measurement unit (IMU) was adopted to adapt the MCPE algorithm to road slope change. In order to estimate the brake pressure accurately, the driving resistance of the vehicle was obtained through a vehicle test under coasting condition. After that, with the active braking function of EHB, the evolution of the BLCF was acquired through extensive real vehicle test under different initial temperatures, different initial vehicle speeds, and different brake pressures. According to the test results, a revised model of the BLCF is proposed. Finally, the performance of the MCPE based on the revised BLCF model was compared with that based on a fixed BLCF model. Vehicle test demonstrates that the former MCPE algorithm is not only more accurate at low vehicle speed than the later, but also robust to road slope change.

Keywords: electro-hydraulic brake system; master cylinder pressure estimation; vehicle longitudinal dynamics; brake linings' coefficient of friction

Citation: Shi, B.; Xiong, L.; Yu, Z. Pressure Estimation Based on Vehicle Dynamics Considering the Evolution of the Brake Linings' Coefficient of Friction. *Actuators* **2021**, *10*, 76. https://doi.org/10.3390/act10040076

Academic Editor: Ronald M. Barrett

Received: 18 March 2021
Accepted: 6 April 2021
Published: 8 April 2021

Publisher's Note: MDPI stays neutral with regard to jurisdictional claims in published maps and institutional affiliations.

1. Introduction

With the development of electric and intelligent vehicles, the conventional brake system (i.e., vacuum booster) cannot meet the new demands any more, and the brake by wire system (BBW) came into being. BBW cannot only maximize the recovery of braking energy through coordinated control with the driven motor for electric vehicles, but for intelligent vehicles, it can also realize high-performance active braking, which is the development trend of automotive brake systems in the future [1,2]. As a branch of BBW, the electro-hydraulic brake system (EHB), which is based on a hydraulic system and activated by electric motors, is superior to the electro-mechanical brake system (EMB) in production inheritance and security reliability [3–12].

Pressure control is the core technology of EHB and has been extensively studied [13–16]. However, as far as the author knows, in addition to some research by the author's team [17–19], all the master cylinder pressure control algorithms in the existing literature adopted the master cylinder pressure sensor as the feedback signal for closed-loop control. The existence of the pressure sensor increased the cost and the risk of sensor failure. As one of the key safety components of automobiles, once the pressure sensor fails, the function of EHB will be seriously affected. Some products adopted two pressure sensors in the master cylinder for mutual inspection as a solution of failure detection and backup, which led to a further increase in cost [20]. For this reason, master cylinder pressure estimation (MCPE) is a promising solution to the above-mentioned problems.

In addition, the motor information of EHB (e.g., motor torque, motor rotational angle, etc.) can increase the possibility of MCPE.

The MCPEs in literatures were mainly based on the relationship between the master cylinder piston position (which can be obtained from the motor rotational angle and the transmission ratio of the reduction mechanism) and the master cylinder pressure. A first-order polynomial, a second-order polynomial, and a look-up table were used to render the pressure–position relationship in [16,21,22], respectively. However, due to the hysteresis and time-varying characteristics of the pressure–position relationship, the above methods were not accurate all the time. For this reason, the extended least squares and the recursive least squares were adopted to update the coefficients of the quadratic polynomial in [23,24], respectively. In [18], the coefficients were further reduced to one and updated by the recursive least squares with a fixed forgetting factor. Although the above algorithms can adjust the pressure–position model online, the coefficients of the polynomials fluctuate violently during the adaptive process due to the significant uncertainty of EHB (e.g., temperature, motor speed, brake pads wear, and so on). Once the pressure sensor fails, the values of coefficients at that moment are fixed for MCPE, resulting in an inaccurate pressure estimation with large uncertainty. Furthermore, if the EHB is only operated within a small pressure region, the pressure–position model may be over-fitted to this region. This would be a common occurrence in road vehicles since most instances of braking in daily driving involve only low decelerations. To this end, Ref. [25] proposed a bin-least-square algorithm, in which the measured (θ, P) data points were allocated according to the P value into n_b "bins", each of which corresponded to a small window of P. These windows were non-overlapping and distributed over the operating range of P as shown in Figure 1. The data points allocated to each bin were aggregated over time into single θ and P values; the (θ, P) pairs from all n_b bins were then imported to the least square algorithm. Although this method improved the stability of polynomial coefficients, it also deteriorated the accuracy of MCPE by reducing the sensitivity of polynomial coefficients to system pressure-position uncertainty.

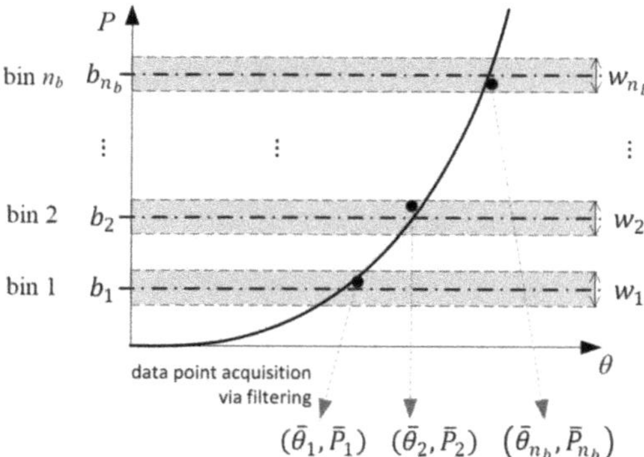

Figure 1. Schemes of the bin least square.

It is worth pointing out that all the above MCPE algorithms were based on the pressure sensors and cannot be used in EHB unequipped with pressure sensors. To this end, Ref. [26] proposed an interconnected pressure estimation method in which the key characteristic parameter of the pressure–position curve, namely, the nonlinearly parameterized perturbations, could be estimated via EHB's dynamics based on the LuGre friction model. The problem is that the friction itself in EHB is time-varying, and this method needs to be demonstrated through extensive real vehicle verifications in the future.

The above-mentioned methods considered only the actuator characteristics (e.g., pressure–position model, friction model of EHB) and depended on the model accuracy. Inspired by the wheel cylinder pressure estimation algorithms [27], Ref. [28] proposed a MCPE algorithm based on vehicle longitudinal dynamics and wheel dynamics for the first time. Real vehicle test demonstrated that the MCPE outperformed that proposed in [26]. However, the brake linings' coefficient of friction (BLCF) was regarded as constant. In fact, the BLCF is greatly affected by vehicle speed, brake pressure and brake linings' temperature [29].

Summarized by the above literature, the MCPE for EHB requires further improvement on accuracy and robustness, and the BLCF needs to be studied further. Two main contributions make this work distinctive from the previous studies: (1) the MCPE in [28] has been expanded in this article to adapt it to more working conditions, such as slope condition, based on inertial measurement unit (IMU), which is easily accessible for vehicles equipped with an electronic stability control system (ESC); (2) a revised model of BLCF is proposed based on extensive real vehicle tests, which contributes to a more accurate MCPE. The rest of this article is organized as follows. The vehicle platform and the EHB prototype under consideration are introduced in Section 2. The MCPE is proposed based on the longitudinal dynamics of the vehicle in Section 3. The driving resistance is tested through real vehicle tests in Section 4. The effect of different initial temperatures, different brake pressures, and different initial vehicle speeds on the BLCF is studied through extensive real vehicle tests, and a revised model of the BLCF is proposed in Section 5. Real vehicle tests under normal driving conditions, including flat road and slope road, are conducted to verify the proposed MCPE in Section 6. Section 7 concludes this article.

2. Test Vehicle and the EHB Prototype

2.1. Test Vehicle

A front-wheel-drive electric vehicle equipped with EHB is selected as the test vehicle, as shown in Figure 2. In this work, the regenerative brake function of the driven motor was invalid in braking, and all the braking force was supplied by the EHB. The test vehicle was equipped with the anti-lock brake system (ABS) and an additional IMU needed to be installed on the test vehicle.

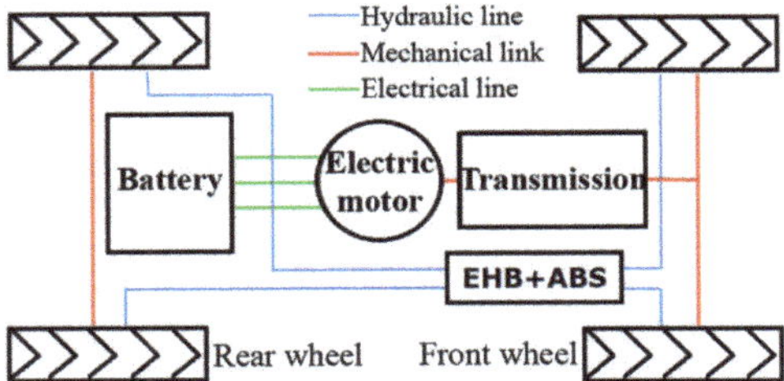

Figure 2. Schematic of the test vehicle.

The scheme of the vehicle platform is displayed in Figure 3; messages could be transferred from one node to another through controller area network (CAN). The electric control unit (ECU) of EHB received signals from the ABS, IMU, and sensors equipped in EHB and generated the control demand to the electric motor of EHB, which drove the reduction gear that directly pushed the master cylinder to generate pressure. A laptop was used for online calibration and observation via corresponding tools. There were two working modes of EHB: normal mode and X-by-wire mode. In normal mode, the EHB tracked the target pressure generated by the brake pedal, while in the X-by-wire mode, the EHB tracked the target pressure generated by the laptop.

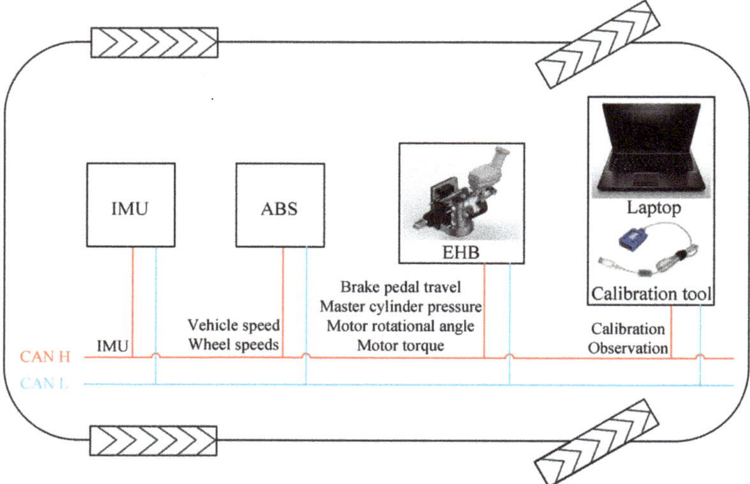

Figure 3. Scheme of the vehicle platform.

2.2. EHB Prototype

The scheme of EHB is shown in Figure 4. The EHB consisted of four parts: brake pedal unit, motor driven unit, brake execution unit, and ECU. The brake pedal unit, which included a brake pedal and a pedal feel simulator, provided the driver with a good pedal feel. The motor driven unit was the power source of the system, including a permanent magnet synchronous motor (PMSM) and reduction gear. The brake execution unit had the same structure as the conventional hydraulic brake system and included the master cylinder, the brake pipelines, and the ABS. A decoupling gap was designed to realize system decoupling, that is, the brake pedal and the master cylinder were not directly connected. In normal mode, the driver depressed the brake pedal, and the brake pedal rod compressed the pedal feel emulator to generate a brake feel. ECU analyzed the driver's braking intention according to the pedal stroke signal and controlled the PMSM to generate corresponding torque; therefore, there was no mechanical connection between the brake pedal and the master cylinder [12]. The pressure sensor was adopted as a feed-back signal for master cylinder pressure control and played no role in MCPE.

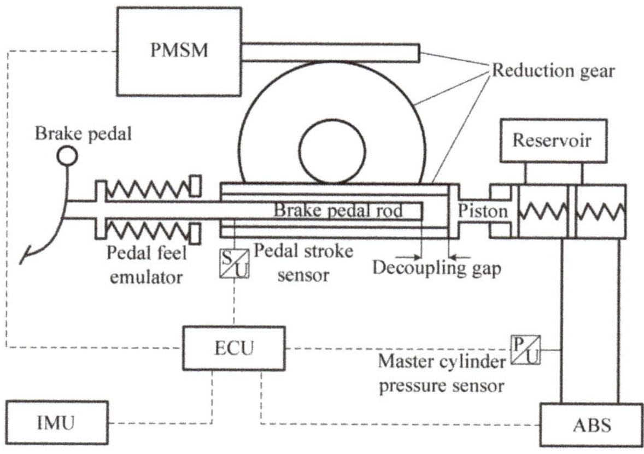

Figure 4. Scheme of the electro-hydraulic brake system (EHB).

3. MCPE Algorithm Design

3.1. Assumptions

In this work, the following assumptions are considered:

1. This work investigates the MCPE in the ordinary braking scenarios. The ABS must not work;
2. The master cylinder pressure is the same as the wheel cylinder pressure. In other words, the pressure wave propagation dynamics are ignored;
3. In the vehicle longitudinal dynamics, the longitudinal tire slip ratio is ignored;
4. The vehicle mass is assumed to be known or could be obtained by using the estimation method during the acceleration process [30];
5. All wheels share the same rolling radius, which is a reasonable assumption for most vehicles [28,31];
6. The vehicle lateral dynamics are ignored.

3.2. MCPE Based on Vehicle Longitudinal Dynamics

Under braking conditions, the vehicle longitudinal dynamics can be expressed by Equation (1) [32].

$$(m + m_\delta)a_x = F_b + F_f + F_w + F_i, \tag{1}$$

where m denotes the vehicle mass, kg; m_δ denotes the vehicle rotational mass, kg; a_x denotes the vehicle longitudinal deceleration, m/s^2; F_b denotes the braking force, N; F_f denotes the rolling resistance, N; F_w denotes the wind resistance, N; F_i denotes the slope resistance, N, as shown in Figure 5.

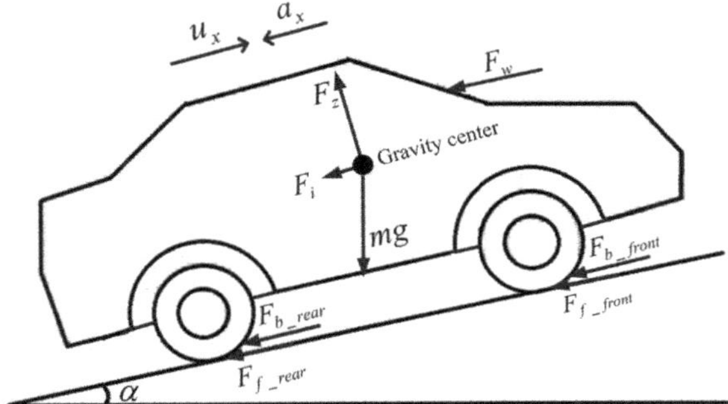

Figure 5. Scheme of the vehicle longitudinal dynamics.

The vehicle longitudinal dynamics of Equation (1) already include the wheel rotational dynamics so the braking force of each wheel can be expressed by Equation (2).

$$F_{bi} = \frac{T_{bi}}{r} \tag{2}$$

where the subscript i ($i = 1, 2, 3, 4$) denotes the front left wheel, the front right wheel, the rear left wheel, and the rear right wheel, respectively. T_{bi} denotes the braking torque of a certain wheel, $N \cdot m$; r denotes the wheel rolling radius, m, as shown in Figure 6.

The braking torque of each wheel can be expressed by Equation (3).

$$T_{bi} = pA_{wci}f_iR_{ei} \tag{3}$$

where p denotes the pressure of the hydraulic circuit, bar; A_{wci}, f_i and R_{ei} denote the wheel cylinder piston area, the BLCF, and the effective friction radius of each wheel, respectively.

Both A_{wci} and R_{ei} are constant. To simplify the problem, a new variable is defined to render the characteristic of BLCF as Equation (4).

$$k_i = A_{wci} f_i R_{ei} \tag{4}$$

where k_i denotes the pressure–torque factor of each wheel, Nm/bar.

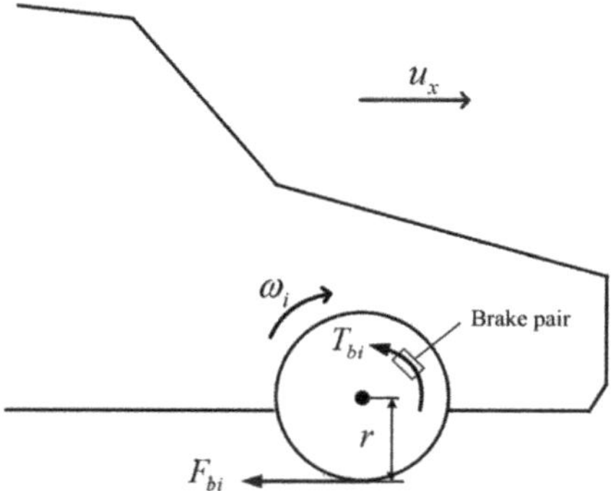

Figure 6. Scheme of the wheel rotational dynamics.

Substituting Equations (2)–(4) into Equation (1), the pressure can be calculated by Equation (5):

$$p = \frac{\left[(m + m_\delta)a_x - F_f - F_w - F_i\right] r}{\sum\limits_{i=1}^{4} k_i} \tag{5}$$

Some variables in Equation (5) can be further expressed as follows:

$$F_i = mg\sin\alpha \tag{6}$$

$$F_f = fmg\cos\alpha \tag{7}$$

where g denotes the acceleration of gravity, m/s^2; α denotes the road slope, rad; f denotes the rolling resistance coefficient.

The signal of the IMU can be expressed by Equation (8) according to its working principle.

$$a_{IMU} = -a_x + g\sin\alpha \tag{8}$$

A new variable is defined to render the equivalent characteristic of all the BLCFs as Equation (9).

$$K_e = \sum_{i=1}^{4} k_i \tag{9}$$

Substituting Equations (6), (8), and (9) into Equation (5), the MCPE algorithm can be expressed as follows:

$$p = \frac{\left(m_\delta a_x - ma_{IMU} - F_f - F_w\right) r}{K_e} \tag{10}$$

It should be noted that m_δ renders all the rotational mass of the vehicle, which mainly includes the wheels and the rotor of the driven motor, so it can be expressed by Equation (11).

$$m_\delta = \frac{\sum\limits_{i=1}^{4} J_{wi}}{r^2} + \frac{Jm i_g^2 \eta_T}{r^2} \tag{11}$$

where J_{wi} denotes the moment of inertia of a single wheel, $kg \cdot m^2$; J_m denotes the moment of inertia of the driven motor's rotor, $kg \cdot m^2$; i_g and η_T denote the transmission ratio and transmission efficiency of the vehicle transmission system. Table 1 provides the specifications of the test vehicle, in which all the vehicle parameters have been calibrated off line.

Table 1. Specifications of the test vehicle.

Item	Value
m	1580 kg
r	0.3183 m
J_{w1}, J_{w2}	0.67 kg·m^2
J_{w3}, J_{w4}	0.76 kg·m^2
i_g	7.65
J_m	0.053 kg·m^2
η_T	0.97

According to Equation (11) and Table 1, we see that $m_\delta/m = 3.7\%$, so the value of $m_\delta a_x$ is too small to be ignored compared to ma_{IMU}. Furthermore, the signal of a_x, which is obtained by vehicle speed or wheels speeds, is full of noise [28]. Considering the above reasons, $m_\delta a_x$ is ignored in Equation (11), and the MCPE algorithm is finally designed as follows:

$$p = \frac{\left(-ma_{IMU} - F_f - F_w\right)r}{K_e} \tag{12}$$

According to Equation (12), in order to estimate the brake pressure, we must first determine $\left(F_f + F_w\right)$, i.e., the driving resistance, and K_e, i.e., the sum of the pressure–torque factors of all wheels.

4. Driving Resistance

Driving resistance of the vehicle, including the rolling resistance and the wind resistance, can be expressed as follows:

$$F_d = F_f + F_w = fmgcos\alpha + F_w \tag{13}$$

The driving resistance is affected by the slope. In fact, the slope of normal road is not large, that is, $\cos \alpha \doteq 1$. Therefore, the influence of slope change on driving resistance is ignored.

Driving resistance is generally obtained through real vehicle tests. Under the coasting condition, the vehicle longitudinal dynamics can be expressed by Equation (14).

$$(m + m_\delta)a_x = F_f + F_w + F_i \tag{14}$$

Substituting Equation (8) into Equation (14) and ignoring $m_\delta a_x$, driving resistance can be acquired by Equation (15).

$$F_f + F_w = -ma_{IMU} \tag{15}$$

Generally, a special road is required to conduct the coasting test. Due to the limitation of test conditions, this article adopted the method of segmented testing, that is, the coasting test was broken down into multiple different vehicle speed segments to be tested separately, and finally, the test data are integrated and fitted by a quadratic polynomial, as shown in Figure 7.

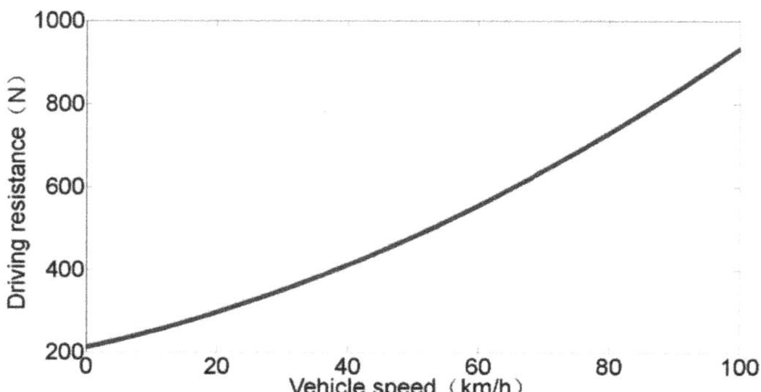

Figure 7. Vehicle driving resistance.

The analytical model of the driving resistance is shown in Equations (16) and (17).

$$F_f + F_w = A + Bu_x + Cu_x^2 \tag{16}$$

$$\begin{cases} A = 211.3 \\ B = 3.529 \\ C = 0.03681 \end{cases} \tag{17}$$

where u_x denotes the vehicle speed, km/h.

5. Revised Model of the BLCF

The BLFC has been widely studied in literature and is affected by several phenomena: fading [33,34]; bedding [35]; hysteresis against the pressure [36]; hysteresis against the speed [37], wear [38,39], and aging [35]; and variation in the environmental conditions [40]. The behavior of a pad–disc coupling is also dependent on the chemical composition and mechanical properties of each component [41]. Therefore, the BLCF can range between 0.3 and 0.6 [41,42], with peaks up to 0.8 and down to 0.1 [36,43].

There are mainly two methods in the literature to estimate the BLCF:

- Model based analytical approach: which strives for a physical understanding and analytical description of the friction behavior. Accurate BLCF models usually incorporate a temperature model, which is solved by the finite element method (FEM). However, owing to the high computational burden, it is not possible to use this approach for an online estimation of the brake temperature. For this reason, it is not feasible to use the FEM, along with a friction model, for estimation purposes [43–45].
- Neural-networks: which found an extensive application in friction modelling in recent years due to the capability of accurately modeling complex nonlinear phenomena with several inputs. The main limitation of the neural-network approach is the high experimental burden for the training/learning processes. Furthermore, the approach based on the neural-network is purely black box; therefore, it is not able to describe the actual phenomenology of the tribological contact [46–48].

The most recent research put forth a semi-empirical dynamic model of BLCF resulting from a thorough experimental campaign conducted on a brake dynamometer. The model rendered the rotor speed, rotor temperature, and contact area dynamics by means of a set of three differential equations and validated for three passenger cars' brake systems [29]. Though the state-of-the-art BLCF model can account for several tribological phenomena, parameter calibration requires lots of experiments.

As far as the author knows, all the above-mentioned methods are based on brake dynamometers, in other words, none of them are based on vehicle test. Furthermore, in normal braking conditions, the variation range of influencing factors of BLCF, such as temperature, may not be that large.

In this article, to estimate brake pressure, K_e needs to be identified. Although K_e is the sum of the pressure–torque factor of all wheels and not the same as BLCF of each wheel, K_e can render the equivalent characteristics of sum of the front and rear BLCFs for both A_{wci} and R_{ei}, which are constant. In this sense, K_e and BLCF have similarities in characteristics. Therefore, the characteristics of BLCF can also be used to explain and analyze the characteristics of K_e.

According to Equation (12), K_e can be measured by the following equation based on a vehicle test:

$$K_e = \frac{\left(-ma_{IMU} - F_f - F_w\right)r}{p} \tag{18}$$

where the brake pressure p can be obtained by pressure sensor. When p is 0, the above equation diverges so that the value of p is set to not less than 2 bar.

5.1. Error Analysis

There are two things to point out: (1) $m_\delta a_x$ is ignored in Equation (12). (2) $m_\delta a_x$ is also ignored in Equation (15) when identifying the driving resistance. That is, the ignored $m_\delta a_x$ is balanced in Equations (12) and (18). In other words, Equations (12) and (18) are the exact formula to calculate p and K_e, respectively.

Tables 2 and 3 provide the specifications of the IMU and the master cylinder pressure sensor, respectively.

Table 2. Specifications of the inertial measurement unit (IMU).

Item	Value
Measuring range	± 4.9 g
Sensitivity	0.00015 g
Accuracy	0.8%

Table 3. Specifications of the master cylinder pressure sensor.

Item	Value
Measuring range	0–300 bar
Sensitivity	0.125 bar
Accuracy	1%

It can be roughly calculated from the sensors' specifications that the error between the K_e calculated by Equation (18) and the actual value should be within $\pm 1.8\%$.

5.2. The Effect of Temperature on K_e

5.2.1. The Effect of Initial Disc Temperature on the Evolution of K_e

Although there are many factors affecting BLCF, the most important are the temperature, brake pressure, and vehicle speed [29]. During the braking process, kinetic energy of the vehicle is converted into heat, and the temperature of the friction pair rises sharply. For organic friction material, which is the most widely used in brakes at present, the BLCF increases first and then decreases with disc temperature. The turning point (critical temperature) varies with different specific ingredients and their ratio. The experimental results in Ref. [49] show that the critical temperature of BLCF under different Sb_2S_3 and $ZrSiO_4$ ratios ranges from 230 °C to 330 °C. Other literature shows that the critical temperature of BLCF is generally around 230 °C [29,50]. For the friction material of the test vehicle in this article, the author only knows that it is organic friction material, but the specific composition and ratio are difficult to find due to proprietary reasons.

In this work, a contact temperature sensor was adopted to measure the disc's temperature, as shown in Figure 8. When the vehicle was static, the probe of the temperature sensor was touched to the surface of the brake disc, and the temperature on the display instrument was stable in 3–5 s.

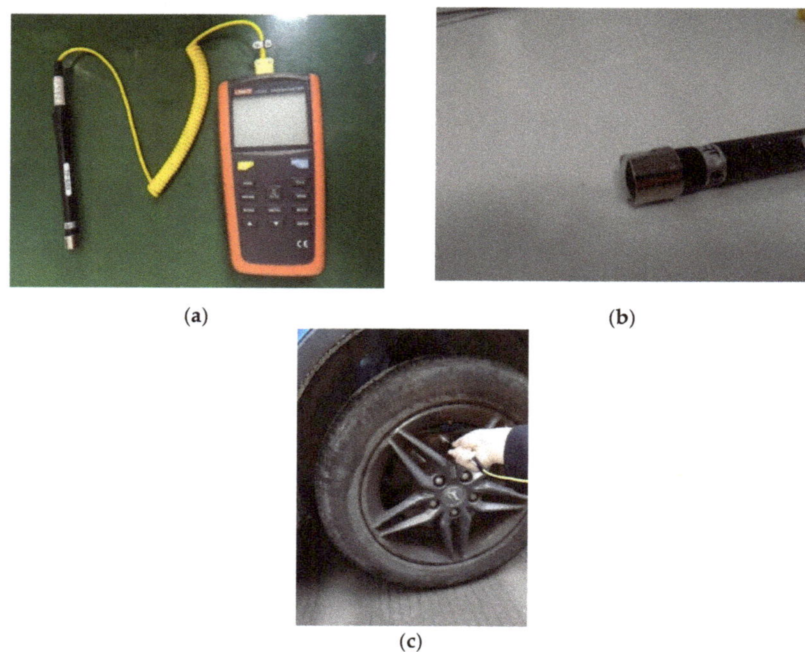

(a)

(b)

(c)

Figure 8. Picture of the contact temperature sensor: (**a**) Picture of the temperature sensor and display instrument; (**b**) Picture of the probe of the temperature sensor; (**c**) Picture of the temperature sensor being used.

The test process was as follows: when the vehicle was static, we, first, measured the temperature of the brake discs, then, accelerated the vehicle to a predetermined speed, and finally, braked. The temperature could only be measured when the vehicle was static. Therefore, we tried to speed up the vehicle as quickly as possible in the test to reduce the temperature change during this period. Six groups of tests with different initial temperatures of the brake discs were conducted, as shown in Table 4. Test results (i.e., evolution of K_e) are shown in Figure 9.

Table 4. Initial temperatures of the brake discs.

Group	Initial Temperatures of the Front Brake Discs (°C)	Initial Temperatures of the Rear Brake Discs (°C)
1	28	22
2	50	31
3	100	70
4	130	90
5	200	150
6	300	230

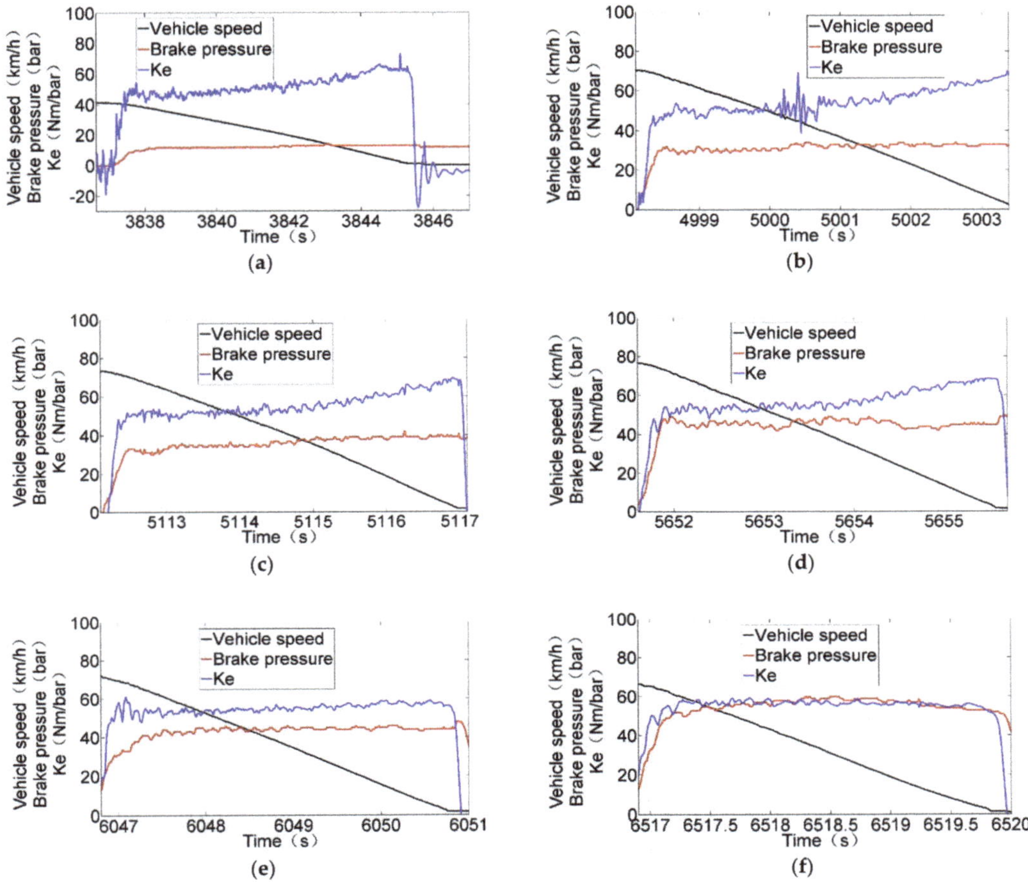

Figure 9. Evolution of K_e under different initial temperatures of the brake discs: (**a**–**f**) represent different initial disc temperatures of test groups 1–6, shown in Table 2, respectively.

In Figure 9a, K_e stayed around zero at the beginning when there was no brake pressure and quickly dropped and converged to a negative value after the vehicle speed was reduced to zero, which verifies the correctness of Equation (18) and the accuracy of driving resistance identification. When braking, K_e rose quickly and converged, indicating that there was a small delay between the brake pressure and the vehicle deceleration (50–100 ms). When braking under a constant pressure, K_e became larger and larger with time because the temperature of the friction pair rose sharply (but did not reach the critical temperature). In addition, the decrease in vehicle speed during braking also led to an increase in K_e, which was the so called "Stribeck" effect.

We heated the brake disc by repeated accelerations and brakings; the temperature of the disc was increased. Figure 9b–f show the constant pressure braking test with the initial vehicle speed of 65–80 km/h and the brake pressure of 35–60 bar, but the initial braking temperature is different. We can conclude that, when the initial temperature of the brake disc is within 130 °C, the temperature has little effect on the evolution of K_e, but the effect is greater when the temperature is above 200 °C, where the temperature of the brake pair reaches the critical value. In addition, the violent fluctuation between 5000 and 5000.5 s in Figure 9b was caused by the speed bump on the road.

5.2.2. Statistics of Initial Disc Temperature

Although the effect of temperature on K_e was studied in Section 5.2.1, the temperature of the brake disc is usually not very high in practice. Generally speaking, the thermal balance of the brake disc is maintained at about 100 °C during low-intensity braking, which is common in city driving conditions [48,50,51].

Additionally, this article recorded statistics of the front brake disc temperature at the end of several regular driving trips, as shown in Figure 10. Due to the different driving styles of the drivers, the most "prudent" driver and the most "adventurous" driver were selected for testing. The results are shown in Tables 5 and 6.

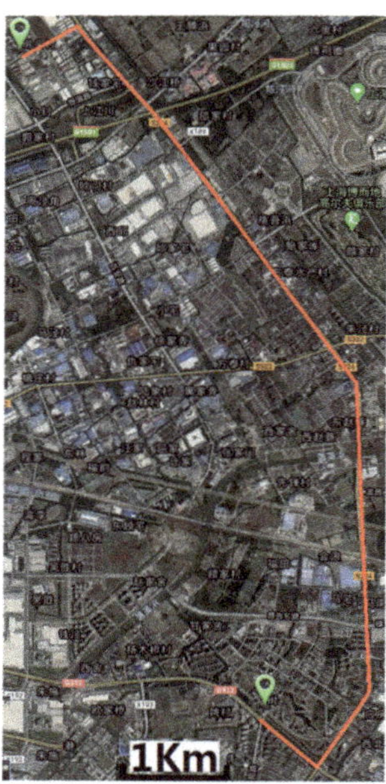

Figure 10. Test route in Google Maps.

Table 5. Static disc temperature under normal driving conditions (prudent driver).

Group	Ambient Temperature (°C)	Temperature of the Front Brake Disc (°C)
1	3	52
2	5	38
3	8	51
4	9	52
5	9	59
6	10	51
7	10	55
8	10	58
9	12	55
10	14	62
11	15	76
12	17	60

Table 6. Static disc temperature under normal driving conditions (adventurous driver).

Group	Ambient Temperature (°C)	Temperature of the Front Brake Disc (°C)
1	8	102
2	10	122
3	10	110
4	14	116
5	15	132
6	15	131
7	16	126
8	17	124

The temperature of the brake disc after each trip was related to the driving style, traffic condition, and ambient temperature. Most of the statistics were within 130 °C and the average was about 90 °C.

It can be concluded from the above that the influence of the initial disc temperature on the evolution of K_e can be ignored under normal driving conditions.

5.3. The Effect of Brake Pressure on K_e

The influence of brake pressure on the BLCF was related to the material of the friction pair, and specific tests were required. Ref. [52] pointed out that, for organic friction materials, BLCF first increases and then decreases with the increase of brake pressure; for powder metallurgy friction materials, BLCF decreases with the increase of braking pressure.

This article carried out tests with initial vehicle speed of 60 km/h and brake pressure of 10 bar, 20 bar, 30 bar, 40 bar, and 50 bar based on the X-by-wire function of EHB. The initial temperature of the brake disc was set to 90 °C each time at the beginning of the test. The test results are shown in Figure 11.

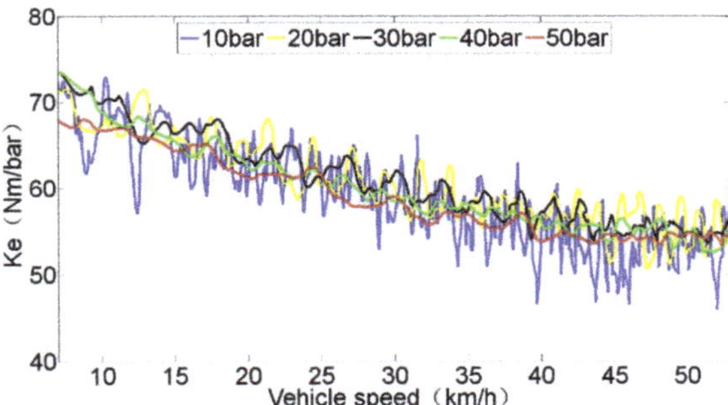

Figure 11. Evolution of K_e under different brake pressure.

From the perspective of the entire vehicle speed range, the average value of K_e first increased and then decreased with the increase of brake pressure, but the overall change was not large (especially within the range of normal brake pressure). Therefore, the effect of brake pressure on K_e is ignored in this article.

5.4. The Effect of Vehicle Speed on K_e

The BLCF was affected by the speed of the vehicle and obeyed the Stribeck characteristic [53–56], that is, the BLCF was greatly affected by speed.

Under normal driving conditions, the brake pressure was within 30 bar. Vehicle tests with the brake pressure of 15 bar and initial vehicle speed of 20 km/h, 40 km/h, 60 km/h, and 80 km/h were carried out based on the X-by-wire function of EHB with initial disc temperature of 90 °C. Test results are shown in Figure 12.

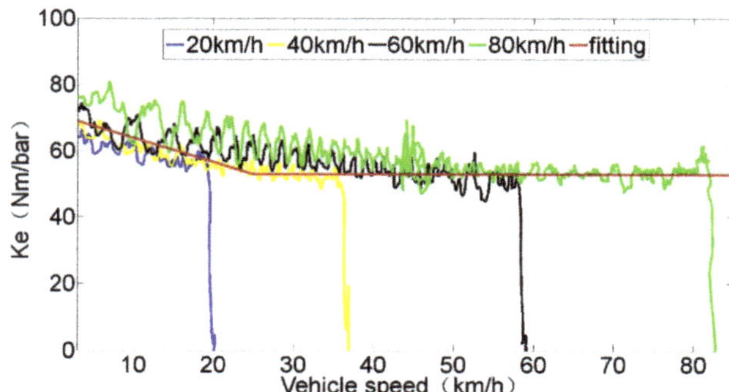

Figure 12. Evolution of K_e under different initial vehicle speeds.

K_e had a Stribeck effect with the vehicle speed and increased when the vehicle speed was under the critical speed. Specifically, the critical speed was about 30 km/h, 45 km/h and 60 km/h with the initial vehicle speed of 40 km/h, 60 km/h, and 80 km/h, respectively (phenomenon 1). The evolution of K_e with different initial vehicle speeds did not coincide. The greater the initial vehicle speed, the greater the K_e at the end of braking (phenomenon 2).

The explanation of the above two phenomena is that the temperature of the friction pair increased during braking (especially when the initial braking speed was high), which made the BLCF increase. The conclusion in Section 5.2 "The influence of the initial temperature on the evolution of K_e under normal driving conditions is negligible" is based on the condition "at the same initial vehicle speed". However, when the initial vehicle speed is different, due to the different braking temperature evolution in the process, even if the initial temperature is the same, the evolution of K_e will be different.

It should be noted that, in normal driving conditions, it is rare to decelerate the vehicle from 80 km/h to zero all at once. The more common situation is to decelerate the vehicle from "80 km/h to 60 km/h", from "60 km/h to 40 km/h", from "40 km/h to 20 km/h", and from "20 km/h to zero" in a braking process. From this point of view, the revised BLCF model was defined as a piecewise linear function according to the trend of K_e in the above several speed ranges. That is, when the vehicle speed was lower than a certain critical speed, K_e increased as the vehicle speed decreased; when the vehicle speed was above the critical speed, K_e was fixed, as shown in Equation (19).

$$K_e = \begin{cases} K_1 - \frac{K_1 - K_0}{u_0} u_x, & u_x \leq u_0 \\ K_0, & u_x > u_0 \end{cases} \tag{19}$$

where u_0 denotes the critical vehicle speed, km/h; K_1 denotes the K_e when the vehicle speed is zero, Nm/bar; K_0 denotes the K_e when the vehicle speed exceeds the critical speed, Nm/bar.

There was a certain degree of subjectivity when dividing the speed zone. In addition, defining the revised BLCF model as a piecewise linear function approximated the test results. Based on the above reasons, the three parameters in Equation (19) can be calibrated more accurately in real vehicle tests. The calibration result of this article is shown in Equation (20).

$$K_e = \begin{cases} 70 - \frac{70 - 53}{25} u_x, & u_x \leq 25 \\ 53, & u_x > 25 \end{cases} \tag{20}$$

6. MCPE Based on the Revised BLCF Model

6.1. Flat Road

The master cylinder pressure can be estimated by Equations (12) and (20). The MCPE was verified by real vehicle tests under normal driving conditions. In order to highlight the superiority of the revised BLCF model, it was compared with the fixed BLCF ($K_e \doteq 53$). Test results are shown in Figure 13. In the legend, "fixed BLCF" refers to "pressure estimated based on fixed BLCF model", and the legend "revised BLCF" refers to "pressure estimated based on revised BLCF model" in Figure 13.

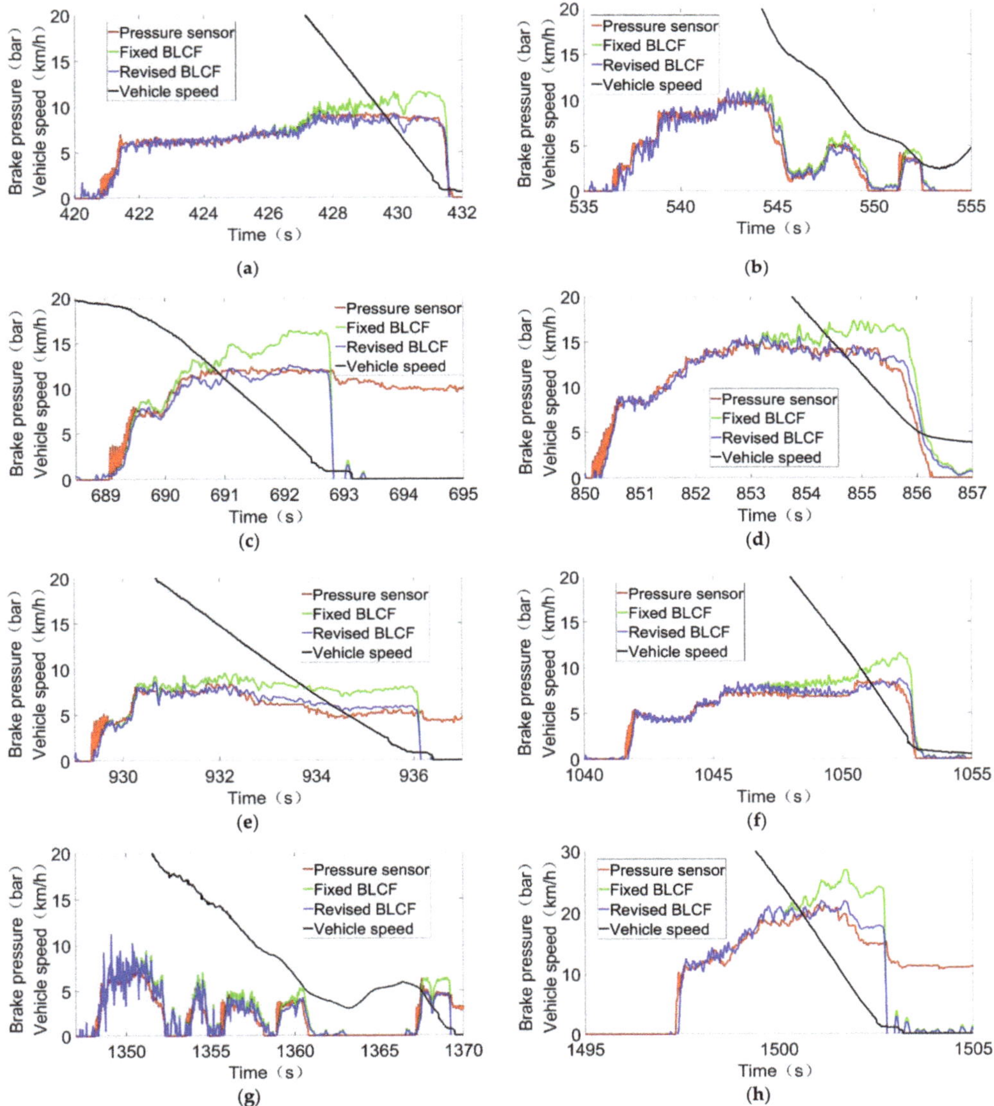

Figure 13. Test results of master cylinder pressure estimation (MCPE): (**a**–**h**) represent different brake conditions.

Thanks to the revised model of BLCF, the pressure estimation algorithm proposed in this article was much more accurate when the vehicle speed is low, and the root mean square error (RMSE) was 0.9182 bar. It was much smaller than 1.8248 bar of the MCPE with a fixed BLCF model.

6.2. Slope Road

Section 6.1 proves the superiority of the MCPE based on revised BLCF model, while this section tries to prove slope adaptability of the proposed method.

The MCPE algorithm, based on longitudinal deceleration, which ignored the road slope, can be expressed by Equation (21) as proposed in [28].

$$p = \frac{\left(ma_x - F_f - F_w\right)r}{K_e} \tag{21}$$

Vehicle tests were conducted on a road with slope of 5.5°; test results are shown in Figures 14 and 15.

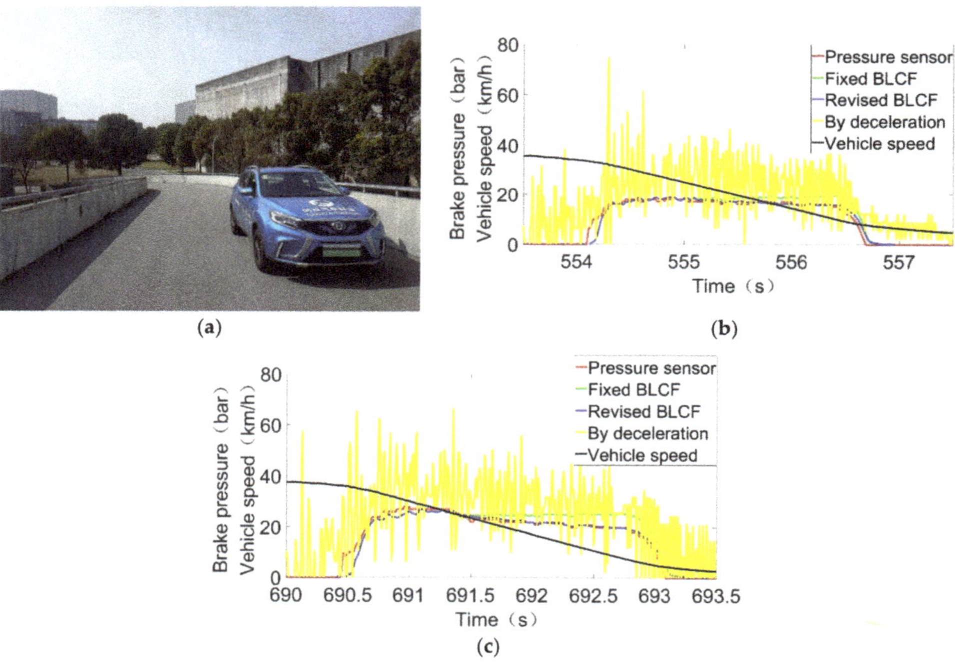

Figure 14. Test results of MCPE on the uphill: (**a**) Test vehicle on the uphill, (**b**,**c**) show the test results.

The error between the MCPE based on vehicle longitudinal deceleration and the actual pressure can be derived by comparing Equations (21) and (12):

$$p_{error} = \frac{mg\sin\alpha}{K_e}r \tag{22}$$

We could conclude from Equation (22) that the estimated pressure was higher than the actual pressure when the vehicle was going uphill and lower than the actual pressure when the vehicle was going downhill, which was consistent with the test results. Furthermore, the error was greatly affected by the slope; even a slope of 5.5° could cause a pressure estimation error of about 9 bar. In addition, the deceleration signal was obtained from the difference between the vehicle speeds at different times and fluctuated sharply, which led to a lot of noise in the estimated pressure.

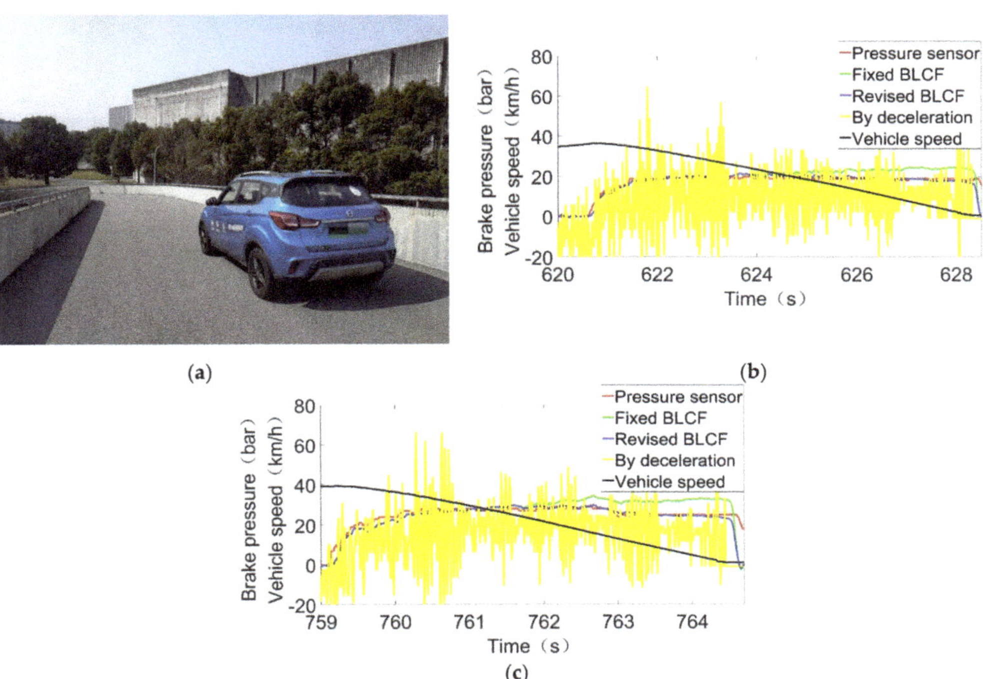

Figure 15. Test results of MCPE on the downhill: (**a**) Test vehicle on the downhill, (**b**,**c**) show the test results.

7. Conclusions

Aiming at the problems of low accuracy and poor robustness of the MCPE algorithm, based on EHB's own sensor information, a MCPE algorithm based on vehicle information is proposed. Compared with the existing literature, the innovation of this article lies in the fact that the BLCF is affected by temperature, brake pressure, and vehicle speed. Additionally, a revised BLCF model is proposed based on a thorough experimental campaign, which is finally verified by real vehicle tests. Compared with the MCPE based on a fixed friction factor, the accuracy is greatly improved. In addition, by adopting IMU information, pressure can be accurately estimated on slopes. In short, the proposed MCPE algorithm can provide EHB with an accurate, robust feedback signal that can be used for pressure control, which can save EHB costs and reduce the risk of pressure sensor failure.

Future works can further study how to integrate different pressure estimation algorithms, such as the MCPE proposed in this work and the MCPEs based on EHB's own information, to further improve the accuracy and robustness of the MCPE algorithm. Furthermore, the effect of the variability of disc thickness, block thickness, etc. on K_e and the MCPE, can be studied in future works.

Author Contributions: Conceptualization, B.S.; Data curation, B.S.; Formal analysis, B.S.; Funding acquisition, L.X. and Z.Y.; Investigation, B.S.; Methodology, B.S.; Project administration, L.X. and Z.Y.; Resources, B.S.; Software, B.S.; Supervision, L.X. and Z.Y.; Validation, B.S.; Visualization, B.S.; Writing—original draft, B.S.; Writing—review and editing, B.S. All authors have read and agreed to the published version of the manuscript.

Funding: This research was funded by the Research on Development of Electronic Control Chassis System and Active Control Technology (Grant No. 20511104601).

Institutional Review Board Statement: Not applicable.

Informed Consent Statement: Not applicable.

Data Availability Statement: Detailed data are contained within the article. More data that support the findings of this study are available from the author B.S. upon reasonable request.

Acknowledgments: The authors are thankful for the support of the School of Automotive Studies and the IIV (Institute of Intelligent Vehicle) of Tongji University and Tongyu Automobile Technology Co., Ltd.

Conflicts of Interest: The authors declare no conflict of interest.

Abbreviations

MCPE	Master cylinder pressure estimation
EHB	Electro-hydraulic brake system
BLCF	Brake linings' coefficient of friction
IMU	Inertial measurement unit
BBW	Brake by wire system
ESC	Electronic stability control system
ABS	Anti-lock brake system
CAN	Controller area network
ECU	Electric control unit
PMSM	Permanent magnet synchronous motor
FEM	Finite element method

References

1. Schenk, D.E.; Wells, R.L.; Miller, J.E. Intelligent Braking for Current and Future Vehicles. *Int. Congr. Expo.* **1995**. [CrossRef]
2. Jonner, W.; Winner, H.; Dreilich, L.; Schunck, E. Electrohydraulic Brake System-The First Approach to Brake-by-Wire Technology. *SAE Trans.* **1996**, *105*, 1368–1375. [CrossRef]
3. Reuter, D.F.; Lloyd, E.W.; Zehnder, J.W.; Elliott, J.A. Hydraulic Design Considerations for EHB Systems. *Braking Syst.* **2003**. [CrossRef]
4. D'alfio, N.; Morgando, A.; Sorniotti, A. Electro-hydraulic brake systems: Design and test through hardware-in-the-loop simulation. *Veh. Syst. Dyn.* **2006**, *44*, 378–392. [CrossRef]
5. Sorniotti, A.; Repici, G. Hardware in the Loop with Electro-Hydraulic Brake Systems. In Proceedings of the 9th WSEAS International Conference on Systems, Athens, Greece, 11 July 2005.
6. Duval-Destin, M.; Kropf, T.; Abadie, V. Effects of an Electric Actuator to the Brake System. *ATZ Automob. Tech. Mag.* **2011**, *113*, 638–643. [CrossRef]
7. Aoki, Y.; Suzuki, K.; Nakano, H.; Akamine, K.; Shirase, T.; Sakai, K. Development of Hydraulic Servo Brake System for Cooperative Control with Regenerative Brake. *SAE Tech. Pap.* **2007**. [CrossRef]
8. Leiber, T.; Köglsperger, C.; Unterfrauner, V. Modular brake system with integrated functionalities. *Auto. Tech. Rev.* **2012**, *1*, 24–29. [CrossRef]
9. Feigel, H.J. Integrated Brake System without Compromises in Functionality. *Atz. Worldw.* **2012**, *114*, 46–50. [CrossRef]
10. Xiong, L.; Yuan, B.; Guang, X.; Xu, S. Analysis and Design of Dual-Motor Electro-Hydraulic Brake System. *SAE Tech. Pap.* **2014**. [CrossRef]
11. Wang, Z.; Yu, L.; Wang, Y.; You, C.; Ma, L.; Song, J. Prototype of Distributed Electro-Hydraulic Braking System and Its Fail-Safe Control Strategy. *SAE Tech. Pap.* **2013**. [CrossRef]
12. Yu, Z.; Xu, S.; Xiong, L.; Han, W. An Integrated-Electro-Hydraulic Brake System for Active Safety. *SAE Tech. Pap.* **2016**. [CrossRef]
13. Liu, Q.H.; Rong, Y.C. Research on Electro-Hydraulic Hybrid Brake System Combined with ABS. *Appl. Mech. Mater.* **2014**, *543–547*, 1525–1528. [CrossRef]
14. Yang, X.; Li, J.; Miao, H.; Shi, Z.T. Hydraulic Pressure Control and Parameter Optimization of Integrated Electro-Hydraulic Brake System. *Brake Colloq. Exhib. Annu.* **2017**. [CrossRef]
15. Xiong, Z.; Pei, X.; Guo, X.; Zhang, C. Model-Based Pressure Control for an Electro Hydraulic Brake System on RCP Test Environment. *SAE Tech. Pap.* **2016**. [CrossRef]
16. Yong, J.W.; Gao, F.; Ding, N.G.; He, Y.P. Pressure-tracking control of a novel electro-hydraulic braking system considering friction compensation. *J. Cent. South Univ.* **2017**, *24*, 1909–1921. [CrossRef]
17. Han, W.; Xiong, L.; Yu, Z. Braking Pressure Tracking Control of a pressure sensor unequipped Electro-Hydraulic Booster based on a Nonlinear Observer. *SAE Tech. Pap. Ser.* **2018**. [CrossRef]

18. Han, W.; Xiong, L.; Yu, Z. Braking pressure control in electro-hydraulic brake system based on pressure estimation with nonlinearities and uncertainties. *Mech. Syst. Signal Process.* **2019**, *131*, 703–727. [CrossRef]

19. Han, W.; Xiong, L.; Yu, Z. Interconnected Pressure Estimation and Double Closed-Loop Cascade Control for an Integrated Electro-Hydraulic Brake System. *IEEE ASME Trans. Mechatron.* **2020**, *99*, 1. [CrossRef]

20. Ohkubo, N.; Matsushita, S.; Ueno, M.; Akamine, K.; Hatano, K. Application of Electric Servo Brake System to Plug-In Hybrid Vehicle. *SAE Int. J. Passeng. Cars* **2013**, *6*, 255–260. [CrossRef]

21. Todeschini, F.; Formentin, S.; Panzani, G.; Corno, M.; Savaresi, S.M.; Zaccarian, L. Deadzone compensation and anti-windup design for brake-by-wire systems. In Proceedings of the 2014 American Control Conference, Portland, OR, USA, 4–6 June 2014.

22. Zhao, J.; Hu, Z.; Zhu, B. Pressure Control for Hydraulic Brake System Equipped with an Electro-Mechanical Brake Booster. *SAE Tech. Pap. Ser.* **2018**. [CrossRef]

23. Todeschini, F.; Corno, M.; Panzani, G.; Savaresi, S.M. Adaptive position–pressure control of a brake by wire actuator for sport motorcycles. *Eur. J. Control* **2014**, *20*, 79–86. [CrossRef]

24. Todeschini, F.; Corno, M.; Panzani, G.; Fiorenti, S.; Savaresi, S.M. Adaptive Cascade Control of a Brake-By-Wire Actuator for Sport Motorcycles. *IEEE ASME Trans. Mechatron.* **2015**, *20*, 1310–1319. [CrossRef]

25. Ho, L.M.; Satzger, C.; de Castro, R. Fault-Tolerant Control of an Electrohydraulic Brake Using Virtual Pressure Sensor. In Proceedings of the 2017 International Conference on Robotics and Automation Sciences (ICRAS), Hong Kong, China, 26–29 August 2017; pp. 76–82.

26. Han, W.; Xiong, L.; Yu, Z. Pressure Estimation Algorithms in Decoupled Electro-Hydraulic Brake System Considering the Friction and Pressure-Position Relationship. *SAE Tech. Pap.* **2019**. [CrossRef]

27. Jiang, G.; Miao, X.; Wang, Y.; Chen, J.; Li, D.; Liu, L.; Muhammad, F. Real-time estimation of the pressure in the wheel cylinder with a hydraulic control unit in the vehicle braking control system based on the extended Kalman filter. *Proc. Inst. Mech. Eng. Part D J. Automob. Eng.* **2017**, *10*. [CrossRef]

28. Han, W.; Xiong, L.; Yu, Z.; Xu, S. Vehicle Validation for Pressure Estimation Algorithms of Decoupled EHB Based on Actuator Characteristics and Vehicle Dynamics. *WCX SAE World Congr. Exp.* **2020**. [CrossRef]

29. Ricciardi, V.; Travagliati, A.; Schreiber, V.; Klomp, M.; Ivanov, V.; Augsburg, K.; Faria, C. A novel semi-empirical dynamic brake model for automotive applications. *Tribol. Int.* **2020**, *146*, 106223. [CrossRef]

30. McIntyre, M.; Ghotikar, T.; Vahidi, A.; Song, X.; Dawson, D. A Two-Stage Lyapunov-Based Estimator for Estimation of Vehicle Mass and Road Grade. *IEEE Trans. Veh. Technol.* **2009**, *58*, 3177–3185. [CrossRef]

31. Ricciardi, V.; Acosta, M.; Augsburg, K.; Kanarachos, S.; Ivanov, V. Robust brake linings friction coefficient estimation for enhancement of ehb control. In Proceedings of the 2017 XXVI International Conference on Information, Communication and Automation Technologies (ICAT), Sarajevo, Bosnia and Herzegovina, 26–28 October 2017.

32. Mitschke, M.; Wallentowitz, H. *Dynamik der Kraftfahrzeuge*; Springer: Berlin/Heidelberg, Germany, 2004.

33. Balotin, J.; Neis, P.; Ferreira, N. Analysis of the influence of temperature on the friction coefficient of friction materials. *ABCM Symp. Ser. Mechatron.* **2010**, *10*, 774–785.

34. Talati, F.; Jalalifar, S. Analysis of heat conduction in a disk brake system. *Heat Mass. Transf.* **2009**, *45*, 1047–1059. [CrossRef]

35. Eriksson, M.; Bergman, F.; Jacobson, S. On the nature of tribological contact in automotive brakes. *Wear* **2002**, *252*, 26–36. [CrossRef]

36. Shyrokau, B.; Wang, D.; Augsburg, K.; Ivanov, V. Vehicle Dynamics with Brake Hysteresis. *Proc. Inst. Mech. Eng. Part D J. Automob. Eng.* **2013**, *227*, 139–150. [CrossRef]

37. Hess, D.; Soom, A. Friction at a lubricated line contact operating at oscillating sliding velocities. *J. Tribol.* **1990**, *112*, 147–152. [CrossRef]

38. Thuresson, D. Thermomechanical analysis of friction brakes. *SAE Tech. Pap.* **2000**, 149–159. [CrossRef]

39. Blau, P.J. Embedding wear models into friction models. *Tribol. Lett.* **2009**, *34*, 75–79. [CrossRef]

40. Chan, D.; Stachowiak, W. Review of automotive brake friction materials. *Proc. Inst. Mech. Eng. Part D J. Automob. Eng.* **2004**, *218*, 953–963. [CrossRef]

41. Nagesh, S.; Siddaraju, C.; Prakash, S.; Ramesh, M. Characterization of brake pads by variation in composition of friction materials. *Proc. Mater. Sci.* **2014**, *5*, 295–302. [CrossRef]

42. Kim, M. Development of the braking performance evaluation technology for high speed brake dynamometer. *Int. J. Syst. Appl. Eng. Dev.* **2012**, *6*, 122–129.

43. Ostermeyer, G. On the dynamics of the friction coefficient. *Wear* **2003**, *254*, 852–858. [CrossRef]

44. Wei-Yi, L.; Hecht Basch, R.; Li, D.; Sanders, P. Dynamic Modelling of Brake Friction Coefficients. *SAE Tech.* **2001**, *110*, 2627–2636.

45. Zhishuai, W.; Xiandong, L.; Haixa, W.; Yingchun, S. Research on the Time-Varying Properties of Brake Friction. *IEEE Access* **2018**, *6*, 69742–69749.

46. Frangu, L.; Ripa, M. Artificial Neural Networks Applications in Tribology—A Survey. In Proceedings of the Advanced Study Institute on Neural Networks for Instrumentation, Measurement, and Related Industrial Applications: Study Cases, Crema, Italy, 9–20 October 2001.

47. Argatov, I. Artificial Neural Networks (ANNs) as a Novel Modeling Technique in Tribology. *Front. Mech. Eng.* **2019**, *5*, 1–9. [CrossRef]

48. Aleksendric, D.; Barton, D. Neural Network Prediction of Disc Brake Performance. *Tribol. Int.* **2009**, *42*, 1074–1080. [CrossRef]

49. Jang, H.; Kim, S.J. The effects of antimony trisulfide (Sb_2S_3) and zirconium silicate ($ZrSiO_4$) in the automotive brake friction material on friction characteristics. *Wear* **2000**, *239*, 229–236. [CrossRef]
50. Chu, L.; Ma, W.; Cai, J.; Liu, D.; Zhang, Y.; Wei, W. Realtime Disc Brake Temperature Model Based on Vehicle Speed. *Automot. Eng.* **2016**, *38*, 61–64.
51. Cirovic, V.; Aleksendric, D. Dynamic Modeling of Disc Brake Contact Phenomena. *FME Trans.* **2020**, *39*, 177–183.
52. Xiao, X.; Yin, Y.; Bao, J.; Lu, L.; Feng, X. Review on the friction and wear of brake materials. *Adv. Mech. Eng.* **2016**, *8*, 8. [CrossRef]
53. Jang, H.; Lee, J.S.; Fash, J.W. Compositional effects of the brake friction material on creep groan phenomena. *Wear* **2001**, *251*, 1477–1483. [CrossRef]
54. Senatore, A.; D'Agostino, V.; Di Giuda, R.; Petrone, V. Experimental investigation and neural network prediction of brakes and clutch material frictional behaviour considering the sliding acceleration influence. *Tribol. Int.* **2011**, *44*, 1199–1207. [CrossRef]
55. Oumlktem, H.; Uygur, I. Advanced Friction–Wear Behavior of Organic Brake Pads Using a Newly Developed System. *Tribol. Trans.* **2019**, *62*, 51–61.
56. Martínez, C.M.; Velenis, E.; Tavernini, D.; Gao, B.; Wellers, M. Modelling and estimation of friction brake torque for a brake by wire system. In Proceedings of the 2014 IEEE International Electric Vehicle Conference (IEVC), Florence, Italy, 17–19 December 2014. [CrossRef]

Article

Mono-Vision Based Lateral Localization System of Low-Cost Autonomous Vehicles Using Deep Learning Curb Detection

Junwei Yu * and Zhuoping Yu

School of Automotive Studies, Tongji University, Shanghai 200092, China; yuzhuoping@tongji.edu.cn
* Correspondence: 13818102200@139.com

Abstract: The localization system of low-cost autonomous vehicles such as autonomous sweeper requires a highly lateral localization accuracy as the vehicle needs to keep a near lateral-distance between the side brush system and the road curb. Existing methods usually rely on a global navigation satellite system that often loses signal in a cluttered environment such as sweeping streets between high buildings and trees. In a GPS-denied environment, map-based methods are often used such as visual and LiDAR odometry systems. Apart from heavy computation costs from feature extractions, they are too expensive to meet the low-price market of the low-cost autonomous vehicles. To address these issues, we propose a mono-vision based lateral localization system of an autonomous sweeper. Our system relies on a fish-eye camera and precisely detects road curbs with a deep curb detection network. Curbs locations are then referred to as straightforward marks to control the lateral motion of the vehicle. With our self-recorded dataset, our curb detection network achieves 93% pixel-level precision. In addition, experiments are performed with an intelligent sweeper to prove the accuracy and robustness of our proposed approach. Results demonstrate that the average lateral distance error and the maximum invalid rate are within 0.035 m and 9.2%, respectively.

Keywords: curb detection; intelligent vehicles; autonomous driving

Citation: Yu, J.; Yu, Z. Mono-Vision Based Lateral Localization System of Low-Cost Autonomous Vehicles Using Deep Learning Curb Detection. *Actuators* **2021**, *10*, 57. https://doi.org/10.3390/act10030057

Academic Editors: Peng Hang, Xin Xia and Xinbo Chen

Received: 21 January 2021
Accepted: 8 March 2021
Published: 11 March 2021

Publisher's Note: MDPI stays neutral with regard to jurisdictional claims in published maps and institutional affiliations.

1. Introduction

Over the few years, autonomous vehicles have gained a lot of attention and witnessed remarkable progress. Companies such as Google and Tesla have the same goal toward fully L5 self-driving cars although they differ in approach from a design and engineering philosophy. Despite great success achieved by these companies, it will still take a rather long time before autonomous cars are widespread on the public roads in any weather and under any condition. As a result, there are a bunch of universities and companies concentrating on developing low-speed and low-cost autonomous vehicles that run in a limited, tightly controlled environment.

Among all the fundamental components (e.g., perception, decision-making, motion planning and localization) in the field of autonomous vehicle, localization is one of the most important and challenging problems. There are always inevitable contradictions of the highly precise localization systems and low-cost hardware requirements, especially for a low-cost autonomous vehicle such as a sweeper.

The easiest way to obtain the location of vehicles is using a global navigation satellite system (GNSS) with an inertial navigation system (INS), which is widely used for autonomous vehicles running in an open area such as on a highway [1,2]. The drawbacks are obvious. Firstly, the cost of a highly accurate GNSS/INS system is almost of equal value to a low-cost vehicle, which is certainly unacceptable. In addition, in a cluttered environment such as streets inside high buildings and trees, or in a GPS-denied environment such as a parking garage, GNSS signals are not feasible. To overcome this problem, several map-based methods are developed, where the features extracted from the environments using LiDARs, cameras, or other sensors are matched to the HD digital map to aid localization [3–11]. Apart from heavy computation costs from feature extraction and data association, they are too expensive to meet the low-price requirements of the low-cost autonomous vehicles.

To address these issues, we propose a fish-eye mono-vision based lateral localization system of an autonomous sweeper, which is highly efficient, low cost, and less complex compared to existing solutions. The framework is illustrated in Figure 1. Our system relies on a monocular fish-eye camera and precisely detects the road curbs with our proposed deep learning model. Curb locations are then referred as straightforward marks to control the lateral motion of the autonomous vehicle. At the heart of our work, we propose a deep curb detection network which serves as a key component to ensure a near lateral-distance (e.g., 0.2 m) between the side brush system and the sweeping road curb.

It is worth noting that, although curb detection is a traditional problem in the field of autonomous vehicles [12–16], most of the existing works utilize a front-facing camera or 3D LiDAR to detect curbs and search the road boundary, and further segment the travelable regions. They differ from our work in three aspects. Firstly, we are using a fish-eye side-facing camera to detect the road curbs while they often use a front-facing camera to detect curbs and lanes to segment the travelable region. Secondly, travelable region segmentation requires a pretty-low accuracy of curb detection as vehicles are often far away from the road boundary. Thirdly, for the expensive LiDAR-based method or stereo camera-based method, curbs are often detected based on strong assumptions such as the height of road curbs. These methods are generally not applicable to detect the curbs without obvious geometric features.

In contrast, our deep curb detection network is designed for a high-precision localization system of an autonomous sweeper. Our network consists of three key modules: a road scene classification module acts as the pre-processing procedure to classify the images as *Scene with Obstacles, Scene with Curbs or Scene with Intersection*. A curb region of interest (CRoI) module is utilized to obtain the curb region of interest. Subsequently, a semantic segmentation module is developed to accurately segment the curbs in CRoI. We combined U-Net [17] and SCNN [18] into our model. U-Net has an excellent performance in semantic segmentation problems and the slice-by-slice convolution in SCNN helps to make better use of spatial information. We evaluate our deep curb detection with a self-recorded dataset and achieve 93% pixel-level precision. Apart from offline experiments, we also perform online experiments on the autonomous sweeper developed at Tongji University, and compare our mono-vision based lateral localization system with the LiDAR-based localization system. The experiment results demonstrate that the average lateral distance error and the maximum invalid rate are within 0.035 m and 9.2%, respectively, and thus our system gets a better performance in terms of the robustness of the localization system compared to the LiDAR-based method.

The main contributions of this work are as follows:

- A mono-vision based lateral localization system of a low-cost autonomous vehicle is developed. Our system relies on a monocular fish-eye camera that is much cheaper than LiDARs.
- A CRoI module is proposed to obtain the curb region of interest to crop the image for the following semantic segmentation module, which improves the efficiency of the proposed method.
- We propose a novel semantic segmentation module based on the combination of U-Net and SCNN.

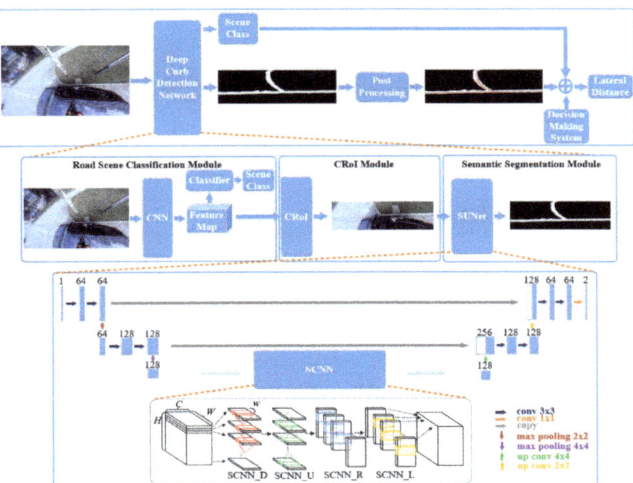

Figure 1. Hierarchical structure of the proposed mono-vision based lateral localization system. The upper layer (the first row) is the lateral localization system. Depending on the curb detection results, the vehicle calculates the lateral distance between itself and the nearest road curb which is then sent to the vehicle control system. The middle layer (the second row) is the core component of our system, which is a deep curb detection network. It consists of three important modules: road scene classification module, curb region of interest module, and semantic segmentation module. The bottom layer (the third row) shows the network architecture of our semantic segmentation module, which is built based on U-Net [17] and SCNN [18].

2. Related Works

The localization system is one of the most important components in the autonomous vehicle. Many efforts have been invested in this research topic. One of the most common solutions is using GNSS/INS. However, the accuracy of the traditional GNSS/INS method cannot meet the requirements of autonomous vehicles in cluttered environments and GPS-denied scenarios. To improve accuracy and robustness of localization, several map-based methods are proposed. For example, in [6], vertical corner features are extracted from the scan data of 3D LiDAR and then matched to pre-built corner map to correct the vehicle position. Similarly, the framework proposed in [5] adopts semantic and distinctive physical objects such as trees, traffic signs, or street lamps as landmarks and the vehicle pose is obtained via the combination of these features and an offline map.

Except for the global localization, lateral localization is also a research focus because of its remarkable assistance for localization. In [7], lateral and orientation information of lane markings is extracted from a video camera to enhance lateral localization accuracy. In [8], two lateral cameras are used to detect road markings and provide the lateral distance between the vehicle and the road borders for lane change. In addition, the information from the camera is combined with a digital map of the road markings to aid the localization from GNSS/INS. In [9], an algorithm which produces the distance of the vehicle to the left and right boundaries of the road-lane is presented. Then, the detected lane-markings are used as measurements for a Bayes filter to obtain the lateral position of the vehicle.

In addition to lane markings, road curb is another important feature for the improvement of lateral localization. In [10], a curb detection algorithm using 3D-LiDAR is performed, and the detection result matches the high-precision map. Then, the map matching result is fused with the localization of GPS and INS via a Kalman filter. In [13], the point cloud data from a 3D-LiDAR sensor are processed to distinguish on-road and off-road areas. A subsequent sliding-beam method can segment the road, then the position of curbs is obtained via a search-based method for each road segment. In [11], curb detection results obtained from a 3D LiDAR are adopted to correct the lateral errors in localization from GNSS/IMU/DMI(Distance Measuring Instruments). In [19], a deep learning-based method is used to detect visible curbs and occluded curbs. In [20], a Conditional Random Field (CRF) is used to assign the 3D points measured by stereo camera to different parts of a 3D environment model in order to reconstruct the surfaces and in particular the curb. In [21], an ultrasonic sensor-based method is proposed for curb detection. However, the detection method has requirements for the height of the curb and can not perform detection on curbs with low height.

Different from the works mentioned above, a fish-eye mono-vision based lateral localization system of an autonomous sweeper is developed in this paper, which is highly efficient, low cost, and less complex compared to existing solutions. Our system relies on a monocular fish-eye camera and precisely detects the road curbs with our proposed deep learning model. Curb locations are then referred to as straightforward marks to control the lateral motion of the autonomous vehicle.

3. Mono-Vision Based Lateral Localization System

In this work, a fish-eye mono-vision based lateral localization system of a low-cost autonomous vehicle is proposed. A hierarchical structure of the framework is shown in Figure 1. The upper layer of our framework (the first row of Figure 1) depicts the overall work-flow of the lateral localization system. Depending on the curb detection results, the vehicle calculates the lateral distance between itself and the nearest road curb, which is then sent to the vehicle control system. The middle layer of our framework (the second row of Figure 1) is the core component of our system which is a deep curb detection network. It consists of three important modules: road scene classification module, curb region of interest module, and semantic segmentation module. The road scene classification module classifies the road scenes into three classes: *Scene with Obstacles*, *Scene with Curbs* and *Scene with Intersection*. The CRoI module detects the interested region of curbs. As the input image of the semantic segmentation module shrinks, it improves the following semantic segmentation module's efficiency. Our semantic segmentation module is built based on U-Net [17] and SCNN [18].

The overall lateral localization system is described in Algorithm 1. A road scene image recorded by our side-facing fish-eye camera is firstly entered into our system. The road scene classification module outputs the class label. In the case of the label *Scene with Obstacles*, there will be no further processing procedure such as semantic segmentation. An obstacle encountering message is transmitted to the decision-making system of the autonomous vehicle. In the case of the label *Scene with Curbs* and *Scene with Intersection*, the CRoI is firstly detected, and then a precise segmentation result of the CRoI is obtained. The road curbs' locations are extracted from the segmentation results. For the road scene classified as *Scene with Curbs*, a curve is fitted based on the curbs' locations. For the road scenes classified as *Scene with Intersection*, there are three possibilities: (1) if the vehicle goes forward, then we fit a straight line with curbs; (2) if the vehicle turns right, then we fit a right-turn curve with curbs; (3) if the vehicle turns left, the localization will be based on the low-cost GPS and the lateral localization accuracy will be less important in this case.

4. Deep Curb Detection Network

In this section, we describe the deep curb detection network which is the core component of the proposed mono-vision based lateral localization system. It consists of three modules: road scene classification module, CRoI module, and semantic segmentation module.

Algorithm 1 Lateral localization system

Require: road scene image
1: Enter the image into curb detection network, obtain the *road scene label* and segmentation results;
2: **if** *road scene label* is *Scene with Obstacles* **then**
3:     Transmit an obstacle encountering message to the decision-making system of the vehicle;
4: **else if** *road scene label* is *Scene with Curbs* **then**
5:     Fit a curve based on the curbs' locations;
6: **else if** *road scene label* is *Scene with Intersection* **then**
7:     The decision-making system decide to go forward or turn right;
8:     **if** The vehicle goes forward **then**
9:         Fit a straight line;
10:     **else if** The vehicle turns right **then**
11:         Fit a right-turn curve;
12:     **else**The vehicle turns left
13:         Use low-cost GPS for localization.
14:     **end if**
15: **end if**

4.1. Road Scene Classification

The road scene classification model acts as a pre-processing procedure of the lateral localization system. It also serves as a basic module for the following CRoI module and semantic segmentation module. We annotate each road scene image recorded by the side-facing fish-eye camera with three labels as *Scene with Obstacles*, *Scene with Curbs* and *Scene with Intersection* (see Figure 2). The road scene classification model is implemented with a pre-trained convolutional neural network VGG-16 [22]. The feature map generated by the VGG-16 model is also passed to the CRoI module. The classified road scene is not only used by the lateral localization system but also transmitted to the motion planning system of the autonomous vehicle.

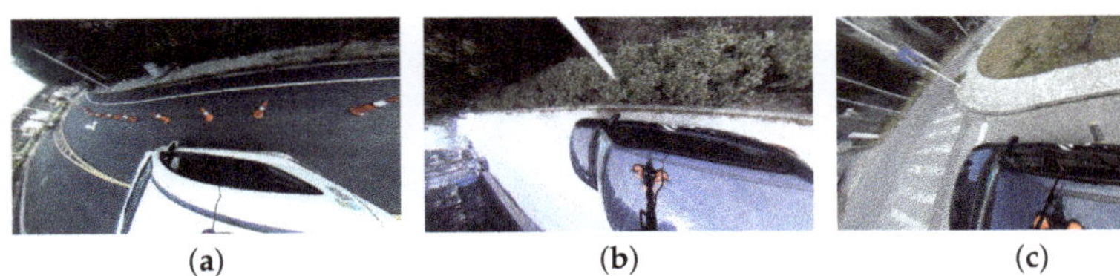

(a)   (b)   (c)

Figure 2. Road scene samples of three classes: (**a**) *road scene with obstacles*; (**b**) *road scene with curbs*; (**c**) *road scene with intersection*.

4.2. CRoI

As shown in Figure 2, curbs only occupy a long and narrow region of the full road scene pictures. To speed up the semantic segmentation module, we decide to detect the curb region of interests (CRoI) before further processing. Inspired by the region proposals widely used by object detection networks [23–25], we develop our CRoI module based on region proposal networks (RPN) introduced in Faster-RCNN [25]. The difference between our CRoI module and other RPN networks is that our CRoI module only needs to generate one curb region proposal that is fast and effective. It is worth noting that the road scene classification module is different from the classification step in Fast-RCNN. The latter cannot substitute the former because the road scene classification module we utilize in this work pays more attention to the global description of a road scene image. Additionally, the VGG-16 model used in the road scene classification module shares the same feature map with the RPN network in the CRoI module to avoid redundant computations. As illustrated in the second row of Figure 1, following the pre-trained CNN, there are two parallel branches: feature map to road scene classifier and to the CRoI module.

4.3. Semantic Segmentation

The semantic segmentation module is designed to precisely segment curbs from the CRoI module. We propose our SUNet segmentation model that is a combination of the U-Net [17] and SCNN [18]. The structure of the SUNet can be seen in the third row of Figure 1. Thanks to the contracting path between high resolution features and the upsampled output in U-Net, the successive convolution layer can get both information with a large receptive field from deeper layers and detailed information from the shallower layer. Consequently, U-Net has shown excellent performance in semantic segmentation problems. Therefore, we choose it as the backbone of our semantic segmentation module. The depth of the original U-Net is reduced to reduce the computational complexity.

As mentioned above, a curb is generally a long and narrow structure. The appearance clues are relatively less and the curbs are often interrupted and occluded. Fortunately, the curb is a highly structured feature and a strong spatial relationship exists between curb pixels. Intuitively, if the spatial information of curb can be better utilized, the algorithm should achieve better performance. Based on the above considerations, we adopt the SCNN structure, which is proposed in [18]. In the SCNN structure, traditional layer-by-layer convolutions are replaced by slice-by-slice convolutions within feature maps. As shown in the third row in Figure 1, SCNN_D, SCNN_U, SCNN_R, and SCNN_L represent four directions that slice-by-slice convolutions are applied: downward, upward, rightward, leftward. For instance, in SCNN_D, the feature map with size $C \times H \times W$ is split into H slices. The first slice is sent into a convolution layer with kernels of size $C \times w$, and the output is added to the next slice to generate a new slice. This process continues until the bottom slice. The processing procedure of other modules can be learned by analogy. Consequently, the spatial information can be propagated across rows and columns in a layer so that the structure is particularly suitable for structured objects like curbs. As mentioned in [18], SCNN can be flexibly applied to any place of a network. Generally, it should be added after a layer that contains richer information. Thus, we choose to apply SCNN at the bottom of U-Net. It is found that the computational efficiency of SCNN is highly dependent on the size of its input layer. In order to reduce computing time, a max pooling layer is added before the SCNN to reduce the size of the input layer.

5. Experiments

This section describes the details and results of experiments of our mono-vision based lateral localization system. The experiments are divided into two parts, offline experiments with the self-recorded dataset and online experiments with an autonomous sweeper. We evaluate the performance of our deep curb detection network with offline experiments. The localization accuracy and robustness are evaluated with an autonomous sweeper developed at Tongji University.

5.1. Deep Curb Detection Network Implementations

5.1.1. Road Scene Classification and Curb Region of Interest

The road scene classification module is implemented with a pre-trained convolutional neural network VGG-16. Firstly, pre-trained VGG-16 has a restriction of the input image's size and the original designed image size is 224×224. However, the resolution of the fish-eye camera is 1920×1080, so the images in our data set must be resized to fit the requirement of VGG-16. We resize the image to 300×168; in addition, this process would not change the height-width ratio of the raw image. In addition, the resizing is a trade-off to achieve a balance between the resolution and the memory usage of GPU.

Secondly, the CNN of our network is composed of the first 30 layers of VGG-16, and the FC (fully connected) layer is discarded temporarily. When it comes to the classification module, the FC layer is implemented again; however, because of the resize process, dimensions of the tensor here should be handled with care. In contrast to the original VGG-16, the dimension of FC layer is $23,040(512*9*5) \times 4096$. In addition, the RPN is adopted from Faster-RCNN [25].

In the train procedure, the initial learning rate is set to 0.001 and the learning rate is decayed by a factor of 0.1 every 10 epochs. We adapt cross-entropy loss as the classification loss, and it is incorporated with the losses in RPN. Except for the 30 frozen layers of VGG-16, all of the new layers of the model are initialized from a zero-mean Gaussian distribution with a standard deviation of 0.01. Since the road scene classification module and curb region of interest module share the same feature map, the networks of these modules are trained simultaneously.

5.1.2. Semantic Segmentation

Since the input image is downsampled and upsampled several times in the semantic segmentation network, in order to ensure the consistency of the input and output image size, we extend the size of irregular CRoI that is generated by the CRoI module to the power of 2, such as 1024×256.

In the training procedure, the initial learning rate is set to 0.001 and decayed by 0.9 every epoch. We also adopt cross-entropy loss as the loss function here. In addition, due to the imbalance of the number of pixels between background and curbs, we set the weight of the loss to be 0.8 for curbs and 0.2 for the background.

The whole network is trained and validated on an Nvidia GTX 1080Ti GPU (NVIDIA Corporation, Santa Clara, CA, USA) and implemented using PyTorch [26].

5.2. Offline Experiments

5.2.1. Dataset

To evaluate the performance of our deep curb detection network, we establish the first-ever road curb detection dataset dedicated to a lateral localization system of a low-speed autonomous vehicle. We use a fish-eye camera with a $180°$ angle of view. The resolution of the fish-eye camera is 1920×1080. The camera is mounted on the right side of the vehicle, and is facing to the right side of the road. In total, our dataset has 7000 images that are recorded at different locations during daytime, which is very challenging. Regarding the annotation, each image has three labels. The first one is the class of the road scene. The second label is a rectangle of the ground truth of the CRoI. The third label is a fitting curve of the curb in the road scene, which results in a pixel-level mask annotation of the curb.

5.2.2. Experimental Results

We evaluate three modules of our deep curb detection network separately with a self-recorded dataset. We consider the road scene classification module as a three-class classification problem. The classification accuracy with our dataset is 96.5%. For the CRoI module, we adopt average precision (AP) to evaluate the model; it is expressed as:

$$AP = \frac{1}{11} \sum_{r \in \{0,0.1,\ldots,1\}} \max_{\tilde{r}:\tilde{r} \geq r} p(\tilde{r}) \tag{1}$$

where $\tilde{r}$ represents recall, p denote precision, AP is the area between precision–recall curve and axis, thus $AP = \int_0^1 P(r)dr$, but to simplify the computation, we set $r \in \{0,0.1,\ldots,1\}$, so we replace the integration with a sum of $p_{\text{interp}}(r)$. On a validation set, the AP of CRoI module is 0.904. For the semantic segmentation module, the performance is evaluated by a parameter called $Ppre$ (i.e., pixel-level precision), which is calculated by $Ppre = N_c/N_{Pred}$, where N_c is the number of correct curb pixels and N_{pred} is the number of all curb pixels detected by our network. We compare our SUNet with the full U-Net. The results are displayed in Table 1, which shows that our SUNet achieves a higher $Ppre$ than U-Net with similar computing time. The detection results of our deep curb detection network are shown in Figure 3.

Table 1. Experiment results of semantic segmentation.

Models	*Ppre*	Computing Time (ms)
Our algorithm	93.86	92
Full U-Net	90.34	91

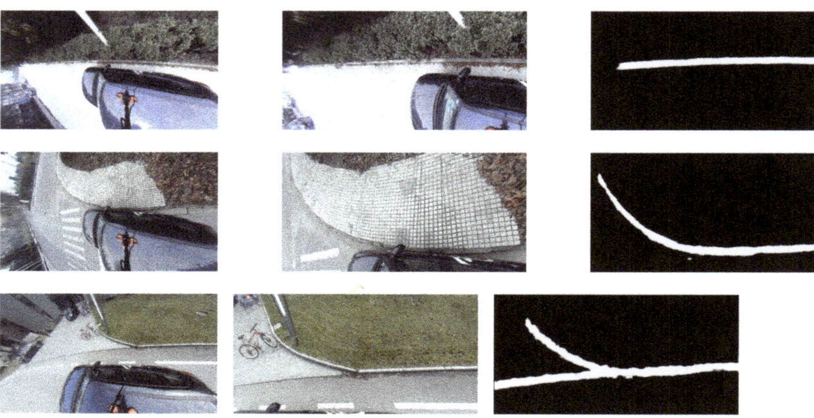

Figure 3. Detection results of deep curb detection network. Left column: image samples from the dataset; Middle Column: enlarged curb regions of interests extracted by CRoI module; Right column: curb segmentation results.

5.3. Experiments with Autonomous Sweeper

5.3.1. Experiment Vehicle

Our experiment vehicle is an intelligent sweeper developed at Tongji University (see Figure 4). The computing platform of this vehicle is a Nvidia Jetson TX2. A LiDAR-based lateral localization system is used by this vehicle, which is described in detail in Section 5.3.2. Two 16-layer LiDARs are equipped and mounted at the bottom of both sides of the vehicle (on top of the side brushes, see Figure 4). Our fish-eye camera is mounted on the right roof of the vehicle, which is the only sensor used by our proposed mono-vision based lateral localization system.

Figure 4. The intelligent sweeper of Tongji University.

5.3.2. LiDAR-Based Lateral Localization System

The LiDAR-based lateral localization system is designed to keep a fixed close distance between the side brush of the sweeper and the curb. The distance is obtained via a LiDAR-based curb detection algorithm. The algorithm first selects candidates in the region of interest from the 3D point cloud generated by LiDARs. The region of interest here is within 1.5 m to the right and 2.5 m to the front of the center of the front axle of the vehicle. The heights of the selected candidates are in the range of 0.09 m to 0.11 m. After that, the algorithm selects the points closest to the vehicle on each row and fits them to a straight line using least squares. Then, the distance between the fitted straight line and the point 1.5 m ahead of the center of the front axle of the vehicle is calculated. The final output of the algorithm is the aforementioned distance minus one offset. An important assumption in this algorithm is that the height of curb is within a certain range, which is also widely used in other LiDAR-based curb detection methods. Consequently, the performance of the LiDAR-based lateral localization system could be greatly affected when the aforementioned assumption is not applicable.

5.3.3. Experimental Results

We select three representative testing routes (route with continuous straight curbs, route with intermittent straight curbs, route with curving curbs) for the autonomous sweeper. Figures 5 and 6 show example scenes of these three routes. The curb detection results are shown in Figure 5. Based on the curb detection results of each frame (image frame for camera and point cloud frame for LiDAR), the lateral distance between the curbs and the sweepers is calculated. Experiments results with our proposed method, LiDAR-based method, and the ground truth are shown in Figure 6

We evaluate the experiment results with two parameters: Average Error and Invalid Rate. The Average Error is defined as the average deviation of the calculated lateral distance value and the manually labeled ground truth. The Invalid Rate is the ratio of the number of failed curb detection frames to the number of all data frames in a testing route. If the deviation of the lateral distance calculated based on the curb detection result and the ground truth exceeds 0.1 m, the curb detection of the current frame is failed. The threshold 0.1 m is decided according the lateral localization accuracy of the intelligent sweeper. We show the experiment results in Tables 2–4.

Table 2. Average error and invalid rate of the testing route with continuous straight curbs.

Evaluation Metrics	Proposed Method	LiDAR-Based Method
Average Error (m)	0.020	0.022
Invalid Rate	0%	0.7%

Table 3. Average error and invalid rate of the testing route with intermittent straight curbs.

Evaluation Metrics	Proposed Method	LiDAR-Based Method
Average Error (m)	0.035	0.038
Invalid Rate	9.2%	72.5%

Table 4. Average error and invalid rate of the testing route with curving curbs.

Evaluation Metrics	Proposed Method	LiDAR-Based Method
Average Error (m)	0.023	0.032
Invalid Rate	0.9%	12.1%

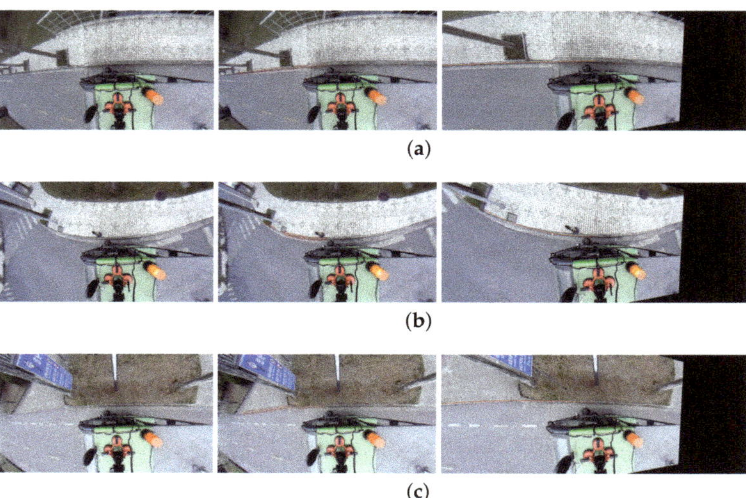

Figure 5. Experimental results of the deep curb detection network with autonomous sweeper. (**a**) curb detection results of the testing route with continuous straight curbs. Left image is a sample. Middle image shows the curb detection results which are highlighted by red color. Right image shows a fitted curve based on the curb detection results. Right image is converted into bird-view; (**b**) curb detection results of the testing route with intermittent straight curbs; (**c**) curb detection results of the testing route with winding road curbs.

(1) Testing route with continuous straight curbs

Table 2 shows that both the LiDAR-based method and the proposed method achieve high accuracy in terms of the Average Error. The Average Error is 0.022 m and 0.020 m for the LiDAR-based method and proposed method, respectively. In terms of the Invalid Rate, our proposed method performs better than the LiDAR-based method. The Invalid Rate is 0.7% and 0% for the LiDAR-based method and proposed method, respectively. It is likely that there are sudden height changes of the road curbs (e.g., at the timestamp of 40 s and 80 s in Figure 6a), which cause the failed curb detection of the LiDAR-based method while our proposed method is not affected.

(2) Testing route with intermittent straight curbs

In the testing route with intermittent straight curbs, the road curbs are intermittent. Typical scenes are from the intersections of the road and the trail, as shown in Figure 6b. As shown in Table 3, compared to the testing route with continuous straight curbs, the Average Error for LiDAR-based method and proposed method increases, and reaches 0.035 m and 0.038 m, respectively. In terms of the Invalid Rate, our proposed method achieves better performance than the LiDAR-based method. The main reason is that the LiDAR-based method assumes a certain height of the road curbs, while this assumption is not applied to the scenes such as road curbs at the intersections. The Invalid Rate for LiDAR-based method is up to 72.5%, which means that the results completely deviate (e.g., in the time interval of [10 s, 30 s] in Figure 6b). In comparison, the Invalid Rate of our proposed method is controlled at 9.2%.

(3) Testing route with winding road curbs

In testing route with winding road curbs, the road curbs are curved. Typical scenes are from the corners of the road, which are shown in Figure 6c. In Table 4, our proposed method achieves better performance in Average Error and Invalid Rate than the LiDAR-based method. The Average Error for the LiDAR-based method and proposed method is 0.032 m and 0.023 m, respectively. The Invalid Rate for our proposed method is 0.9%, which is much better than 12.1% for the LiDAR-based method. The reason for this phenomenon is that the LiDAR-based method has poor fitting performance on the curve (e.g., in the time interval of (22 s, 26 s) in Figure 6c).

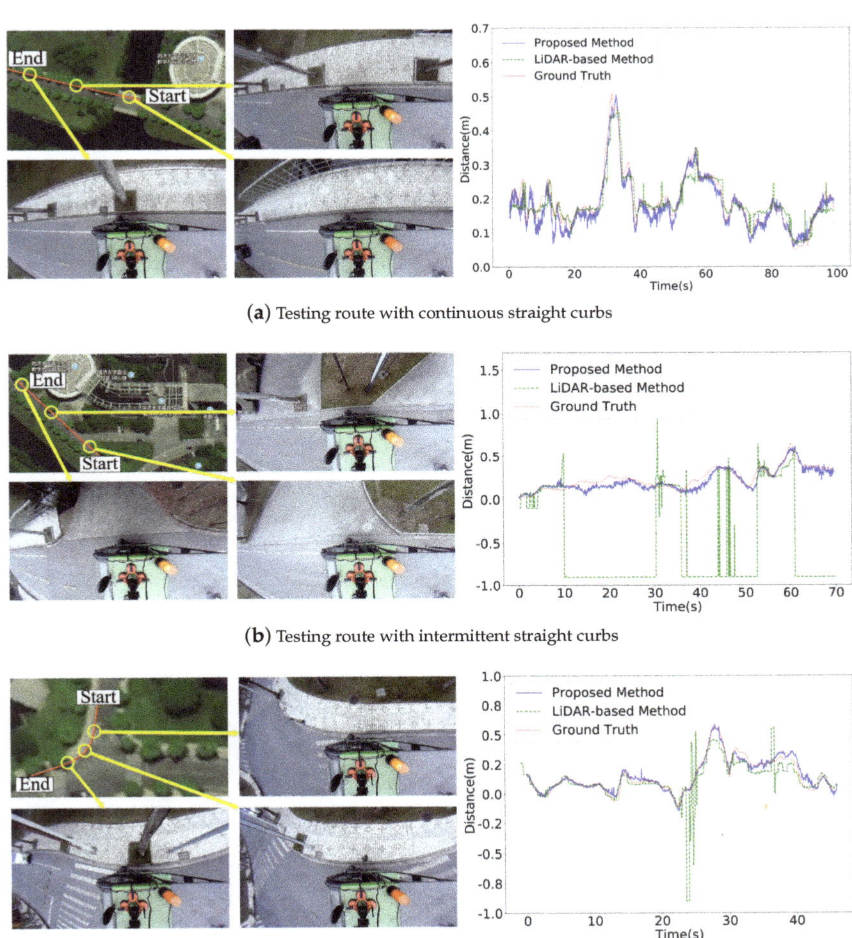

(**a**) Testing route with continuous straight curbs

(**b**) Testing route with intermittent straight curbs

(**c**) Testing route with winding road curbs

Figure 6. Experimental results with autonomous sweeper. (**a**) testing route with continuous straight curb. The left figure shows the satellite images and sampled scenes. The right figure shows the lateral localization results of the proposed method, LiDAR-based method and ground truth; (**b**) testing route with intermittent straight curbs; (**c**) testing route with winding road curbs.

6. Conclusions

We propose a mono-vision based lateral localization system of low-cost autonomous vehicles. Our system relies on a side-facing monocular fish-eye camera that precisely detects the road curbs with the proposed deep curb detection network. Compared with existing methods such as the global navigation satellite system and the LiDAR-based method, a monocular fish-eye camera is cheap, and our solution meets the low-price requirement of a low-speed low-cost autonomous vehicle such as sweepers. We conduct two experiments to evaluate the accuracy and robustness of our mono-vision based lateral localization. Our deep curb detection network achieves 93% pixel-level precision. Our experiment with the intelligent sweeper developed at Tongji University demonstrates that the average lateral distance error of our method is controlled within 0.035 m, and the maximum invalid rate is controlled within 9.2%.

In future work, several directions are worth investigating. Vision-based detection methods generally have problems that are easily affected by environmental factors such as the lighting. Thus, the combination of low-cost LiDAR (e.g., single layer laser scanner) and monocular camera could be a better solution. The future work will focus on how to efficiently fuse the data from the LiDAR and camera to develop a highly efficient and robust lateral localization system.

Author Contributions: J.Y. designs the method and performs the experiments; Z.Y. supervises the manuscripts. All authors have read and agreed to the published version of the manuscript.

Funding: The research leading to these results has partially received funding from the Shanghai Automotive Industry Sci-Tech Development Program according to Grant Agreement No. 1838, from the Shanghai AI Innovative Development Project, and from the State Key Laboratory of Advanced Design and Manufacturing for Vehicle Body according to Grant Agreement No.31815005.

Conflicts of Interest: The authors declare no conflict of interest.

References

1. Zhou, J.; Yang, Y.; Zhang, J.; Edwan, E.; Loffeld, O.; Knedlik, S. Tightly-coupled INS/GPS using Quaternion-based Unscented Kalman filter. In Proceedings of the AIAA Guidance, Navigation, and Control Conference, Portland, Oregon, 8–11 August 2011; p. 6488.
2. Enkhtur, M.; Cho, S.Y.; Kim, K.H. Modified unscented Kalman filter for a multirate INS/GPS integrated navigation system. *Etri J.* **2013**, *35*, 943–946. [CrossRef]
3. Suhr, J.K.; Jang, J.; Min, D.; Jung, H.G. Sensor fusion-based low-cost vehicle localization system for complex urban environments. *IEEE Trans. Intell. Transp. Syst.* **2017**, *18*, 1078–1086. [CrossRef]
4. Vu, A.; Ramanandan, A.; Chen, A.; Farrell, J.A.; Barth, M. Real-time computer vision/DGPS-aided inertial navigation system for lane-level vehicle navigation. *IEEE Trans. Intell. Transp. Syst.* **2012**, *13*, 899–913. [CrossRef]
5. Sefati, M.; Daum, M.; Sondermann, B.; Kreisköther, K.D.; Kampker, A. Improving vehicle localization using semantic and pole-like landmarks. In Proceedings of the 2017 IEEE Intelligent Vehicles Symposium (IV), Los Angeles, CA, USA, 11–14 June 2017; pp. 13–19.
6. Im, J.H.; Im, S.H.; Jee, G.I. Vertical corner feature based precise vehicle localization using 3D LIDAR in urban area. *Sensors* **2016**, *16*, 1268. [CrossRef] [PubMed]
7. Tao, Z.; Bonnifait, P.; Fremont, V.; Ibanez-Guzman, J. Mapping and localization using GPS, lane markings and proprioceptive sensors. In Proceedings of the 2013 IEEE/RSJ International Conference on Intelligent Robots and Systems, Tokyo, Japan, 3–7 November 2013; pp. 406–412.
8. Gruyer, D.; Belaroussi, R.; Revilloud, M. Accurate lateral positioning from map data and road marking detection. *Expert Syst. Appl.* **2016**, *43*, 1–8. [CrossRef]
9. Seo, Y.W.; Rajkumar, R. Tracking and estimation of ego-vehicle's state for lateral localization. In Proceedings of the 17th International IEEE Conference on Intelligent Transportation Systems (ITSC), Qingdao, China, 8–11 October 2014; pp. 1251–1257.
10. Wang, L.; Zhang, Y.; Wang, J. Map-based localization method for autonomous vehicles using 3D-LIDAR. *IFAC-PapersOnLine* **2017**, *50*, 276–281. [CrossRef]
11. Meng, X.; Wang, H.; Liu, B. A robust vehicle localization approach based on gnss/imu/dmi/lidar sensor fusion for autonomous vehicles. *Sensors* **2017**, *17*, 2140. [CrossRef] [PubMed]
12. Zhao, G.; Yuan, J. Curb detection and tracking using 3D-LIDAR scanner. In Proceedings of the 2012 19th IEEE International Conference on Image Processing, Orlando, FL, USA, 30 September–3 October 2012; pp. 437–440.
13. Zhang, Y.; Wang, J.; Wang, X.; Dolan, J.M. Road-segmentation-based curb detection method for self-driving via a 3D-LiDAR sensor. *IEEE Trans. Intell. Transp. Syst.* **2018**, *19*, 3981–3991.
14. Fernández, C.; Llorca, D.F.; Stiller, C.; Sotelo, M. Curvature-based curb detection method in urban environments using stereo and laser. In Proceedings of the 2015 IEEE Intelligent Vehicles Symposium (IV), Seoul, Korea, 28 June–1 July 2015; pp. 579–584.
15. Cheng, M.; Zhang, Y.; Su, Y.; Alvarez, J.M.; Kong, H. Curb detection for road and sidewalk detection. *IEEE Trans. Veh. Technol.* **2018**, *67*, 10330–10342. [CrossRef]
16. Kellner, M.; Hofmann, U.; Bouzouraa, M.E.; Stephan, N. Multi-cue, model-based detection and mapping of road curb features using stereo vision. In Proceedings of the 2015 IEEE 18th International Conference on Intelligent Transportation Systems, Gran Canaria, Spain, 15–18 September 2015; pp. 1221–1228.
17. Ronneberger, O.; Fischer, P.; Brox, T. U-net: Convolutional networks for biomedical image segmentation. In *International Conference on Medical Image Computing and Computer-Assisted Intervention*; Springer: Berlin, Germany, 2015; pp. 234–241.

18. Pan, X.; Shi, J.; Luo, P.; Wang, X.; Tang, X. Spatial as deep: Spatial cnn for traffic scene understanding. In Proceedings of the Thirty-Second AAAI Conference on Artificial Intelligence, New Orleans, LA, USA, 15–18 February 2018.
19. Suleymanov, T.; Amayo, P.; Newman, P. Inferring road boundaries through and despite traffic. In Proceedings of the 2018 21st International Conference on Intelligent Transportation Systems (ITSC), Maui, HI, USA, 4–7 November 2018; pp. 409–416.
20. Siegemund, J.; Pfeiffer, D.; Franke, U.; Förstner, W. Curb reconstruction using conditional random fields. In Proceedings of the 2010 IEEE Intelligent Vehicles Symposium, La Jolla, CA, USA, 21–24 June 2010; pp. 203–210.
21. Rhee, J.H.; Seo, J. Low-Cost Curb Detection and Localization System Using Multiple Ultra sonic Sensors. *Sensors* **2019**, *19*, 1389. [CrossRef] [PubMed]
22. Simonyan, K.; Zisserman, A. Very deep convolutional networks for large-scale image recognition. *arXiv* **2014**, arXiv:1409.1556.
23. Girshick, R.; Donahue, J.; Darrell, T.; Malik, J. Rich feature hierarchies for accurate object detection and semantic segmentation. In Proceedings of the IEEE conference on computer vision and pattern recognition, Columbus, OH, USA, 23–28 June 2014, pp. 580–587.
24. Girshick, R. Fast r-cnn. In Proceedings of the IEEE International Conference on Computer Vision, Santiago, Chile, 7–13 December 2015; pp. 1440–1448.
25. Ren, S.; He, K.; Girshick, R.; Sun, J. Faster r-cnn: Towards real-time object detection with region proposal networks. Advances in neural information processing systems. *arXiv* **2015**, arXiv:1506.01497.
26. Paszke, A.; Gross, S.; Chintala, S.; Chanan, G.; Yang, E.; DeVito, Z.; Lin, Z.; Desmaison, A.; Antiga, L.; Lerer, A. *Automatic Differentiation in PyTorch*; Long Beach, CA, USA, 2017. Available online: https://openreview.net/forum?id=BJJsrmfCZ (accessed on 11 March 2021).

 *actuators*

Article

UWB Based Relative Planar Localization with Enhanced Precision for Intelligent Vehicles

Mingyang Wang, Xinbo Chen, Pengyuan Lv, Baobao Jin, Wei Wang * and Yong Shen ***

Institute of Intelligent Vehicles, School of Automotive Studies, Tongji University, No. 4800 Cao'an Highway, Jiading District, Shanghai 201804, China; wangmingyang@sina.cn (M.W.); austin_1@163.com (X.C.); lv_Pengyuan@tongji.edu.cn (P.L.); jin_baobao2019@163.com (B.J.)
* Correspondence: lazguronwang@gmail.com (W.W.); shenyong@tongji.edu.cn (Y.S.)

Abstract: Along with the rapid development of advanced driving assistance systems for intelligent vehicles, essential functions such as forward collision warning and collaborative cruise control need to detect the relative positions of surrounding vehicles. This paper proposes a relative planar localization system based on the ultra-wideband (UWB) ranging technology. Three UWB modules are installed on the top of each vehicle. Because of the limited space on the vehicle roof compared with the ranging error, the traditional triangulation method leads to significant positioning errors. Therefore, an optimal localization algorithm combining homotopy and the Levenberg–Marquardt method is first proposed to enhance the precision. The triangular side lengths and directed area are introduced as constraints. Secondly, a UWB sensor error self-correction method is presented to further improve the ranging accuracy. Finally, we carry out simulations and experiments to show that the presented algorithm in this paper significantly improves the relative position and orientation precision of both the pure UWB localization system and the fusion system integrated with dead reckoning.

Keywords: ultra-wideband; relative localization; enhanced precision; clock self-correction; homotopy; Levenberg–Marquardt

Citation: Wang, M.; Chen, X.; Lv, P.; Jin, B.; Wang, W.; Shen, Y. UWB Based Relative Planar Localization with Enhanced Precision for Intelligent Vehicles. *Actuators* **2021**, *10*, 144. https://doi.org/10.3390/act10070144

Academic Editor: Hai Wang

Received: 10 May 2021
Accepted: 24 June 2021
Published: 26 June 2021

Publisher's Note: MDPI stays neutral with regard to jurisdictional claims in published maps and institutional affiliations.

1. Introduction

The intelligent vehicle has become one of the most concerning social hot spots and academic research directions. The demand for autonomous vehicles is expected to grow in the coming decades, and the development of autonomous driving technology is followed by the prevalence of the advanced driving assistance system (ADAS). Many vital functions in ADAS, such as blind-spot detection, forward collision avoidance, collaborative cruise control, and collaborative merge assist, require estimation of the relative position among vehicles [1–3]. At present, the relative positioning technology is mainly divided into two types of techniques: (1) Calculating the relative position by the absolute position of each vehicle; and (2) detecting the relative position of the target by radar, camera, and other sensors.

In the first type of technique, relative positioning relies on absolute positioning. There have already been a variety of absolute positioning technologies, however, they are all flawed. Global Navigation Satellite Systems (GNSS) is the most common choice for absolute positioning. However, the accuracy of consumer GNSS is around 10 m [4]. Besides, satellite signals are usually disrupted or blocked in urban canyons, rural tree canopies, and tunnels, leading to degradation or interruption in the positioning information [5]. Many solutions have been proposed on this issue. Integrating the inertial navigation system (INS) with GNSS is a common workaround, which was costly in the past [6]. Only low-cost inertial measurement units (IMU) based on micro-electro-mechanical systems (MEMS) technology are affordable for large-scale promotion [7]. However, due to the low quality of MEMS IMU, positioning errors explode when the GPS signal is unreachable [8], which is also the inevitable problem for INS. In order to fundamentally solve the problem of positioning in GPS blind areas, the wireless sensor network (WSN) can be applied in positioning [9], which relies on wireless technologies such as Bluetooth, WIFI, radio frequency identification (RFID), and Zigbee, etc. [10]. Sensors with known locations are used to locate the sensors with unknown locations. However, the WSN positioning systems are limited by the coverage of base stations; it costs too much to construct base stations widely.

Since all the absolute localization technologies mentioned above are difficult to cover all zones, sensor-based relative positioning is a better choice under certain scenarios. Sensor-based systems use laser, radar, or camera to acquire the relative positions of surrounding vehicles [11–14]. Under favorable road and weather conditions, these systems can facilitate many critical ADAS functions well. However, the relative positioning technologies based on radar and laser are still affected by factors such as weather. Similarly, most vision-based systems work well under adequate lighting and road conditions, but it is not the same case when the environment is dark or lane markings are worn out [15–20]. Although advanced image processing algorithms have been proposed to improve performance at night or under poor lighting and road conditions, it is still very challenging to implement these techniques in real scenarios [21,22].

Moreover, in certain scenarios, such as collision prevention and intelligent fleet following, the two vehicles need to communicate, which means combining traditional positioning technologies with vehicle-to-vehicle (V2V) communication. Shen et al. [23] proposed a tightly-coupled relative positioning method, which used a low-cost IMU and dedicated short-range communications (DSRC) to improve the system's accuracy and robustness. Ponte et al. [24] presented a collaborative positioning method combining radar for the relative positioning of road vehicles. Pinto Neto et al. [25] developed a cooperative GNSS positioning system (CooPS), which used V2V communications to cooperatively determine the absolute and relative position of the ego-vehicle with enough precision. However, localization and communication are accomplished separately in the existing positioning systems, which will affect the real-time performance.

To deal with this problem, a relative positioning system using UWB can accomplish positioning and communication simultaneously without delay. UWB-based relative positioning technology is more adaptable to the environment and has all the advantages of cooperative positioning systems compared to the traditional positioning technologies mentioned above. UWB is a wireless carrier communication technology that uses nanosecond to microsecond non-sine wave narrow pulses to achieve data transmission and high-precision ranging [26]. It also belongs to WSN positioning technologies but provides much higher positioning accuracy than other wireless sensors because of its high temporal resolution. In recent years, UWB technology has been increasingly used in the transportation field, but in most cases, UWB anchors need to be installed on the roadside to locate the absolute position of the vehicle [27,28]. However, the problem is installing UWB anchors on a massive scale costs too much, and the deployment of anchors is quite complicated.

There are also some related studies that used onboard UWB modules for relative positioning between vehicles. Monica et al. [29] used UWB modules installed on the automated guided vehicle and the target node to perform real-time ranging to avoid collisions in the warehouse. In the proposed system, positioning still relied on roadside-based stations. Pittokopiti et al. [30] proposed a UWB based collision avoidance system for miners, which used the distance measured by UWB as the relative position between the worker and the mining vehicle. In other words, it is just a line localization system instead of a planar localization system. Zhang et al. [31] used two UWB tags on the car to calculate the coordinates of the front tag but did not calculate the relative orientation, and the horizontal error was as large as 1 m. Ernst-Johann Theussl et al. [32] proposed a measurement method of the relative position and orientation (RPO) using UWB. They weighted the distances ranged by UWB in different directions to minimize the dilution of precision and to get more accurate results. However, their application scenario was confined to the mobile machinery that did not move fast in a large range because the weights in different directions are hard to be calibrated entirely. Ehab Ghanem et al. [33] proposed a method to estimate vehicular RPO based on multiple UWB ranges and improved the precision using an extended Kalman filter (EKF). Their work was simply an application of UWB in relative positioning for vehicles but did not make improvements in the algorithm. Their experiments were only conducted at a constant vehicle speed and a short vehicle distance.

Generally, current studies on UWB based relative localization for vehicles are relatively less and not thorough enough. One of the essential reasons that limit UWB in the application of relative planar localization for vehicles is the horizontal dilution of precision (HDOP) [34]. Because of the limited space on a vehicle, UWB modules have to be installed closely. Without improvements to the algorithm, the positioning accuracy will decrease drastically with the increase of vehicle distance. However, most of the existing research on UWB relative vehicular localization stays in basic applications without in-depth study of the algorithm. In this paper, a UWB based relative planar localization system is designed with three UWB modules on each vehicle. An improved homotopy-Levenberg–Marquardt (HOMO-LM) algorithm with triangular side length and directed area constraints is proposed, which significantly improves the RPO accuracy.

This paper is organized as follows: In Section 2, the UWB based relative localization system is established, including an improved HOMO-LM positioning algorithm, a timing error self-correction method, and a simple fusion model with DR. Section 3 validates the superiority of the HOMO-LM algorithm by simulation. In Section 4, experiments are conducted to compare the RPO accuracy with and without sensor correction in two conditions, pure UWB mode and fusion mode integrating UWB with dead reckoning (DR).

2. UWB Based Relative Planar Localization System

To design a high precision relative localization system, we should first identify the factors that affect the UWB positioning accuracy. Figure 1 shows multiple sources of the UWB positioning error.

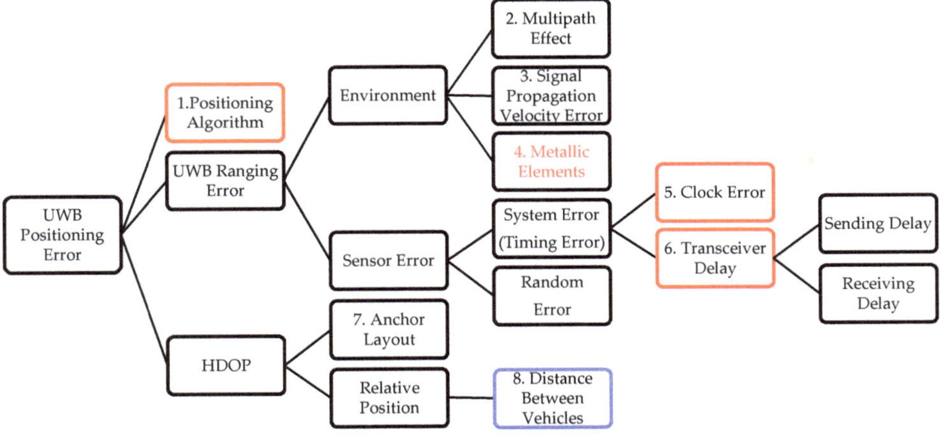

Figure 1. Sources of UWB Positioning error. Red boxes show the factors that we will address in the proposed system. The blue box shows the variable that we will control in the experiments.

As shown in Figure 1, the UWB positioning error is determined by seven factors.

1. The positioning algorithm has profound implications for positioning precision.
2. The multipath effect error can be easily identified due to the high temporal resolution of UWB signals.
3. The signal propagation velocity in the air is almost constant.

4. Metallic elements will affect the UWB systems that range using received signal strength indicators (RSSI) [35]. The proposed system, which range using time of flight (TOF), will not encounter this problem.

5. The clock error is one of the most critical factors to the ranging accuracy. In the UWB system, the distance is calculated by multiplying the time of flight (TOF) of a UWB signal from a module to another by the speed of light, which means one nanosecond clock error will lead to a 30 cm ranging error.

6. The transceiver delay, including sending and receiving delay, is also reflected as a timing error and has a similar effect as the clock error.

7. The anchor layout cannot be significantly adjusted in the proposed system, limited by the space on top of the vehicle.

8. The distance between two vehicles depends on the particular driving scenarios and is not determined by the system.

Therefore, the factors that need to be considered while designing the system consist of positioning algorithm and timing error, including clock error and antenna delays. In this section, an improved HOMO-LM positioning algorithm is proposed, and a timing error self-correction method is presented. Vehicle positioning is usually not completed in only one way but through multi-sensor integration. For extending the application value of the proposed UWB system, we establish a simple UWB/DR fusion system to validate its contribution to the fusion accuracy.

2.1. UWB Relative Planar Localization Algorithm

2.1.1. The Classic Triangulation Algorithm

The positioning and directing model is shown in Figure 2. Three UWB modules are installed on the roof of each vehicle. A, B, and C represent the UWB modules on vehicle 1, while E, F, and G represent those on vehicle 2. O_1 and O_2 represent the centroids of the two vehicles. We define $X_K^{(g)} = \left[x_K^{(g)}, y_K^{(g)}\right]^T$ as the position of point K under the global coordinates system, where $K = (A, B, C, E, F, G, O_1, O_2)$. $\varphi_i^{(g)}$ denotes the heading angle of vehicle i under the global coordinates system, where $i = (1, 2)$. Similarly, $X_K^{(1)} = \left[x_K^{(1)}, y_K^{(1)}\right]^T$ and $\varphi_i^{(1)}$ denote the position and orientation under the coordinate system of vehicle 1. $X_K^{(2)} = \left[x_K^{(2)}, y_K^{(2)}\right]^T$ and $\varphi_i^{(2)}$ denote the position and orientation under the coordinate system of vehicle 2.

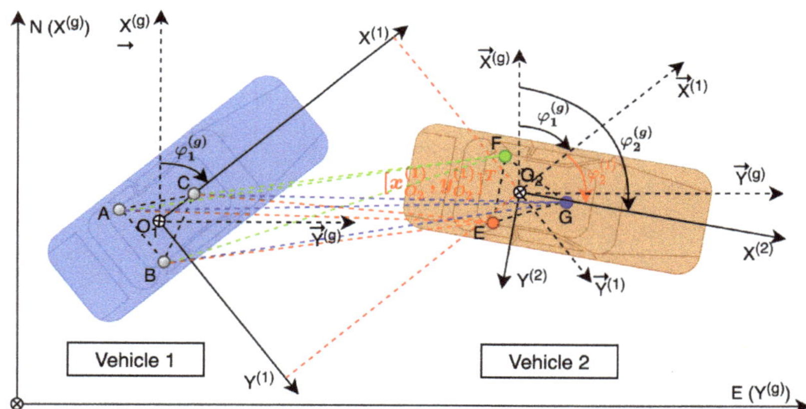

Figure 2. The relative positioning and directing model.

After the UWB modules were installed, $X_A^{(1)}$, $X_B^{(1)}$, $X_C^{(1)}$, $X_{O_1}^{(1)}$, $X_D^{(2)}$, $X_E^{(2)}$, $X_F^{(2)}$ and $X_{O_2}^{(2)}$ were confirmed. UWB measures distances between modules on vehicle 1 and those on vehicle 2. $D_{j,k}$ denotes the distance between module j and module k, where $j = (A,\ B,\ C)$ and $k = (E,\ F,\ G)$. In the proposed system, what we want to know is the relative position $X_{O_2}^{(1)}$ and the relative orientation $\varphi_2^{(1)}$. Equation (1) can be established using the known parameters;

$$D_{j,k} = \| X_j^{(1)} - X_k^{(1)} \|, \ (j = A, B, C;\ k = E, F, G) \tag{1}$$

$X_E^{(1)}$, $X_F^{(1)}$, $X_G^{(1)}$ can be solved from (1), as shown in (2).

$$X_k^{(1)} = P^{-1} N_k, (k = E, F, G) \tag{2}$$

where

$$P = 2\left[X_B^{(1)} - X_A^{(1)}, X_C^{(1)} - X_B^{(1)} \right]^T,$$

$$N_k = \begin{bmatrix} D_{A,k}^2 - D_{B,k}^2 - \| X_A^{(1)} \|^2 + \| X_B^{(1)} \|^2 \\ D_{B,k}^2 - D_{C,k}^2 - \| X_B^{(1)} \|^2 + \| X_C^{(1)} \|^2 \end{bmatrix}.$$

Then $\varphi_2^{(1)}$ can be derived by (3).

$$\varphi_2^{(1)} = \frac{1}{3} \sum_{j,k} \left[\mathrm{atan2}\left(y_j^{(1)} - y_k^{(1)}, x_j^{(1)} - x_k^{(1)} \right) - \mathrm{atan2}\left(y_j^{(2)} - y_k^{(2)}, x_j^{(2)} - x_k^{(2)} \right) \right], (j, k = E, F; F, G; E, G) \tag{3}$$

where,

$$\mathrm{atan2}(y, x) = \mathrm{sgn}(x)^2 \arctan\left(\frac{y}{x}\right) + \frac{1 - \mathrm{sgn}(x)}{2}\left(1 + \mathrm{sgn}(y) - \mathrm{sgn}(y)^2\right)\pi,$$

$$\mathrm{sgn}(x) = \begin{cases} 1 & x > 0 \\ 0 & x = 0 \\ -1 & x < 0 \end{cases}.$$

$X_{O_2}^{(1)}$ can be derived by (4);

$$X_{O_2}^{(1)} = \overline{X_{EFG}^{(1)}} - R\overline{X_{EFG}^{(2)}} \tag{4}$$

where

$$R = \begin{bmatrix} \cos\left(\varphi_2^{(1)}\right) & -\sin\left(\varphi_2^{(1)}\right) \\ \sin\left(\varphi_2^{(1)}\right) & \cos\left(\varphi_2^{(1)}\right) \end{bmatrix}.$$

The overlines represent the mean values of the coordinates of the three modules, such that:

$$\overline{X_{EFG}^{(1)}} = \frac{1}{3} \sum_k X_k^{(1)} \ (k = E, F, G), \quad \overline{X_{EFG}^{(2)}} = \frac{1}{3} \sum_k X_k^{(2)} \ (k = E, F, G).$$

2.1.2. An Improved HOMO-LM Localization Algorithm

According to the error distribution of triangulation, the positioning and directing error of the classic triangulation method will be extensive when two vehicles are far away. Therefore, an improved HOMO-LM method is proposed for better solutions.

With $X_E^{(1)}$, $X_F^{(1)}$, $X_G^{(1)}$ derived from (2), the side lengths of triangle $\triangle_{EFG}$, i.e., $D_{E,F}$, $D_{F,G}$, and $D_{G,E}$ can be calculated by

$$D_{i,j} = \| X_i^{(1)} - X_j^{(1)} \| \ (j, k = E, F; F, G; E, G) \tag{5}$$

However, the real side lengths, $D_{E,F}$, $D_{F,G}$, and $D_{E,G}$ are determined by $X_E^{(2)}$, $X_F^{(2)}$, $X_G^{(2)}$, as shown in (6).

$$D_{i,j}^{real} = \| X_i^{(2)} - X_j^{(2)} \| \quad (j,k = E,F;F,G;E,G) \tag{6}$$

Since ranging error always exists, $D_{E,F}^{real} \neq D_{E,F}$, $D_{F,G}^{real} \neq D_{F,G}$, and $D_{G,E}^{real} \neq D_{G,E}$.

For more accurate solutions, (5) and (6) can be combined as a constraint, i.e., triangular side length constraint, as shown in (7).

$$\begin{cases} \| X_E^{(2)} - X_F^{(2)} \| - \| X_E^{(1)} - X_F^{(1)} \| = 0 \\ \| X_F^{(2)} - X_G^{(2)} \| - \| X_F^{(1)} - X_G^{(1)} \| = 0 \\ \| X_E^{(2)} - X_G^{(2)} \| - \| X_E^{(1)} - X_G^{(1)} \| = 0 \end{cases} \tag{7}$$

Combining (1) and (7), function p can be established:

$$p(x) = \begin{bmatrix} D_{AE} - \| X_A^{(1)} - X_E^{(1)} \| \\ D_{BE} - \| X_B^{(1)} - X_E^{(1)} \| \\ D_{CE} - \| X_C^{(1)} - X_E^{(1)} \| \\ D_{AF} - \| X_A^{(1)} - X_F^{(1)} \| \\ D_{BF} - \| X_B^{(1)} - X_F^{(1)} \| \\ D_{CF} - \| X_C^{(1)} - X_F^{(1)} \| \\ D_{AG} - \| X_A^{(1)} - X_G^{(1)} \| \\ D_{BG} - \| X_B^{(1)} - X_G^{(1)} \| \\ D_{CG} - \| X_C^{(1)} - X_G^{(1)} \| \\ \| X_E^{(2)} - X_F^{(2)} \| - \| X_E^{(1)} - X_F^{(1)} \| \\ \| X_F^{(2)} - X_G^{(2)} \| - \| X_F^{(1)} - X_G^{(1)} \| \\ \| X_E^{(2)} - X_G^{(2)} \| - \| X_E^{(1)} - X_G^{(1)} \| \end{bmatrix} = 0 \tag{8}$$

The least-square (LS) solution of (8) will be more accurate than the solution of (2). However, as the positioning error grows extensive, the triangle composed of E, F, and G may flip. The LS solution of (8) will also encounter significant directing errors, as shown in Figure 3. $\triangle_{EFG}$ represents the real triangle determined by the real relative positions of module E, F, and G. $\triangle_{E_0 F_0 G_0}$ represents the triangle determined by the relative positions of module E, F, and G calculated by (2). $\triangle_{E_1^* F_1^* G_1^*}$ represents the triangle determined by the relative positions of module E, F, and G derived from (8). $\triangle_{E_2^* F_2^* G_2^*}$ represents the triangle determined by the relative positions of module E, F, and G derived from (11). φ denotes the real relative orientation. φ_0, φ_1^*, and φ_2^* denote the relative orientations determined by $\triangle_{E_0 F_0 G_0}$, $\triangle_{E_1^* F_1^* G_1^*}$, and $\triangle_{E_2^* F_2^* G_2^*}$, respectively.

As shown in Figure 3a, because of the large positioning error of the classic triangulation method, the triangle $\triangle_{E_0 F_0 G_0}$, which is constructed by the calculated module coordinates, is seriously deformed. E, F, and G were originally arranged clockwise, but become counterclockwise under the influence of positioning errors. Therefore, the corresponding relative orientation is apparently inaccurate. The shape of the triangle $\triangle_{E_1^* F_1^* G_1^*}$, which is constructed by the LS solutions of (8) with the side length constraint, is similar to the real triangle $\triangle_{EFG}$. However, the rotation direction of the three points was still opposite to the real situation, which means that the triangle flipped, as shown in Figure 3b. The error of relative orientation was still large.

The side length constraint cannot deal with this issue. To suppress the triangle flipping, another constraint, i.e., directed area constraint, is necessary. As shown in Figure 3c, with the introduction of the directed area constraint, both the shape and the direction of $\triangle_{E_2^* F_2^* G_2^*}$

are approximate to the real triangle $\triangle_{EFG}$. Therefore, the relative orientation is much more accurate. Equation (9) is established to express the directed area constraint.

$$S\left(\triangle_{EFG}^{(2)}\right) - S\left(\triangle_{EFG}^{(1)}\right) = 0 \tag{9}$$

where S represents the directed area of the triangle, as shown in (10).

$$S\left(\triangle_{EFG}^{(2)}\right) = \frac{1}{2}\begin{vmatrix} x_E^{(2)} & x_F^{(2)} & x_G^{(2)} \\ y_E^{(2)} & y_F^{(2)} & y_G^{(2)} \\ 1 & 1 & 1 \end{vmatrix}, \quad s\left(\triangle_{EFG}^{(1)}\right) = \frac{1}{2}\begin{vmatrix} x_E^{(1)} & x_F^{(1)} & x_G^{(1)} \\ y_E^{(1)} & y_F^{(1)} & y_G^{(1)} \\ 1 & 1 & 1 \end{vmatrix}. \tag{10}$$

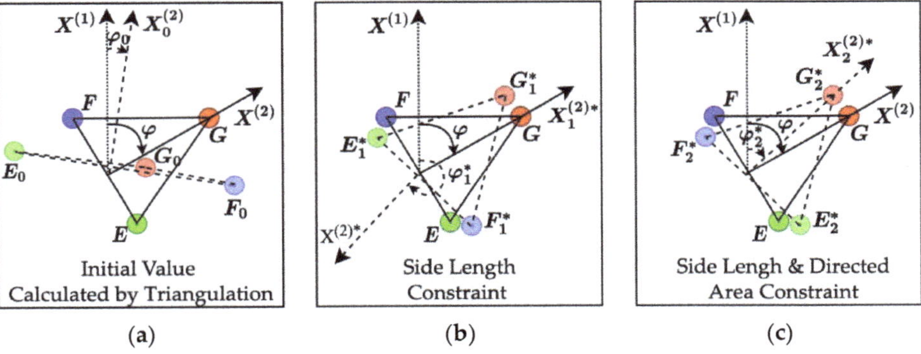

Figure 3. The real triangle and the calculated triangles using different methods: (**a**) the classic triangulation method; (**b**) the LS method with side length constraint; (**c**) the LS method with side length and directed area constraint.

The directed area is a signed area, which can also be described as half the cross products of triangular edge-vectors. According to the basic properties of the cross product, its sign indicates the rotation direction of the triangle vertices. It should be noted that (9) not only limits the triangle flip but also further constrains the triangle shape. Combining (1), (7), and (9), function l is established:

$$L = l(x) = \begin{bmatrix} D_{AE} - \| X_A^{(1)} - X_E^{(1)} \| \\ D_{BE} - \| X_B^{(1)} - X_E^{(1)} \| \\ D_{CE} - \| X_C^{(1)} - X_E^{(1)} \| \\ D_{AF} - \| X_A^{(1)} - X_F^{(1)} \| \\ D_{BF} - \| X_B^{(1)} - X_F^{(1)} \| \\ D_{CF} - \| X_C^{(1)} - X_F^{(1)} \| \\ D_{AG} - \| X_A^{(1)} - X_G^{(1)} \| \\ D_{BG} - \| X_B^{(1)} - X_G^{(1)} \| \\ D_{CG} - \| X_C^{(1)} - X_G^{(1)} \| \\ \| X_E^{(2)} - X_F^{(2)} \| - \| X_E^{(1)} - X_F^{(1)} \| \\ \| X_F^{(2)} - X_G^{(2)} \| - \| X_F^{(1)} - X_G^{(1)} \| \\ \| X_E^{(2)} - X_G^{(2)} \| - \| X_E^{(1)} - X_G^{(1)} \| \\ \det\left(\triangle_{EFG}^{(2)}\right) - \det\left(\triangle_{EFG}^{(1)}\right) \end{bmatrix} = 0 \tag{11}$$

where $x = \left[x_E^{(1)}, y_E^{(1)}, x_F^{(1)}, y_F^{(1)}, x_G^{(1)}, y_G^{(1)}\right]^T$.

Define $x^* = \left[x_E^{(1)*}, y_E^{(1)*}, x_F^{(1)*}, y_F^{(1)*}, x_G^{(1)*}, y_G^{(1)*} \right]^T$ as the LS solution of (11). Then x^* satisfy

$$x^* = \mathrm{argmin}(\| L \|). \tag{12}$$

The localization problem can be transformed into a nonlinear least square (NLLS) optimization problem. To address this problem, a HOMO-LM algorithm is proposed in this section. The LM method is an improved Gauss-Newton (GN) and gradient descent (GD) method. LM method has a faster convergence rate than the GD method and can solve the problem with a singular Jacobian matrix, whereas the GN method cannot. To further improved the convergence rate, we integrated the LM method with the Armijo search [36]. The optimized objective function is l, which has been defined in (11). The initial value x_0 is solved by (2). Define J_L as the Jacobian matrix of function l, which can be expressed as shown in (13).

$$J_L = \frac{\partial l}{\partial x} = \begin{bmatrix} J_E & 0 & 0 \\ 0 & J_F & 0 \\ 0 & 0 & J_G \\ J_{EFG_1} & J_{EFG_2} & J_{EFG_3} \end{bmatrix} \tag{13}$$

where

$$J_i = \begin{bmatrix} \dfrac{x_A^{(1)} - x_i^{(1)}}{\| X_A^{(1)} - X_i^{(1)} \|} & \dfrac{y_A^{(1)} - y_i^{(1)}}{\| X_A^{(1)} - X_i^{(1)} \|} \\ \dfrac{x_B^{(1)} - x_i^{(1)}}{\| X_B^{(1)} - X_i^{(1)} \|} & \dfrac{y_B^{(1)} - y_i^{(1)}}{\| X_B^{(1)} - X_i^{(1)} \|} \\ \dfrac{x_C^{(1)} - x_i^{(1)}}{\| X_C^{(1)} - X_i^{(1)} \|} & \dfrac{y_C^{(1)} - y_i^{(1)}}{\| X_C^{(1)} - X_i^{(1)} \|} \end{bmatrix} (i = E, F, G), \quad J_{EFG_1} = \begin{bmatrix} \dfrac{x_F^{(1)} - x_E^{(1)}}{\| X_F^{(1)} - X_E^{(1)} \|} & \dfrac{y_F^{(1)} - y_E^{(1)}}{\| X_F^{(1)} - X_E^{(1)} \|} \\ \dfrac{x_G^{(1)} - x_E^{(1)}}{\| X_G^{(1)} - X_E^{(1)} \|} & \dfrac{y_G^{(1)} - y_E^{(1)}}{\| X_G^{(1)} - X_E^{(1)} \|} \\ \dfrac{y_G^{(1)} - y_F^{(1)}}{2} & \dfrac{x_F^{(1)} - x_G^{(1)}}{2} \end{bmatrix},$$

$$J_{EFG_2} = \begin{bmatrix} \dfrac{x_E^{(1)} - x_F^{(1)}}{\| X_E^{(1)} - X_F^{(1)} \|} & \dfrac{y_E^{(1)} - y_F^{(1)}}{\| X_E^{(1)} - X_F^{(1)} \|} \\ \dfrac{x_G^{(1)} - x_F^{(1)}}{\| X_G^{(1)} - X_F^{(1)} \|} & \dfrac{y_G^{(1)} - y_F^{(1)}}{\| X_G^{(1)} - X_F^{(1)} \|} \\ 0 & 0 \\ \dfrac{y_E^{(1)} - y_G^{(1)}}{2} & \dfrac{x_G^{(1)} - x_E^{(1)}}{2} \end{bmatrix}, \quad J_{EFG_3} = \begin{bmatrix} \dfrac{10}{1)} & \\ \dfrac{x_F^{(1)} - x_G^{(1)}}{\| X_F^{(1)} - X_G^{(1)} \|} & \dfrac{y_F^{(1)} - y_G^{(1)}}{\| X_F^{(1)} - X_G^{(1)} \|} \\ \dfrac{x_E^{(1)} - x_G^{(1)}}{\| X_E^{(1)} - X_G^{(1)} \|} & \dfrac{x_E^{(1)} - x_G^{(1)}}{\| X_E^{(1)} - X_G^{(1)} \|} \\ \dfrac{y_F^{(1)} - y_E^{(1)}}{2} & \dfrac{x_E^{(1)} - x_F^{(1)}}{2} \end{bmatrix}.$$

Let ε denotes the iteration termination threshold of the optimization algorithm. Define ρ, $\sigma \in (0,1)$ as the regulatory factors of the Armijo search, μ as the regulatory factor of the LM method. Assume k_{max} and m_{max} as the maximum iterations of the LM method and the Armijo search, respectively. Table 1 shows the pseudocode of the LM method with the Armijo search.

Like many other fitting algorithms, the optimization result of the LM method relies on the initial value. LM may only find a local minimum instead of the global minimum or even diverge without the proper initial value. The positioning error increases with the increase of the distance between two vehicles, which also means the error of the initial value x_0 increases. Therefore, the homotopy method is introduced for searching for an optimal initial value in a broader range. Assume $f(x) = 0$ is the equation we want to solve, and f_0 is a known function with an available zero solution x^*, i.e., $f_0(x^*) = 0$. We conduct a depending parameter function

$$h(x,s) = sf(x) + (1-s)f_0(x) \quad s \in [0,1]. \tag{14}$$

$h(x,0) = 0$ is the problem with a known solution x^*, and $H(x,1) = 0$ is the original problem $f(x) = 0$.

In the proposed system, we define $f(x) = l(x)$, $f_0(x) = l(x) - l(x^*)$. This gives the homotopy function

$$H = h(x,s) = l(x) + (s-1)l(x^*) \quad s \in [0,1]. \tag{15}$$

Table 1. The pseudocode of the LM algorithm with the Armijo search.

Step	Pseudocode
Step 1	Define $L = l(x)$ as (11) and $J_L = \frac{\partial l}{\partial X}$ as (13);
Step 2	Set $\varepsilon = 10^{-4}$, $k_{max} = 200$, $\rho = 0.5$, $\sigma = 0.5$, $m_{max} = 20$;
Step 3	Compute the initial value $x_0 = \left[X_E^{(1)^T}, X_F^{(1)^T}, X_G^{(1)^T} \right]^T$ by (2); Set $x_0^* = x_0$; $\mu_0 = \| x_0 \|$;
Step 4	$k = 0$;
Step 5	**while** $(k < k_{\max})$ **do**
Step 6	Compute $L_k = l(x_k)$, J_{L_k};
Step 7	$g_k = J_{L_k}^T L_k$; $d_k = \left(-J_{L_k}^T J_{L_k} + \mu_k I \right)^{-1} g_k$;

Step 8	**if** $\| g_k \| < \varepsilon$ **do**
	     break;
	 end if

Step 9	$m = 0$;
	 while $(m < m_{max})$ **do**
	     if $\left(\| l(x_k + \rho^m d_k) \| < \| l(x_k) \| + \sigma \rho^m g_k^T d_k \right)$ **do**
	         $m = m + 1$;
	     end if
	 end while
	 $\alpha = \rho^m$;

Step 10	$x_{k+1} = x_k + \alpha d_k$;
Step 11	$k = k + 1$;
Step 12	$\mu_k = \| x_k \|$;
Step 13	**end while**
Step 14	$x^* = x_k$;
Step 15	**Output x^* as the optimal solution of equation $l(x) = 0$**

As the solution of $h(x, s) = 0$ depends on s, we denote it by $x^*(s)$. s can be discretized into $0 = s_0 < s_1 < s_2 < \cdots < s_n = 1$. Then the optimization of (11) can be transformed into solving a sequence of nonlinear equations with the LM method such that

$$h(x, s_i) = 0. \tag{16}$$

Each iteration is started with the solution $x^*(s_{i-1})$. Table 2 shows the pseudocode of the proposed HOMO-LM algorithm.

After the LS solution of x^* of (2) being calculated, positions of E, F, G are confirmed. Define

$$
\begin{aligned}
X_{EFG}^{(1)*} &= \left[X_E^{(1)*}, X_F^{(1)*}, X_G^{(1)*} \right] \\
X_{EFG}^{(2)} &= \left[X_E^{(2)}, X_F^{(2)}, X_G^{(1)} \right] \\
\overline{X_{EFG}^{(1)*}} &= \frac{1}{3} \sum_k X_k^{(1)*}, \ (k = E, F, G) \\
\overline{X_{EFG}^{(2)}} &= \frac{1}{3} \sum_k X_k^{(2)}, \ (k = E, F, G)
\end{aligned}
\tag{17}
$$

Table 2. The pseudocode of the HOMO-LM algorithm.

Step	Pseudocode
Step 1	Define $L = l(x)$ as (11) and derive $J_L = \frac{\partial l}{\partial X}$ as (13); Define $H = h(x, s, x^*) = l(x) + (s-1)l(x^*)$ and $J_H = J_L$;
Step 2	Set $\varepsilon = 10^{-4}$, $k_{max} = 200$, $s_0 = 0$, $\Delta s = 0.1$, $s_{end} = 1$, $\rho = 0.5$, $\sigma = 0.5$, $m_{max} = 20$;
Step 3	Compute the initial value $x_0 = \left[X_E^{(1)T}, X_F^{(1)T}, X_G^{(1)T} \right]^T$ by (2); Set $x_0^* = x_0$; $\mu_0 = \parallel x_0 \parallel$;
Step 4	**for** $n = 1 : \frac{(s_{end} - s_0 + \Delta s)}{\Delta s}$ **do**
Step 5	$s_n = s_0 + (n-1)\Delta s$;
Step 6	$k = 0$;
Step 7	**while** $(k < k_{\max})$ **do**
Step 8	Compute $H_k = h\left(x_k, s_n, x_{n-1}^*\right)$, J_{H_k};
Step 9	$g_k = J_{H_k}^T H_k$; $d_k = \left(-J_{H_k}^T J_{H_k} + \mu_k I\right)^{-1} g_k$;
Step 10	**if** $\parallel g_k \parallel < \varepsilon$ **do** **break**; **end if**
Step 11	$m = 0$; **while** $(m < m\_max)$ **do** **if** $(\parallel f(x_k + \rho^m d_k) \parallel < \parallel f(x_k) \parallel + \sigma \rho^m g_k^T d_k)$ **do** $m = m + 1$; **end if** **end while** $\alpha = \rho^m$;
Step 12	$x_{k+1} = x_k + \alpha d_k$;
Step 13	$k = k + 1$;
Step 14	$\mu_k = \parallel x_k \parallel$;
Step 15	**end while**
Step 16	$x_n^* = x_k$;
Step 17	**end for**
Step 18	**Output** $x^* = x_n^*$ **as the optimal solution of equation** $f(x) = 0$

Then the LS solutions of the relative position $X_{O_2}^{(1)*}$ and orientation $\varphi_2^{(1)*}$ can be derived using singular value decomposition (SVD) [37]. Define M as

$$M = \left[X_{EFG}^{(2)} - \overline{X_{EFG}^{(2)}}[1, 1, 1] \right]^T \left[X_{EFG}^{(1)*} - \overline{X_{EFG}^{(1)*}}[1, 1, 1] \right] \tag{18}$$

Take SVD of M:

$$M = U\Sigma V^T \tag{19}$$

where U, Σ, and V represent the three decomposed matrixes of M. Then we have

$$R = VU^T \tag{20}$$

$$\begin{cases} X_{O_2}^{(1)*} = \overline{X_{EFG}^{(1)*}} - R\overline{X_{EFG}^{(2)}} \\ \varphi_2^{(1)*} = \text{atan2}(R_{2,1}, R_{1,1}) \end{cases} \tag{21}$$

2.2. A UWB Timing Error Self-Correction Method

Figure 4 shows the typical two-way ranging (TWR) system. Assume d_{ij} is the original measurement of the distance between module i and j without correction, $\hat{d}_{ij}$ is the corrected measurement and d_{ij}^{real} is the real distance. Other adopted symbols are illustrated in Table 3. Then we have

$$d_{ij}^{real} = ct_{ij}^{f},$$

$$d_{ij} = \frac{c\left(\Delta t_{ij} - t_{j}^{p}\right)}{2}, \tag{22}$$

$$\hat{d}_{ij} = \frac{c\left(k_{i}\Delta t_{ij} - k_{j}t_{j}^{p} - t_{i}^{s} - t_{i}^{r} - t_{j}^{s} - t_{j}^{r}\right)}{2}.$$

d_{ij} ignores several important parameters including t_{i}^{s}, t_{i}^{r}, t_{j}^{s}, t_{j}^{r}, k^{i} and k_{j}.

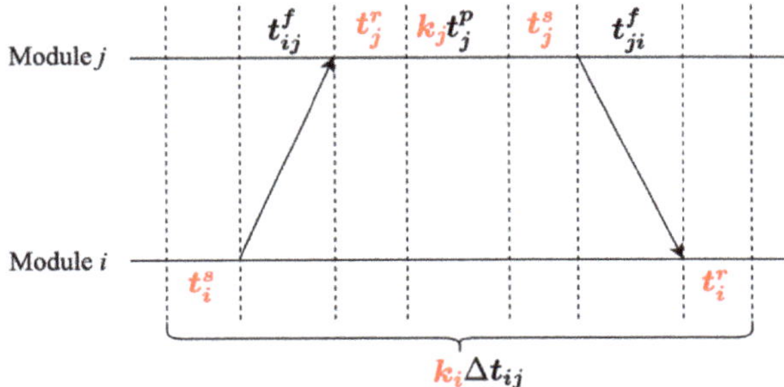

Figure 4. Two-way ranging model. Module i sends a signal to module j firstly. Then module j receives the signal and sends a signal back to module i immediately. In the ranging process, the sending and receiving delays and the clock errors of the two modules will affect the ranging accuracy.

Table 3. The symbols used in the calibration mode.

Symbols	Title 2
$k_{i}\ (i = A, B, C)$	The clock correction coefficient, which is unknown.
$t_{i}^{s}\ (i = A, B, C)$	The sending delay of module i, which is unknown.
$t_{i}^{r}\ (i = A, B, C)$	The receiving delay of module i, which is unknown.
$t_{i}^{d}\ (i = A, B, C)$	The antenna delay. $t_{i}^{d} = t_{i}^{s} + t_{i}^{r}$.
$t_{i}^{p_{j}}\ (i = A, B, C; j = 1, 2, 3, 4)$	The time spent by module i from receiving to sending a signal, which is measured by the crystal oscillator inside module i. As t_{i}^{p} is not a constant, number j is added to distinguish the measurements at different times.
$t_{ij}^{f}\ (ij = AB, BC, AC)$	The signal propagation time from module i to j. It is a particular value after modules being installed. That is $t_{ij}^{f} = \frac{\lVert X_{i}^{(1)} - X_{j}^{(1)} \rVert}{c}$, where c is the speed of light.
$\Delta t_{i}^{sr}\ (i = A, B, C)$	The interval time from module i sending a signal to receiving a signal, measured by the crystal oscillator inside module i.
$\Delta t_{i}^{rr}\ (i = A, B, C)$	The interval time for module i from receiving a signal to another, measured by the crystal oscillator inside module i. Number j is added to distinguish the measurements at different times.
$k_{i}\ (i = A, B, C)$	The clock correction coefficient.

UWB modules A, B, and C on vehicle 1 are taken as an example to show the correction process. The symbols used in the correction method are interpreted in Table 3.

Since $t_i^{p_j}$, Δt_i^{sr}, and Δt_i^{rr} are measured by the crystal oscillator inside UWB module i, which do not equal the real interval time, because of the crystal oscillation frequency error. The real values are

$$\hat{t}_i^{p_j} = k_i t_i^{p_j}, \Delta \hat{t}_i^{sr} = k_i \Delta t_i^{sr}, \Delta \hat{t}_i^{rr} = k_i \Delta t_i^{rr}. \tag{23}$$

Two calibration modes are designed to implement the correction algorithm, as described as follows.

1. Circulation Mode: UWB signals transmit among the modules in turn. All the correction parameters are encountered in this mode, as shown in Figure 5.
2. Differential Mode: UWB signals transmit from a module to another in two paths. The sending delay of the sending module and the receiving delay of the receiving module are eliminated, as shown in Figure 6.

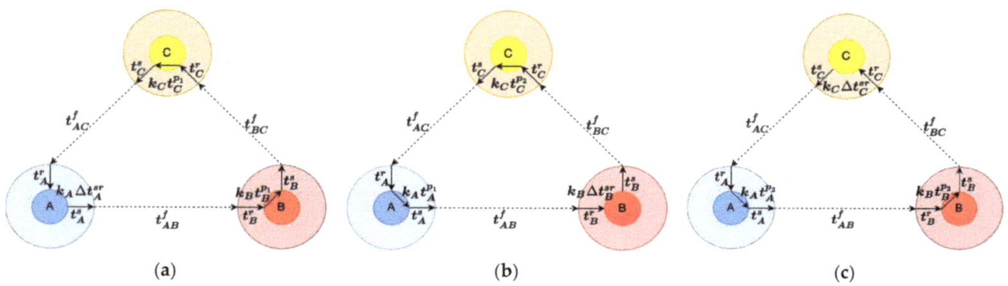

Figure 5. The circulation calibration mode: (**a**) Module A firstly send a signal, and BC send signals in turn after receiving signals from the former module; (**b**) Module B firstly sends a signal and CA sends signals in turn after receiving signals from the former module; (c) Module A firstly send a signal, and BC send signals in turn after receiving signals from the former module.

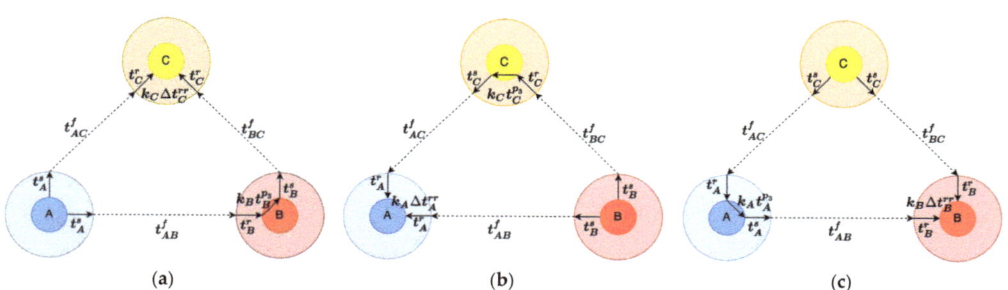

Figure 6. The differential calibration mode: (**a**) signals transmit from A to C in two paths; (**b**) signals transmit from B to A in two paths; (**c**) signals transmit from C to B in two paths.

k_i and t_i^d are the parameters needed to correct the ranging error. Equation (24) is established according to the circulation mode;

$$\begin{cases} k_A \Delta t_A^{sr} = t_A^s + t_{AB}^f + t_B^r + k_B t_B^{p_1} + t_B^s + t_{BC}^f + t_C^r + k_C t_C^{p_1} + t_C^s + t_{AC}^f + t_A^r \\ k_B \Delta t_B^{sr} = t_B^s + t_{BC}^f + t_C^r + k_C t_C^{p_2} + t_C^s + t_{AC}^f + t_A^r + k_A t_A^{p_1} + t_A^s + t_{AB}^f + t_B^r \alpha \\ k_C \Delta t_C^{sr} = t_C^s + t_{AC}^f + t_A^r + k_A t_A^{p_2} + t_A^s + t_{AB}^f + t_B^r + k_B t_B^{p_2} + t_B^s + t_{BC}^f + t_C^r \end{cases} \tag{24}$$

Equation (25) is established according to the differential mode.

$$
\begin{cases}
k_C \Delta t_C^{rr} = \left(t_A^s + t_{AB}^f + t_B^r + k_B t_B^{p3} + t_B^s + t_{BC}^f + t_C^r \right) - \left(t_A^s + t_{AC}^f + t_C^r \right) \\
k_A \Delta t_A^{rr} = \left(t_B^s + t_{BC}^f + t_C^r + k_C t_C^{p3} + t_C^s + t_{AC}^f + t_A^r \right) - \left(t_B^s + t_{AB}^f + t_A^r \right) \\
k_B \Delta t_B^{rr1} = \left(t_C^s + t_{AC}^f + t_A^r + k_A t_A^{p3} + t_A^s + t_{AB}^f + t_B^r \right) - \left(t_C^s + t_{BC}^f + t_B^r \right)
\end{cases}
\tag{25}
$$

Define $t = \left[k_A, k_B, k_C, t_A^d, t_B^d, t_C^d \right]^T$. Equation (26) can be established combining (24) and (25). t_i^s and t_i^r are eliminated because $t_i^d = t_i^s + t_i^r$, which has been introduced in Table 3.

$$
Q_1 t =
\begin{bmatrix}
\Delta t_A^{sr} & -t_B^{p1} & -t_C^{p1} & -1 & -1 & -1 \\
-t_A^{p1} & \Delta t_B^{sr} & -t_C^{p2} & -1 & -1 & -1 \\
-t_A^{p2} & t_B^{p2} & \Delta t_C^{sr} & -1 & -1 & -1 \\
0 & -t_B^{p3} & \Delta t_C^{rr} & 0 & -1 & 0 \\
\Delta t_A^{rr} & 0 & -t_C^{p3} & 0 & 0 & -1 \\
-t_A^{p3} & \Delta t_B^{rr} & 0 & -1 & 0 & 0
\end{bmatrix}
\begin{bmatrix}
k_A \\ k_B \\ k_C \\ t_A^d \\ t_B^d \\ t_C^d
\end{bmatrix}
=
\begin{bmatrix}
t_{AB}^f + t_{BC}^f + t_{AC}^f \\
t_{AB}^f + t_{BC}^f + t_{AC}^f \\
t_{AB}^f + t_{BC}^f + t_{AC}^f \\
t_{AB}^f + t_{BC}^f - t_{AC}^f \\
t_{AB}^f - t_{BC}^f + t_{AC}^f \\
-t_{AB}^f + t_{BC}^f + t_{AC}^f
\end{bmatrix}
= b_1
\tag{26}
$$

The correction process needs repeating for a while to decrease the influence of random noise. That means

$$
Qt = \left[Q_1^T, Q_2^T, \cdots, Q_n^T \right]^T t = \left[b_1^T, b_2^T, \cdots, b_n^T \right]^T = b
\tag{27}
$$

Then the LS solution $t^* = \left[k_A^*, k_B^*, k_C^*, t_A^{*s}, t_B^{*s}, t_C^{*s}, t_A^{*r}, t_B^{*r}, t_C^{*r} \right]^T$ of (27) can be expressed as

$$
t^* = \left(Q^T Q \right)^{-1} Q^T b.
\tag{28}
$$

Besides, we also preprocess the range measurements before positioning to further improve the positioning accuracy. Figure 4 shows a typical two-way ranging system. In fact, the UWB system can range in high frequency up to thousands of times per second. The positioning system does not always need that high data refresh rate. Therefore, we can average multiple measurements, which reduces the interference of random errors without introducing too much latency.

2.3. An Intergrating Model of UWB and DR

The DR system consists of wheel-speed sensors. Each vehicle is equipped with four wheel-speed sensors. According to the Ackerman steering principle, the instantaneous centers of four wheels coincide at point P, as shown in Figure 7. In the DR model, v_{fr}, v_{fl}, v_{rr}, and v_{rl} are the speeds of the wheels measured by wheel-speed sensors. To obtain the longitudinal velocity v_x, lateral velocity v_y, and yaw rate ω, (29) can be established.

$$
\begin{cases}
v_{fr} = \left\| \left[v_x + \omega W_{fr}, v_y + \omega L_f \right]^T \right\| \\
v_{fl} = \left\| \left[v_x - \omega W_{fl}, v_y + \omega L_f \right]^T \right\| \\
v_{rr} = \left\| \left[v_x + \omega W_{rr}, v_y - \omega L_r \right]^T \right\| \\
v_{rl} = \left\| \left[v_x - \omega W_{rr}, v_y - \omega L_r \right]^T \right\|
\end{cases}
\tag{29}
$$

Equation (29) has a similar form as (11), so the proposed HOMO-LM algorithm in Section 2.1.2 is also suitable here.

After v_x, v_y, and ω solved, the UWB/DR fusion model can be established based on the relative kinematic model shown in Figure 8.

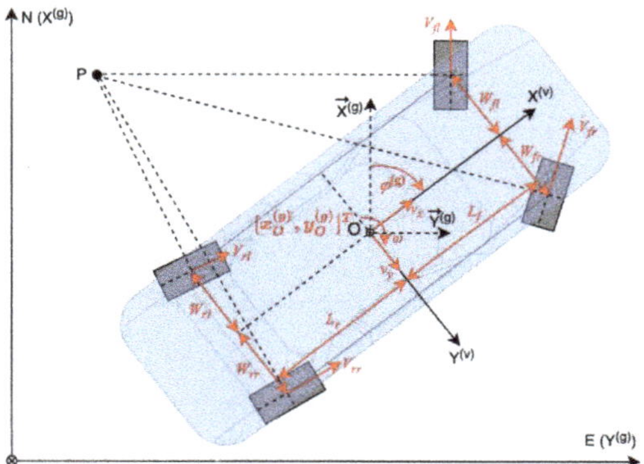

Figure 7. The DR model based on four wheel-speed sensors and IMU.

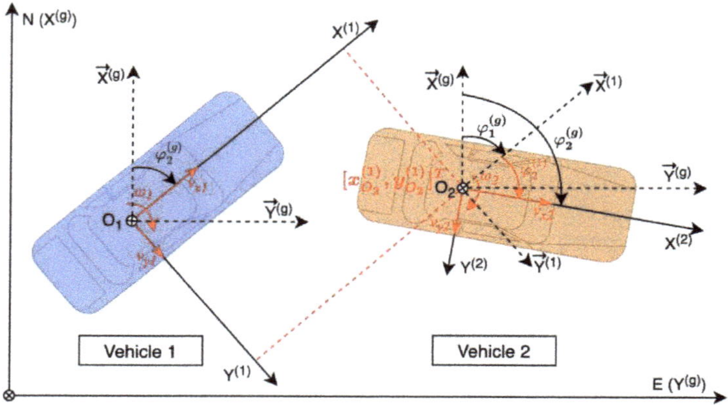

Figure 8. The relative kinematic model.

Let X_k denotes the state vector, which contains the relative position and orientation of vehicle 2 in the coordinate system of vehicle 1 $\left[x_k^{(1)}, y_k^{(1)}, \varphi_k^{(1)}\right]^T$, yaw rates $\left[\omega_{1k}, \omega_{2k}\right]$, and velocities $\left[v_{x1k}, v_{y1k}, v_{x2k}, v_{y2k}\right]^T$ of two vehicles. That is

$$X_k = \left[x_k^{(1)}, y_k^{(1)}, \varphi_k^{(1)}, \omega_{1k}, \omega_{2k}, v_{x_1 1k}, v_{y1k}, v_{x2k}, v_{y2k}\right]^T. \tag{30}$$

The state at time k can be predicted by function f with reference to the state at time $k-1$. Δt denotes the update interval. The state prediction equation can be expressed as

$$X_k = \begin{bmatrix} X'\cos\left(\omega_{1_{k-1}}\Delta t\right) + Y'\sin\left(\omega_{1_{k-1}}\Delta t\right) \\ -X'\sin\left(\omega_{1_{k-1}}\Delta t\right) + Y'\cos\left(\omega_{1_{k-1}}\Delta t\right) \\ \varphi_{k-1}^{(1)} - \omega_{1_{k-1}}\Delta t + \omega_{2_{k-1}}\Delta t \\ \omega_{1_{k-1}} + W_{\omega_1}\Delta t \\ \omega_{2_{k-1}} + W_{\omega_2}\Delta t \\ v_{x1_{k-1}} + W_{v_{x1}}\Delta t \\ v_{y1_{k-1}} + W_{v_{y1}}\Delta t \\ v_{y2_{k-1}} + W_{v_{x2}}\Delta t \\ v_{x2_{k-1}} + W_{v_{y2}}\Delta t \end{bmatrix} \tag{31}$$

where

$$X' = x_{k-1} + v_{x2k-1} \cos \varphi_{k-1} \Delta t - v_{y2k-1} \sin \varphi_{k-1} \Delta t - v_{x1k-1} \Delta t,$$

$$Y' = y_{k-1} + v_{x2k-1} \sin \varphi_{k-1} \Delta t + v_{y2k-1} \cos \varphi_{k-1} \Delta t - v_{y1k-1} \Delta t,$$

W_{ω_1}, W_{ω_2}, $W_{v_{x1}}$, $W_{v_{y1}}$, $W_{v_{x2}}$, and $W_{v_{y2}}$ denote the noise in the prediction process.

Because all the state values can be measured directly or being calculated, the observation vector is

$$z_k = \left[Z_{x_k^{(1)}}, Z_{y_k^{(1)}}, Z_{\varphi_k^{(1)}}, Z_{\omega_{1k}}, Z_{\omega_{2k}}, Z_{v_{x1k}}, Z_{v_{y1k}}, Z_{v_{x2k}}, Z_{v_{y2k}} \right]^T. \tag{32}$$

The observation equation can be expressed as

$$z_k = I X_k \tag{33}$$

where I is an identity matrix.

Since the fusion system is nonlinear and the Jacobian matrix of the state prediction equation is complex, the unscented Kalman filter (UKF) [38,39] method has advantages in dealing with this kind of problem. The UKF model is not difficult to build based on (31) and (33), so the process will not be elaborated in this paper.

3. Simulations

As timing errors cannot be eliminated entirely, we create a virtual environment to validate the feasibility and necessity of the proposed HOMO-LM localization algorithm by simulation, regardless of the influence of clock errors and antenna delays. The driving scenario is established in the Driving Scenario Designer of MATLAB, as shown in Figure 9. The parameter setups of the two vehicles are shown in Tables 4 and 5, respectively. In the Driving Scenario Designer, the coordinates of the waypoints are selected randomly. The velocities of the vehicles can be updated only at the waypoints. The values of velocities are adjusted to avoid vehicle collisions during the simulation. The wait time means the duration that a vehicle stays at a waypoint. The sample interval of the system is set to 0.01 s.

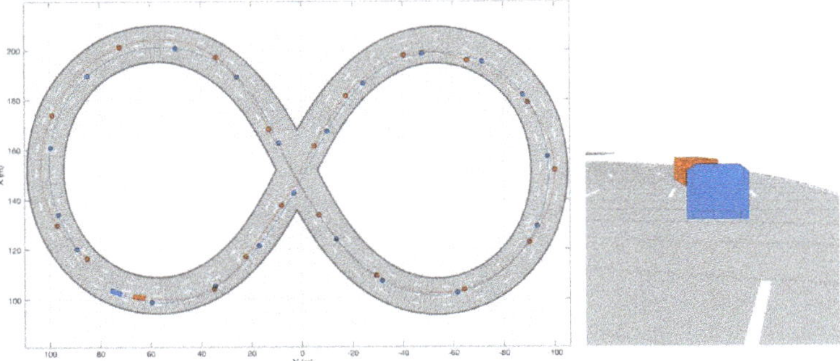

Figure 9. The virtual scenario in the Driving Scenario Designer. The blue cube represents vehicle 1. The red cube represents vehicle 2. The blue/red dots indicate the waypoints of the two vehicles defined in the Driving Scenario Designer.

The simulation results are shown in Figure 10. The proposed method significantly improves the longitudinal positioning accuracy, whether with or without the directed area constraint. As for the lateral positioning accuracy, it cannot be improved by the side length constraint alone. Besides, the directing accuracy is improved by the side length constraint in most cases, although some jumping points still exist. All abnormal orientation data is eliminated with the introduction of the directed area constrain.

Table 4. Parameters of vehicle 1 in the Driving Scenario Designer.

Sequence Number of Waypoints	Coordinate (m)	Velocity (m/s)	Wait Time (s)
1	[103.5; 74.9]	0.0	1.0
2	[99.0; 59.6]	5.0	0.0
3	[105.4; 34.4]	20.0	0.0
4	[121.5; 17.0]	20.0	0.0
5	[142.6; 3.2]	30.0	0.0
6	[167.3; −10.1]	30.0	0.0
7	[186.6; −24.3]	30.0	0.0
8	[198.5; −47.7]	30.0	0.0
9	[195.3; −71.6]	30.0	0.0
10	[182.0; −87.6]	30.0	0.0
11	[157.2; −97.3]	30.0	0.0
12	[129.3; −93.1]	30.0	0.0
13	[102.6; −61.5]	30.0	0.0
14	[107.4; −31.7]	30.0	0.0
15	[124.1; −13.6]	30.0	0.0
16	[162.7; 9.2]	30.0	0.0
17	[189.4; 25.7]	30.0	0.0
18	[200.8; 50.0]	30.0	0.0
19	[189.8; 84.9]	30.0	0.0
20	[160.9; 99.6]	30.0	0.0
21	[134.0; 96.5]	20.0	0.0
22	[120.3; 89.3]	0.0	2.0

Table 5. Parameters of vehicle 2 in the Driving Scenario Designer.

Sequence Number of Waypoints	Coordinate (m)	Velocity (m/s)	Wait Time (s)
1	[101.2; 66.0]	0.0	1.0
2	[104.2; 34.9]	10.0	0.0
3	[117.0; 22.4]	15.0	0.0
4	[137.5; 8.1]	20.0	0.0
5	[161.4; −4.9]	20.0	0.0
6	[181.5; −17.6]	20.0	0.0
7	[197.9; −40.5]	40.0	0.0
8	[195.9; −65.5]	30.0	0.0
9	[178.8; −89.5]	30.0	0.0
10	[151.8; −100.2]	30.0	0.0
11	[122.7; −90.1]	30.0	0.0
12	[103.9; −64.2]	30.0	0.0
13	[109.4; −29.5]	50.0	0.0
14	[133.7; −6.8]	30.0	0.0
15	[168.1; 13.0]	30.0	0.0
16	[197.2; 33.9]	30.0	0.0
17	[201.3; 72.0]	30.0	0.0
18	[173.8; 98.9]	30.0	0.0
19	[129.6; 97.0]	30.0	0.0
20	[116.4; 85.2]	0.0	2.0

We also conduct a simulation of the UWB/DR fusion system, with results shown in Figure 11. HOMO-LM represents the algorithm with both triangular side length and directed area constraint hereafter.

In comparison to Figure 10, the positioning accuracy is further improved by integrating DR. However, the accuracy improvement was limited after fusion using the traditional triangulation method, especially for longitudinal positioning accuracy. The RPO calculated by the HOMO-LM algorithm lead to apparent better fusion accuracy, especially for longitudinal positioning accuracy. Table 6 shows the quantitative comparison results. The root mean square error (RMSE) is recommended to indicate the positioning error. ρ represents the Euclidean distance from the measurement to the real position. $RMSE_\rho$ represents the absolute positioning error, which is

$$RMSE_\rho = \sqrt{\frac{\sum_{i=1}^{n}\left[(x_{measure} - x_{real})^2 + (y_{measure} - y_{real})^2\right]}{n}} \tag{34}$$

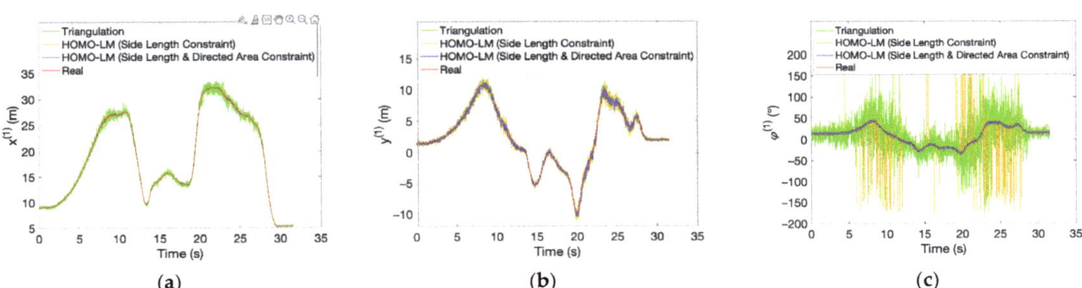

Figure 10. Simulation results for the proposed HOMO-LM method: (**a**) the relative longitudinal position; (**b**) the relative lateral position; (**c**) the relative orientation.

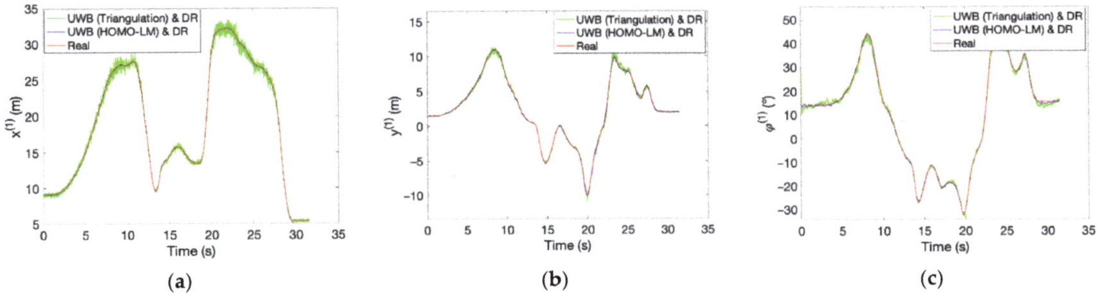

Figure 11. Simulation results of UWB and DR fusion: (**a**) Relative longitude position; (**b**) Relative lateral position; (**c**) Relative orientation.

Table 6. RMSE of position and orientation in simulation.

Algorithm	$RMSE_x$ (m)	$RMSE_y$ (m)	$RMSE_\rho$ (m)	$RMSE_\varphi$ (°)
UWB (Triangulation)	0.48	0.50	0.69	32.83
UWB (HOMO-LM)	0.08	0.28	0.29	1.85
UWB (Triangulation) and DR	0.42	0.21	0.47	2.97
UWB (HOMO-LM) and DR	0.02	0.12	0.12	0.38

Table 6 shows the same conclusion as Figure 11. The HOMO-LM algorithm provides noticeable better results and contributes to better fusion accuracy as well. The enhanced rate of the RPO accuracy can be computed as

$$Enhanced\ Rate = \frac{RMSE_{Classic\ Algorithm} - RMSE_{Prposed\ Algorithm}}{RMSE_{Classic\ Algorithm}} \tag{35}$$

In the pure UWB mode, the proposed HOMO-LM algorithm improved RPO accuracy by 83, 44, 58, and 94%, respectively, in the longitudinal position, lateral position, absolute position, and orientation. In the UWB/DR fusion mode, the proposed algorithm improved the fusion accuracy by 95, 43, 74, and 87%, respectively, in the longitudinal position, lateral position, absolute position, and orientation. The simulation results provided vital support to the proposed algorithm. Therefore, the improvement we made in the localization algorithm was necessary and feasible.

4. Experiments

According to the sources of UWB positioning error we discussed in Section 2, vehicle distance is the only factor that we did not consider in this paper because it is determined by real driving scenarios. Therefore, in this section, the experiments are designed under different vehicle distances. Simulations in Section 3 proved the immense superiority of the proposed HOMO-LM positioning algorithm to the traditional triangulation method. In the experiments, the traditional algorithm was abandoned and would not be verified repeatedly. Nevertheless, the actual timing error cannot be precisely simulated, so the effectiveness of the timing error self-correction method was validated in the experiments.

4.1. Experiment Environment and Equipment

The experimental area and driving routes are shown in Figure 12. Vehicles were driven through a similar route in different experiments but in different vehicle distances.

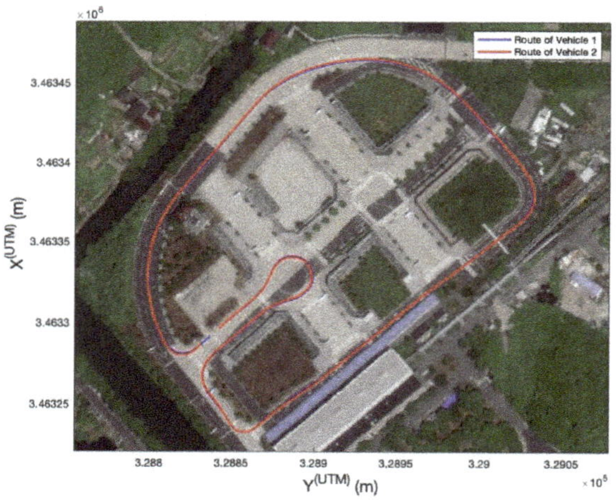

Figure 12. The routes of two vehicles in the experiments.

The equipment installed on the vehicles is shown in Figure 13. Two vehicles were necessary, and three UWB modules were installed on top of each vehicle. A high-precision real-time kinematic (RTK)-GPS/INS, which has the positioning accuracy of 1–2 cm, was recognized as the actual reference. Experimental results are compared to the RTK-GPS/INS. A long-range radio (LoRa) antenna was used to receive differential signals from the RTK base station, which was installed in the testing ground.

4.2. Experiment Results

Limited by the size of the testing ground, the two vehicles needed to be closer when turning at corners to keep UWB modules in line of sight. The distance between the two vehicles could not be kept to a constant, so we only guaranteed the maximum vehicle distances during different experiments. The online data of UWB, DR, and RTK-GPS were recorded into the computer and processed in MATLAB/Simulink offline. Figures 14–19 show the comparison results of the experiments at maximum vehicle distances of 17 m, 37 m, and 70 m. The quantified positioning deviation from the high precision RTK-GPS/INS is shown in Tables 7–9. Similar to simulations, the results of the experiments were also compared in two conditions, the pure UWB mode and UWB/DR fusion mode.

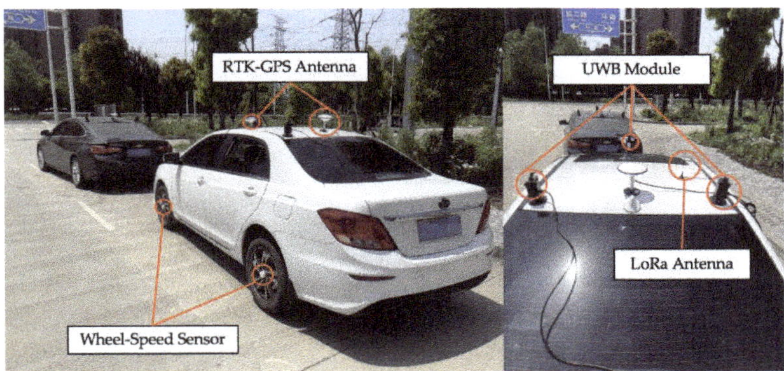

Figure 13. Experimental equipment.

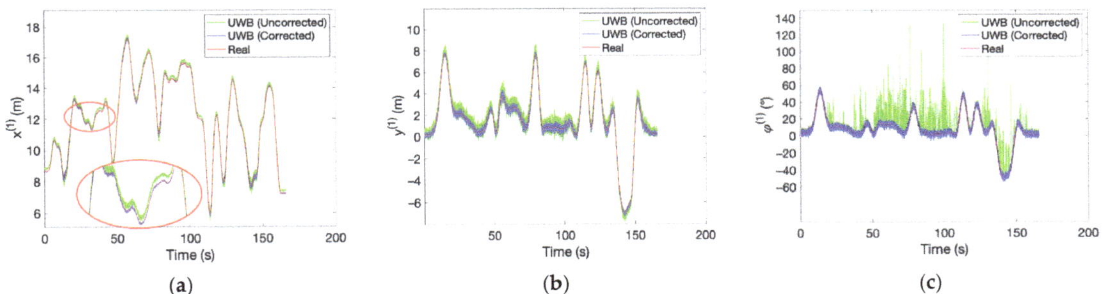

Figure 14. Comparison of UWB localization data with and without correction in the experiment with maximum vehicle distance of 17 m: (**a**) the relative longitudinal position; (**b**) the relative lateral position; (**c**) the relative orientation.

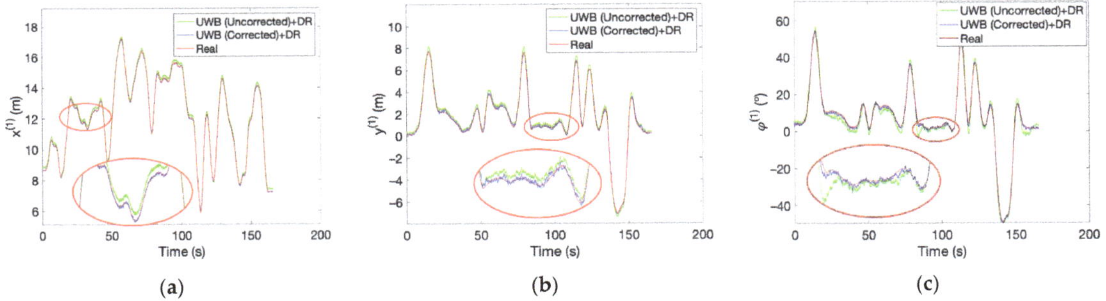

Figure 15. Comparison of fusion localization with and without correction in the experiment with maximum vehicle distance of 17 m: (**a**) the relative longitudinal position; (**b**) the relative lateral position; (**c**) The relative orientation.

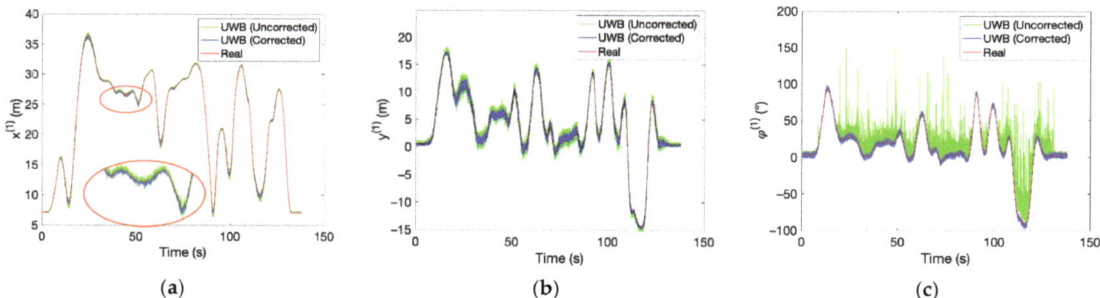

Figure 16. Comparison of UWB localization data with and without correction in the experiment with maximum vehicle distance of 37 m: (**a**) the relative longitudinal position; (**b**) the relative lateral position; (**c**) the relative orientation.

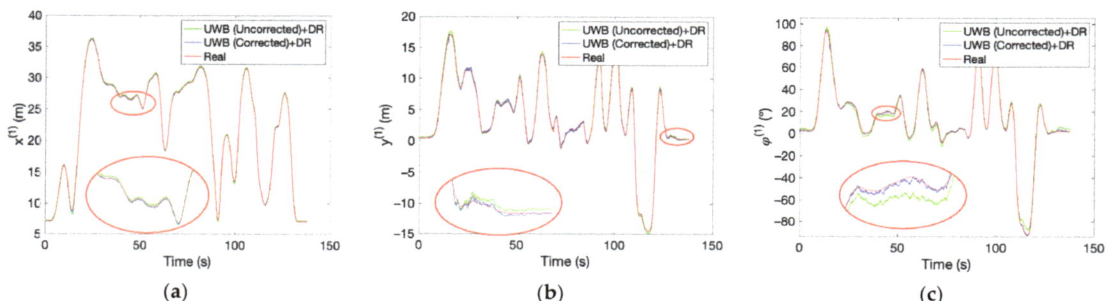

Figure 17. Comparison of fusion localization with and without correction in the experiment with maximum vehicle distance of 37 m: (**a**) the relative longitudinal position; (**b**) the relative lateral position; (**c**) the relative orientation.

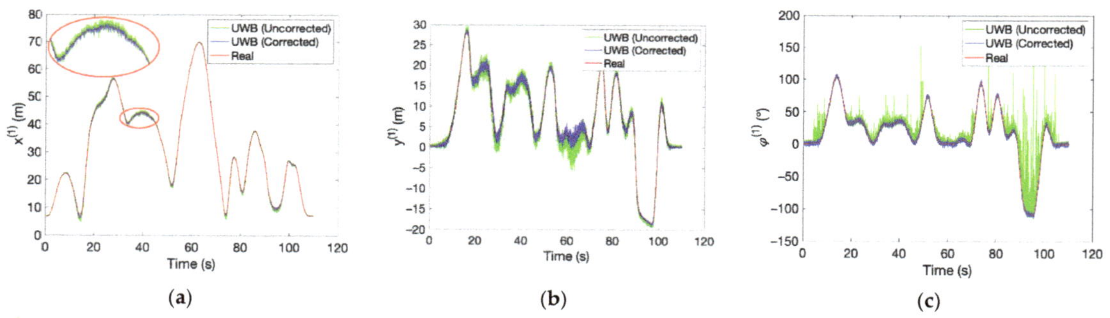

Figure 18. Comparison of UWB localization data with and without correction in the experiment with maximum vehicle distance of 70 m: (**a**) the relative longitudinal position; (**b**) the relative lateral position; (**c**) the relative orientation.

4.3. Results Analysis

Figures 14, 16 and 18 show the comparison results in pure UWB mode. Significant migration exists compared to the real values. As for the orientation, the data was always jumping. For comparison, the results in the UWB/DR fusion mode, shown in Figures 15, 17 and 19, display that the data curves with and without correction are smoother than those in the pure UWB mode, but apparent migration still exists without correction. According to Tables 7–9, the effectiveness of the timing error self-correction method is very noticeable.

The accuracy of the corrected UWB is even better than that of the uncorrected UWB/DR fusion. In pure UWB mode, the RPO accuracy enhanced rates in the three experiments are computed and shown in Table 10, and that in UWB/DR fusion mode is shown Table 11. The RPO accuracy increased substantially either in the pure UWB mode or in the UWB/DR fusion mode. The RPO accuracy improvement with timing error correction was more noticeable in the fusion mode because the UKF eliminates part of the random errors but cannot dispose of system errors, i.e., timing errors, so the influence of timing errors appears more visible.

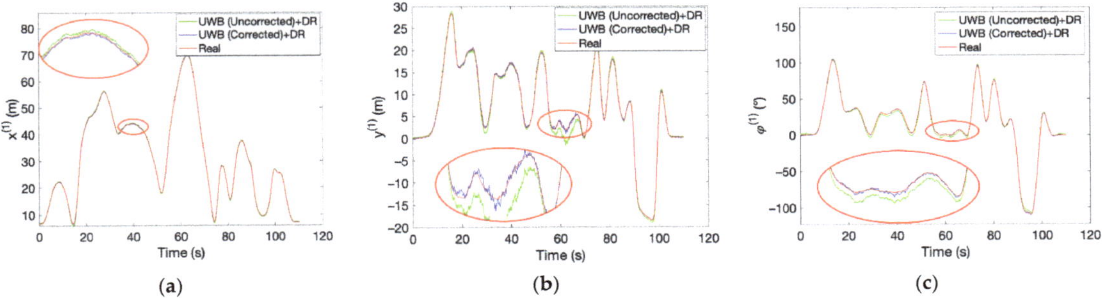

Figure 19. Comparison of fusion localization with and without correction in the experiment with maximum vehicle distance of 70 m: (**a**) the relative longitudinal position; (**b**) the relative lateral position; (**c**) the relative orientation.

Table 7. RMSE of the experiment with the maximum distance of 17 m.

Algorithm	$RMSE_x$ (m)	$RMSE_y$ (m)	$RMSE_\rho$ (m)	$RMSE_\varphi$ (°)
UWB (Uncorrected)	0.17	0.40	0.43	4.80
UWB(Uncorrected) + DR	0.15	0.30	0.34	2.50
UWB (Corrected)	0.04	0.17	0.17	1.84
UWB (Corrected) + DR	0.04	0.13	0.14	0.66

Table 8. RMSE of RPO in the second experiment with the maximum distance of 37 m.

Algorithm	$RMSE_x$ (m)	$RMSE_y$ (m)	$RMSE_\rho$ (m)	$RMSE_\varphi$ (°)
UWB (Uncorrected)	0.25	0.63	0.68	11.64
UWB(Uncorrected) + DR	0.18	0.30	0.35	2.36
UWB (Corrected)	0.10	0.32	0.34	1.83
UWB (Corrected) + DR	0.05	0.16	0.17	0.49

Table 9. RMSE of RPO in the third experiment with the maximum distance of 70 m.

Algorithm	$RMSE_x$ (m)	$RMSE_y$ (m)	$RMSE_\rho$ (m)	$RMSE_\varphi$ (°)
UWB (Uncorrected)	0.38	1.11	1.17	10.26
UWB(Uncorrected) + DR	0.27	0.72	0.77	2.11
UWB (Corrected)	0.16	0.48	0.51	1.83
UWB (Corrected) + DR	0.08	0.22	0.23	0.45

Table 10. RPO accuracy enhanced rate in pure UWB mode in different experiments.

Maximum Vehicle Distance	x	y	ρ	φ
17 m	76%	58%	60%	62%
37 m	60%	49%	50%	84%
70 m	58%	57%	56%	82%

Table 11. RPO accuracy enhanced rate in UWB/DR fusion mode in different experiments.

Maximum Vehicle Distance	x	y	ρ	φ
17 m	73%	57%	59%	74%
37 m	72%	47%	51%	79%
70 m	70%	69%	70%	79%

As shown in Tables 7–9, the positioning accuracy decreases with the increase of vehicle distance. However, even in the third experiment with the maximum vehicle distance of 70 m, the calibrated system using the proposed algorithm provided a positioning error of 0.48 m RMSE in the lateral position and 0.2 m RMSE in the longitudinal position. Besides, the relative orientation error was always within 1.85° in all three experiments. It is a significant improvement under the condition that UWB anchors were installed in such a limited space, and the positioning target was so far away. In addition, with the integration of DR, the RPO error decreased to 0.08 m RMSE, 0.22 m RMSE, and 0.45° RMSE, respectively.

5. Conclusions

In this paper, a relative planar localization system with enhanced precision is proposed. We firstly analyze the UWB positioning error sources and confirm that the influencing factors consist of the positioning algorithm, timing errors, and the vehicle distance. Then, a HOMO-LM optimal positioning algorithm is proposed with the triangular side length and directed area constraints, and a UWB timing error calibration method is presented to correct the clock error and antenna delay. Furthermore, a UWB/DR fusion model is established to extend the application scope of the proposed system and evaluate the contribution of the proposed system to integrated positioning accuracy. Finally, simulations and experiments are conducted to validate our work. The main conclusions are as follows:

1. The proposed HOMO-LM method significantly improves the localization precision comparing to the traditional triangulation method. As expected, the side length constraint ensures the positioning accuracy, and the directed area constraint guarantees the stability of the relative orientation. According to the simulation results, in the pure UWB mode, the proposed HOMO-LM algorithm improves RPO accuracy by 87, 44, 58, and 94%, respectively, in the longitudinal position, lateral position, absolute position, and orientation. As for the UWB/DR fusion mode, the proposed algorithm improves the fusion accuracy by 95, 43, 74, and 87%, respectively, in the longitudinal position, lateral position, absolute position, and orientation.

2. The proposed timing error calibration method improves the positioning accuracy significantly. Experimental results show that in the pure UWB mode, the RPO accuracy with timing error correction improves by at least 58, 49, 50, and 62%, respectively, in the longitudinal position, lateral position, absolute position, and orientation. In the UWB/DR fusion mode, the enhanced rate is 70, 47, 50, and 74%, respectively.

3. With the improved algorithm and corrected sensors, even in the third experiment under the maximum vehicle distance at 70 m, our system provided the RPO error within 0.16 m RMSE, 0.48 m RMSE, 0.51 m RMSE, and 1.85° RMSE in the longitudinal position, lateral position, absolute position, and orientation, respectively. Integrated with DR, the RPO error even decreased to 0.08 m RMSE, 0.22 m RMSE, 0.23 m RMSE, and 0.45° RMSE.

Author Contributions: Conceptualization, M.W. and Y.S.; funding acquisition, X.C. and Y.S.; investigation, W.W., P.L. and B.J.; formal analysis, W.W.; methodology M.W.; software, M.W. and P.L.; visualization: M.W.; validation, X.C. and M.W.; supervision, W.W. and X.C.; writing—original draft, M.W.; writing—review and editing, W.W. and Y.S. All authors have read and agreed to the published version of the manuscript.

Funding: This research was funded by the National Key R&D Program of China (Grant No. 2018YFB0104802) and the Industry University Research Project of Shanghai Automotive Industry Science and Technology Development Foundation (Grant No. 1705).

Institutional Review Board Statement: Not applicable.

Informed Consent Statement: Informed consent was obtained from all subjects involved in the study.

Acknowledgments: The authors appreciate the reviewers and editors for their helpful comments and suggestions in this study.

Conflicts of Interest: The authors declare no conflict of interest.

References

1. Selloum, A.; Betaille, D.; Le Carpentier, E.; Peyret, F. Lane Level Positioning Using Particle Filtering. In Proceedings of the 2009 12th International IEEE Conference on Intelligent Transportation Systems, St. Louis, MO, USA, 4–7 October 2009; pp. 1–6.
2. Lee, S.; Kim, S.-W.; Seo, S.-W. Accurate Ego-Lane Recognition Utilizing Multiple Road Characteristics in a Bayesian Network Framework. In Proceedings of the 2015 IEEE Intelligent Vehicles Symposium (IV), Seoul, Korea, 28 June–1 July 2015; pp. 543–548.
3. Casapietra, E.; Weisswange, T.H.; Goerick, C.; Kummert, F.; Fritsch, J. Building a Probabilistic Grid-Based Road Representation from Direct and Indirect Visual Cues. In Proceedings of the 2015 IEEE Intelligent Vehicles Symposium (IV), Seoul, Korea, 28 June–1 July 2015. [CrossRef]
4. Noureldin, A.; Karamat, T.B.; Georgy, J. *Fundamentals of Inertial Navigation, Satellite-Based Positioning and Their Integration*; Springer Science & Business Media: Berlin/Heidelberg, Germany, 2012; ISBN 978-3-642-30465-1.
5. Abdelfatah, W.F.; Georgy, J.; Iqbal, U.; Noureldin, A. 2d Mobile Multi-Sensor Navigation System Realization Using FPGA-Based Embedded Processors. In Proceedings of the 2011 24th Canadian Conference on Electrical and Computer Engineering (CCECE), Niagara Falls, ON, Canada, 8–11 May 2011; pp. 001218–001221.
6. Berdjag, D.; Pomorski, D. DGPS/INS Data Fusion for Land Navigation. *J. Dong Hua Univ.* **2004**, *21*, 88–93.
7. Barbour, N.; Schmidt, G. Inertial Sensor Technology Trends. *IEEE Sens. J.* **2001**, *1*, 332–339. [CrossRef]
8. Toledo-Moreo, R.; Zamora-izquierdo, M.; Gomez-skarmeta, A. IMM-EKF Based Road Vehicle Navigation with Low Cost GPS/INS. In Proceedings of the 2006 IEEE International Conference on Multisensor Fusion and Integration for Intelligent Systems, Heidelberg, Germany, 3–6 September 2006; pp. 433–438.
9. Lv, Y.; Liu, Y.; Hua, J. A Study on the Application of WSN Positioning Technology to Unattended Areas. *IEEE Access* **2019**, *7*, 38085–38099. [CrossRef]
10. Katiyar, V.; Kumar, P.; Chand, N. An Intelligent Transportation Systems Architecture Using Wireless Sensor Networks. *IJCA* **2011**, *14*, 22–26. [CrossRef]
11. Li, Q.; Chen, L.; Li, M.; Shaw, S.-L.; Nuchter, A. A Sensor-Fusion Drivable-Region and Lane-Detection System for Autonomous Vehicle Navigation in Challenging Road Scenarios. *IEEE Trans. Veh. Technol.* **2014**, *63*, 540–555. [CrossRef]
12. Zhao, H.; Chiba, M.; Shibasaki, R.; Katabira, K.; Cui, J.; Zha, H. Driving Safety and Traffic Data Collection—A Laser Scanner Based Approach. In Proceedings of the 2008 IEEE Intelligent Vehicles Symposium, Eindhoven, The Netherlands, 4–6 June 2008. [CrossRef]
13. Sherer, T.P.; Martin, S.M.; Bevly, D.M. Radar Probabilistic Data Association Filter/Dgps Fusion for Target Selection and Relative Range Determination for a Ground Vehicle Convoy. *NAVIGATION J. Inst. Navig.* **2019**, *66*, 441–450. [CrossRef]
14. Patra, S.; Van Hamme, D.; Veelaert, P.; Calafate, C.T.; Cano, J.-C.; Manzoni, P.; Zamora, W. Detecting Vehicles' Relative Position on Two-Lane Highways through a Smartphone-Based Video Overtaking Aid Application. *Mob. Netw. Appl.* **2020**, *25*, 1084–1094. [CrossRef]
15. Sakjiraphong, S.; Pinho, A.; Dailey, M.N. Real-Time Road Lane Detection with Commodity Hardware. In Proceedings of the 2014 International Electrical Engineering Congress (iEECON), Chonburi, Thailand, 19–21 March 2014; pp. 1–4.
16. Hang, P.; Lv, C. Human-Like Decision Making for Autonomous Driving: A Noncooperative Game Theoretic Approach. *IEEE Trans. Intell. Transp. Syst.* **2021**, *22*, 2076–2087. [CrossRef]
17. Baek, S.; Kim, H.; Boo, K. Robust Vehicle Detection and Tracking Method for Blind Spot Detection System by Using Vision Sensors. In Proceedings of the 2014 Second World Conference on Complex Systems (WCCS), Agadir, Morocco, 10–12 November 2014; pp. 729–735.
18. Kim, S.G.; Kim, J.E.; Yi, K.; Jung, K.H. Detection and Tracking of Overtaking Vehicle in Blind Spot Area at Night Time. In Proceedings of the 2017 IEEE International Conference on Consumer Electronics (ICCE), Las Vegas, NV, USA, 8–10 January 2017; pp. 47–48.
19. Hang, P.; Lv, C.; Huang, C.; Cai, J.; Hu, Z.; Xing, Y. An Integrated Framework of Decision Making and Motion Planning for Autonomous Vehicles Considering Social Behaviors. *IEEE Trans. Veh. Technol.* **2020**, *69*, 14458–14469. [CrossRef]
20. Ma, H.; Ma, Y.; Jiao, J.; Bhutta, M.U.M.; Bocus, M.J.; Wang, L.; Liu, M.; Fan, R. Multiple Lane Detection Algorithm Based on Optimised Dense Disparity Map Estimation. In Proceedings of the 2018 IEEE International Conference on Imaging Systems and Techniques (IST), Krakow, Poland, 16–18 October 2018; pp. 1–5.

21. Amimi, O.; Mansouri, A.; Dose Bennani, S.; Ruichek, Y. Stereo Vision Based Advanced Driver Assistance System. In Proceedings of the 2017 International Conference on Wireless Technologies, Embedded and Intelligent Systems (WITS), Fez, Morocco, 19–20 April 2017; pp. 1–5.
22. Peng, Z.; Hussain, S.; Hayee, M.I.; Donath, M. Acquisition of Relative Trajectories of Surrounding Vehicles Using GPS and SRC Based V2V Communication with Lane Level Resolution. In *Proceedings of the 3rd International Conference on Vehicle Technology and Intelligent Transport Systems*; SCITEPRESS—Science and Technology Publications: Porto, Portugal, 2017; pp. 242–251.
23. Shen, F.; Cheong, J.W.; Dempster, A.G. A DSRC Doppler/IMU/GNSS Tightly-Coupled Cooperative Positioning Method for Relative Positioning in VANETs. *J. Navig.* **2017**, *70*, 120–136. [CrossRef]
24. de Ponte Muller, F.; Diaz, E.M.; Rashdan, I. Cooperative Positioning and Radar Sensor Fusion for Relative Localization of Vehicles. In Proceedings of the 2016 IEEE Intelligent Vehicles Symposium (IV), Gothenburg, Sweden, 19–22 June 2016; pp. 1060–1065.
25. Pinto Neto, J.B.; Gomes, L.C.; Ortiz, F.M.; Almeida, T.T.; Campista, M.E.M.; Costa, L.H.M.K.; Mitton, N. An Accurate Cooperative Positioning System for Vehicular Safety Applications. *Comput. Electr. Eng.* **2020**, *83*, 106591. [CrossRef]
26. Sakkila, L.; Tatkeu, C.; Boukour, F.; El Hillali, Y.; Rivenq, A.; Rouvean, J.-M. UWB Radar System for Road Anti-Collision Application. In Proceedings of the 2008 3rd International Conference on Information and Communication Technologies: From Theory to Applications, Damascus, Syria, 7–11 April 2008; pp. 1–6.
27. Shi, D.; Mi, H.; Collins, E.G.; Wu, J. An Indoor Low-Cost and High-Accuracy Localization Approach for AGVs. *IEEE Access* **2020**, *8*, 50085–50090. [CrossRef]
28. Xue, W.-W.; Jiang, P. The Research on Navigation Technology of Dead Reckoning Based on UWB Localization. In Proceedings of the 2018 Eighth International Conference on Instrumentation & Measurement, Computer, Communication and Control (IMCCC), Harbin, China, 19–21 July 2018; pp. 339–343.
29. Monica, S.; Ferrari, G. Low-Complexity UWB-Based Collision Avoidance System for Automated Guided Vehicles. *ICT Express* **2016**, *2*, 53–56. [CrossRef]
30. Pittokopiti, M.; Grammenos, R. Infrastructureless UWB Based Collision Avoidance System for the Safety of Construction Workers. In Proceedings of the 2019 26th International Conference on Telecommunications (ICT), Hanoi, Vietnam, 8–10 April 2019; pp. 490–495.
31. Zhang, R.; Song, L.; Jaiprakash, A.; Talty, T.; Alanazi, A.; Alghafis, A.; Biyabani, A.A.; Tonguz, O. Using Ultra-Wideband Technology in Vehicles for Infrastructure-Free Localization. In Proceedings of the 2019 IEEE 5th World Forum on Internet of Things (WF-IoT), Limerick, Ireland, 15–18 April 2019; pp. 122–127.
32. Theussl, E.-J.; Ninevski, D.; O'Leary, P. Measurement of Relative Position and Orientation Using UWB. In Proceedings of the 2019 IEEE International Instrumentation and Measurement Technology Conference (I2MTC), Auckland, New Zealand, 20–23 May 2019; pp. 1–6.
33. Ghanem, E.; O'Keefe, K.; Klukas, R. Testing Vehicle-to-Vehicle Relative Position and Attitude Estimation Using Multiple UWB Ranging. In Proceedings of the 2020 IEEE 92nd Vehicular Technology Conference (VTC2020-Fall), Victoria, BC, Canada, 18 November–16 December 2020; pp. 1–5.
34. Akopian, D.; Melkonyan, A.; Chen, C.L.P. Validation of HDOP Measure for Impact Detection in Sensor Network-Based Structural Health Monitoring. *IEEE Sens. J.* **2009**, *9*, 1098–1102. [CrossRef]
35. Bonnin-Pascual, F.; Ortiz, A. UWB-Based Self-Localization Strategies: A Novel ICP-Based Method and a Comparative Assessment for Noisy-Ranges-Prone Environments. *Sensors* **2020**, *20*, 5613. [CrossRef] [PubMed]
36. Zhang, L.; Zhou, W.; Li, D. Global Convergence of a Modified Fletcher–Reeves Conjugate Gradient Method with Armijo-Type Line Search. *Numer. Math.* **2006**, *104*, 561–572. [CrossRef]
37. Sorkine-Hornung, O.; Rabinovich, M. Least-Squares Rigid Motion Using SVD. *Computing* **2017**, *1*, 1–5.
38. Julier, S. A General Method for Approximating Nonlinear Transformations of Probability Distributions. *Citeseer* **1996**. [CrossRef]
39. Wan, E.A.; Van Der Merwe, R.; Nelson, A.T. Nelson Dual Estimation and the Unscented Transformation. *NIPS* **1999**, *29*, 666–672.

MDPI

St. Alban-Anlage 66

4052 Basel

Switzerland

Tel. +41 61 683 77 34

Fax +41 61 302 89 18

www.mdpi.com

Actuators Editorial Office

E-mail: actuators@mdpi.com

www.mdpi.com/journal/actuators